AF786194

Modern Aspects of Electrochemistry

Volume 63

Series Editors

Constantinos G. Vayenas, Department of Chemical Engineering, University of Patras, Patras, Greece

Ralph E. White, Department of Chemical Engineering, University of South Carolina, Columbia, SC, USA

This well-respected series commenced publication in the early 1950s. Over the years it has earned an excellent reputation by offering high-quality monographs and reviews of current research into all areas of electrochemistry and electrochemical engineering and their applications. The international magazine Chemistry & Industry notes that this series "deserves a place in electrochemistry libraries and should prove useful to electrochemists and related workers." Starting with Number 43 in the series, two features have been introduced:

- Each volume is thematic and focuses on a new development in the field or reviews a well-established area that is regaining interest.
- Each volume is guest-edited. As in the past, all chapters are topical reviews.

Subjects covered in recent volumes include:

1. Modeling and Numerical Simulations
2. Progress in Corrosion Science and Engineering
3. Electrodeposition: Theory and Practice
4. Electrocatalysis: Theory and Experiments
5. Interfacial Phenomena in Fuel Cell Electrocatalysis
6. Diagnostics and Modeling of Polymer Electrolyte Fuel Cells
7. Applications of Electrochemistry and Nanotechnology in Biology and Medicine

Kumar Raju · Charl J. Jafta · Eric Lichtfouse
Editors

Hybrid Nanostructures for Advanced Supercapacitors

Springer

Editors
Kumar Raju
Department of Engineering
University of Cambridge
Cambridge, Cambridgeshire, UK

Charl J. Jafta
BASF (United States)
Beachwood, OH, USA

Eric Lichtfouse
State Key Laboratory of Multiphase Flow
in Power Engineering
Xi'an Jiaotong University
Xi'an, Shaanxi, China

ISSN 0076-9924 ISSN 2197-7941 (electronic)
Modern Aspects of Electrochemistry
ISBN 978-3-032-04592-8 ISBN 978-3-032-04593-5 (eBook)
https://doi.org/10.1007/978-3-032-04593-5

© The Editor(s) (if applicable) and The Author(s), under exclusive license to Springer Nature Switzerland AG 2026

This work is subject to copyright. All rights are solely and exclusively licensed by the Publisher, whether the whole or part of the material is concerned, specifically the rights of translation, reprinting, reuse of illustrations, recitation, broadcasting, reproduction on microfilms or in any other physical way, and transmission or information storage and retrieval, electronic adaptation, computer software, or by similar or dissimilar methodology now known or hereafter developed.
The use of general descriptive names, registered names, trademarks, service marks, etc. in this publication does not imply, even in the absence of a specific statement, that such names are exempt from the relevant protective laws and regulations and therefore free for general use.
The publisher, the authors and the editors are safe to assume that the advice and information in this book are believed to be true and accurate at the date of publication. Neither the publisher nor the authors or the editors give a warranty, expressed or implied, with respect to the material contained herein or for any errors or omissions that may have been made. The publisher remains neutral with regard to jurisdictional claims in published maps and institutional affiliations.

This Springer imprint is published by the registered company Springer Nature Switzerland AG
The registered company address is: Gewerbestrasse 11, 6330 Cham, Switzerland

If disposing of this product, please recycle the paper.

Preface

Climate change and the rising energy demand have recently strongly aggravated energy supply worldwide, thus calling for the development of green fuels and powerful energy storage. Supercapacitors have recently found wide applications in managing renewable energy sources. Despite the availability of wind and solar energy, most energy consumption for daily life and industrial production still relies on fossil fuels. In this context, supercapacitors are viewed as new tools for rapid energy storage and release. Because supercapacitors are versatile and have long-lasting charge cycles, they are ideal for storing and releasing energy efficiently in various applications such as mobile devices, vehicles, and building power supplies. Supercapacitors' ability to deliver quick bursts of energy, coupled with a longer lifespan compared to batteries, is garnering special attention. However, conventional supercapacitors face limitations in terms of energy density, which is a critical factor for high-performance applications. Therefore, research has recently focused on advanced supercapacitors made of hybrid nanostructures, which are reviewed in this book.

Hybrid nanostructures bring a transformative edge to supercapacitor technology by combining different materials at the nanoscale. These materials synergize to create a unique hybrid, offering enhanced properties and capabilities such as the customization of supercapacitors. For instance, hybrid nanostructures enable a remarkable increase in energy density, allowing supercapacitors to store more energy efficiently. Additionally, the synergy of materials in hybrid nanostructures facilitates rapid charge and discharge, resulting in superior power density compared to traditional supercapacitors. Also, by carefully selecting and combining materials, researchers can fine tune the properties of hybrid nanostructures to meet specific application requirements. The unique characteristics of hybrid nanostructures contribute to prolonged cycle life, ensuring sustained performance over an extended period and making them suitable for a wide range of applications, from wearable electronics to electric vehicles.

Carbon-based hybrid nanomaterial. This artistic figure was drawn by the authors

In Chap. 1, we present the fundamentals of hybrid supercapacitors. The groundwork laid here will serve as a foundation for the subsequent chapters, creating a better understanding of these materials and hybrid supercapacitors. Chapter 2 reviews hybrid nanostructures with diverse morphologies and interfaces and their potential to achieve better electrochemical properties for high energy density supercapacitors. Chapter 3 focuses on the synergy between conducting polymers and hybrid nanostructures. Chapter 4 reviews graphene-metal oxide hybrid nanostructures and their synergistic effects in hybrid supercapacitor performances. Chapters 5 and 6 expand our examination of lithium and sodium-ion capacitors, addressing the challenges and solutions involving hybrid nanostructures, along with their integration into flexible and wearable devices. Chapter 7 explains the role of conducting polymers in hybrid nanostructures, notably with metal oxides and sulfides for enhancing the cycle stability and rate capability. Chapter 8 presents the synthesis and characterization of various hybrid nanocomposites and their advantages such as tuning the morphology and phase structures for boosting the supercapacitor properties. Chapter 9 reviews the advanced nanostructured materials in hybrid ion capacitors, notably with the importance of structural and compositional optimization, incorporation of heteroatoms, and pore size manipulation for improving the performance of hybrid capacitors. Chapter 10 reviews the advancements in computational modeling of hybrid supercapacitors.

As editors, we extend our gratitude to the authors for their invaluable contributions and hope this compilation serves as a source of inspiration and knowledge for researchers, students, and practitioners.

Cambridge, UK — Kumar Raju
Beachwood, USA — Charl J. Jafta
Xi'an, China — Eric Lichtfouse

Contents

Chapter 1
Hybrid Supercapacitors

Mpho S. Ratsoma, Katlego Makgopa, Kabir O. Oyedotun, Kwena D. Modibane, Kumar Raju, and Ncholu Manyala

Abstract Hybrid supercapacitors have gained widespread attention due to their electrochemical performance features, which include high-energy and high-power densities, resulting from the combination of electric double-layer capacitance and battery-type electrodes. Thus, these devices play a crucial role in bridging the gap between high-power-density conventional supercapacitors and high-energy density batteries. In this chapter, we provide a review of the progress made with hybrid supercapacitor devices, with an emphasis on the electrode materials, their synthesis, and their applications in hybrid supercapacitors. We first highlight the differences between the various types of electrodes commonly used in supercapacitors, as classified by their charge storage mechanisms. We then give detailed accounts of some of the electric double-layer capacitance-type and battery-type electrode materials used in hybrid supercapacitors and summarize the numerous works reported on hybrid supercapacitors based on such materials.

Keywords Hybrid supercapacitors · Asymmetric design · Nanostructures · Electric double-layer capacitance · Faradaic redox reactions · High-energy density · High-power density

M. S. Ratsoma · K. Makgopa (✉)
Department of Chemistry, Faculty of Science, Tshwane University of Technology (Arcadia Campus), Pretoria, South Africa
e-mail: MakgopaK@tut.ac.za

K. O. Oyedotun · N. Manyala
Department of Physics, Institute of Applied Materials, SARChI Chair in Carbon Technology and Materials, University of Pretoria, Pretoria, South Africa

K. D. Modibane
Nanotechnology Research Lab, Department of Chemistry, School of Physical and Mineral Sciences, University of Limpopo (Turfloop), Polokwane, South Africa

K. Raju
Energy Centre, Council for Scientific and Industrial Research (CSIR), Pretoria, South Africa

© The Author(s), under exclusive license to Springer Nature Switzerland AG 2026
K. Raju et al. (eds.), *Hybrid Nanostructures for Advanced Supercapacitors*, Modern Aspects of Electrochemistry 63, https://doi.org/10.1007/978-3-032-04593-5_1

1.1 Introduction

The advent of supercapacitors as promising electrochemical energy storage systems has led to a continuous requirement for advancing their energy storage capabilities to endorse their application in renewable energy technology, as well as in the rapidly evolving portable electronic sector, transportation sector, and utility in power grid energy storage. Supercapacitors are characterized by a high-power density, fast charge/discharge capability, long cycle lifespan, good operational safety, and environmental friendliness, which collectively give them an advantage over other electrochemical storage and conversion systems such as batteries and fuel cells [1–5]. Supercapacitors usually store charge either through ion adsorption using carbon-based electrode materials, i.e. electrochemical double-layer capacitors (EDLCs) or through fast surface redox reactions using transition metal compounds and/or conducting polymers i.e., pseudocapacitors [6]. The supercapacitor devices have proven to play a key role in bridging the energy and power density gap between the power density electrolytic capacitors and the high-energy density batteries [7] and can be used to augment batteries or extend their lifetime by balancing temporary power peaks in hybrid systems. Supercapacitors have been used for fast charge/discharge and uninterrupted power supply applications [8–10]. However, the low energy density of supercapacitors hinders their potential widespread application, opening a window for researchers to design and develop supercapacitors with enhanced energy densities without forgoing the power densities [5, 11].

Recently, hybrid supercapacitors have received a far-reaching research expansion as they have been demonstrated to have promising potential in fabricating novel charge storage devices that present reliability, high-energy density, fast charging, and sustainability. Hybrid supercapacitors also known as supercapatteries can accumulate charge both electrostatically, i.e. formation of an electrical double and electrochemically, i.e. Faradaic redox reactions [12–20]. The hybrid supercapacitors are assembled through the combination of electric double-layer capacitor-type and pseudocapacitive/battery-type electrode materials in a single cell configuration [21–25]. The synergy of the capacitor-like and pseudocapacitive/battery-like electrodes results in a device that possesses high energy and power densities. The setup used in these hybrid supercapacitors can effectively overcome the low energy and power densities associated with conventional supercapacitors and batteries, respectively [12, 26–30]. The Ragone plot in Fig. 1.1a demonstrates the higher energy density of hybrid supercapacitors compared to symmetric supercapacitors, as well as their higher power density compared to batteries. Due to their high-power and energy densities, as well as the prevalence of flexible electrode materials of different morphologies, hybrid supercapacitors have found application in fields that include portable electronics, energy harvesting, smart cloth, smart grids, and electric vehicles (see Fig. 1.1b). Lower-self discharge and longer cycle life (> 100,000) are some of the advantages obtained from hybrid supercapacitors as compared to other conventional supercapacitors [31, 32].

Normally, the EDLC-type materials store charge by forming a Helmholtz double layer with electrolyte ions through adsorption and desorption of these ions on the

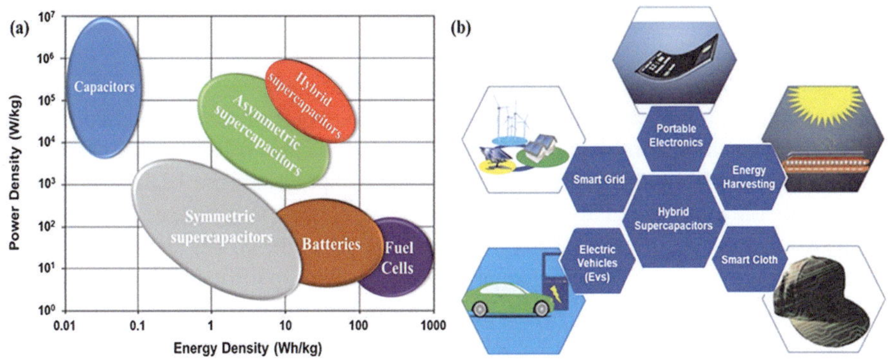

Fig. 1.1 **a** Ragone plot comparing the power and energy density capabilities of various energy storage devices. **b** Prospective application areas for hybrid supercapacitors. Reprinted with permission of Wiley from [33]

surface of the electrode. These EDLC materials have been demonstrated to have high cycling stability [34]. Pseudocapacitive materials such as MnO_2, RuO_2, and Mn_3O_4 utilize fast faradaic redox reactions to store energy by transfer of electrons within a given potential, owing to the ability of the metal atom in these compounds to change oxidation states in flexible mechanisms during the electrochemical measurements (e.g., Mn^{3+} and Mn^{4+}) [29, 35]. On the other hand, battery-type materials, which include transition metal oxides/hydroxides, sulphides, nitrides, and selenides mainly store energy storage through dynamic redox reactions with electrolyte ions at a given potential [36, 37]. As such, battery-type electrode materials are better candidates for application as hybrid supercapacitor cathodes because they can store higher electrochemical energy in comparison to pseudocapacitive materials.

Moreover, scientists have adopted a distinctive approach to separate devices built with pseudocapacitive electrodes from those with battery-type electrodes, the former designated as plainly "asymmetric supercapacitors" and the latter recognized as hybrid supercapacitors, which are labelled as a "special case" of asymmetric supercapacitors [14, 20, 38]. Figure 1.1a further shows the slightly higher energy and power densities of hybrid supercapacitors compared to asymmetric supercapacitors. For that reason, this chapter follows a similar trend, and we discuss the recent progress made on hybrid supercapacitors with emphasis on the preparation methods and application of carbon-based EDLC-type and battery-type electrode materials and as well as their composites in hybrid supercapacitors.

An enhanced energy density in hybrid supercapacitors is realized as a result of an increased operating potential arising from the synergetic combination of the complementary potential windows of the two dissimilar electrodes [19, 39, 40]. Furthermore, the overall performance of hybrid supercapacitors, like any other electrochemical energy storage and conversion devices, is intrinsically dependent on the nature of the electrode materials, which are vital for the construction of an energy storage device that has a long cycle life and high energy and power densities [41–43]. However, even though hybrid supercapacitors show potential improvement along the diagonal of Ragone plots in terms of achieving higher power and energy density at the same time, several challenges are encountered with these devices. These include poor cycling behaviour, the struggle to find a well-matched pair of electrode materials, as well as overcoming the imbalance between the specific capacity and reaction kinetics of the electrodes [44–49]. The rather low specific capacity of the EDLC-type electrode limits the energy density of the hybrid supercapacitor, whereas the relatively sluggish reactions of the battery-type electrode stifle the power density [50–53]. Therefore, improving the specific capacitance and rate capability of the EDLC- and battery-type electrodes, respectively, has become important for the construction of hybrid supercapacitors with enhanced electrochemical performances [54]. Herein, we give a review of the progress made on the EDLC- and battery-type electrode materials, as well as the performance of the hybrid supercapacitor devices built using combinations of the respective anode and cathode materials that possess unique and novel nano-architectures.

1.2 Supercapacitors

1.2.1 Design and Types of Electrodes Used for Supercapacitors

Supercapacitors are usually built with two sandwiching electrodes separated by liquid or gel electrolytes and ion-porous separator material in one of two designs, either in a symmetric or asymmetric configuration. In symmetric supercapacitors, both electrodes are identical in every aspect like material, weight, thickness and essentially have the same capacitance [14]. Symmetric supercapacitors are known for their higher cycle life as compared to their asymmetric counterparts, as well as a higher power density in comparison to the asymmetric devices due to their ability to be discharged to and held at 0 V with fast response time. Pseudocapacitors and EDLCs are typical symmetric supercapacitors. However, symmetric supercapacitors suffer from lower energy densities as compared to the asymmetric devices. On the other hand, asymmetric supercapacitors, as adjudged by Lukatskaya et al. [14], arise from any difference in the aspects of the electrodes, but most commonly from different electrode materials.

Asymmetric devices possess essentially different capacitance values for each electrode [6, 12, 30]. Typically, the cycle life is lower than that of symmetric supercapacitors, and unlike the latter, they mostly cannot be discharged to and held at 0 V potential and usually show longer response time. Asymmetric electrode designs are mostly associated with the combination of electrode materials with different storage mechanisms, ideally, EDLC-type electrodes in conjunction with pseudocapacitive material or as in the scope of this discourse, with battery-type faradaic electrode material in hybrid supercapacitors. In hybrid devices, which are considered a special case of asymmetric supercapacitors, the battery-type electrode provides high-energy density while the EDLC electrode furnishes the system with a high-power capability [12, 55]. Figure 1.2a shows the different types of supercapacitors categorized according to their design and storage mechanism, and (b) explaining the charge storage behaviour of different electrodes and their storage mechanisms such as electric double-layer capacitors, pseudocapacitors, and asymmetric and hybrid supercapacitors. The electrochemical characteristics such as cell voltage, capacitance, and the energy and power density make these energy storage systems different from each other. Compared to EDLCs and pseudocapacitors, the hybrid supercapacitor stores more energy, last longer (over 100,000 cycles), and operate at higher voltages (up to 0.5 V).

With an increasing development of electrode materials for example transition metal oxides, hydroxides, sulphides, carbides, nitrides, and conducting polymers that present an electrochemical behaviour that is intermediate of the capacitive and Faradaic charge storage mechanisms, distinguishing the charge storage modalities of electrode materials used in electrochemical systems necessary so at to avert confusion. In essence, it is usually apparent what materials can be classified as EDLCs-type, as well as how these materials differ from the battery-type and pseudocapacitive materials, both of which store charge using faradaic reactions [56]. Most often, indistinctness is encountered when researchers try to make a distinction between battery-type and pseudocapacitive electrodes materials, despite there being guidelines that were proposed for distinguishing between these two types of energy storage materials [14, 20, 56].

Figure 1.3 illustrates the cyclic voltametric and current discharge curves proposed by Gogotsi and Penner as a guideline for making a distinction between the three types of electrochemical energy storage materials [35]. It is apparent from the cyclic voltammetric curves that the EDLC-type, pseudocapacitive, and battery-type electrode materials respond differently when subjected to potential scanning. The EDLC-type materials usually exhibit a potential-independent capacitance and consequently, an independent current (Fig. 1.3a, b). On the other hand, the battery-type materials exhibit largely protruding and widely separated peaks that are characteristic of the reduction and oxidation of the metal centres involved in the Faradaic reactions (Fig. 1.3g, h). Pseudocapacitive materials tend to show behaviour that is intermediate of EDLC-type and battery-type electrode materials (Fig. 1.3d, e). Further distinction can be made using the galvanostatic discharge curves of these electrode materials. At a constant current discharge, EDLC-type electrodes display a linear E versus t curve (Fig. 1.3c), whereas in the case of battery-type materials, an extremely nonlinear curve is observed.

The nonlinear E versus t plot of the battery-type electrode materials is characterized by plateaus of near constant potential that are equivalent to the potentials at which the Faradaic redox reactions of the metal centres is occurring (Fig. 1.3i). Moreover, the peak current (i) of the EDLC-type materials displays a linear response to the scanning rate, whereas that of the battery-like materials show dependency on the square root of the scanning rate. A more particular distinction is made in

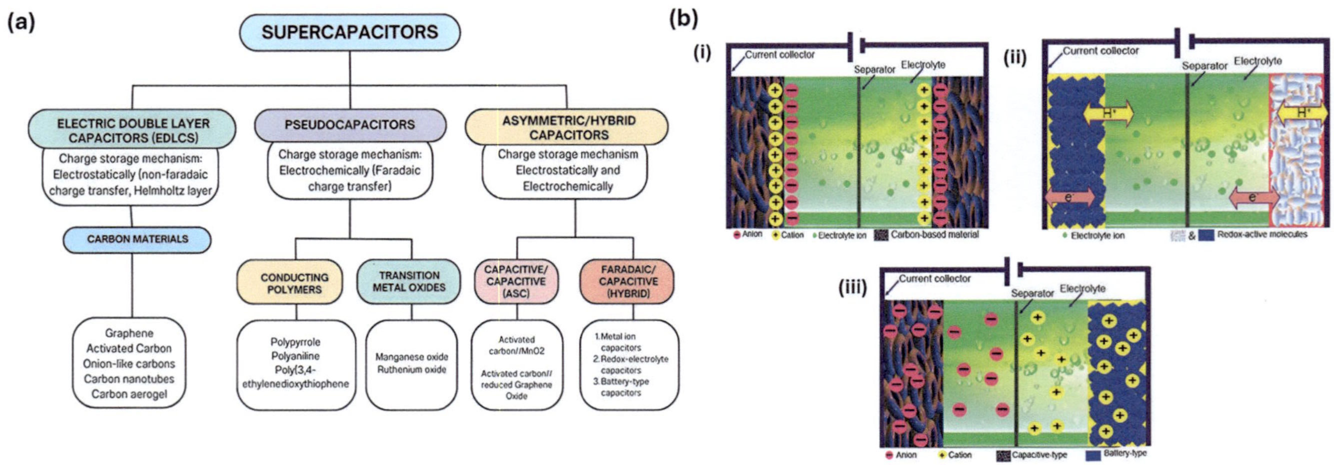

Fig. 1.2 **a** Flow chart explaining the three different types of supercapacitor devices as categorized according to electrode materials and their respective storage mechanisms. Suitable electrode materials are mentioned to distinguish the charge storage mechanisms. **b** The working principle of (i) electrical double-layer capacitors (ii) pseudocapacitors and (iii) hybrid supercapacitors, explaining the activities of electrode materials and devices with their respective charge storage mechanisms

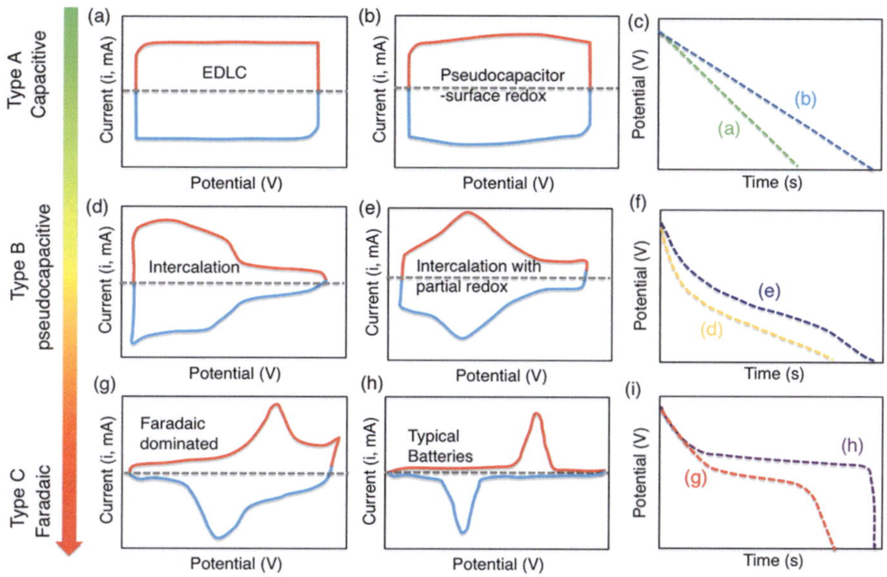

Fig. 1.3 Cyclic voltammetric profiles (**a**, **b**, **d**, **e**, **g**, **h**) and corresponding galvanostatic discharge curves (**c**, **f**, **i**) for various kinds of electrode materials. Reprinted with permission of American Chemical Society from [35]. (EDLC- Electrochemical double-layer capacitors)

Table 1.1 where we discuss the different battery-type electrode materials used in hybrid supercapacitors.

Table 1.1 Classification of electrode materials according to their different charge storage mechanisms [57]

Charge storage mechanism	Typical electrode materials
Surface redox reaction-based pseudocapacitance	RuO_2, Mn_3O_4, MnO_2, MnOOH, FeOOH, Fe_3O_4, Fe_2O_3 (0–0.8 V versus Ag/AgCl in neutral electrolytes)
Intercalation dominated pseudocapacitance	V_2O_5, WO_3, MoO_3, Nb_2O_5, TiO_2, $Na_2Ti_2O_{5-x}$, MoS_2, WS_2, VS_2, TiS_2
Battery-type electrode materials	Ni, Co, Cu, and Cd based oxides/hydroxides, sulphides, selenides and phosphides: $Co(OH)_2$, CoOOH, CoO_x ($x =$ 1–2), $Ni(OH)_2$, NiOOH, NiO, Fe_2O_3, γ–FeOOH, α–NiS, NiCu LDH, NiMn LDH, CoAl-LDH, NiAl LDH, Ni–Co LDH, $NiCo_2O_4$, $NiMoO_4$, $ZnFe_2O_4$, $FeCo_2O_4$, $Ni_3V_2O_8$, $NiCo_2S_4$, Ni–Mn sulphides, Co–Mn sulphides, Zn–Co sulphides, Zn–Ni sulphides, NiV_2S_4, Ni–Mo sulphides, $NiGa_2S_4$, NiCoP, $Co_{0.85}Se$, $(Ni, Co)_{0.85}Se$
	(LDH—layered double hydroxide)

1.2.2 Performance Evaluation in Supercapacitors

The capacitive performance evaluation for supercapacitors is established from the generic EDLC setup that comprises of two sandwiching electrodes made of carbon materials that are embedded on a current collector material and separated by a porous separator material immersed in an electrolyte. This setup is equivalent to two capacitors connected in series, and the resulting capacitance (C), which is obtained after polarizing the electrodes by applying potential difference (voltage) between them, can be expressed through Eq. 1.1:

$$\frac{1}{C_{\text{cell}}} = \frac{1}{C_+} + \frac{1}{C_-}, \tag{1.1}$$

where C_{cell} represents the capacitance (in Farad = Coulomb/Volt) of the resulting device or cell, and C_+ and C_- are the capacitances of the positive and the negative electrodes, respectively. The capacitance (C_{cell}) probed from the cyclic voltametric curve, as well as from the slope of the discharge curve of the galvanostatic charge–discharge graph, is acquired through Eq. 1.2:

$$C_{\text{cell}}(F) = \frac{i}{\Delta V/\Delta t}, \tag{1.2}$$

where i (A) is the applied current, $\Delta V/\Delta t$ (V s^{-1}) is the slope of the discharge curve after the initial iR drop. For symmetrical devices, where the two electrodes (positive and negative electrode) are identical in all aspects, i.e. material, weight, thickness, and the same capacitance is obtained from the electrodes ($C_+ = C_-$), the definition of the cell capacitance (C_{cell}) in Eq. 1.2 adapts to Eq. 1.3:

$$C_{\text{cell}} = \frac{C_{\text{e}}}{2}, \tag{1.3}$$

where $C_{\text{e}} = C_+ = C_-$ The electrode capacitance (C_{e}) can be valued by rearranging Eq. 1.3 to form Eq. 1.4:

$$C_{\text{e}} = 2C_{\text{cell}}. \tag{1.4}$$

Generally, a more efficient comparison of supercapacitor electrode materials is achieved by effecting a specific capacitance, $C_{\text{e, sp}}$ (F g^{-1}), which is related to the capacitance obtained from a single electrode. Therefore, the specific capacitance is calculated by dividing Eq. 1.4 by the mass of the single electrode to obtain Eq. 1.5:

$$C_{\text{e,sp}}\left(\text{Fg}^{-1}\right) = \frac{2C_{\text{cell}}}{m_{\text{e}}}, \tag{1.5}$$

where m_e (g) is the mass of the single electrode. The evaluation of a single electrode in a *three-electrode* (half-cell) setup usually shows a higher capacitance value than the actual cell capacitance obtained from a *two-electrode* (full-cell) system. In a symmetric system, the specific (gravimetric) capacitance of the two electrodes (C_{sp}) is obtained from Eq. 1.6:

$$C_{sp}\left(\mathrm{Fg}^{-1}\right) = \frac{4C_{cell}}{M}, \tag{1.6}$$

where M is the total mass of the active materials of the two electrodes (i.e., $M = 2m_e$ since the weight of each electrode is the same). The multiplier of 4 is used as an operation to relate the cell capacitance (C_{cell}) and the combined weight of two electrodes (M) to the capacitance (C_e) and mass (m_e) of a single electrode [58, 59].

Besides the specific capacitance, the energy and power densities of a supercapacitor are also relevant measures of the capacitive performance of the device [60]. These two parameters are used to gauge the amount of energy that can be stored (energy density) and how long can it be discharged (power density). The energy (E) stored in the EDLC is related to the charge (Q) at each electrode–electrolyte interface, as well as to the applied voltage between the two electrodes. Consequently, the energy density shows a directly proportional dependency on the cell capacitance as shown by Eq. 1.7:

$$E_{sp}\left(\mathrm{Wh\ kg}^{-1}\right) = \frac{C_{cell}V^2}{2M}, \tag{1.7}$$

where M (kg) is the mass of the electrodes and V (V) is the maximum voltage of electrochemical stability. The maximum power of the device is elucidated using the following Eq. 1.8:

$$P_{max}\left(W.\mathrm{kg}^{-1}\right) = \frac{V^2}{4R_sM}, \tag{1.8}$$

where R_s is the internal resistance, which is calculated from the voltage drop at the beginning of a discharge curve as shown in Eq. 1.9:

$$R_s(\Omega) = \frac{\Delta V_{IR}}{2i}, \tag{1.9}$$

where ΔV_{IR} is the voltage drop between the first two points obtained at the start of the discharge curve.

Equations 1.1–1.9 remain relevant when probing the performance of an asymmetric cell, but before they are implemented on these devices, the "*mass-balancing*" from the *three-electrode* evaluation should be affected by obtaining the electrochemical performance of each electrode. Contrary to the equal split of the applied voltage in symmetric systems because of the use of identical electrodes, voltage splitting in asymmetric devices is dependent on the capacitance of the respective active material in each electrode. In general, the capacitance is associated with the mass and the measured specific capacitance of the active material in each electrode [61]. Thus, to affect an equal voltage split, the mass balance between the two electrodes must be optimized using the relationship Eq. 1.10:

$$q_+ = q_-, \tag{1.10}$$

where q_+ and q_- represent the charges stored at the positive and negative electrode, respectively. The stored charge is expressed using Eqs. 1.11 or 1.12:

$$q = C_{\text{sp}}.m.\Delta V \tag{1.11}$$

or:

$$\frac{m_+}{m_-} = \frac{C_{\text{sp}-}}{C_{\text{sp}+}} \times \frac{\Delta V_-}{\Delta V_+}, \tag{1.12}$$

where m, C_{sp}, and ΔV represent the working electrode-respective mass, specific capacitance, and potential range obtained from the charging/discharging process of three-electrode configuration.

However, in the case of the asymmetric hybrid supercapacitors, the definition of capacitance does not apply to battery-type electrode materials, which show a plateau during the discharging process (Fig. 1.3i). Instead, the capacity (Q) is implemented as a measure of the performance of the battery-type electrode. It is calculated using Eq. 1.13:

$$Q(\text{coulomb } (C) \text{ or mAh}) = I \cdot \Delta t, \tag{1.13}$$

where I is the constant discharge current and Δt is the time of discharge (s). Comparison with the ELDC-type electrode can be made by translating the constant capacitance of the carbon electrode into C using the width of the potential window ($\mathit{\Delta V}$) that is traversed when cycling the device as shown in Eq. 1.14:

$$Q\left(C\,x^{-1}\text{or mAh}\,x^{-1}\right) = \frac{C_e.\Delta V}{3.6.x}, \tag{1.14}$$

where C_{e} is the capacitance of the EDLC-type electrode, $\mathit{\Delta V}$ is the width of the potential window ($V_{\text{max}} - V_{\text{min}}$), and x represents either the mass (m), area (a) or volume (v) of the electrode material for normalizing the capacity to either one of the aforementioned parameters of the electrode for better comparison to literature normalizing to area or volume also applies to capacitance of the electrode as is the mass.

1.3 Electrode Materials in Hybrid Supercapacitor Applications

The development of electro-active materials that can endorse energy storage and conversion devices to meet modern energy requirements is ultimately supreme, since generally in supercapacitors, the role of active materials in energy generation and storage is of paramount importance. The choice of electrode dictates and limits the electrical properties of the device [62], and ideal supercapacitor electrodes should possess exceptional thermal and chemical stability, a high specific surface area, high corrosion resistance, high electrical conductivity, and adequate surface wettability. Moreover, they should also be low-cost and environmentally friendly. The electrodes competency for charge transfer is vital for improving the devices' capacitance performance; besides the surface area, other crucial factors such as pore structures, shapes, and size distribution as well as their accessibility to electrolyte ions also affect the capacitance.

In addition, as the asymmetric electrodes of the hybrid supercapacitor device afford the device with a higher working potential, an improved electrochemical performance (in terms of energy density) can be realized by further improving the working potential. The selection of the electrode materials is important in determining the potential window of the device. Therefore, appropriately matching the high-power EDLC-type electrode with the high-energy battery-type electrodes results in a much-improved energy density due to the improved voltage window. In this section, we look at EDLC-type and battery-type materials as electrodes for hybrid supercapacitors.

1.3.1 Carbon Materials as Electrodes in Hybrid Supercapacitors

Carbon materials play a remarkable role as they exist in various and structurally stable forms due to their flexible electronic configuration [63]. Excellent properties, which include cost-effectiveness, natural abundance, nontoxicity, and environmentally friendly nature, as well as high electronic conductivity and excellent mechanical stability, have afforded carbon-based electrode materials a preferential role as electrode material in hybrid supercapacitors [7, 64]. An EDLC capacitance is usually attained from these materials, which present a tailorable porous structure, large surface area and high electrical conductivity, as well as favourable surface chemistry. These features, together with a range of shapes, sizes, and pore size distributions, are exploited extensively to develop electrode materials with optimum electrochemical energy storage performance [65]. Moreover, their thermal stability as well as chemical and electrochemical stability in various electrolytes (acidic to basic media), symmetrical galvanostatic charge–discharge profiles and the ideal rectangular cyclic voltammetric curves validate the use of carbon-based materials in supercapacitors. A typical reaction at the negative carbon electrode in a hybrid supercapacitor is shown in Eq. 1.15:

$$\text{Carbon} + M^{+} + e^{-} \underset{\text{Discharging}}{\overset{\text{Charging}}{\rightleftarrows}} \text{Carbon}/M, \tag{1.15}$$

where M^+ is a cation [33]. In this section, we look at the properties of activated carbon, carbon nanotubes graphene, and metal–organic framework-derived carbon materials as some of the carbon-based EDLC-type materials used as electrodes in hybrid supercapacitors.

1.3.1.1 Activated Carbons

Owing to qualities such as large theoretical surface area (~ 3000 $m^2\ g^{-1}$), robustly tunable porous structure (pore size), low cost, and scalable production, activated carbons have gained popularity as the first choice and most used carbon-based electrode material for supercapacitors in research and commercial purposes [66–70]. They are majorly prepared via the oxidation of bulk carbon in water vapour, KOH, or CO_2, otherwise known as an activation process. The process creates a three-dimensional porous network (see Fig. 1.4c) with micropores (pore size < 2 nm), mesopores (2–50 nm), and macropores (> 50 nm), giving rise to the high specific surface area associated with these carbon materials. Low-cost activated carbons can be produced from carbonization and activation of natural carbon precursors such as wood, coconut shells, palm kernel shell, or anthracite, which are amorphous with a network structure majorly consisting of sp^2 and some sp^3 bonded atoms [71]. Also, metal–organic frameworks such as silica, zeolite, and MgO templates have been demonstrated to be effective in modifying the pore size in activated carbons [72, 73].

However, a decrease in the usable surface area, which leads to a decreased specific capacitance is usually encountered with activated carbons because a large portion of the micropores is inaccessible to the electrolyte. This is due to the incompatibility of the pores of the activated carbon materials with the size of electrolyte ions (Figs. 1.4a and 1.5a), limiting the specific surface area to between 1000 and 2000 $m^2\ g^{-1}$ [76]. Thus, the capacitance of activated carbons is optimized with the use of compatible electrolytes, with activated carbons with pore sizes in the range of 0.4–0.7 nm, and ~ 0.8 nm proven to be well suited for aqueous, and most organic electrolytes, respectively [67, 68, 74]. A study by Largeot et al. [74] demonstrated the relation of the activated carbons pore sizes to the size of the electrolyte ions, and how it impacts on the capacitance of activated carbons. The group showed that activated carbons with pore sizes that are similar to the size of electrolyte ions displayed optimum EDL capacitance, as seen in Fig. 1.4a, b), whereas attenuation of capacitance is observed when using activated carbons with pore sizes that are different from the electrolyte-ion sizes.

Moreover, an enhanced capacitive performance is realized by introducing surface functionalities on activated carbons. The introduction of heteroatoms such as oxygen,

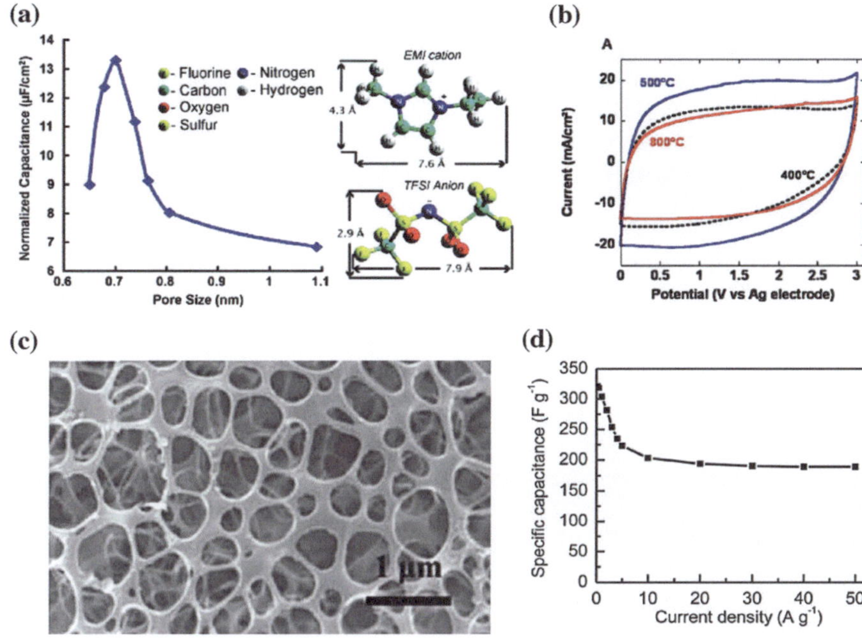

Fig. 1.4 **a** Activated carbon capacitance as a function of pore size. The images on the right show the size of the electrolyte ions that are within the range of the activated carbon pore size. The maximum capacitance was obtained when the pore size was like the size of the ions used. **b** Cyclic voltammetric curve showing typical EDLC behaviour in activated carbon. Reprinted with permission of American Chemical Society from [74]. **c** Scanning electron microscopic image showing the 3D hierarchical porous carbon electrode structure. **d** C_{sp} as a function of current density ranging from 0.5 to 50 A g^{-1}. Reprinted with permission of Wiley from [33], Reprinted with permission of Royal Society of Chemistry from [75]. (EDLC—Electrochemical double-layer capacitors, EMI–TFSI—1-ethyl-3-methylimidazolium-bis(trifluoromethylsulfonyl)imide)

nitrogen, boron, and sulphur into the porous carbon network makes the activated carbons hydrophilic, resulting in improved adsorption of ions and bequeathing activated carbons Faradaic charge transfer reactions that result in about 5–10% increase in the C_{sp} [78–81]. Apart from the introduction of surface functionalities and using compatible electrolytes, the electrochemical performance of the activated carbons could be improved through several different strategies. These include the Fermi level position modification of the activated carbons using ultrasonic radiation [82], and combining activated carbons with other carbon nanomaterials, e.g., carbon nanotubes, carbon nanofibres or conducting polymers to form composite electrodes [83–85].

Wang et al. [86] fabricated a hybrid supercapacitor using activated carbon as the negative electrode and Ni_3S_4 as the positive electrode. The activated carbon//Ni_3S_4 device showed energy power densities of 18.625 Wh kg^{-1} and 1500.2 W kg^{-1}, respectively. The hybrid supercapacitor device showed a maximum C_{sp} of 59.6 F g^{-1}

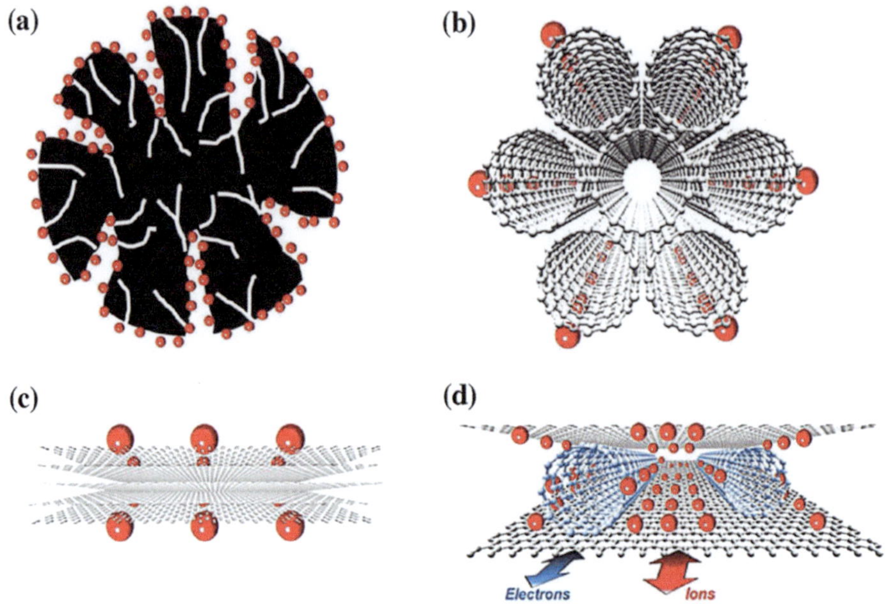

Fig. 1.5 Intrinsic limitations of different carbon materials. **a** Activated carbon has a large surface area, but many pores are inaccessible by electrolyte ions. **b** Carbon nanotubes: form bundles and only the outer surface is available for electrolyte–carbon nanotube interaction. **c** Graphene sheets: tend to agglomerate due to Van der Waal interactions and thereby electrolyte ions refrain from reaching small pores. **d** Graphene/carbon nanotube composite: carbon nanotubes act as spacers and binders from graphene sheets, as well as provide large conduction pathways to electrons and ions. Reprinted with permission of the Royal Society of Chemistry from [77]

at a current density of 0.2 A g^{-1}, as well as an excellent cycling performance with retention of about 93% of the capacitance over 5000 cycles. Other examples of hybrid supercapacitor devices that employed activated carbons as well as the succeeding carbon-based materials as electrode materials will be relayed as we progress with this discourse in the next section where we will discuss the different battery-type electrodes.

1.3.1.2 Carbon Nanotubes

Carbon nanotubes have found use as EDLC-type electrodes in hybrid supercapacitors as an alternative to the activated carbons that suffer from poor electrical conductivity which leads to a higher internal resistance and the inaccessibility of electrolyte ions to reach into their micropores and/or interior atoms at higher scan rates [47, 87, 88]. The one-dimensional ("quasi 1D") carbon nanotubes present unique and favourable properties that include an accessible external structure having a high and specific internal mesoporous network, chemical and thermal stability, high electrical conductivity, and low mass densities. In addition, their mechanical flexibility and open framework afford carbon nanotubes with a link to active material, greater interconnection, and the ability to form a continuous pathway that leads to maximum utilization of the specific surface area. Despite having reasonably lower theoretical surface areas than activated carbons (~ 50–1315 $m^2\ g^{-1}$), carbon nanotubes have demonstrated higher capacitances compared to other activated carbons as a result of their unique tubular structures with a high density of mesopores that allow for fast charge/ion transport [89, 90]. Multi-walled carbon nanotubes (MWCNTs), the randomly oriented and entangled variant of the carbon nanotubes, were the first to be used as electrode materials in supercapacitors and were able to achieve a C_{sp} in the range of 102–135 F g^{-1} [91, 92]. On the other hand, single-walled nanotubes (SWNTs) display superior electrochemical performances because of their large specific surface area (~ 1600 $m^2\ g^{-1}$), high aspect ratio, and improved accessibility for the electrolyte ions compared to MWCNTs, [93–95].

Usually, carbon nanotubes are synthesized through the catalytic decomposition of hydrocarbon precursor material, with control of parameters such as crystalline orders to influence the conformation of the resulting carbon nanotubes to be either single- or multi-walled [64]. The single-walled carbon nanotubes tend to form bundles, which reduces their surface area and restricts the interaction with the electrolyte ions to the outermost surface of the substrate (see Fig. 1.5b) [77]. Manipulation of the morphology in the synthesis process circumvents the problems associated with the tangled carbon nanotubes by producing aligned carbon nanotubes. The latter provides unbundled structures that offer more mesopores and accessible surfaces. Besides, they display a reduced contact resistance during the charging-discharging process, which leads to enhanced energy and power densities [96, 97].

Chen et al. [98] synthesized highly ordered MWCNT-array electrodes using the anodic aluminium oxide template, which showed a high specific capacitance of 365 F g^{-1}. Lu et al. [99] used a template-free chemical vapour deposition method to grow aligned MWCNTs electrodes that showed an improved C_{sp} of 440 F g^{-1}. Wang's group assembled a hybrid supercapacitor device comprised of hierarchical porous nitrogen-doped carbon nanotubes (HPNCNTs) as the negative electrode and NiO nanoparticles grown on highly conductive three-dimensional graphene foam /NiO as the positive electrode [51]. The device was able to achieve a specific capacitance of 116 F g^{-1} at a current density of 1 A g^{-1} and retained 61 F g^{-1} at a 60 A g^{-1} current density achieving a full charge–discharge within 2.8 s and showed an energy density of 32 W h kg^{-1} at a power density of 700 W kg^{-1} in the voltage range from 0.0 to 1.4 V. Moreover, the NiO/GF//HPNCNT "asymmetric" supercapacitor showed outstanding cycling stability, with a retention of 94% of the specific capacitance after 2000 cycles.

Although there have been extensive research efforts aimed at exploiting carbon nanotubes as supercapacitor electrode material, the lack of effective and scalable synthesis methods restricts their widespread implementation in commercial devices. In addition, bundling of the carbon nanotubes poses a big challenge as the bundling leads to a decreased ionic transport efficacy, and the high cost associated with the purification of carbon nanotubes is another factor limiting their introduction into the marketplace [33].

1.3.1.3 Graphene

Graphene is a one-atom-thick 2D layer comprised of sp^2-bonded carbon atoms arranged in a honeycomb-like crystal lattice [10, 100–103]. This “wonder material” presents excellent features that include a remarkably large theoretical surface area (2630 $m^2\ g^{-1}$), great mechanical strength, high electrical and thermal conductivities, high chemical stability, as well as a high carrier mobility [104–109]. Moreover, using the 2D structure of graphene as the electro-active material reduces the thickness of the electrode, which results in an improved potential window, superior flexibility, as well as high chemical and thermal stability [109, 110]. These outstanding properties make graphene a favourable electrode material candidate for energy storage and conversion applications [103]. Moreover, the high inherent electrochemical double-layer capacitance of about ~ 21 $\mu F\ cm^{-2}$ of graphene can deliver a large specific capacitance as high as 550 $F\ g^{-1}$, which is far higher than that of most EDLC-type electrode materials, including activated carbons, carbon nanotubes, mesoporous carbon, and xerogels [64, 111, 112].

Several routes have been developed to synthesize various types of graphene, which include micromechanical exfoliation of graphite, chemical vapour deposition, epitaxial growth, unzipping of carbon nanotubes, electrochemical and chemical approaches, arch discharge method and intercalation techniques in graphite [113–115]. Demonstrated in Fig. 1.6 is one such method, which entails the synthesis of N-doped porous graphene through chemical treatment. The porous graphene was obtained by oxidizing the epoxy groups on the surface of reduced graphene oxide through treatment with H_2O_2, as well as from the hydrolysis of hydroxyl functional groups by breaking parallel C–C bonds [116, 117].

Graphene and its derivatives exhibit an impressive combination of favourable features such as conductivity, surface area, and mechanical strength, and this results in enhanced electrochemical performance in comparison to activated carbons and carbon nanotubes [112, 119]. Nonetheless, graphene sheets are prone to forming irreversible agglomerates of graphite via weak van der Waals forces attraction. As seen in Fig. 1.5c, this leads to a substantial reduction of the specific surface area, which ultimately restricts ion diffusion and the specific capacitance [120–122]. To address this issue, scientists have suggested methods such as gas–solid reduction processes, the use of weak reducing agents, and adding a spacer between graphene sheets to prevent them from restacking [123–125].

Liu and company reported on the synthesis of graphene nanosheets with a unique curved morphology that prevents them from restacking [105]. The curved morphology was also able to endorse the formation of mesopores that are accessible to ionic electrolytes, which bolstered the energy density to 90 $W\ h\ kg^{-1}$. Cheng’s group suggested the use of carbon nanotubes as spacers in between graphene sheets as a means of preventing agglomeration and lowering internal resistances, as

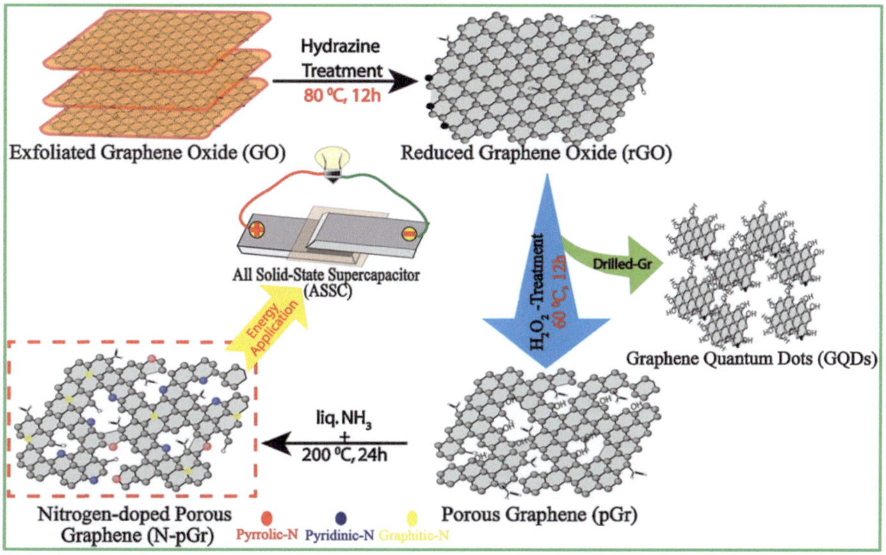

Fig. 1.6 Synthesis of nitrogen-doped porous graphene and graphene quantum dots and their potential application for the fabrication of all-solid-state asymmetric supercapacitor devices. Reprinted with permission of Elsevier from [118]

well as broadening the diffusion pathway for electrolyte ions (Fig. 1.5d) [77]. The carbon nanotubes acted as a binder for the graphene nanosheets, providing them with improved mechanical stability and a longer cycle life.

Generally, chemical reduction or functionalisation of graphene produces a substrate having poor electrical conductivity (100–200 S m^{-1}), which has resulted in several instances where dire electrochemical performances were observed [126]. Lately, graphene-based 3D frameworks such as foams, sponges, hydrogels, and aerogels have emerged as promising materials for the fabrication of ultra-lightweight electrode materials due to their stability, large surface area, interconnected micro-/macroporous structures, and excellent electrical conductivity [127–130]. Their unique 3D structural design not only prevents the agglomeration of graphene nanosheets with high efficacy but also enables fast transportation of the electrolyte ions. Moreover, their freestanding nature bypasses the need for electrochemically inactive binders that suppress the ionic diffusion within the electrode material by blocking the electro-active sites [131–133].

In a bid to enhance the specific capacitances of cobalt sulphide and reduced graphene oxide for application as electrode materials in a hybrid supercapacitor, Xu's group modified the electrode materials with nitrogen-doped carbon dots The as-assembled CoS/N-CDs//rGO/N-CDs hybrid supercapacitor device achieved an energy density of 36.6 Wh kg^{-1} at a power density of 800 W kg^{-1}, and even at a higher power density of 16,000 W kg^{-1}, it could still maintain 25.6 Wh kg^{-1}. The device also showed good cycling stability with 85.9% capacitance retention after 10,000 cycles [134].

1.3.1.4 Metal–Organic Framework Derived Carbon Materials

Metal–organic frameworks have always been considered preferable templates for synthesizing highly porous functionalized electrode materials that include carbons, metal compounds, and many composite materials because of their dual organic–inorganic nature. In the case of hybrid supercapacitors, this is particularly attractive as there is still difficulty in producing densely porous electrode materials. Hence, there has been an intensive use of metal–organic frameworks in the production of both EDLC-type and battery-type hybrid supercapacitor electrode materials in recent years. Usually, various metal–organic frameworks are custom tailored through careful consideration of the reactant metal salts and organic ligands to meet particular requirements [135].

The metal–organic frameworks are usually employed as self-sacrificial templates to produce nanoporous carbons that possess favourable structural porosity features such as a reasonable surface area and competent pore structures. In comparison to nanoporous carbons produced via conventional precursors (such as biomass), metal–organic framework-derived nanoporous carbons present conveniences that include facile synthesis, high surface area and porosity, as well as easy control over pore size and volumes [136]. Liu's group reported one of the pioneering works that used metal–organic frameworks as a precursor material for the preparation of nanoporous carbons [137]. The group impregnated the pores of MOF-5 with furfuryl-alcohol and carbonized the mixture at 1000 °C in the Ar atmosphere. The as-prepared nanoporous carbons presented a high specific area, as well as an impressive electrochemical performance when exploited as an EDLC-type electrode material.

Yi et al. [138] used ZIF-8 polyhedrons to produce porous carbon polyhedrons for application as EDLC—48 + HU-type electrode material. The team assembled a hybrid supercapacitor device using the EDLC—48 + HU-type porous carbon polyhedrons as an anode and core–shell structural carbon nanotubes-nickel oxide nanosheet composites as a cathode. The as-assembled carbon nanotubes/NiO// porous carbon polyhedrons device was able to reach energy and power densities of 25.4 Wh kg^{-1} and 400 W kg^{-1}, respectively. Furthermore, the device displayed a full charge/discharge time of 4.4 s and was able to retain the energy and power densities at 9.8Wh kg^{-1} and 16 kW kg^{-1}, respectively, with a working potential window range of 0–1.6 V. The hybrid supercapacitors further showed excellent cycling stability, with 93% retention of the initial capacity after 10,000 cycles at a current density of 6 A g^{-1}.

Tang et al. [139] prepared nanoporous carbon anode material by treating MOX–Al xerogels derived from MIL-100 (Al) metal–organic frameworks with NaOH to augment the surface area and to also eliminate the inactive Al atoms inside the carbon substrate. The nanoporous carbon anode, designated MOXC–Al, was assembled with a MOCX–Fe (Fe_3O_4/Fe/C), a cathode material derived from MOX–Fe metal–organic framework to fabricate a hybrid supercapacitor device. The Fe_3O_4/Fe/C// nanoporous carbons device showed a C_{sp} of 202.5 F g^{-1} at a current density of 0.5 A g^{-1}, as well as energy and power densities of 17.496 Wh kg^{-1} and 388.8 W kg^{-1}, respectively. Other works on hybrid supercapacitors based on metal–organic framework-derived carbon material include but are not limited to transition metal-nanorods//transition metal—nanoporous carbons [136], Carbon//Co_3O_4 [140], NCMO// nanoporous carbons [141].

1.3.2 Battery-Type Materials as Electrodes in Hybrid Supercapacitors

Generally, battery-type electrode materials comprise the redox-active oxide, hydroxide, sulphide, and selenide compounds of the Ni, Co, Cu, Cd, Mn^{II}, and Fe transition metals. In alkaline media, these electrode materials undergo a reaction with the OH^- ion in which the transition metals are oxidized from the lower to the higher valance states, releasing electrons in the charging process. During the reverse discharging process, the transition metal is reduced to its original lower valance state. As a result, at least one phase change can be observed with the low valance compounds during each charge and discharge process, which results in the battery-like charge storage behaviour. These electrode materials are promising for hybrid supercapacitor applications since they demonstrate comparable kinetics to the EDLC-type materials in special device designs, as well as complementary high electrochemical performances as a result of the battery-like charge storage mechanism [57]. Some examples of the battery-type electrode materials, together with some typical pseudocapacitive electrode materials, are listed in Table 1.1. The application of pseudocapacitive electrode materials in supercapacitors is reviewed in refs. [142] and [143].

1.3.2.1 Transition Metal Oxides

Binary transition metal oxides such as Fe_2O_3, NiO, CoO, and Co_3O_4 have received extensive research attention as hybrid supercapacitor electrode materials mostly due to their exceptionally high theoretical specific capacitances, natural abundance and low cost, and facile synthetic methodologies for preparing nanomaterials with tailored morphologies, structures, sizes, and composition. These transition metal oxide electrode materials have been demonstrated to achieve electrochemical stability in long-term charge–discharge cycling. For example, Ding et al. [144] reported on flower-like Co_3O_4 nanostructures made up of mesoporous hexagonal nanosheets, which showed 96.07% capacitance retention after 10,000 cycles at 5 A g^{-1}.

On the other hand, it is generally difficult to achieve such high cycling performance as these battery-type transition metal oxides portray a semiconductive behaviour, and as a result, suffer from poor electrical conductivity and attenuated electrochemical activity in comparison to hydroxides [57]. Thus, intensive research devotion has been aimed at improving the electrochemical capacitive performance of transition metal oxides for hybrid supercapacitor applications. In the case of Co_3O_4, many efforts have been devoted to preparing different nano-morphologies such as nanosheets [145], nanotubes [146, 147], nanowires [148, 149], nanobelts [150], and nanocapsules [151]. Nanomaterials are used to replace bulk electrode material in a bid to overcome the large volume change of Co_3O_4 during the charge and discharging process that results in a reduced cycle life. Generally, nanostructures can cushion the strain from swelling of the electrodes and avert pulverization, to enhance the cycling performance. In addition, improved electrochemical performance is realized from the nanosized electrode material as they afford a large specific surface area, as well as reduced ion diffusion and electron transport pathways, which lead to fast reaction kinetics.

Sivakumar et al. [152] reported on the solvothermal-calcination synthesis of Co_3O_4 nanohorn for application as a cathode for hybrid supercapacitors. Using activated carbon as the anode, the Co_3O_4 nanohorn//activated carbon hybrid supercapacitor showed a specific capacitance of 101 F g^{-1} at a current density of 0.5 A g^{-1}. The device further showed high cyclic stability with 91.73% retention of the initial capacitance after 350 000 cycles at a current density of 20 A g^{-1}, as well as high coulombic efficiency approximating 100% throughout cycling tests. The system also showed a maximum energy density of ~ 31.70 Wh kg^{-1} with a power density of ~ 424.29 W kg^{-1}, as well as a maximum power density of ~ 16 710 W kg^{-1} with an energy density of ~ 25.52 Wh kg^{-1}. The authors ascribed the hybrid device's electrochemical performance to the unique structure of the Co_3O_4 nanohorn. The structural and electrochemical results are shown in Fig. 1.7.

Conversely, immense attention has been given to ternary metal oxides, which use multiple metal cations. This kind of oxide usually offers improved electric conductivity and electrochemical performance compared to binary metal oxides. As an example, the spinel-structured ternary $NiCo_2O_4$ has a Ni cation occupying the octahedral sites, whereas Co cations occupy both the octahedral and tetrahedral sites [153]. Both the Ni and Co ions are observed to have mixed valence states of + 2 and + 3, consequently giving the oxide a general formula of $Co^{2+}_{1-x}Co^{3+}_{x}(Co^{3+}Ni^{2+}_{x}Ni^{3+}_{1-x})O_4$ $(0 < x < 1)$ [154]. As a result, an enhanced electrical conductivity is realized from the multivalences of transition metals in the $NiCo_2O_4$ ternary compound. Liu's group reported a conductivity value reaching 62 S cm^{-1} for a single crystal $NiCo_2O_4$ sample, as well as 0.6 S cm^{-1} for a polycrystalline sample at 300 °C [155]. These results are at least two orders of magnitude higher compared to those of binary metal oxides of NiO and Co_3O_4. Moreover, the bimetallic $NiCo_2O_4$ affords hybrid supercapacitor devices with improved electrochemical activity. Song's group assembled hybrid supercapacitor devices using activated carbon as the anode and $NiCo_2O_4$ nanosheet arrays grown in situ on nickel foam as the cathode. Using PVA-CMC-KOH as the electrolyte, the flexible solid-state $NiCo_2O_4$/nickel foam//activated carbon hybrid supercapacitor showed a specific capacitance of 186.7 F g^{-1} at a current density of 1 A g^{-1}, with a rate performance of 69.6% at a current density of 30 A g^{-1}. The device was able to retain 73.7% and 81.1% of the initial capacitance for the

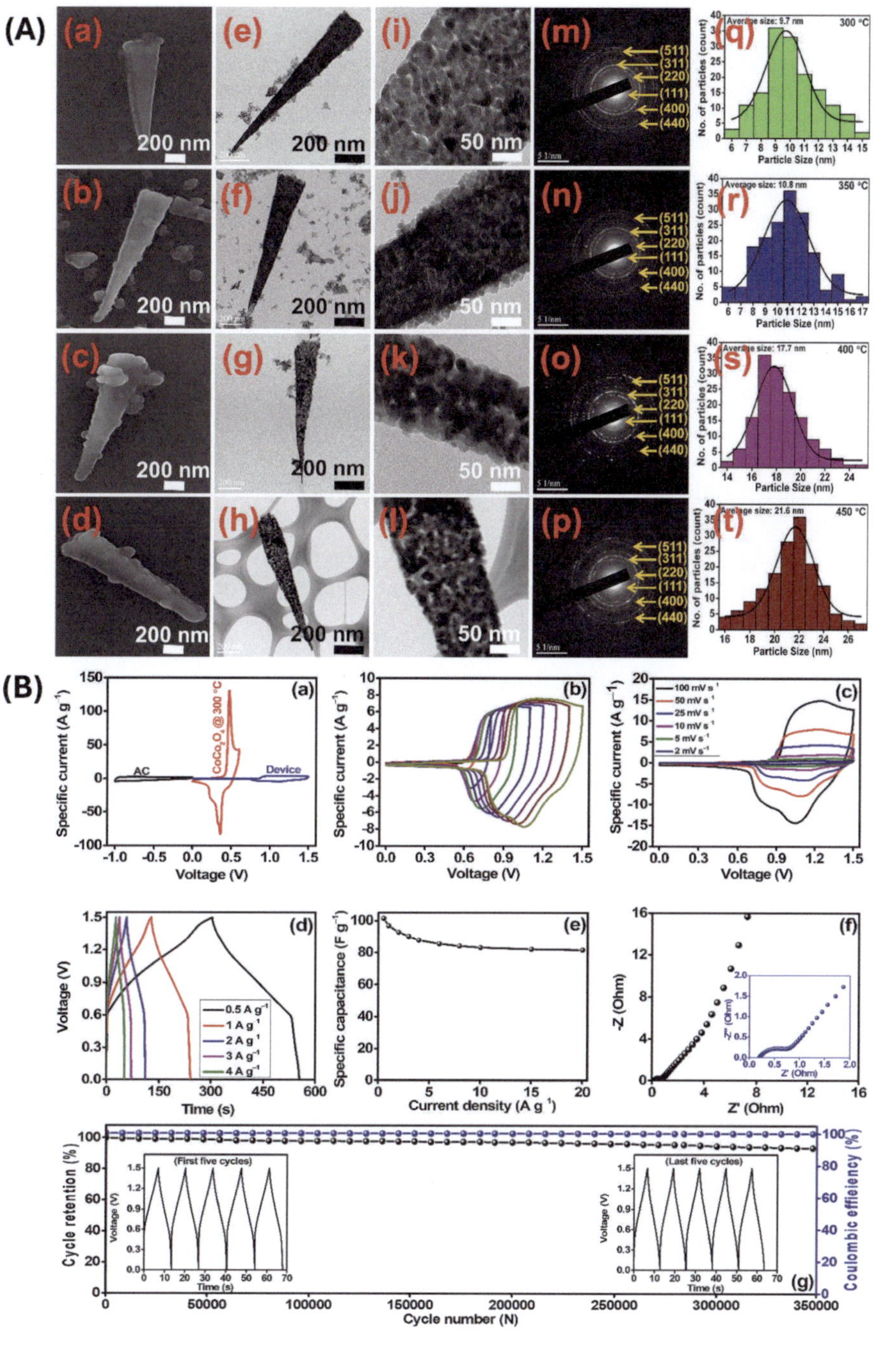
(A)
(a)
(b)
(c)
(d)
(e)
(f)
(g)
(h)
(i)
(j)
(k)
(l)
(m)
(n)
(o)
(p)
(q)
(r)
(s)
(t)
200 nm
50 nm
(511)
(311)
(220)
(111)
(400)
(440)
Average size: 9.7 nm
300 °C
Average size: 10.8 nm
350 °C
Average size: 17.7 nm
400 °C
Average size: 21.6 nm
450 °C
No. of particles (count)
Particle Size (nm)
(B)
Specific current (A g⁻¹)
Voltage (V)
AC
Device
CoCo2O4@300 °C
100 mV s⁻¹
50 mV s⁻¹
25 mV s⁻¹
10 mV s⁻¹
5 mV s⁻¹
2 mV s⁻¹
Time (s)
0.5 A g⁻¹
1 A g⁻¹
2 A g⁻¹
3 A g⁻¹
4 A g⁻¹
Specific capacitance (F g⁻¹)
Current density (A g⁻¹)
-Z (Ohm)
Z' (Ohm)
Cycle retention (%)
Coulombic effiency (%)
(First five cycles)
(Last five cycles)
Cycle number (N)

◂**Fig. 1.7** **A** Co_3O_4 nanohorns at 300, 350, 400, and 450 °C: **a–d** high-resolution transmission electron microscopic image, **e–l** transmission electron microscopic (low and high magnification) images, **m–p** selected area electron diffraction patterns, and **q–t** histogram of the particle size distribution, estimated from transmission electron microscope. **B**: **a** cyclic voltammetric curves of the activated carbon, Co_3O_4@300 °C, and hybrid supercapacitor, **b** cyclic voltammetric curves at different potential windows, **c** Cyclic voltammetric curves at different scan rates (0–1.5 V), **d** galvanostatic charge–discharge plots at different current densities, **e** specific capacitances at different current densities, **f** nyquist plots of electrochemical impedance spectroscopy, and **g** cycle stabilities and coulombic efficiency of the Co_3O_4@300 °C nanohorn//activated carbon hybrid supercapacitor. Reprinted with permission of Elsevier from [152]

charge and discharge processes, respectively, after 5000 cycling routines at a current density of 1 A g^{-1}. It further showed a 1 of 55.8 Wh kg^{-1} with a specific power density of 733.7 W kg^{-1}, as well as a maximum specific power density of 22,500 W kg^{-1} with a specific energy density of 40.6 Wh kg^{-1} [156].

Table 1.2 further lists and summarizes some of the work reported on transition metal oxide-based hybrid supercapacitors. The various EDLC-type materials are also indicated as used in the different reported hybrid supercapacitor devices. The same trend is adopted in the sub-sections to follow, which cover hybrid supercapacitor devices based on transition metal hydroxides and chalcogenides.

1.3.2.2 Transition Metal Hydroxides

Transition metal hydroxides usually store charges through deprotonation/protonation reactions with the OH^- ion following a reaction path outlined in Eq. 1.16, with some compounds undergoing secondary reactions as outlined in Eq. 1.17 for $Co(OH)_2$ [56, 168]:

$$TM(OH)_2 + OH^- \leftrightarrow TMOOH + H_2O + e^- \quad (1.16)$$

$$CoOOH + OH^- \leftrightarrow CoO_2 + H_2O + e^-. \quad (1.17)$$

In general, transition metal hydroxides are found with either α- or β- phase crystal structures. The α-phase hydroxides consist of a hydrotalcite-like and turbostratic-disordered structure that is composed of positively charged hydroxyl-deficient $M(OH)_{2-x}$ layers intercalated by anions or water. The existence of intercalating species in the structure of the α-phase hydroxides leads to enhanced electro-activity as a result of the larger interlayer spacing (> 7.0 Å), compared to that of the β-phase counterpart materials (4.6 Å). On the other hand, the α-phase material is structurally unstable and is easily retorted to the β-phase in alkaline electrolyte. The easy-to—synthesize β-phase hydroxides have a brucite-like structure that comprises hexagonal closed-packed unit cells having a stoichiometric composition of the hydroxide anions and metal cations that occupy alternate rows of octahedral sites. The β-phase hydroxides suffer from poor electrical conductivity due to their p-type semiconductor

Table 1.2 Summary of hybrid supercapacitor devices built with transition metal oxide-based battery-type electrode materials

Hybrid supercapacitor device fabrication	Electrolyte (KOH)	V_{max} (V)	Specific capacitance (F g^{-1})	E_{max} (Wh kg^{-1})	P_{max} (kW kg^{-1})	References
AC//ZNACO	2 M	1.5	232	72.4	24.77	[43]
HPNCNT//NiO/GF	1 M	1.4	116	32	42	[51]
(APDC)//Ni–Co oxide	2 M	1.6	202	71.7	16	[157]
NF–AC// $Ni_xCo_{3-x}O_4$	2 M	1.6	105	37.4	4.1	[158]
AC//KOH/β–$Ni(OH)_2$/NF	6 M	1.6	105.8	36.2	687.3	[159]
AC//$NiCo_2O_4$–rGO	2 M	1.3	99.4	23.3	12.99	[160]
Co_3O_4–rGO//AC	6 M	1.5	114.1	35.7	3.75	[161]
$ZnCo_2O_4$//F–RGO	2 M	1.6	66	84.48	4.0	[162]
AC/NF//Co_3O_4–rGO/NF	1 M	1.6	83.68	30	1.2	[163]
AC//P–$Co_{0.21}Ni_{0.79}MoO_4$	2 M	1.5	158.5	49.2	–	[164]
AC//Ni–Co oxide	1 M	1.2	60	12	1.903	[165]
Co_3O_4/RGONR//RGONR	1 M	1.6	–	64.2	8.3	[166]
$NiCo_2O_4$/NF//AC	PVA-CMC-KOH	1.5	186.7	55.8	22.5	[156]
CA//$NiCo_2O_4$	2 M	1.6	–	47.5	13.0	[167]

KOH—Potassium hydroxide, AC—Activated Carbon, NF—Nano foam, RGO—Reduced graphene oxide, PVA—Polyvinyl alcohol, CMC—Carboxymethyl-cellulose nature and also have metastable crystal structures [169]. These drawbacks lead to a lethargic charge storage kinetics, and hence a poor capacity retention rate and short cycle lifetime.

Many attempts have been made to bolster the capacity retention and cycling stability of the transition metal hydroxides by fabricating nanostructured electrode materials of these compounds. Taking $Ni(OH)_2$ as an example, Lang et al. [170] reported on a hybrid supercapacitor device assembled with activated carbon as an anode and Al-stabilized α-$Ni(OH)_2$ as the cathode. The device showed a specific capacitance of 127 F g^{-1} at a current density of 5 mA cm^{-2} and also showed a maximum specific energy density of 42 Wh kg^{-1}, as well as a maximum specific power density of 1100 W kg^{-1}. The device further showed high cycling stability, with 82% retention of the initial capacitance after 1000 cycles at a current density of 10 mA cm^{-2}.

Doping of metal cations into the crystal phase brings about cyclic stability for the transition metal hydroxide electrode materials. Xie et al. [171] further demonstrated the cycling stability gained by the α–$Ni(OH)_2$ crystal phase when doped with Mg cations. A hybrid supercapacitor built with Mg–$Ni(OH)_2$ as a cathode and activated carbon as the anode showed a 95% retention of the initial capacitance after 10,000 cycles at a current density of 10 A g^{-1}. Alternatively, the pure $Ni(OH)_2$-based hybrid supercapacitor only achieved a 51% retention of the initial capacitance after merely 3000 cycles. The Mg–$Ni(OH)_2$//AC hybrid supercapacitor also showed a maximum specific capacitance of 167 F g^{-1}. Furthermore, the Mg–$Ni(OH)_2$-based device exhibited maximum energy and power densities of 57.9 Wh kg^{-1} and 26000 W kg^{-1}, respectively, which were than those obtained from the pure $Ni(OH)_2$-based device ($E_d = 33.9$ Wh kg^{-1}, $P_{max} = 24$ kW kg^{-1}). The morphological and electrochemical results of the study are shown in Fig. 1.8.

A special type of transition metal hydroxide compound that has also gained traction for application in hybrid supercapacitors is the bimetallic cation-based layered double hydroxide materials. These layered double hydroxide compounds have a hydrotalcite-like crystalline structure with a general formula of $\left[M^{2+}_{1-x}M^{3+}_{x}(OH)_2\right]_{x+}\left[A^{n-}_{x/n}\right]_{x-}$ mH_2O, where M^{2+} and M^{3+} are the metal cations that form octahedral host layers, An^- is the charge-balancing anionic species such as CO_3^{2-}, NO_3^{2-}.. The water molecules are found in the interlayers of the crystal structure. Improved electrochemical performance is achieved from layered double hydroxide in comparison to single metal hydroxide material since the coexistence of the two metal cations brings about two benefits: (1) stabilization of the layered double hydroxide structure to enhance the cycling stability and rate capability; (2) generation of robust synergistic effects in augmenting the electrochemical performance of the bimetallic layered double hydroxide materials (Table 1.3).

The Ni–Co layered double hydroxide exhibits a robust synergistic effect from the coexistence of Ni and Co that leads to improved charge storage ability and long-cycling life. Other transition metal-based layered double hydroxide compounds

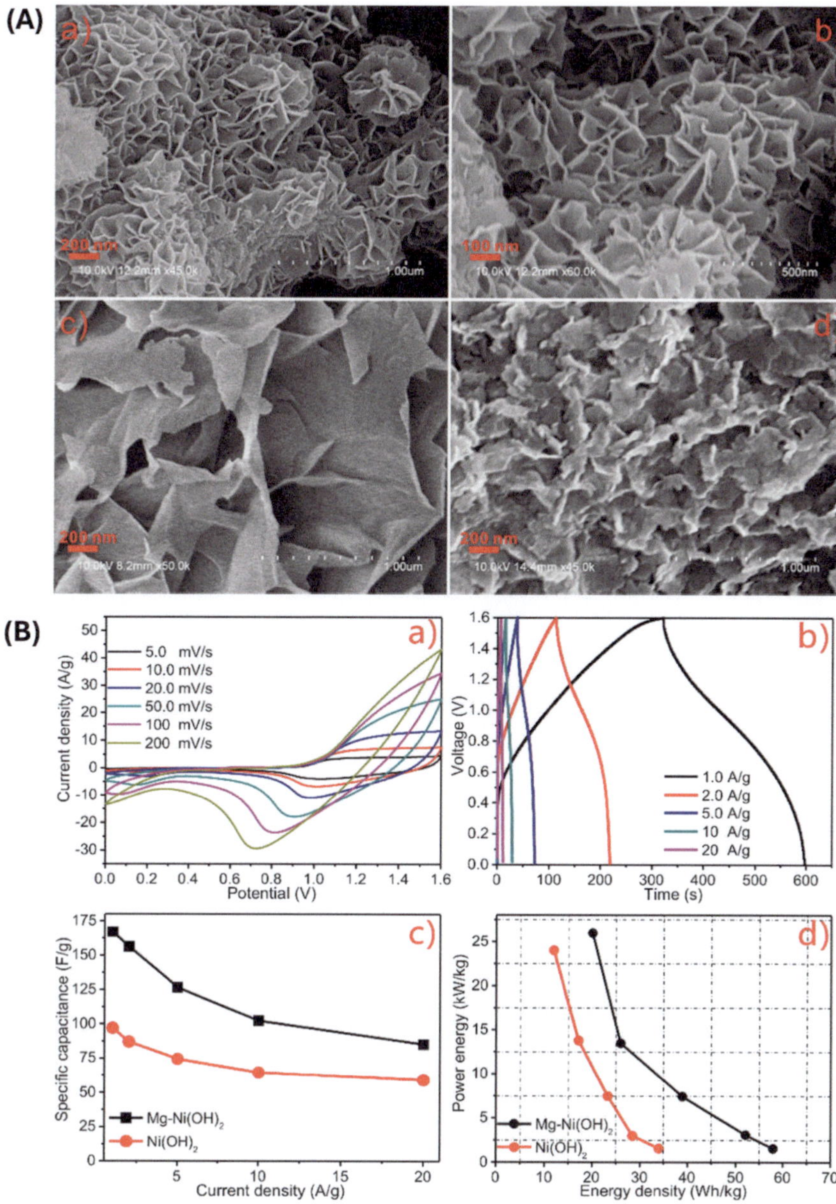

Fig. 1.8 **A** Scanning electron microscopic images of the two nickel hydroxides before and after the long-cycling test at a current density of 10.0 A/g. Scanning electron microscopic image of Mg-Ni$(OH)_2$ before **a** and after **b** 10 000 cycles test; scanning electron microscopic images of dopant-free Ni$(OH)_2$ before **c** and after **d** 3000 cycles test. **B** Electrochemical performance of asymmetric supercapacitor based on Mg–Ni$(OH)_2$ and dopant-free Ni$(OH)_2$ electrodes in 6.0 M KOH. **a** Cyclic voltammograms of Mg–Ni$(OH)_2$ at various scan rates. **b** Charge/discharge curves of Mg–Ni$(OH)_2$ at different current densities. **c** Specific capacitances versus current densities. **d** Ragone plots of the two asymmetric cells. Reprinted with permission of the American Chemical Society from [171]

Table 1.3 Summary of hybrid supercapacitor devices built with transition metal hydroxide-based battery-type electrode materials

Hybrid supercapacitor device fabrication	Electrolyte (KOH)	V_{max} (V)	Specific, areal capacitance (F g^{-1}, *F cm^{-2}) or **specific capacity (C g^{-1})	E_{max} (Wh kg^{-1}, ˠmW h cm^{-3})	P_{max} (kW kg^{-1}, ˠmW cm^{-3})	References
AC//$NiCo_2S_4$@$Ni(OH)_2$	2 M	1.7	132.7	53.3	6.42	[16]
AC//Al–α–$Ni(OH)_2$	2 M	1.6	127	42.3	1.10	[170]
CNT@CF// $Ni(OH)_2$@CF	6 M	1.4	151	41.1	3.5	[172]
CNT@CF// $Co(OH)_xCO_3$@CF	6 M	1.4	123	33.45	3.5	[172]
AC// C@NiMn–OH–Ni_3S_2/ NF	3 M	1.7	112.8	45.3	12.75	[173]
EG//$Ni(OH)_2$/af–GQDs	1 M	1.8	103.3	46.5	9	[4]
AC//$NiCo_2S_4$@$Ni(OH)_2$	2 M	1.7	132.7	53.3	6.42	[16]
AC// MnO_2–$Ni(OH)_2$–NF/AC	1 M	1.6	5.99*	3.62ˠ	43.23ˠ	[174]
PPD/rGO// $Co_xNi_{1-x}(OH)_2$–rGO	2 M	1.6	–	72	16.7	[41]
AC// NiFe–LDH–RGO–CNFs	6 M	1.57	98.4	33.7	11.78	[175]
AC/CoNiFe–LDH/CNFs	6 M	1.6	84.9	30.2	–	[176]
rGO//CoAl–LDH/ $NiCo_2S_4$	2 M	1.6	201**	44.67	4.0	[177]
AC//$NiCo_2$Al–LDH	2 M	1.7	226.7	91	22.7	[178]
AC//$Co(OH)_2$	2 M	1.6	72.4	92.7	2.395	[179]
AC//CuO@CoFe–LDH	1 M	1.2	9.38*	1.857ˠ	914.5ˠ	[180]
RGO//RGO–$Ni(OH)_2$	1 M	1.6	–	75	40	[181]

AC—Activated Carbon, CNF—Carbon nanofibres, LDH—Layered double hydroxide, CF—Carbon fibre, NF—Nickel foam, KOH—Potassium hydroxide, rGO—Reduced graphene oxide

such as Ni–Mn layered double hydroxide, CoAl layered double hydroxide, and Ni–Al layered double hydroxide have also been proven to achieve long-term cycling stability of over 10,000 cycles [174].

1.3.2.3 Transition Metal Chalcogenides

The transition metal chalcogenide compounds, especially sulphides and selenides, have gained intense research attention due to their variable electronic structures and physical properties that make them attractive for energy storage devices such as hybrid supercapacitors [182–184]. A study by Chhowalla et al. [183] showed an apparent effect brought on by the introduction of chalcogen atoms on the electronic structures of metal compounds, which relates to the broadening of the d bands as well as the correlating decrease of the band gap that occurs going down the chalcogen group. Hence, it would come as no surprise that the transition metal chalcogenides may exhibit greater conductivity compared to analogous oxides [185].

Transition metal sulphides have found application as battery-type electrode materials for hybrid supercapacitors since they show favourable properties such as greater electric conductivity, and enhanced electrochemical activity, as well as a diversity of stoichiometric chemistries in comparison to metal oxides. Mostly, the sulphides are prepared by using solution-based methods, thus eliminating the annealing step generally used in the preparation of oxides. As a result, low crystallinity is obtained from the elimination of the annealing step, and hence a high specific surface area that affords more electro-active sites and good hydrophilic properties for aqueous electrolytes. Moreover, many transition metal sulphides have demonstrated higher conductivity. Some of them like $NiCo_2S_4$ and α–NiS even possess metallic conductivity features, which is very important to boost the power performance. Much like transition metal oxides, transition metal sulphides store charge through reversible surface redox reactions of transition metal ions with OH^-. The process is expressed in Eq. 1.18:

$$TMS + OH^- \leftrightarrow TMSOH + e^-. \tag{1.18}$$

Even though transition metal sulphides are considered vital for energy storage because of their theoretical specific capacity, only electrode materials have shown cycling stability that reaches beyond 10,000 cycling routines [186–189]. To improve the electrochemical performance of the sulphide-based electrode materials for hybrid supercapacitor applications, various attempts have been made to overcome these barriers. One such attempt involves the fabrication of micro-nano transition metal sulphide structures. Chen et al. [190] reported on the synthesis of a sea-urchin-like $NiCo_2S_4$ nanostructure. The as-synthesized nanomaterial showed significantly enhanced electro-capacitive properties that include high electric conductivity, high electro-capacitive activity, as well as improved cycling and rate capabilities compared to single metal Ni or Co sulphide. The material was able to retain 91.4% of its initial specific capacity after 5000 charge–discharge cycles at a current density of 20 A g^{-1}, demonstrating its potential as an ultra-long lifespan supercapacitor material.

Wang's group assembled hybrid supercapacitors using $NiCo_2S_4$ supported on graphene as a cathode and reduced graphene hydrogels as the anode [189]. The $NiCo_2S_4$@G electrode was synthesized using a one-step solvothermal process. The $NiCo_2S_4$@G//RHG device showed good cycling stability, with 84.4% retention of its initial capacitance after 15,000 cycles at a current density of 10 A g^{-1}. Furthermore, it achieved a specific capacitance of 111.8 F g^{-1}, as well as high energy and power densities of 39.5 Wh kg^{-1} at a power density of 1778 W kg^{-1}. The morphological and electrochemical results of the study are shown in Fig. 1.9.

Transition metal selenide presents greater prospects for application in hybrid capacitors since selenium has a lower electronegativity than oxygen and sulphur, making the bond between the Se atom and the bonding electrons weaker. Thus, transition metal selenides present greater electrochemical activity compared to metal hydroxides, oxides, and sulphides. Furthermore, the inherent higher electronic conductivity of selenium (1×10^{-3} S m^{-1}) means the electrode of transition metal selenides is likely to display a higher electric conductivity than transition metal-hydroxides, oxides, and sulphides as the generated selenium can play the role of a conductive additive and play a desirable role during cycling [191–194].

Xia et al. [195] observed a linear increase in resistivity of $Co_{0.85}Se$ with an increase in temperature, which highlighted the metallic nature of the material. However, introducing Ni as a dopant was found to further increase the electric conductivity of $Co_{0.85}Se$. The (Ni, $Co)_{0.85}Se$ material shows a much-improved electric conductivity that is comparable to that of platinum (1.67×10^6 S m^{-1} for $(Ni,Co)_{0.85}Se$ and 9.1 $\times 10^6$ S m^{-1} for Pt). Zhai et al. [196] reported on a mixed metal selenide Co_9Se_8/CdSe nanorods prepared on nickel foam via the selenization of Co_3O_4/CdO. The as-synthesized Co_9Se_8/CdSe electrode showed much higher capacity and better cycling performance compared Co_3O_4/CdO electrodes. In addition to the coexistence of two active components and the integrated nanorod structures, the impressive performance of the Co_9Se_8/CdSe electrode was also credited to the superior capacity properties of selenide species.

Nagaraju's group reported on the synthesis of cauliflower-like nickel selenide nanostructures supported on carbon cloth (Ni_3Se_2@CC) using a single-step electrochemical deposition method for application in hybrid supercapacitors [36]. The device, which employed ac coated on carbon cloth as an anode and the as-synthesized Ni_3Se_2@CC as a cathode, showed a maximum specific capacity of 48.4 mA h g^{-1} at a current density of 1 A g^{-1}. The device was also able to maintain a specific capacity of 36.6 mA h g^{-1} at a current density of 10 A g^{-1}. Furthermore, the device showed a maximum energy density of 32.8 W h kg^{-1} at a power density of 677.03 W kg^{-1}, as well as a maximum power density of 6027.3 W kg^{-1} at an energy density of 22.1 W h kg^{-1}. The hybrid supercapacitor further showed a 98% retention of the initial after 5000 cycles at a current density of 5 A g^{-1} (Table 1.4).

1.3.2.4 Metal–Organic Frameworks

Alternatively, the crystalline materials metal–organic frameworks, in which organic ligands are repeatedly coordinated to metal cations, have received growing interest for application as battery-type electrode material. Much attention into these materials is sparked by their attractive combination of nanotechnology and energy storage,

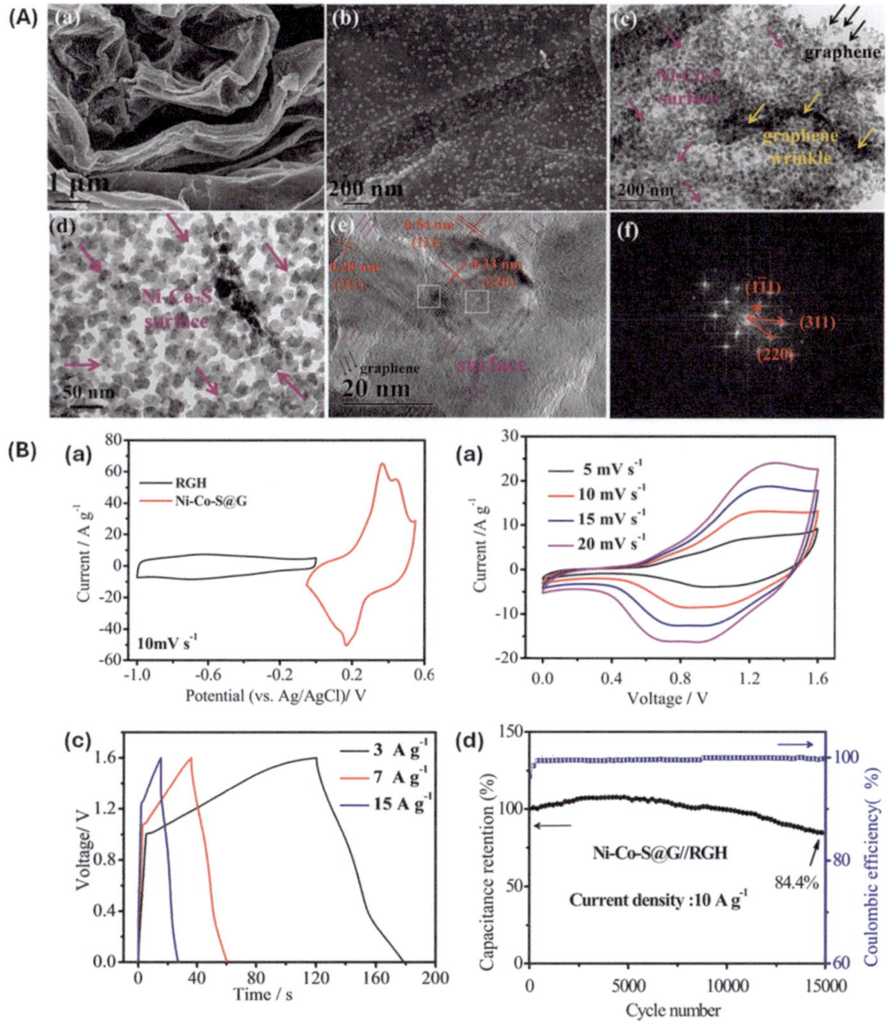

Fig. 1.9 **A**: Typical (**a**, **b**) images and (**c**, **d**) transmission electron microscopic images of the Ni-Co-S@G hybrids with different magnifications; **e** high-resolution transmission electron microscopic image of the Ni-Co-S@G hybrids; **f** Fourier transform image obtained from the white square regions in (**e**). **B**: **a** cyclic voltammetric curves of Ni-Co-S@G hybrids and reduced graphene hydroxide in a three-electrode system at 10 mV s^{-1}, **b** cyclic voltammetric curves of the Ni-Co-S@G/reduced graphene hydroxide hybrid supercapacitor at various scan rates from 5 to 20 mV s^{-1}, **c** the charge–discharge curves of the hybrid supercapacitor at different current densities, and **d** cycling performance and coulombic efficiency of asymmetric supercapacitor devices at a current density of 10 A g^{-1}. Reprinted with permission of Elsevier from [189]

Table 1.4 Hybrid supercapacitor devices built with transition metal chalcogenides-based battery-type electrode materials

Hybrid supercapacitor device fabrication	Electrolyte (KOH)	V_{max} (V)	Specific, areal capacitance or capacity	E_{max} (Wh kg^{-1}, $^{\gamma}$mW h cm^{-3})	P_{max} (kW kg^{-1}, $^{\gamma}$mW cm^{-3})	References
rGO//$NiCo_2S_4$	6 M	1.6	4.68*	31.5	156.6	[197]
AC//NiS NTs/NF	1 M	1.7	1.4*	0.472$^{\gamma}$	21.5$^{\gamma}$	[198]
PG@CC//Ni–C–S@CC	1 M	1.8	13	60.0	28.8	[199]
AC//$NiCo_2S_4$/NF–1	2 M	1.6	111.3	34.7	15.0	[200]
AC// Co_9S_8–NSA	1 M	1.7	52.7	20.0	8.3	[201]
NG//Ni_1Co_1–S/NF	2 M	1.6	164.9	58.1	25.1	[202]
rGO// Fe–Co–S/NF	1 M	1.6	131.0	43.6	5.5	[203]
α–CA//$NiCo_2S_4$/CA	6 M	1.6	95.0	33.8	7.3	[204]
CC//Ni–$Bi_{1.5}$–S/Ni	1 M	1.8	13.28*	2.96$^{\gamma}$	133.3$^{\gamma}$	[205]
AC//$NiCo_2S_4$	3 M	1.6	54.0	19.2	–	[206]
AC//$NiCo_2S_4$@PPy–50	3 M	1.6	97.4	34.6	–	[206]
AC// $Ni_{1-x}Zn_xS$/NF	3 M	1.6	127.7	38.9	6.0	[207]
AC//$ZnCo_2S_4$	3 M		160.0**	101.4	57.5	[208]
AC// C@NiMn–OH–Ni_3S_2/NF	3 M	1.7	112.8	45.3	12.8	[173]
AC//$CoNiSe_2$	3 M	1.6	0.0014*	50.7	42.8	[209]
AC//H–$NiCoSe_2$	6 M	1.5	112	35.0	3.8	[210]
AC//CoSe	2 M	1.5	–	18.6	18.8	[211]
PCNF// Co–Zn–Se–1@CNT–CNF	6 M	1.6	285.2**	61.4	15.4	[212]
AC// NiSe–G	6 M	1.6	141.0	50.1	7.9	[213]
AC// Ni_3Se_2	3 M	1.6	131.1	23.3	2.0	[214]
PC//NiSe NNs	3 M	1.7	78.4**	66.6	8.5	[5]
N–G//NF@rGO/Ag/ Ni_3S_2	2 M	1.7	71.5	28.7	67.9	[24]
PC//$CoNi_2S_4$–rGO	2 M	1.6	152	51.7	11.7	[215]
rGO//CoAl–LDH/ $NiCo_2S_4$	2 M	1.6	201**	44.67	4.0	[177]
CoS/N–CDs//rGO/N–CDs	3 M	1.7	–	36.6	16.0	[216]

Specific, areal capacitance (F g^{-1}, * F cm^{-2}), or Specific, areal capacity (** C g^{-1}, *** cm^{-2}), NF—Nickel foam, rGO—Reduced graphene oxide, G—graphene, CNF—Carbon nanofibre, AC—Activated Carbon, KOH—Potassium hydroxide, CNT—Carbon nanotubes, CD—Carbon dots

owing to their inherent properties that include a high surface area, abundant pores, tuneable nano-architectures, and the presence of multiple functionalities [217–222].

Research work on metal–organic frameworks has largely focused on enhancing the conductivity of metal–organic frameworks, as well as reducing the steric hindrance and variance between the electrolyte and metal–organic frameworks, by developing metal–organic frameworks with uniquely tailored nanostructures. Usually, the conductivity of the metal–organic frameworks is improved by modifying the linker structures inside the metal–organic frameworks, which offer high charge transfer capability inside the framework, and by promoting the redox activity of the metal sites within the metal–organic frameworks to provide electron transportation ways [223]. Alternatively, composite electrode materials of metal–organic frameworks with materials such as carbon materials, metals, and conducting polymers can be used to improve chemical durability and poor conductivity.

In a practical application of metal–organic framework-based electrode material in hybrid supercapacitors, Teffu et al. [224] assembled a hybrid supercapacitor using nanofibres-rGO/MOF nanocomposite as the positive electrode and activated carbon as the negative electrode. When studied in 3.0 M KOH electrolyte, the as-assembled devices exhibited a specific capacity of about 48.81 C g^{-1} at a current density of 0.4 A g^{-1}, with a maximum energy density of 11.0 Wh kg^{-1} and a maximum power density of 640.45 W kg^{-1}. The device further showed good cycling stability by achieving a 70% retention of the initial capacity after 2000 cycles. Other work on metal–organic framework-based hybrid supercapacitors includes Ni–MOF//Activated carbon [225–227], Co-MOF//Activated carbon [228], Ni–MOF/CNTs//rGO/g–C_3N_4 [229], Ni-pPDA//glassy carbon [230], Mn–pPDA// glassy carbon [230], Ni/Co–MOF//activated carbon [231], Zr–MOF//porous carbon [232].

1.4 Perspectives on Hybrid Supercapacitors

As we have briefly shown in this chapter, numerous research efforts have been and continue to be done by researchers in the quest to develop efficient and state-of-the-art energy storage and conversion devices that will be able to meet the energy demands of this rapidly changing modern world. Hybrid supercapacitors emerge as a standout amongst energy storage devices as they present desirable properties that are intermediate to the high-energy density batteries and the high-power density conventional supercapacitors. Thus, the hybrid supercapacitor design, which incorporates the use of an EDLC-type electrode that is countered by a battery-type electrode, comes as a revelation because it has enabled the construction of a single energy storage device that possesses both high-energy and high-power densities. However, challenges such as the lack of low-cost, scalable, and effective procedures that provide more control of the structural features, e.g. porosity and morphology of the electrode materials still hinder the construction of hybrid supercapacitors [233].

Research work around the improvement of hybrid supercapacitors has largely focused on enhancing the electrochemical performance of the devices through material engineering and developing electrodes with robust designs, morphologies, and compositions. Nevertheless, as we highlighted in this chapter, a profound understanding of the charge storage modalities offered by different electrode materials is even more essential in the process of developing new electrode materials and overall devices with specialized designs, enhanced energy and power densities, as well good stability [168]. Furthermore, a good understanding of the effect electrolytes used in hybrid supercapacitors have on the charge storage mechanisms of electrode materials is vital for developing speciality devices such as hybrid supercapacitors. Electrode materials, especially transition metal-based compounds, have been observed to display different charge storage modalities in different electrolytes [233].

With modern technology rapidly changing, hybrid supercapacitors are posed with the benefit of finding applications in numerous sectors since their asymmetric configuration allows them to have features such as flexibility and miniaturization. Such properties are favourable for application in wearable technologies [57], and such applications require the use of solid electrolytes for consumer convenience and comfort. So, more work can be channelled at developing new, as well as improving on the already existing solid-state hybrid supercapacitor devices.

Overall, hybrid supercapacitors present highly attractive attributes that are promising for applications such as temporary to long-term energy storage for power grids, portable electronics, as well as in electric vehicles. A major promising direction in hybrid supercapacitor research is their integration with renewable energy harvesting technologies such as solar cells and wind turbines for private and commercial use, as well as in industrial activities requiring small to medium loads of energy. Overcoming issues such as the cycling life span of the already impressive and reasonable unparalleled energy storage devices through the development of stable and efficient electrode materials can enable the movement of hybrid supercapacitor fabrication from the research laboratory phase to the large-scale commercialisation phase.

1.5 Conclusion

In this chapter, we reviewed and discussed recent progress made on hybrid supercapacitors, with emphasis on the electrode material development and the electrochemical performance thereof. Termed a special case of asymmetric supercapacitors, the hybrid supercapacitors have been shown in this discourse to possess remarkable features that are desirable from energy storage devices. We particularly delved into the properties and performance of the electrode materials used in hybrid supercapacitors, highlighting their strengths and weaknesses, as well as the efforts taken to improve their performance to perfect the overall device. We also set out to give a clear distinction between the types of electrode materials about the charge storage mechanism, especially between the battery-type and pseudocapacitive electrode materials, to avoid confusion with the type of electrode materials used in hybrid supercapacitors. We gave detailed accounts of the synthesis and application of typical electric double-layer capacitive-type materials used in hybrid supercapacitors such as activated carbons, carbon nanotubes, graphene, and metal–organic framework-derived carbon nanofibres. A pedigree of performance improvement was deduced from moving from activated carbon to carbon nanotubes to graphene. We also discussed in detail the different battery-type electrodes of hybrid supercapacitors that encompass transition metal oxides, hydroxides, and chalcogenides such as sulphides and selenide where we saw an improvement in the electrochemical performance moving in the respective manner of listing. In conclusion, we have relayed the various work done by researchers to develop unique electrode materials with designs and compositions, in a bid to fabricate hybrid supercapacitors that offer high energy and power densities, as well as stability.

Acknowledgements KM and KDM would like to thank the National Research Foundation (NRF-Thuthuka) (Grant Nos. 117984 and 118113), Tshwane University of Technology, South Africa, University of Limpopo (Research Development Grants R202 and R232), South Africa, for financial support. KO and NM acknowledge financial support from the South African Research Chairs Initiative of the Department of Science and Technology and the National Research Foundation (NRF) of South Africa (Grant No. 61056).

References

1. L. Niu, Y. Wang, F. Ruan, C. Shen, S. Shan, M. Xu, Z. Sun, C. Li, X. Liu, Y. Gong, In situ growth of $NiCo_2S_4@Ni_3V_2O_8$ on Ni foam as a binder-free electrode for asymmetric supercapacitors. J. Mater. Chem. A Mater. **4**, 5669–5677 (2016). https://doi.org/10.1039/c6ta00078a
2. H. Chen, S. Chen, Y. Zhu, C. Li, M. Fan, D. Chen, G. Tian, K. Shu, Synergistic effect of Ni and Co ions on molybdates for superior electrochemical performance. Electrochim. Acta **190**, 57–63 (2016). https://doi.org/10.1016/j.electacta.2015.12.212
3. D. Feng, T. Lei, M.R. Lukatskaya, J. Park, Z. Huang, M. Lee, L. Shaw, S. Chen, A.A. Yakovenko, A. Kulkarni, J. Xiao, K. Fredrickson, J.B. Tok, X. Zou, Y. Cui, Z. Bao, Robust and conductive two-dimensional metal-organic frameworks with exceptionally high volumetric and areal capacitance. Nat. Energy **3**, 30–36 (2018). https://doi.org/10.1038/s41560-017-0044-5
4. L. Zheng, L. Guan, J. Song, H. Zheng, Rational design of a sandwiched structure $Ni(OH)_2$ nanohybrid sustained by amino-functionalized graphene quantum dots for outstanding capacitance. Appl. Surf. Sci. **480**, 727–737 (2019). https://doi.org/10.1016/j.apsusc.2019.02.243
5. D. Yu, Z. Li, G. Zhao, H. Zhang, H. Aslan, J. Li, F. Sun, L. Zhu, B. Du, B. Yang, W. Cao, Y. Sun, F. Besenbacher, M. Yu, Porous ultrathin NiSe nanosheet networks on nickel foam for high-performance hybrid supercapacitors. Chemsuschem **13**, 260–266 (2020). https://doi.org/10.1002/cssc.201901766
6. P. Simon, Y. Gogotsi, Materials for electrochemical capacitors, in *Nanoscience and Technology: A Collection of Reviews* (Nature Journals, Co-Published with Macmillan Publishers Ltd, UK, 2009), pp. 320–329. https://doi.org/10.1142/9789814287005_0033
7. E. Frackowiak, F. Béguin, Carbon materials for the electrochemical storage of energy in capacitors. Carbon N Y **39**, 937–950 (2001). https://doi.org/10.1016/S0008-6223(00)00183-4
8. J.R. Miller, Valuing reversible energy storage. Science **335**(2012), 1312–1313 (1979). https://doi.org/10.1126/science.1219134
9. A. Lahyani, P. Venet, A. Guermazi, A. Troudi, Battery/supercapacitors combination, **28**, 1509–1522 (2013)
10. Y. Bin Tan, J.M. Lee, Graphene for supercapacitor applications, J. Mater. Chem. A Mater. **1**, 14814–14843 (2013). https://doi.org/10.1039/c3ta12193c

11. S. Vijayakumar, S. Nagamuthu, K.S. Ryu, In situ preparation of $MgCo_2O_4$ nanosheets on Ni-foam as a binder-free electrode for high performance hybrid supercapacitors. Dalton Trans. **47**, 6722–6728 (2018). https://doi.org/10.1039/c8dt00591e
12. D.P. Dubal, O. Ayyad, V. Ruiz, P. Gómez-Romero, Hybrid energy storage: the merging of battery and supercapacitor chemistries. Chem. Soc. Rev. **44**, 1777–1790 (2015). https://doi.org/10.1039/c4cs00266k
13. M. Salanne, B. Rotenberg, K. Naoi, K. Kaneko, P.-L. Taberna, C.P. Grey, B. Dunn, P. Simon, Efficient storage mechanisms for building better supercapacitors. Nat. Energy **1**, 16070 (2016). https://doi.org/10.1038/nenergy.2016.70
14. M.R. Lukatskaya, B. Dunn, Y. Gogotsi, Multidimensional materials and device architectures for future hybrid energy storage. Nat. Commun. **7**, 1–13 (2016). https://doi.org/10.1038/ncomms12647
15. V. Aravindan, J. Gnanaraj, Y.S. Lee, S. Madhavi, Insertion-type electrodes for nonaqueous Li-ion capacitors. Chem. Rev. **114**, 11619–11635 (2014). https://doi.org/10.1021/cr5000915
16. Y. Yang, D. Cheng, S. Chen, Y. Guan, J. Xiong, Construction of hierarchical $NiCo_2S_4$@$Ni(OH)_2$ core-shell hybrid nanosheet arrays on Ni foam for high-performance aqueous hybrid supercapacitors. Electrochim. Acta **193**, 116–127 (2016). https://doi.org/10.1016/j.electacta.2016.02.053
17. F.X. Ma, L. Yu, C.Y. Xu, X.W. Lou, Self-supported formation of hierarchical $NiCo_2O_4$ tetragonal microtubes with enhanced electrochemical properties. Energy Environ. Sci. **9**, 862–866 (2016). https://doi.org/10.1039/c5ee03772g
18. B. Akinwolemiwa, C. Peng, G.Z. Chen, Redox electrolytes in supercapacitors. J. Electrochem. Soc. **162**, A5054–A5059 (2015). https://doi.org/10.1149/2.0111505jes
19. L. Yu, G.Z. Chen, High energy supercapattery with an ionic liquid solution of $LiClO_4$. Faraday Discuss. **190**, 231–240 (2016). https://doi.org/10.1039/c5fd00232j
20. T. Brousse, D. Bélanger, J.W. Long, To be or not to be pseudocapacitive? J. Electrochem. Soc. **162**, A5185–A5189 (2015). https://doi.org/10.1149/2.0201505jes
21. J.S. Lee, D.H. Shin, W. Kim, J. Jang, Highly ordered, polypyrrole-coated $Co(OH)_2$ architectures for high-performance asymmetric supercapacitors. J. Mater. Chem. A Mater. **4**, 6603–6609 (2016). https://doi.org/10.1039/c6ta01419d
22. S.T. Senthilkumar, N. Fu, Y. Liu, Y. Wang, L. Zhou, H. Huang, Flexible fiber hybrid supercapacitor with $NiCo_2O_4$ nanograss@carbon fiber and bio-waste derived high surface area porous carbon. Electrochim. Acta **211**, 411–419 (2016). https://doi.org/10.1016/j.electacta.2016.06.059
23. L. Yu, B. Yang, Q. Liu, J. Liu, X. Wang, D. Song, J. Wang, X. Jing, Interconnected NiS nanosheets supported by nickel foam: soaking fabrication and supercapacitors application. J. Electroanal. Chem. **739**, 156–163 (2015). https://doi.org/10.1016/j.jelechem.2014.12.031
24. J. Qi, Y. Chang, Y. Sui, Y. He, Q. Meng, F. Wei, Y. Ren, Y. Jin, Facile Synthesis of Ag-decorated Ni_3S_2 Nanosheets with 3D bush structure grown on rGO and its application as positive electrode material in asymmetric supercapacitor. Adv. Mater. Interfaces **5**, 1–9 (2018). https://doi.org/10.1002/admi.201700985
25. G. Nagaraju, S.C. Sekhar, B. Ramulu, L.K. Bharat, G.S.R. Raju, Y.K. Han, J.S. Yu, Enabling redox chemistry with hierarchically designed bilayered nanoarchitectures for pouch-type hybrid supercapacitors: A sunlight-driven rechargeable energy storage system to portable electronics. Nano Energy **50**, 448–461 (2018). https://doi.org/10.1016/j.nanoen.2018.05.063
26. P. Zhang, B.Y. Guan, L. Yu, X.W.D. Lou, Formation of double-shelled zinc-cobalt sulfide dodecahedral cages from bimetallic zeolitic imidazolate frameworks for hybrid supercapacitors. Angew. Chemie—International Ed. **56**, 7141–7145 (2017). https://doi.org/10.1002/anie.201702649
27. B. Senthilkumar, Z. Khan, S. Park, K. Kim, H. Ko, Y. Kim, Highly porous graphitic carbon and $Ni_2P_2O_7$ for a high performance aqueous hybrid supercapacitor. J. Mater. Chem. A Mater. **3**, 21553–21561 (2015). https://doi.org/10.1039/c5ta04737d
28. G. Nagaraju, S. Chandra Sekhar, L. Krishna Bharat, J.S. Yu, Wearable fabrics with self-branched bimetallic layered double hydroxide coaxial nanostructures for hybrid supercapacitors. ACS Nano **11**, 10860–10874 (2017). https://doi.org/10.1021/acsnano.7b04368

29. B. Akinwolemiwa, C. Wei, G.Z. Chen, Mechanisms and designs of asymmetrical electrochemical capacitors. Electrochim. Acta **247**, 344–357 (2017). https://doi.org/10.1016/j.electacta.2017.06.088
30. G.G. Amatucci, F. Badway, A. Du Pasquier, T. Zheng, An asymmetric hybrid nonaqueous energy storage cell. J. Electrochem. Soc. **148**, A930 (2001). https://doi.org/10.1149/1.1383553
31. J. Ding, H. Wang, Z. Li, K. Cui, D. Karpuzov, X. Tan, A. Kohandehghan, D. Mitlin, Peanut shell hybrid sodium ion capacitor with extreme energy-power rivals lithium ion capacitors. Energy Environ. Sci. **8**, 941–955 (2015). https://doi.org/10.1039/c4ee02986k
32. H. Chen, T.N. Cong, W. Yang, C. Tan, Y. Li, Y. Ding, Progress in electrical energy storage system: a critical review. Prog. Nat. Sci. **19**, 291–312 (2009). https://doi.org/10.1016/j.pnsc.2008.07.014
33. N. Choudhary, C. Li, J. Moore, N. Nagaiah, L. Zhai, Y. Jung, J. Thomas, Asymmetric supercapacitor electrodes and devices. Adv. Mater. **29** (2017). https://doi.org/10.1002/adma.201605336.
34. A. Jayakumar, R.P. Antony, R. Wang, J.M. Lee, MOF-derived hollow cage $NixCo_3-xO_4$ and their synergy with graphene for outstanding supercapacitors. Small **13**, 1–12 (2017). https://doi.org/10.1002/smll.201603102
35. Y. Gogotsi, R.M. Penner, Energy storage in nanomaterials—capacitive, pseudocapacitive, or battery-like? ACS Nano **12**, 2081–2083 (2018). https://doi.org/10.1021/acsnano.8b01914
36. G. Nagaraju, S.M. Cha, S.C. Sekhar, J.S. Yu, Metallic layered polyester fabric enabled nickel selenide nanostructures as highly conductive and binderless electrode with superior energy storage performance. Adv. Energy Mater. **7**, 1–13 (2017). https://doi.org/10.1002/aenm.201601362
37. W. Jiang, F. Hu, Q. Yan, X. Wu, Investigation on electrochemical behaviors of $NiCo_2O_4$ battery-type supercapacitor electrodes: the role of an aqueous electrolyte. Inorg. Chem. Front. **4**, 1642–1648 (2017). https://doi.org/10.1039/c7qi00391a
38. Y. Gogotsi, P. Simon, True performance metrics in electrochemical energy storage. Science **334**(2011), 917–918 (1979). https://doi.org/10.1126/science.1213003
39. W.G. Pell, B.E. Conway, Peculiarities and requirements of asymmetric capacitor devices based on combination of capacitor and battery-type electrodes. J. Power. Sources **136**, 334–345 (2004). https://doi.org/10.1016/j.jpowsour.2004.03.021
40. Q. Abbas, P. Babuchowska, E. Frąckowiak, F. Béguin, Sustainable AC/AC hybrid electrochemical capacitors in aqueous electrolyte approaching the performance of organic systems. J. Power. Sources **326**, 652–659 (2016). https://doi.org/10.1016/j.jpowsour.2016.03.094
41. B. Zhao, D. Chen, X. Xiong, B. Song, R. Hu, Q. Zhang, B.H. Rainwater, G.H. Waller, D. Zhen, Y. Ding, Y. Chen, C. Qu, D. Dang, C.P. Wong, M. Liu, A high-energy, long cycle-life hybrid supercapacitor based on graphene composite electrodes. Energy Storage Mater. **7**, 32–39 (2017). https://doi.org/10.1016/j.ensm.2016.11.010
42. S. Peng, L. Li, H. Bin Wu, S. Madhavi, X.W. Lou, Controlled growth of $NiMoO_4$ nanosheet and nanorod arrays on various conductive substrates as advanced electrodes for asymmetric supercapacitors. Adv Energy Mater **5**, 1–7 (2015). https://doi.org/10.1002/aenm.201401172
43. Q. Zhang, B. Zhao, J. Wang, C. Qu, H. Sun, K. Zhang, M. Liu, High-performance hybrid supercapacitors based on self-supported 3D ultrathin porous quaternary Zn-Ni-Al-Co oxide nanosheets. Nano Energy **28**, 475–485 (2016). https://doi.org/10.1016/j.nanoen.2016.08.049
44. Z. Liang, C. Qu, W. Zhou, R. Zhao, H. Zhang, B. Zhu, W. Guo, W. Meng, Y. Wu, W. Aftab, Q. Wang, R. Zou, Synergistic effect of Co–Ni hybrid phosphide nanocages for ultrahigh capacity fast energy storage. Adv. Sci. **6** (2019). https://doi.org/10.1002/advs.201802005
45. C. Qu, Z. Liang, Y. Jiao, B. Zhao, B. Zhu, D. Dang, S. Dai, Y. Chen, R. Zou, M. Liu, One-for-all strategy in fast energy storage: production of pillared MOF nanorod-templated positive/negative electrodes for the application of high-performance hybrid supercapacitor. Small **14**, 1–8 (2018). https://doi.org/10.1002/smll.201800285
46. B. Zhao, L. Zhang, Q. Zhang, D. Chen, Y. Cheng, X. Deng, Y. Chen, R. Murphy, X. Xiong, B. Song, C.P. Wong, M.S. Wang, M. Liu, Rational design of nickel hydroxide-based nanocrystals

on graphene for ultrafast energy storage. Adv. Energy Mater. **8**, 1–10 (2018). https://doi.org/10.1002/aenm.201702247
47. H.B. Li, M.H. Yu, F.X. Wang, P. Liu, Y. Liang, J. Xiao, C.X. Wang, Y.X. Tong, G.W. Yang, Amorphous nickel hydroxide nanospheres with ultrahigh capacitance and energy density as electrochemical pseudocapacitor materials. Nat. Commun. **4**, 1–7 (2013). https://doi.org/10.1038/ncomms2932
48. K. Naoi, S. Ishimoto, J.I. Miyamoto, W. Naoi, Second generation "nanohybrid supercapacitor": evolution of capacitive energy storage devices. Energy Environ. Sci. **5**, 9363–9373 (2012). https://doi.org/10.1039/c2ee21675b
49. H. Wang, C. Guan, X. Wang, H.J. Fan, A high energy and power Li-ion capacitor based on a TiO_2 nanobelt array anode and a graphene hydrogel cathode. Small **11**, 1470–1477 (2015). https://doi.org/10.1002/smll.201402620
50. B. Li, F. Dai, Q. Xiao, L. Yang, J. Shen, C. Zhang, M. Cai, Nitrogen-doped activated carbon for a high energy hybrid supercapacitor. Energy Environ. Sci. **9**, 102–106 (2016). https://doi.org/10.1039/c5ee03149d
51. H. Wang, H. Yi, X. Chen, X. Wang, Asymmetric supercapacitors based on nano-architectured nickel oxide/graphene foam and hierarchical porous nitrogen-doped carbon nanotubes with ultrahigh-rate performance. J. Mater. Chem. A Mater. **2**, 3223–3230 (2014). https://doi.org/10.1039/c3ta15046a
52. G.W. Lee, M.S. Kim, J.H. Jeong, H.K. Roh, K.C. Roh, K.B. Kim, Comparative study of $Li_4Ti_5O_{12}$ composites prepared withpristine, oxidized, and surfactant-treated multiwalled carbon nanotubes for high-power hybrid supercapacitors. ChemElectroChem **5**, 2357–2366 (2018). https://doi.org/10.1002/celc.201800408
53. R. Satish, V. Aravindan, W.C. Ling, S. Madhavi, Carbon-coated $Li_3V_2(PO_4)_3$ as insertion type electrode for lithium-ion hybrid electrochemical capacitors: an evaluation of anode and cathodic performance. J. Power. Sources **281**, 310–317 (2015). https://doi.org/10.1016/j.jpowsour.2015.01.190
54. J.H. Jeong, G.W. Lee, Y.H. Kim, Y.J. Choi, K.C. Roh, K.B. Kim, A holey graphene-based hybrid supercapacitor. Chem. Eng. J. **378**, 122126 (2019). https://doi.org/10.1016/j.cej.2019.122126
55. T. Brousse, M. Toupin, D. Bélanger, A hybrid activated carbon-manganese dioxide capacitor using a mild aqueous electrolyte. J. Electrochem. Soc. **151**, A614 (2004). https://doi.org/10.1149/1.1650835
56. P. Simon, Y. Gogotsi, B. Dunn, Where do batteries end and supercapacitors begin? Science **343**(2014), 1210–1211 (1979). https://doi.org/10.1126/science.1249625
57. T. Wang, H.C. Chen, F. Yu, X.S. Zhao, H. Wang, Boosting the cycling stability of transition metal compounds-based supercapacitors. Energy Storage Mater **16**, 545–573 (2019). https://doi.org/10.1016/j.ensm.2018.09.007
58. M.D. Stoller, R.S. Ruoff, Best practice methods for determining an electrode material's performance for ultracapacitors. Energy Environ Sci. **3**, 1294–1301 (2010). https://doi.org/10.1039/c0ee00074d
59. D. Qu, H. Shi, Studies of activated carbons used in double-layer capacitors. J. Power. Sources **74**, 99–107 (1998). https://doi.org/10.1016/S0378-7753(98)00038-X
60. M. Zhi, C. Xiang, J. Li, M. Li, N. Wu, Nanostructured carbon-metal oxide composite electrodes for supercapacitors: a review. Nanoscale **5**, 72–88 (2013). https://doi.org/10.1039/c2nr32040a
61. Y. Chen, X. Zhang, D. Zhang, P. Yu, Y. Ma, High performance supercapacitors based on reduced graphene oxide in aqueous and ionic liquid electrolytes. Carbon N Y **49**, 573–580 (2011). https://doi.org/10.1016/j.carbon.2010.09.060
62. A.S. Aricò, P. Bruce, B. Scrosati, J.M. Tarascon, W. Van Schalkwijk, Nanostructured materials for advanced energy conversion and storage devices. Nat. Mater. **4**, 366–377 (2005). https://doi.org/10.1038/nmat1368
63. M. Scarselli, P. Castrucci, M. De Crescenzi, Electronic and optoelectronic nano-devices based on carbon nanotubes. J. Phys. Condens. Matter **24** (2012). https://doi.org/10.1088/0953-8984/24/31/313202

64. A.G. Pandolfo, A.F. Hollenkamp, Carbon properties and their role in supercapacitors. J. Power. Sources **157**, 11–27 (2006). https://doi.org/10.1016/j.jpowsour.2006.02.065
65. K. Makgopa, L.F. Mabena, C.G. Brink, G.N. Chauke, M.D. Teffu, K.D. Modibane, M.J. Hato, Nanostructured carbon-based electrode materials for supercapacitor applications, in: *Carbon Related Materials*, ed. by S. Kaneko, M. Aono, A. Pruna, M. Can, P. Mele, M. Ertugrul, T. Endo (Springer Singapore, Singapore, 2021), pp. 317–355. https://doi.org/10.1007/978-981-15-7610-2_14
66. L.L. Zhang, X.S. Zhao, Carbon-based materials as supercapacitor electrodes. Chem. Soc. Rev. **38**, 2520 (2009). https://doi.org/10.1039/b813846j
67. G. Salitra, A. Soffer, L. Eliad, Y. Cohen, D. Aurbach, Carbon electrodes for double-layer capacitors I. Relations between ion and pore dimensions. J. Electrochem. Soc. **147**, 2486 (2000). https://doi.org/10.1149/1.1393557
68. E. Raymundo-Piñero, K. Kierzek, J. Machnikowski, F. Béguin, Relationship between the nanoporous texture of activated carbons and their capacitance properties in different electrolytes. Carbon N Y **44**, 2498–2507 (2006). https://doi.org/10.1016/j.carbon.2006.05.022
69. Y. Zhai, Y. Dou, D. Zhao, P.F. Fulvio, R.T. Mayes, S. Dai, Carbon materials for chemical capacitive energy storage. Adv. Mater. **23**, 4828–4850 (2011). https://doi.org/10.1002/adma.201100984
70. Z. Yu, L. Tetard, L. Zhai, J. Thomas, Supercapacitor electrode materials: nanostructures from 0 to 3 dimensions. Energy Environ. Sci. **8**, 702–730 (2015). https://doi.org/10.1039/c4ee03229b
71. G. Wang, L. Zhang, J. Zhang, A review of electrode materials for electrochemical supercapacitors. Chem. Soc. Rev. **41**, 797–828 (2012). https://doi.org/10.1039/c1cs15060j
72. L. Ru-Shi (ed.), *Controlled Nanofabrication* (Jenny Stanford Publishing, 2012). https://doi.org/10.1201/b13155
73. C.J. Meyers, S.D. Shah, S.C. Patel, R.M. Sneeringer, C.A. Bessel, N.R. Dollahon, R.A. Leising, E.S. Takeuchi, Templated synthesis of carbon materials from zeolites (Y, Beta, and ZSM-5) and a montmorillonite clay (K10): physical and electrochemical characterization. J. Phys. Chem. B **105**, 2143–2152 (2001). https://doi.org/10.1021/jp0029663
74. C. Largeot, C. Portet, J. Chmiola, P.L. Taberna, Y. Gogotsi, P. Simon, Relation between the ion size and pore size for an electric double-layer capacitor. J. Am. Chem. Soc. **130**, 2730–2731 (2008). https://doi.org/10.1021/ja7106178
75. L. Qie, W. Chen, H. Xu, X. Xiong, Y. Jiang, F. Zou, X. Hu, Y. Xin, Z. Zhang, Y. Huang, Synthesis of functionalized 3D hierarchical porous carbon for high-performance supercapacitors. Energy Environ. Sci. **6**, 2497–2504 (2013). https://doi.org/10.1039/c3ee41638k
76. K. Kierzek, E. Frackowiak, G. Lota, G. Gryglewicz, J. Machnikowski, Erratum: electrochemical capacitors based on highly porous carbons prepared by KOH activation (Electrochimica Acta **49**, 515–523 (2004)), Electrochim Acta **49**, 1169–1170 (2004). https://doi.org/10.1016/j.electacta.2004.01.005
77. Q. Cheng, J. Tang, J. Ma, H. Zhang, N. Shinya, L.C. Qin, Graphene and carbon nanotube composite electrodes for supercapacitors with ultra-high energy density. Phys. Chem. Chem. Phys. **13**, 17615–17624 (2011). https://doi.org/10.1039/c1cp21910c
78. R.R.A. Rios, D.E. Alves, I. Dalmázio, S.F.V. Bento, C.L. Donnici, R.M. Lago, Tailoring activated carbon by surface chemical modification with O, S, and N containing molecules. Mater. Res. **6**, 129–135 (2003). https://doi.org/10.1590/s1516-14392003000200004
79. B. Duong, Z. Yu, P. Gangopadhyay, S. Seraphin, N. Peyghambarian, J. Thomas, High throughput printing of nanostructured carbon electrodes for supercapacitors. Adv. Mater. Interfaces **1**, 1–5 (2014). https://doi.org/10.1002/admi.201300014
80. W. Li, D. Chen, Z. Li, Y. Shi, Y. Wan, J. Huang, J. Yang, D. Zhao, Z. Jiang, Nitrogen enriched mesoporous carbon spheres obtained by a facile method and its application for electrochemical capacitor. Electrochem. Commun. **9**, 569–573 (2007). https://doi.org/10.1016/j.elecom.2006.10.027

81. H. Pan, C.K. Poh, Y.P. Feng, J. Lin, Supercapacitor electrodes from tubes-in-tube carbon nanostructures. Chem. Mater. **19**, 6120–6125 (2007). https://doi.org/10.1021/cm071527e
82. I. Bordun, V. Pohrebennyk, M. Sadowa, V. Ptashnyk, A. Klos-Witkowska, V. Martsenyuk, Impedance spectroscopy of supercapacitors on the basis on modified by the ultrasound activated carbon material, in *Proceedings of the 2017 IEEE 9th International Conference on Intelligent Data Acquisition and Advanced Computing Systems: Technology and Applications, IDAACS 2017*, vol. 1, pp. 86–90 (2017). https://doi.org/10.1109/IDAACS.2017.8095055
83. W.-C. Chen, T.-C. Wen, H. Teng, Polyaniline-deposited porous carbon electrode for supercapacitor. Electrochim. Acta **48**, 641–649 (2003). https://doi.org/10.1016/S0013-4686(02)00734-X
84. F. Cheng, X. Yang, S. Zhang, W. Lu, Boosting the supercapacitor performances of activated carbon with carbon nanomaterials. J. Power. Sources **450**, 227678 (2020). https://doi.org/10.1016/j.jpowsour.2019.227678
85. I.I.G. Inal, Y. Gokce, Z. Aktas, Waste tea derived activated carbon/polyaniline composites as supercapacitor electrodes, in *2016 IEEE International Conference on Renewable Energy Research and Applications, ICRERA 2016*, vol. 5, pp. 458–462 (2017). https://doi.org/10.1109/ICRERA.2016.7884380
86. H. Wang, M. Liang, D. Duan, W. Shi, Y. Song, Z. Sun, Rose-like Ni_3S_4 as battery-type electrode for hybrid supercapacitor with excellent charge storage performance. Chem. Eng. J. **350**, 523–533 (2018). https://doi.org/10.1016/j.cej.2018.05.004
87. L. Qie, W.M. Chen, Z.H. Wang, Q.G. Shao, X. Li, L.X. Yuan, X.L. Hu, W.X. Zhang, Y.H. Huang, Nitrogen-doped porous carbon nanofiber webs as anodes for lithium ion batteries with a superhigh capacity and rate capability. Adv. Mater. **24**, 2047–2050 (2012). https://doi.org/10.1002/adma.201104634
88. C. Niu, E.K. Sichel, R. Hoch, D. Moy, H. Tennent, High power electrochemical capacitors based on carbon nanotube electrodes. Appl. Phys. Lett. **70**, 1480–1482 (1997). https://doi.org/10.1063/1.118568
89. H. Pan, J. Li, Y.P. Feng, Carbon nanotubes for supercapacitor. Nanoscale Res. Lett. **5**, 654–668 (2010). https://doi.org/10.1007/s11671-009-9508-2
90. R.A. Fisher, M.R. Watt, W. Jud Ready, Functionalized carbon nanotube supercapacitor electrodes: a review on pseudocapacitive materials. ECS J. Solid State Sci. Technol. **2**, M3170–M3177 (2013). https://doi.org/10.1149/2.017310jss
91. E. Frackowiak, K. Jurewicz, S. Delpeux, F. Béguin, Nanotubular materials for supercapacitors. J. Power. Sources **97–98**, 822–825 (2001). https://doi.org/10.1016/S0378-7753(01)00736-4
92. E. Frackowiak, K. Metenier, V. Bertagna, F. Beguin, Supercapacitor electrodes from multiwalled carbon nanotubes. Appl. Phys. Lett. **77**, 2421–2423 (2000). https://doi.org/10.1063/1.1290146
93. L. Dai, D.W. Chang, J.B. Baek, W. Lu, Carbon nanomaterials for advanced energy conversion and storage. Small **8**, 1130–1166 (2012). https://doi.org/10.1002/smll.201101594
94. R.H. Baughman, A.A. Zakhidov, W.A. De Heer, Carbon nanotubes—the route toward applications. Science **297**(2002), 787–792 (1979). https://doi.org/10.1126/science.1060928
95. H. Zhang, G. Cao, Y. Yang, Z. Gu, Comparison between electrochemical properties of aligned carbon nanotube array and entangled carbon nanotube electrodes. J. Electrochem. Soc. **155**, K19 (2008). https://doi.org/10.1149/1.2811864
96. M. Saghafi, F. Mahboubi, S. Mohajerzadeh, R. Holze, Preparation of vertically aligned carbon nanotubes and their electrochemical performance in supercapacitors. Synth. Met. **195**, 252–259 (2014). https://doi.org/10.1016/j.synthmet.2014.06.012
97. B. Kim, H. Chung, W. Kim, High-performance supercapacitors based on vertically aligned carbon nanotubes and nonaqueous electrolytes. Nanotechnology **23** (2012). https://doi.org/10.1088/0957-4484/23/15/155401
98. Q.L. Chen, K.H. Xue, W. Shen, F.F. Tao, S.Y. Yin, W. Xu, Fabrication and electrochemical properties of carbon nanotube array electrode for supercapacitors. Electrochim. Acta **49**, 4157–4161 (2004). https://doi.org/10.1016/j.electacta.2004.04.010

99. W. Lu, L. Qu, K. Henry, L. Dai, High performance electrochemical capacitors from aligned carbon nanotube electrodes and ionic liquid electrolytes. J. Power. Sources **189**, 1270–1277 (2009). https://doi.org/10.1016/j.jpowsour.2009.01.009
100. S.N. Alam, N. Sharma, L. Kumar, Synthesis of graphene oxide (GO) by modified hummers method and its thermal reduction to obtain reduced graphene oxide (rGO)*. Graphene **06**, 1–18 (2017). https://doi.org/10.4236/graphene.2017.61001
101. A.K. Geim, Graphene: status and prospects. Science **324**(2009), 1530–1534 (1979). https://doi.org/10.1126/science.1158877
102. A.K. Geim, K.S. Novoselov, The rise of graphene. Nat. Mater. **6**, 183–191 (2007). https://doi.org/10.1038/nmat1849
103. M. Pumera, Graphene-based nanomaterials and their electrochemistry. Chem. Soc. Rev. **39**, 4146–4157 (2010). https://doi.org/10.1039/c002690p
104. H. Wang, Q. Hao, X. Yang, L. Lu, X. Wang, A nanostructured graphene/polyaniline hybrid material for supercapacitors. Nanoscale **2**, 2164–2170 (2010). https://doi.org/10.1039/c0nr00224k
105. C. Liu, Z. Yu, D. Neff, A. Zhamu, B.Z. Jang, Graphene-based supercapacitor with an ultrahigh energy density. Nano Lett. **10**, 4863–4868 (2010). https://doi.org/10.1021/nl102661q
106. M.F. El-Kady, V. Strong, S. Dubin, R.B. Kaner, Laser scribing of high-performance and flexible graphene-based electrochemical capacitors. Science **335**(2012), 1326–1330 (1979). https://doi.org/10.1126/science.1216744
107. K.S. Novoselov, V.I. Fal'Ko, L. Colombo, P.R. Gellert, M.G. Schwab, K. Kim, A roadmap for graphene, Nature **490**, 192–200 (2012). https://doi.org/10.1038/nature11458
108. S. Stankovich, D.A. Dikin, G.H.B. Dommett, K.M. Kohlhaas, E.J. Zimney, E.A. Stach, R.D. Piner, S.T. Nguyen, R.S. Ruoff, Graphene-based composite materials. Nature **442**, 282–286 (2006). https://doi.org/10.1038/nature04969
109. Y. Zhu, S. Murali, W. Cai, X. Li, J.W. Suk, J.R. Potts, R.S. Ruoff, Graphene and graphene oxide: synthesis, properties, and applications. Adv. Mater. **22**, 3906–3924 (2010). https://doi.org/10.1002/adma.201001068
110. M.J. Madito, K. Makgopa, C.B. Mtshali, A. Bello, Graphene-based electrode materials for supercapacitor applications, in: *Electrochemical Devices for Energy Storage Applications* ed. by F.I. Kebede, M.A., Ezema, (Taylor & Francis Group, CRC Press, 2019), pp. 101–130. https://doi.org/10.1201/9780367855116-6
111. H. Xia, C. Hong, B. Li, B. Zhao, Z. Lin, M. Zheng, S.V. Savilov, S.M. Aldoshin, Facile synthesis of hematite quantum-dot/functionalized graphene-sheet composites as advanced anode materials for asymmetric supercapacitors. Adv. Funct. Mater. **25**, 627–635 (2015). https://doi.org/10.1002/adfm.201403554
112. H.A. Becerril, J. Mao, Z. Liu, R.M. Stoltenberg, Z. Bao, Y. Chen, Evaluation of solution-processed reduced graphene oxide films as transparent conductors. ACS Nano **2**, 463–470 (2008). https://doi.org/10.1021/nn700375n
113. S.M. Chen, R. Ramachandran, V. Mani, R. Saraswathi, Recent advancements in electrode materials for the high-performance electrochemical supercapacitors: a review. Int. J. Electrochem. Sci. **9**, 4072–4085 (2014)
114. G.K. Sinniah, M.Z. Shah, H.C. Siong, The needs for changes in travel behaviour towards a low carbon society, in *Proceedings of the 26th International Business Information Management Association Conference—Innovation Management and Sustainable Economic Competitive Advantage: From Regional Development to Global Growth, IBIMA 2015*, 3868–3876 (2015)
115. T. Kuilla, S. Bhadra, D. Yao, N.H. Kim, S. Bose, J.H. Lee, Recent advances in graphene based polymer composites. Progress in Polymer Science (Oxford) **35**, 1350–1375 (2010). https://doi.org/10.1016/j.progpolymsci.2010.07.005
116. E. Paek, A.J. Pak, K.E. Kweon, G.S. Hwang, On the origin of the enhanced supercapacitor performance of nitrogen-doped graphene. J. Phys. Chem. C **117**, 5610–5616 (2013). https://doi.org/10.1021/jp312490q
117. T. Palaniselvam, V. Kashyap, S.N. Bhange, J.B. Baek, S. Kurungot, Nanoporous graphene enriched with Fe/Co-N active sites as a promising oxygen reduction electrocatalyst for anion

exchange membrane fuel cells. Adv. Funct. Mater. **26**, 2150–2162 (2016). https://doi.org/10.1002/adfm.201504765
118. S.K. Singh, V.M. Dhavale, R. Boukherroub, S. Kurungot, S. Szunerits, N-doped porous reduced graphene oxide as an efficient electrode material for high performance flexible solid-state supercapacitor. Appl. Mater. Today, **8**, 141–149 (2017). https://doi.org/10.1016/j.apmt.2016.10.002
119. V.C. Tung, M.J. Allen, Y. Yang, R.B. Kaner, High-throughput solution processing of large-scale graphene. Nat. Nanotechnol. **4**, 25–29 (2009). https://doi.org/10.1038/nnano.2008.329
120. H. Gao, F. Xiao, C.B. Ching, H. Duan, High-performance asymmetric supercapacitor based on graphene hydrogel and nanostructured MnO_2. ACS Appl. Mater. Interfaces **4**, 2801–2810 (2012). https://doi.org/10.1021/am300455d
121. X. Cao, Z. Yin, H. Zhang, Three-dimensional graphene materials: preparation, structures and application in supercapacitors. Energy Environ. Sci. **7**, 1850–1865 (2014). https://doi.org/10.1039/c4ee00050a
122. Y. Gong, S. Yang, L. Zhan, L. Ma, R. Vajtai, P.M. Ajayan, A bottom-up approach to build 3D architectures from nanosheets for superior lithium storage. Adv. Funct. Mater. **24**, 125–130 (2014). https://doi.org/10.1002/adfm.201300844
123. X. Zhang, D. Zhao, Y. Zhao, P. Tang, Y. Shen, C. Xu, H. Li, Y. Xiao, High performance asymmetric supercapacitor based on MnO_2 electrode in ionic liquid electrolyte. J. Mater. Chem. A Mater. **1**, 3706–3712 (2013). https://doi.org/10.1039/C3TA00981E
124. Y. Wang, Z. Shi, Y. Huang, Y. Ma, C. Wang, M. Chen, Y. Chen, Supercapacitor devices based on graphene materials. J. Phys. Chem. C **113**, 13103–13107 (2009). https://doi.org/10.1021/jp902214f
125. Y. Wang, Y. Wu, Y. Huang, F. Zhang, X. Yang, Y. Ma, Y. Chen, Preventing graphene sheets from restacking for high-capacitance performance. J. Phys. Chem. C **115**, 23192–23197 (2011). https://doi.org/10.1021/jp206444e
126. S. Park, R.S. Ruoff, Chemical methods for the production of graphenes. Nat. Nanotechnol. **4**, 217–224 (2009). https://doi.org/10.1038/nnano.2009.58
127. Z.S. Wu, A. Winter, L. Chen, Y. Sun, A. Turchanin, X. Feng, K. Müllen, Three-dimensional nitrogen and boron co-doped graphene for high-performance all-solid-state supercapacitors. Adv. Mater. **24**, 5130–5135 (2012). https://doi.org/10.1002/adma.201201948
128. B.G. Choi, M. Yang, W.H. Hong, J.W. Choi, Y.S. Huh, 3D macroporous graphene frameworks for supercapacitors with high energy and power densities. ACS Nano **6**, 4020–4028 (2012). https://doi.org/10.1021/nn3003345
129. H.P. Cong, X.C. Ren, P. Wang, S.H. Yu, Macroscopic multifunctional graphene-based hydrogels and aerogels by a metal ion induced self-assembly process. ACS Nano **6**, 2693–2703 (2012). https://doi.org/10.1021/nn300082k
130. X.L. Wu, T. Wen, H.L. Guo, S. Yang, X. Wang, A.W. Xu, Biomass-derived sponge-like carbonaceous hydrogels and aerogels for supercapacitors. ACS Nano **7**, 3589–3597 (2013). https://doi.org/10.1021/nn400566d
131. G. Zhou, L.C. Yin, D.W. Wang, L. Li, S. Pei, I.R. Gentle, F. Li, H.M. Cheng, Fibrous hybrid of graphene and sulfur nanocrystals for high-performance lithium-sulfur batteries. ACS Nano **7**, 5367–5375 (2013). https://doi.org/10.1021/nn401228t
132. X. Fan, C. Yu, Z. Ling, J. Yang, J. Qiu, Hydrothermal synthesis of phosphate-functionalized carbon nanotube-containing carbon composites for supercapacitors with highly stable performance. ACS Appl. Mater. Interfaces **5**, 2104–2110 (2013). https://doi.org/10.1021/am303052n
133. Q. Lu, Y. Chen, W. Li, J.G. Chen, J.Q. Xiao, F. Jiao, Ordered mesoporous nickel cobaltite spinel with ultra-high supercapacitance. J. Mater. Chem. A Mater. **1**, 2331–2336 (2013). https://doi.org/10.1039/c2ta00921h
134. Z.S. Wu, W. Ren, L. Xu, F. Li, H.M. Cheng, Doped graphene sheets as anode materials with superhigh rate and large capacity for lithium ion batteries. ACS Nano **5**, 5463–5471 (2011). https://doi.org/10.1021/nn2006249

135. D.G. Wang, Z. Liang, S. Gao, C. Qu, R. Zou, Metal-organic framework-based materials for hybrid supercapacitor application. Coord. Chem. Rev. **404**, 213093 (2020). https://doi.org/10.1016/j.ccr.2019.213093
136. R.R. Salunkhe, Y.V. Kaneti, J. Kim, J.H. Kim, Y. Yamauchi, Nanoarchitectures for metal-organic framework-derived nanoporous carbons toward supercapacitor applications. Acc. Chem. Res. **49**, 2796–2806 (2016). https://doi.org/10.1021/acs.accounts.6b00460
137. B. Liu, H. Shioyama, T. Akita, Q. Xu, Metal-organic framework as a template for porous carbon synthesis. J. Am. Chem. Soc. **130**, 5390–5391 (2008). https://doi.org/10.1021/ja7106146
138. H. Yi, H. Wang, Y. Jing, T. Peng, X. Wang, Asymmetric supercapacitors based on carbon nanotubes@NiO ultrathin nanosheets core-shell composites and MOF-derived porous carbon polyhedrons with super-long cycle life. J. Power. Sources **285**, 281–290 (2015). https://doi.org/10.1016/j.jpowsour.2015.03.106
139. N. Tang, W. Wang, H. You, Z. Zhai, J. Hilario, L. Zeng, L. Zhang, Morphology tuning of porous CoO nanowall towards enhanced electrochemical performance as supercapacitors electrodes. Catal Today, 240–245 (2019). https://doi.org/10.1016/j.cattod.2018.03.024
140. R.R. Salunkhe, J. Tang, Y. Kamachi, T. Nakato, J.H. Kim, Y. Yamauchi, Asymmetric supercapacitors using 3D nanoporous carbon and cobalt oxide electrodes synthesized from a single metal-organic framework. ACS Nano **9**, 6288–6296 (2015). https://doi.org/10.1021/acsnano.5b01790
141. X. Wei, H. Peng, Y. Li, Y. Yang, S. Xiao, L. Peng, Y. Zhang, P. Xiao, In situ growth of zeolitic imidazolate Framework-67-derived nanoporous carbon@K0.5Mn2O4 for high-performance 2.4 V aqueous asymmetric supercapacitors. ChemSusChem **11**, 3167–3174 (2018). https://doi.org/10.1002/cssc.201801439
142. K. Makgopa, A. Bello, K. Raju, K.D. Modibane, M.J. Hato, Nanostructured Metal Oxides for Supercapacitor Applications, in: *Emerging Nanostructured Materials for Energy and Environmental Science*, ed. by B.R.S Rajendran, M.Naushad, K. Raju, (Springer, 2019), pp. 247–303. https://doi.org/10.1007/978-3-030-04474-9_6
143. K. Makgopa, P.M. Ejikeme, K.I. Ozoemena, Nanostructured manganese oxides in supercapacitors, in: *Nanomaterials in Advanced Batteries and Supercapacitors*, pp. 345–376 2016. https://doi.org/10.1007/978-3-319-26082-2_10
144. K. Ding, X. Zhang, P. Yang, X. Cheng, A precursor-derived morphology-controlled synthesis method for mesoporous Co_3O_4 nanostructures towards supercapacitor application. Cryst. Eng. Comm. **18**, 8253–8261 (2016). https://doi.org/10.1039/c6ce01921h
145. Y. Fan, H. Shao, J. Wang, L. Liu, J. Zhang, C. Cao, Synthesis of foam-like freestanding Co_3O_4 nanosheets with enhanced electrochemical activities. Chem. Commun. **47**, 3469–3471 (2011). https://doi.org/10.1039/c0cc05383j
146. L. Zhuo, J. Ge, L. Cao, B. Tang, Solvothermal synthesis of CoO, Co_3O_4, $Ni(OH)_2$ and $Mg(OH)_2$ nanotubes. Cryst. Growth Des. **9**, 1–6 (2009). https://doi.org/10.1021/cg070482r
147. N. Du, H. Zhang, B. Chen, J. Wu, X. Ma, Z. Liu, Y. Zhang, D. Yang, X. Huang, J. Tu, Porous Co_3O_4 nanotubes derived from $Co_4(CO)_{12}$ clusters on carbon nanotube templates: a highly efficient material for Li-battery applications. Adv. Mater. **19**, 4505–4509 (2007). https://doi.org/10.1002/adma.200602513
148. C. Li, X. Yin, L. Chen, Q. Li, T. Wang, Synthesis of cobalt ion-based coordination polymer nanowires and their conversion into porous Co_3O_4 nanowires with good lithium storage properties. Chem. Eur. J. **16**, 5215–5221 (2010). https://doi.org/10.1002/chem.200901632
149. Y. Li, B. Tan, Y. Wu, Mesoporous Co_3O_4 nanowire arrays for lithium ion batteries with high capacity and rate capability. Nano Lett. **8**, 265–270 (2008). https://doi.org/10.1021/nl0725906
150. L. Tian, H. Zou, J. Fu, X. Yang, Y. Wang, H. Guo, X. Fu, C. Liang, M. Wu, P.K. Shen, Q. Gao, Topotactic conversion route to mesoporous quasi-single-crystalline Co_3O_4 nanobelts with optimizable electrochemical performance. Adv. Funct. Mater. **20**, 617–623 (2010). https://doi.org/10.1002/adfm.200901503
151. J. Liu, H. Xia, L. Lu, D. Xue, Anisotropic Co_3O_4 porous nanocapsules toward high-capacity Li-ion batteries. J. Mater. Chem. **20**, 1506–1510 (2010). https://doi.org/10.1039/b923834d

152. P. Sivakumar, M. Jana, M. Kota, M.G. Jung, A. Gedanken, H.S. Park, Controllable synthesis of nanohorn-like architectured cobalt oxide for hybrid supercapacitor application. J. Power. Sources **402**, 147–156 (2018). https://doi.org/10.1016/j.jpowsour.2018.09.026
153. J. Li, S. Xiong, Y. Liu, Z. Ju, Y. Qian, High electrochemical performance of monodisperse $NiCo_2O_4$ mesoporous microspheres as an anode material for Li-ion batteries. ACS Appl. Mater. Interfaces **5**, 981–988 (2013). https://doi.org/10.1021/am3026294
154. C. Yuan, H. Bin Wu, Y. Xie, X.W. Lou, Mixed transition-metal oxides: design, synthesis, and energy-related applications. Angew. Chem.—Int. Ed. **53**, 1488–1504 (2014). https://doi.org/10.1002/anie.201303971
155. S. Liu, L. Hu, X. Xu, A.A. Al-Ghamdi, X. Fang, Nickel cobaltite nanostructures for photoelectric and catalytic applications. Small **11**, 4267–4283 (2015). https://doi.org/10.1002/smll.201500315
156. X. Zhang, F. Yang, H. Chen, K. Wang, J. Chen, Y. Wang, S. Song, In situ growth of 2D ultrathin $NiCo_2O_4$ nanosheet arrays on Ni foam for high performance and flexible solid-state supercapacitors. Small **16**, 1–14 (2020). https://doi.org/10.1002/smll.202004188
157. R. Wang, X. Yan, Superior asymmetric supercapacitor based on Ni-Co oxide nanosheets and carbon nanorods. Sci. Rep. **4**, 1–9 (2014). https://doi.org/10.1038/srep03712
158. X. Wang, C. Yan, A. Sumboja, P.S. Lee, High performance porous nickel cobalt oxide nanowires for asymmetric supercapacitor. Nano Energy **3**, 119–126 (2014). https://doi.org/10.1016/j.nanoen.2013.11.001
159. J. Huang, P. Xu, D. Cao, X. Zhou, S. Yang, Y. Li, G. Wang, Asymmetric supercapacitors based on β–$Ni(OH)_2$ nanosheets and activated carbon with high energy density. J. Power. Sources **246**, 371–376 (2014). https://doi.org/10.1016/j.jpowsour.2013.07.105
160. X. Wang, W.S. Liu, X. Lu, P.S. Lee, Dodecyl sulfate-induced fast faradic process in nickel cobalt oxide-reduced graphite oxide composite material and its application for asymmetric supercapacitor device. J. Mater. Chem. **22**, 23114–23119 (2012). https://doi.org/10.1039/c2jm35307e
161. L.J. Xie, J.F. Wu, C.M. Chen, C.M. Zhang, L. Wan, J.L. Wang, Q.Q. Kong, C.X. Lv, K.X. Li, G.H. Sun, A novel asymmetric supercapacitor with an activated carbon cathode and a reduced graphene oxide-cobalt oxide nanocomposite anode. J. Power. Sources **242**, 148–156 (2013). https://doi.org/10.1016/j.jpowsour.2013.05.081
162. L. Xu, Y. Zhao, J. Lian, Y. Xu, J. Bao, J. Qiu, L. Xu, H. Xu, M. Hua, H. Li, Morphology controlled preparation of $ZnCo_2O_4$ nanostructures for asymmetric supercapacitor with ultrahigh energy density. Energy **123**, 296–304 (2017). https://doi.org/10.1016/j.energy.2017.02.018
163. S. Raj, S.K. Srivastava, P. Kar, P. Roy, In situ growth of Co_3O_4 nanoflakes on reduced graphene oxide-wrapped Ni-foam as high performance asymmetric supercapacitor. Electrochim. Acta **302**, 327–337 (2019). https://doi.org/10.1016/j.electacta.2019.02.010
164. T. Xing, Y. Ouyang, Y. Chen, L. Zheng, C. Wu, X. Wang, P-doped ternary transition metal oxide as electrode material of asymmetric supercapacitor. J. Energy Storage **28**, 101248 (2020). https://doi.org/10.1016/j.est.2020.101248
165. C. Tang, Z. Tang, H. Gong, Hierarchically porous Ni–Co oxide for high reversibility asymmetric full-cell supercapacitors. J. Electrochem. Soc. **159**, A651–A656 (2012). https://doi.org/10.1149/2.074205jes
166. S.K. Ujjain, G. Singh, R.K. Sharma, Co_3O_4@Reduced graphene oxide nanoribbon for high performance asymmetric supercapacitor. Electrochim. Acta **169**, 276–282 (2015). https://doi.org/10.1016/j.electacta.2015.03.141
167. P. Hao, J. Tian, Y. Sang, C.C. Tuan, G. Cui, X. Shi, C.P. Wong, B. Tang, H. Liu, 1D Ni-Co oxide and sulfide nanoarray/carbon aerogel hybrid nanostructures for asymmetric supercapacitors with high energy density and excellent cycling stability. Nanoscale **8**, 16292–16301 (2016). https://doi.org/10.1039/c6nr05385h
168. T. Wang, S. Zhang, X. Yan, M. Lyu, L. Wang, J. Bell, H. Wang, 2-Methylimidazole-derived Ni-Co layered double hydroxide nanosheets as high rate capability and high energy density storage material in hybrid supercapacitors. ACS Appl. Mater. Interfaces **9**, 15510–15524 (2017). https://doi.org/10.1021/acsami.7b02987

169. H. Wang, H.S. Casalongue, Y. Liang, H. Dai, $Ni(OH)_2$ nanoplates grown on graphene as advanced electrochemical pseudocapacitor materials. J. Am. Chem. Soc. **132**, 7472–7477 (2010). https://doi.org/10.1021/ja102267j
170. J.W. Lang, L. Bin Kong, M. Liu, Y.C. Luo, L. Kang, Asymmetric supercapacitors based on stabilized α–$Ni(OH)_2$ and activated carbon. J. Solid State Electrochem. **14**, 1533–1539 (2010). https://doi.org/10.1007/s10008-009-0984-1
171. M. Xie, S. Duan, Y. Shen, K. Fang, Y. Wang, M. Lin, X. Guo, In-situ-grown $Mg(OH)_2$-derived hybrid α–$Ni(OH)_2$ for highly stable supercapacitor. ACS Energy Lett. **1**, 814–819 (2016). https://doi.org/10.1021/acsenergylett.6b00258
172. D. Ghosh, M. Mandal, C.K. Das, Solid state flexible asymmetric supercapacitor based on carbon fiber supported hierarchical $Co(OH)_xCO_3$ and $Ni(OH)_2$. Langmuir **31**, 7835–7843 (2015). https://doi.org/10.1021/acs.langmuir.5b00649
173. T. Xu, G. Li, X. Yang, Z. Guo, L. Zhao, Design of the seamless integrated C@NiMn–OH–Ni_3S_2/Ni foam advanced electrode for supercapacitors. Chem. Eng. J. **362**, 783–793 (2019). https://doi.org/10.1016/j.cej.2019.01.083
174. Y. Xi, G. Wei, J. Li, X. Liu, M. Pang, Y. Yang, Y. Ji, V.Y. Izotov, Q. Guo, W. Han, Facile synthesis of MnO_2–$Ni(OH)_2$ 3D ridge-like porous electrode materials by seed-induce method for high-performance asymmetric supercapacitor. Electrochim. Acta **233**, 26–35 (2017). https://doi.org/10.1016/j.electacta.2017.02.038
175. F. Wang, T. Wang, S. Sun, Y. Xu, R. Yu, H. Li, One-step synthesis of nickle iron-layered double hydroxide/reduced graphene oxide/carbon nanofibres composite as electrode materials for asymmetric supercapacitor. Sci. Rep. **8**, 1–10 (2018). https://doi.org/10.1038/s41598-018-27171-0
176. F. Wang, S. Sun, Y. Xu, T. Wang, R. Yu, H. Li, High performance asymmetric supercapacitor based on cobalt nickle iron-layered double hydroxide/carbon nanofibres and activated carbon. Sci. Rep. **7**, 1–11 (2017). https://doi.org/10.1038/s41598-017-04807-1
177. X. Zhang, S. Wang, L. Xu, T. He, F. Lu, H. Li, J. Ye, Controllable synthesis of cross-linked CoAl-LDH/$NiCo_2S_4$ sheets for high performance asymmetric supercapacitors. Ceram. Int. **43**, 14168–14175 (2017). https://doi.org/10.1016/j.ceramint.2017.07.159
178. Y. Xiao, D. Su, X. Wang, S. Wu, L. Zhou, Z. Sun, Z. Wang, S. Fang, F. Li, Ultrahigh energy density and stable supercapacitor with 2D NiCoAl layered double hydroxide. Electrochim. Acta **253**, 324–332 (2017). https://doi.org/10.1016/j.electacta.2017.09.033
179. L.-B. Kong, M. Liu, J.-W. Lang, Y.-C. Luo, L. Kang, Asymmetric supercapacitor based on loose-packed cobalt hydroxide nanoflake materials and activated carbon. J. Electrochem. Soc. **156**, A1000 (2009). https://doi.org/10.1149/1.3236500
180. Z. Li, M. Shao, L. Zhou, R. Zhang, C. Zhang, J. Han, M. Wei, D.G. Evans, X. Duan, A flexible all-solid-state micro-supercapacitor based on hierarchical CuO@layered double hydroxide core-shell nanoarrays. Nano Energy **20**, 294–304 (2016). https://doi.org/10.1016/j.nanoen.2015.12.030
181. Y. Liu, R. Wang, X. Yan, Synergistic effect between ultra-small nickel hydroxide nanoparticles and reduced graphene oxide sheets for the application in high-performance asymmetric supercapacitor. Sci. Rep. **5**, 1–12 (2015). https://doi.org/10.1038/srep11095
182. M.R. Gao, Y.F. Xu, J. Jiang, S.H. Yu, Nanostructured metal chalcogenides: synthesis, modification, and applications in energy conversion and storage devices. Chem. Soc. Rev. **42**, 2986–3017 (2013). https://doi.org/10.1039/c2cs35310e
183. M. Chhowalla, H.S. Shin, G. Eda, L.J. Li, K.P. Loh, H. Zhang, The chemistry of two-dimensional layered transition metal dichalcogenide nanosheets. Nat. Chem. **5**, 263–275 (2013). https://doi.org/10.1038/nchem.1589
184. C.H. Lai, M.Y. Lu, L.J. Chen, Metal sulfide nanostructures: synthesis, properties and applications in energy conversion and storage. J. Mater. Chem. **22**, 19–30 (2012). https://doi.org/10.1039/c1jm13879k
185. Y. Li, H. Peng, L. Yang, H. Dong, P. Xiao, Investigating the effect of sulfur and selenium on the electrochemical properties of nickel–cobalt oxides: enhanced charge capacity and composition–property relationships. J. Mater. Sci. **51**, 7108–7118 (2016). https://doi.org/10.1007/s10853-016-9968-6

186. F. Tao, Y.Q. Zhao, G.Q. Zhang, H.L. Li, Electrochemical characterization on cobalt sulfide for electrochemical supercapacitors. Electrochem. Commun. **9**, 1282–1287 (2007). https://doi.org/10.1016/j.elecom.2006.11.022
187. M.S. Javed, S. Dai, M. Wang, D. Guo, L. Chen, X. Wang, C. Hu, Y. Xi, High performance solid state flexible supercapacitor based on molybdenum sulfide hierarchical nanospheres. J. Power. Sources **285**, 63–69 (2015). https://doi.org/10.1016/j.jpowsour.2015.03.079
188. B.T. Zhu, Z. Wang, S. Ding, J.S. Chen, X.W. Lou, Hierarchical nickel sulfide hollow spheres for high performance supercapacitors. RSC Adv. **1**, 397–400 (2011). https://doi.org/10.1039/c1ra00240f
189. X. Wang, S.X. Zhao, L. Dong, Q.L. Lu, J. Zhu, C.W. Nan, One-step synthesis of surface-enriched nickel cobalt sulfide nanoparticles on graphene for high-performance supercapacitors. Energy Storage Mater. **6**, 180–187 (2017). https://doi.org/10.1016/j.ensm.2016.11.005
190. H. Chen, J. Jiang, L. Zhang, H. Wan, T. Qi, D. Xia, Highly conductive $NiCo_2S_4$ urchin-like nanostructures for high-rate pseudocapacitors. Nanoscale **5**, 8879–8883 (2013). https://doi.org/10.1039/c3nr02958a
191. P. Xu, W. Zeng, S. Luo, C. Ling, J. Xiao, A. Zhou, Y. Sun, K. Liao, 3D Ni-Co selenide nanorod array grown on carbon fiber paper: towards high-performance flexible supercapacitor electrode with new energy storage mechanism. Electrochim. Acta **241**, 41–49 (2017). https://doi.org/10.1016/j.electacta.2017.04.121
192. X. Liu, Y. Liu, L.Z. Fan, MOF-derived $CoSe_2$ microspheres with hollow interiors as high-performance electrocatalysts for the enhanced oxygen evolution reaction. J. Mater. Chem. A Mater. **5**, 15310–15314 (2017). https://doi.org/10.1039/c7ta04662f
193. S.K. Park, J.K. Kim, Y. Chan Kang, Metal-organic framework-derived $CoSe_2$/(NiCo)Se_2 box-in-box hollow nanocubes with enhanced electrochemical properties for sodium-ion storage and hydrogen evolution. J. Mater. Chem. A Mater. **5**, 18823–18830 (2017). https://doi.org/10.1039/c7ta05571d
194. W. Sun, C. Cai, X. Tang, L.P. Lv, Y. Wang, Carbon coated mixed-metal selenide microrod: bimetal-organic-framework derivation approach and applications for lithium-ion batteries. Chem. Eng. J. **351**, 169–176 (2018). https://doi.org/10.1016/j.cej.2018.06.093
195. C. Xia, Q. Jiang, C. Zhao, P.M. Beaujuge, H.N. Alshareef, Asymmetric supercapacitors with metal-like ternary selenides and porous graphene electrodes. Nano Energy **24**, 78–86 (2016). https://doi.org/10.1016/j.nanoen.2016.04.012
196. Z.B. Zhai, K.J. Huang, X. Wu, Superior mixed Co-Cd selenide nanorods for high performance alkaline battery-supercapacitor hybrid energy storage. Nano Energy **47**, 89–95 (2018). https://doi.org/10.1016/j.nanoen.2018.02.059
197. H. Chen, J. Jiang, L. Zhang, D. Xia, Y. Zhao, D. Guo, T. Qi, H. Wan, In situ growth of $NiCo_2S_4$ nanotube arrays on Ni foam for supercapacitors: maximizing utilization efficiency at high mass loading to achieve ultrahigh areal pseudocapacitance. J. Power. Sources **254**, 249–257 (2014). https://doi.org/10.1016/j.jpowsour.2013.12.092
198. G.S. Rama Raju, E. Pavitra, G. Nagaraju, S.C. Sekhar, S.M. Ghoreishian, C.H. Kwak, J.S. Yu, Y.S. Huh, Y.K. Han, Rational design of forest-like nickel sulfide hierarchical architectures with ultrahigh areal capacity as a binder-free cathode material for hybrid supercapacitors. J. Mater. Chem. A Mater. **6**, 13178–13190 (2018). https://doi.org/10.1039/c8ta02597e
199. W. Chen, C. Xia, H.N. Alshareef, One-step electrodeposited nickel cobalt sulfide nanosheet arrays for high-performance asymmetric supercapacitors. ACS Nano **8**, 9531–9541 (2014). https://doi.org/10.1021/nn503814y
200. F. Wang, G. Li, J. Zheng, J. Ma, C. Yang, Q. Wang, Microwave synthesis of three-dimensional nickel cobalt sulfide nanosheets grown on nickel foam for high-performance asymmetric supercapacitors. J. Colloid Interface Sci. **516**, 48–56 (2018). https://doi.org/10.1016/j.jcis.2018.01.038
201. X. Han, K. Tao, D. Wang, L. Han, Design of a porous cobalt sulfide nanosheet array on Ni foam from zeolitic imidazolate frameworks as an advanced electrode for supercapacitors. Nanoscale **10**, 2735–2741 (2018). https://doi.org/10.1039/c7nr07931a

202. D. Zha, Y. Fu, L. Zhang, J. Zhu, X. Wang, Design and fabrication of highly open nickel cobalt sulfide nanosheets on Ni foam for asymmetric supercapacitors with high energy density and long cycle-life. J. Power. Sources **378**, 31–39 (2018). https://doi.org/10.1016/j.jpowsour.2017.12.020
203. K. Le, M. Gao, W. Liu, J. Liu, Z. Wang, F. Wang, V. Murugadoss, S. Wu, T. Ding, Z. Guo, MOF-derived hierarchical core-shell hollow iron-cobalt sulfides nanoarrays on Ni foam with enhanced electrochemical properties for high energy density asymmetric supercapacitors. Electrochim. Acta **323**, 134826 (2019). https://doi.org/10.1016/j.electacta.2019.134826
204. J. Dong, S. Li, Y. Ding, Anchoring nickel-cobalt sulfide nanoparticles on carbon aerogel derived from waste watermelon rind for high-performance asymmetric supercapacitors. J. Alloys Compd. **845**, 155701 (2020). https://doi.org/10.1016/j.jallcom.2020.155701
205. T. Xiao, F. Chen, W. Zhou, P. Che, S. Wang, X. Chen, X. Tan, P. Xiang, L. Jiang, X. Chen, Ni-Bi-S nanosheets/Ni foam as a binder-free high-performance electrode for asymmetric supercapacitors. Chem. Eng. J. **378**, 122162 (2019). https://doi.org/10.1016/j.cej.2019.122162
206. M. Yan, Y. Yao, J. Wen, L. Long, M. Kong, G. Zhang, X. Liao, G. Yin, Z. Huang, Construction of a hierarchical $NiCo_2S_4$@PPy core-shell heterostructure nanotube array on Ni foam for a high-performance asymmetric supercapacitor. ACS Appl. Mater. Interfaces **8**, 24525–24535 (2016). https://doi.org/10.1021/acsami.6b05618
207. X. Wang, J. Hao, Y. Su, F. Liu, J. An, J. Lian, A Ni1-: XZnxS/Ni foam composite electrode with multi-layers: one-step synthesis and high supercapacitor performance. J. Mater. Chem. A Mater. **4**, 12929–12939 (2016). https://doi.org/10.1039/c6ta04022e
208. A. Pramanik, S. Maiti, T. Dhawa, M. Sreemany, S. Mahanty, High faradaic charge storage in $ZnCo_2S_4$ film on Ni-foam with a hetero-dimensional microstructure for hybrid supercapacitor. Mater. Today Energy **9**, 416–427 (2018). https://doi.org/10.1016/j.mtener.2018.07.007
209. X. Shi, H. Wang, S. Ji, V. Linkov, F. Liu, R. Wang, $CoNiSe_2$ nanorods directly grown on Ni foam as advanced cathodes for asymmetric supercapacitors. Chem. Eng. J. **364**, 320–327 (2019). https://doi.org/10.1016/j.cej.2019.01.156
210. L. Hou, Y. Shi, C. Wu, Y. Zhang, Y. Ma, X. Sun, J. Sun, X. Zhang, C. Yuan, Monodisperse metallic $NiCoSe_2$ hollow sub-microspheres: formation process, intrinsic charge-storage mechanism, and appealing pseudocapacitance as highly conductive electrode for electrochemical supercapacitors. Adv. Funct. Mater. **28**, 1–12 (2018). https://doi.org/10.1002/adfm.201705921
211. Y. Zhu, Z. Huang, Z. Hu, L. Xi, X. Ji, Y. Liu, 3D interconnected ultrathin cobalt selenide nanosheets as cathode materials for hybrid supercapacitors. Electrochim. Acta **269**, 30–37 (2018). https://doi.org/10.1016/j.electacta.2018.02.146
212. L.P. Lv, C. Zhi, Y. Gao, X. Yin, Y. Hu, D. Crespy, Y. Wang, Hierarchical "tube-on-fiber" carbon/mixed-metal selenide nanostructures for high-performance hybrid supercapacitors. Nanoscale **11**, 13996–14009 (2019). https://doi.org/10.1039/c9nr03088c
213. B. Kirubasankar, V. Murugadoss, J. Lin, T. Ding, M. Dong, H. Liu, J. Zhang, T. Li, N. Wang, Z. Guo, S. Angaiah, In situ grown nickel selenide on graphene nanohybrid electrodes for high energy density asymmetric supercapacitors. Nanoscale **10**, 20414–20425 (2018). https://doi.org/10.1039/c8nr06345a
214. S. Jiang, J. Wu, B. Ye, Y. Fan, J. Ge, Q. Guo, M. Huang, Growth of $Ni3Se_2$ nanosheets on Ni foam for asymmetric supercapacitors. J. Mater. Sci. Mater. Electron. **29**, 4649–4657 (2018). https://doi.org/10.1007/s10854-017-8416-y
215. Z. Gao, C. Chen, J. Chang, L. Chen, P. Wang, D. Wu, F. Xu, Y. Guo, K. Jiang, Enhanced cycleability of faradic $CoNi_2S_4$ electrode by reduced graphene oxide coating for efficient asymmetric supercapacitor. Electrochim. Acta **281**, 394–404 (2018). https://doi.org/10.1016/j.electacta.2018.05.194
216. Z. Ji, N. Li, M. Xie, X. Shen, W. Dai, K. Liu, K. Xu, G. Zhu, High-performance hybrid supercapacitor realized by nitrogen-doped carbon dots modified cobalt sulfide and reduced graphene oxide. Electrochim Acta **334** (2020). https://doi.org/10.1016/j.electacta.2020.135632

217. Z. Liang, C. Qu, W. Guo, R. Zou, Q. Xu, Pristine metal-organic frameworks and their composites for energy storage and conversion. Adv. Mater. **30**, 1–39 (2018). https://doi.org/10.1002/adma.201702891
218. R. Zhao, Z. Liang, R. Zou, Q. Xu, Metal-organic frameworks for batteries. Joule **2**, 2235–2259 (2018). https://doi.org/10.1016/j.joule.2018.09.019
219. B. Zhu, D. Xia, R. Zou, Metal-organic frameworks and their derivatives as bifunctional electrocatalysts. Coord. Chem. Rev. **376**, 430–448 (2018). https://doi.org/10.1016/j.ccr.2018.07.020
220. B. Zhu, R. Zou, Q. Xu, Metal-organic framework based catalysts for hydrogen evolution. Adv. Energy Mater. **8**, 1–33 (2018). https://doi.org/10.1002/aenm.201801193
221. H. Wang, Q.L. Zhu, R. Zou, Q. Xu, Metal-organic frameworks for energy applications. Chem **2**, 52–80 (2017). https://doi.org/10.1016/j.chempr.2016.12.002
222. L. Wang, Y. Han, X. Feng, J. Zhou, P. Qi, B. Wang, Metal-organic frameworks for energy storage: batteries and supercapacitors. Coord. Chem. Rev. **307**, 361–381 (2016). https://doi.org/10.1016/j.ccr.2015.09.002
223. A. Morozan, F. Jaouen, Metal organic frameworks for electrochemical applications. Energy Environ. Sci. **5**, 9269–9290 (2012). https://doi.org/10.1039/c2ee22989g
224. D.M. Teffu, M.D. Makhafola, M.M. Ndipingwi, E. Makhado, M.J. Hato, E.I. Iwuoha, K.D. Modibane, K. Makgopa, Interrogation of electrochemical performance of reduced graphene oxide/metal-organic framework hybrid for asymmetric supercabattery application. Electroanalysis **32**, 2827–2837 (2020). https://doi.org/10.1002/elan.202060303
225. Y. Yan, P. Gu, S. Zheng, M. Zheng, H. Pang, H. Xue, Facile synthesis of an accordion-like Ni-MOF superstructure for high-performance flexible supercapacitors. J. Mater. Chem. A Mater. **4**, 19078–19085 (2016). https://doi.org/10.1039/c6ta08331e
226. S. Gao, Y. Sui, F. Wei, J. Qi, Q. Meng, Y. He, Facile synthesis of cuboid Ni–MOF for high-performance supercapacitors. J. Mater. Sci. **53**, 6807–6818 (2018). https://doi.org/10.1007/s10853-018-2005-1
227. P. Du, Y. Dong, C. Liu, W. Wei, D. Liu, P. Liu, Fabrication of hierarchical porous nickel based metal-organic framework (Ni-MOF) constructed with nanosheets as novel pseudo-capacitive material for asymmetric supercapacitor. J. Colloid Interface Sci. **518**, 57–68 (2018). https://doi.org/10.1016/j.jcis.2018.02.010
228. R. Abazari, S. Sanati, A. Morsali, A. Slawin, C.L. Carpenter-Warren, Dual-purpose 3D pillared metal-organic framework with excellent properties for catalysis of oxidative desulfurization and energy storage in asymmetric supercapacitor. ACS Appl. Mater. Interfaces **11**, 14759–14773 (2019). https://doi.org/10.1021/acsami.9b00415
229. P. Wen, P. Gong, J. Sun, J. Wang, S. Yang, Design and synthesis of Ni-MOF/CNT composites and rGO/carbon nitride composites for an asymmetric supercapacitor with high energy and power density. J. Mater. Chem. A Mater. **3**, 13874–13883 (2015). https://doi.org/10.1039/c5ta02461g
230. Y.Y. Kannangara, U.A. Rathnayake, J.K. Song, Hybrid supercapacitors based on metal organic frameworks using p-phenylenediamine building block. Chem. Eng. J. **361**, 1235–1244 (2019). https://doi.org/10.1016/j.cej.2018.12.173
231. C. Chen, M.K. Wu, K. Tao, J.J. Zhou, Y.L. Li, X. Han, L. Han, Formation of bimetallic metal-organic framework nanosheets and their derived porous nickel-cobalt sulfides for supercapacitors. Dalton Trans. **47**, 5639–5645 (2018). https://doi.org/10.1039/c8dt00464a
232. W. Gao, D. Chen, H. Quan, R. Zou, W. Wang, X. Luo, L. Guo, Fabrication of hierarchical porous metal-organic framework electrode for aqueous asymmetric supercapacitor. ACS Sustain. Chem. Eng. **5**, 4144–4153 (2017). https://doi.org/10.1021/acssuschemeng.7b00112
233. P. Forouzandeh, V. Kumaravel, S.C. Pillai, Electrode materials for supercapacitors: a review of recent advances. Catalysts **10**, 1–73 (2020). https://doi.org/10.3390/catal10090969

Chapter 2
Hybrid Nanostructures in High Energy Density Supercapacitors

Karthick Sivalingam Nallathambi, Manik Clinton Franklin, Lekshmi Sunil, Jayashree Chandrasekar, Yogapriya Selvaraj, and Hemalatha Kuzhandaivel

Abstract The degradation of fossil fuels, environmental conservation, and the increase in demand for energy are the factors that address the need for energy storage. The electrochemical energy is stored in an energy storage device. Supercapacitors are energy storage devices that provide high charge and discharge rates, extraordinary cyclic stability, high power density, and moderate energy density. Supercapacitors are applied in various fields, such as power hybrid electric vehicles and other electronic systems. Based on the type of electrode material used, the charge storage mechanism differs, and supercapacitors are classified as electric double-layer capacitors, pseudo capacitors, and hybrid supercapacitors. The combined effect of carbonaceous materials due to their stability and redox materials that improve the charging-discharging process in hybrid supercapacitors creates a new era in electric energy storage devices. The individual materials having specific properties can be tuned for hybrid supercapacitors. The literature review on hybrid nanostructures reveals that various morphology and interfaces create the possibility and flexibility to obtain new properties and functions. The improving properties are mechanical strength, high stability, light emission, and electron conductivity. The hybrid nanostructures possess morphologies such as core–shell nanoparticles, nanowires, nanorods, nanobelts, nanoplates, and nanocubes, offering the highest impacts in electric energy storage devices. These designs depend on the internal and external interfacing capabilities and the physical/chemical compatibilities. The electron transfer between the current collector and the working electrode material is highly productive due to the void volume that leads to easy electrolyte access and fast ion transport. The surface area increases while the void volume increases in the material, hence the storage ability. The electrochemical properties of the individual materials can be doubled or more when the material is composited. The impact of hybrid nanostructures for fabricating efficient high-energy supercapacitors based on

K. Sivalingam Nallathambi (✉) · M. C. Franklin · L. Sunil · J. Chandrasekar
Electrochemical Materials and Devices Lab, Department of Chemistry, Bharathiar University, Coimbatore 641046, Tamil Nadu, India
e-mail: snkarthick@buc.edu.in

Y. Selvaraj · H. Kuzhandaivel
Materials Research and Product Laboratory, Department of Chemistry, Coimbatore Institute of Technology, Coimbatore 641014, Tamil Nadu, India

© The Author(s), under exclusive license to Springer Nature Switzerland AG 2026

K. Raju et al. (eds.), *Hybrid Nanostructures for Advanced Supercapacitors*, Modern Aspects of Electrochemistry 63, https://doi.org/10.1007/978-3-032-04593-5_2

their charge storage mechanism is sorted out. The involvement of an effective hybrid nanostructure will improve the storage device in the near future.

Keywords Supercapacitors · Multi-dimensional hybrid nanostructures · Charge storage mechanism · Power density · Hybrid supercapacitors

2.1 Introduction

Energy has become an essential and crucial part of our daily life. Energy consumed from non-renewable sources like fossil fuels and coal is termed "non-renewable energy." About 80% of the world's energy consumption is from non-renewable sources. The energy produced from renewable sources like solar, wind, tidal, and geothermal is called "renewable energy," but the occurrences are seasonal and vary based on the topography [1]. The produced energy is stored through two primary energy storage devices, i.e., supercapacitors and rechargeable batteries. They are widely used in various applications such as electronic devices, hybrid electric vehicles, communication devices, and several electronic gadgets such as mobile phones, digital cameras, and electric tools. Materials play a significant role in storing energy [2]. Designing hybrid materials at the nanoscale becomes a key factor for energy storage applications. Around the globe, various research communities are trying to develop multifunctional and advanced hybrid materials to improve next-generation storage device's power and energy densities [3]. Micro-sized materials are inefficient in meeting the increasing global energy demands, leading us to focus on hybrid nanomaterials. The quantum properties of the materials get affected as the sizes are reduced. The small size of these materials makes them free from internal imperfections in structures and protects them from mechanical failures. Even thermal and catalytic property changes were observed once the bulk material was broken into the nanoscale. Increasing the surface area to volume ratio makes the material more reactive [4]. These materials can rearrange the crystalline structure upon dispersion. The advantages of using nanomaterials in energy applications are (i) the nano-size offers possibilities of tuning the material's properties effortlessly, (ii) the highly porous nature of nanomaterials leads to a high surface area, which enhances the accessibility of the charges for efficient energy applications. The need for hybrid nanomaterials is to enhance the energy density of the supercapacitor device. The device's high energy density helps to deliver the same amount of energy for a prolonged time. Supercapacitors have high power density but low energy density. This is the main reason batteries are not replaced by supercapacitors effectively. In order to improve the energy density, more focus should be on increasing the cell capacitance and cell voltage,

$$\boldsymbol{E} = \frac{1}{7.2}\boldsymbol{C}.\boldsymbol{V}^2\left(\text{Unit} : \mathbf{Wh\ kg^{-1}}\right)$$

where ***C*** is the capacitance of the cell and ***V*** is the voltage of the cell.

Capacitance and potential window are directly proportional to energy density, so increasing these factors helps to improve the energy density, as shown in Fig. 2.1. Choosing the proper materials helps improve the dual properties, enhancing the energy density. To prepare an excellent hybrid material, there is a need to understand the individual material's properties and storage mechanisms. Carbon-based materials like graphene, graphene oxide, reduced graphene oxide, activated carbon, and carbon nanotubes show electrical double-layer capacitance behavior where ion adsorption and desorption occur on the surface area of the electrode [6]. Materials such as metal oxides, two-dimensional metal chalcogenides, and MXenes show pseudocapacitance behavior, which occurs by charge storage through the surface redox reaction on the electrode. Combining materials with different charge storage mechanisms is called hybrid materials.

The advantage of hybrid materials' combined mechanism of electric double-layer capacitors and faradaic reaction within the single electrode enhances the supercapacitor's overall performance. For example, using lithium titanate ($Li_4Ti_5O_{12}$) as the negative electrode material in batteries has its limitations, such as poor conductivity and cyclic stability due to the degradation of the crystal structure over a prolonged operating period. Therefore, it is hybridized with carbon-based material to improve

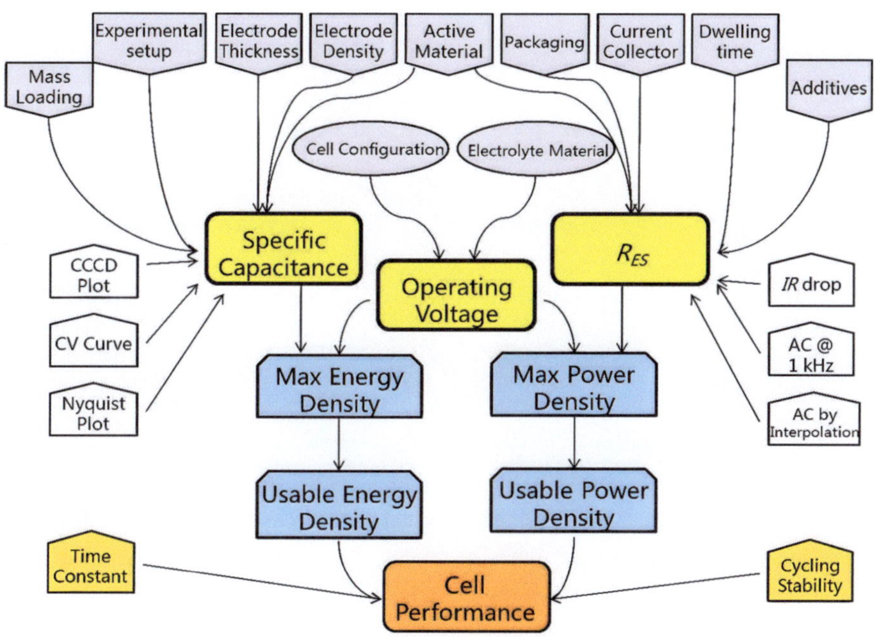

Fig. 2.1 Components of cell such as electrode materials, electrolytes, current collectors, additives, and their electrochemical properties affect the various parameters such as capacitance, energy density, power density, and cyclic stability of the cells. Reprinted with permission of Wiley from ref. [5]

the conductivity and stability of the electrode. Generally, the metal-based structures and their charge-transfer reaction mechanism led to the crystal structure's deformation. Hence, an efficient carbon backbone for metal-based materials can overcome the deformation, increase electrical conductivity with surface area, and provide high pore channels for electrolytic ions accessibility [7].

2.2 Charge Storage Mechanism in Hybrid Nanostructures

The supercapacitor's charge storage mechanism consists of three different types (1) Capacitive non-faradaic charge storage (electrical double-layer capacitance), (2) Capacitive faradaic charge storage (pseudocapacitance), (3) Non-capacitive faradaic charge storage (Battery), as shown in Fig. 2.2.

2.2.1 Capacitive Non-faradaic Charge Storage

Capacitive is a type of ion storage mechanism where the electrolyte ions are electrostatically adsorbed and desorbed on the surface of the electrode material. Commonly,

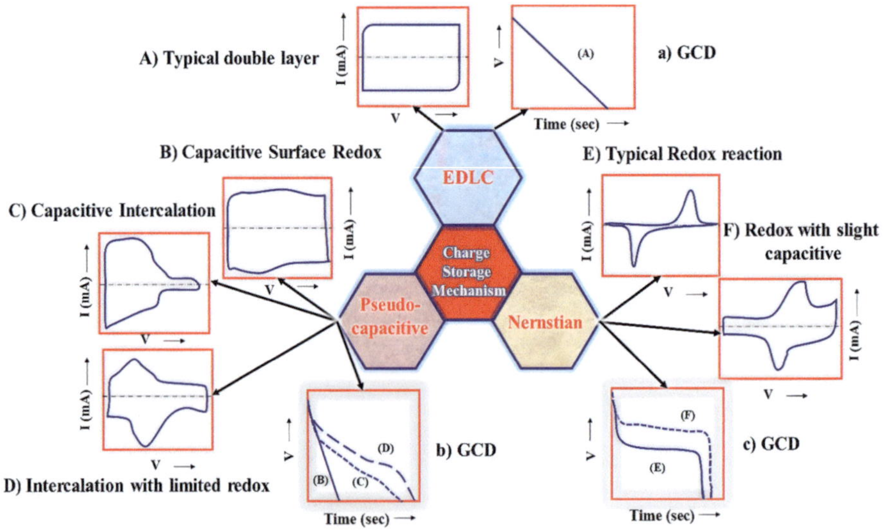

Fig. 2.2 Schematic illustration of the cyclic voltammograms (**A–F**), and galvanostatic charge–discharge profiles (a–c) for the various charge storage mechanisms occur at the electrode–electrolyte interface in energy storage device. Reprinted with permission of the American Chemical Society from ref. [8]. EDLC—Electrochemical double-layer capacitors, GCD—Galvanostatic charge–discharge

the capacitive behaviors are exhibited by the carbon and carbon derivative materials, which are highly capable of providing a large surface area for the accessibility of the ions. Carbon-based supercapacitors have been highly valuable in the market because of their excellent electrical conductivity, stability over mechanical and thermal conditions, specific surface area, and cost-effectiveness. Carbon-based materials like activated carbon, graphene, carbon nanotubes, and graphene-based materials fall under this category. These materials are known for their high surface area, porosity, electrical conductivity, and mechanical stability. The capacitive non-faradaic materials provide a rectangular-type cyclic voltammogram like an ideal capacitor. The integrated area between the anodic and cathodic currents shows the quantity of charges that are capacitively stored at the surface of the electrode. The total charge accommodated on the electrode can be quantified by $Q = C^{*}\Delta E$, where Q is the number of charges accommodated on the electrode surface, C is the capacitance of the electrode material, and ΔE is the potential window. The capacitance can be determined from the expression of $C = \frac{\varepsilon_o \varepsilon \cdot A}{d}$ where ε_o is the vacuum permittivity (8.854×10^{-12} F m^{-1}), ε is the dielectric constant, A is the area of the interfacial surface of the electrode, and d is the double-layer thickness. If the surface of the electrode material is a sheet-like structure, then the charge stored on the surface is two-dimensional, but if these electrode materials also have a porous structure, the charge stored on the surface is 3D dimensional [9, 10]. Pristine raw carbon does not have the above-mentioned properties, which are incapable of storage applications. Once carbon gets activated by various reagents like potassium hydroxide, H_3PO_4, and $ZnCl_2$, it enhances the porosity and surface area for the high accessibility of the electrolyte ions [11]. Even activated carbon suffers from low electrical conductivity and specific capacitance. Carbon nanotubes are used to rectify the issues to increase the electrical conductivity and improve the power density by highly utilizing the surface area of the electrode with electrolytic ions.

The other type of carbon-based material is graphene. The graphene materials are 2D in nature and have high possible access to ions and excellent conductivity compared to activated carbon and carbon nanotubes. Functionalizing the graphene structure using various heteroatoms like oxygen, nitrogen, and sulfur, creates more defects in the sp^2 carbon ring structures, and exfoliation of the graphene sheets leads to excellent specific capacitance, electrical conductivity, and mechanical stability. The theoretical capacitance of graphene is 526 F g^{-1} in aqueous-based electrolytes, making graphene an effective capacitive-based material for supercapacitor applications [12]. These highly efficient graphene materials even have a few drawbacks, like low electrical conductivity. The electrical conductivity of reduced graphene is 10–200 S m^{-1}, which is lower than that of carbon nanotubes, and the graphene sheets restack by van der Waals force of attraction, hindering the performance of the graphene materials. Therefore, researchers are involved in synthesizing hybrid nanomaterials, as shown in Fig. 2.3, by combining the activated carbon with graphene or carbon nanotubes and graphene with carbon nanotubes, which rectifies their disadvantages and produces high-performance hybrid nanostructures for supercapacitors [12, 13].

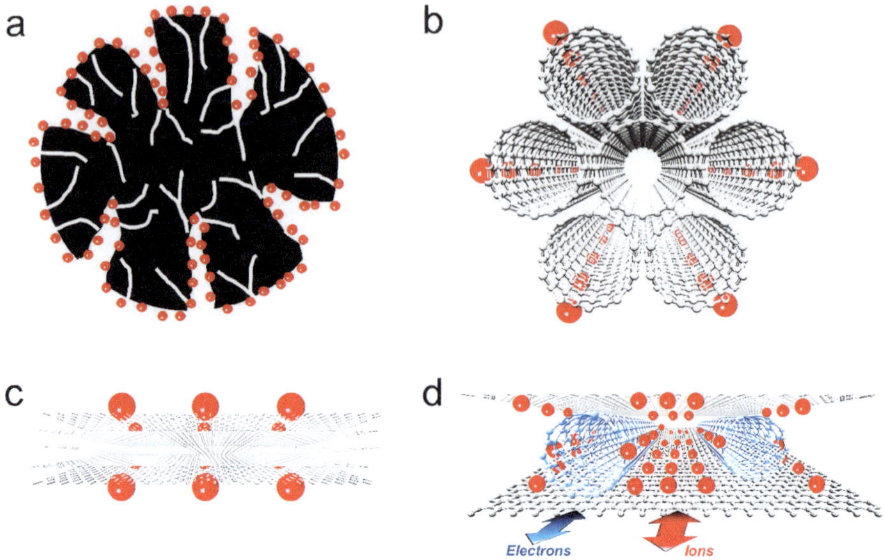

Fig. 2.3 Comparison of the carbon-based materials interacting with the electrolytic ions: **a** activated carbon; **b** pristine single-walled carbon nanotubes; **c** pristine graphene; **d** single-walled carbon nanotubes incorporated into the graphene structure. Reprinted with permission of the Royal Society of Chemistry from ref. [12]

The symmetric devices of activated carbon/carbon nanotube composites exhibit an energy density of 50 Wh kg^{-1} and power density of 22 kW kg^{-1} at a voltage of 4 V using ionic liquid as electrolyte. The device's high performance is due to the combination of large surface area activated carbon and high electrical conductivity of carbon nanotubes, which increases the device's capacitance. The added advantage is using ionic liquid as an electrolyte, which has high thermodynamic stability, leading to an enhanced operating voltage of up to 4 V [14]. The N-doped carbon layer and carbon nanotube composite as a core–shell structure were synthesized using a facile single-step hydrothermal method. The nitrogen-doped carbon layer/carbon nanotube composite electrodes showed a high specific capacitance of 322.1 F g^{-1} at a specific current of 1 A g^{-1}. This is due to the high specific surface area and conductivity of the nitrogen-doped carbon/carbon nanotube composite, and the influence of N–atom doping enhances the performance [15]. A self-standing electrode's fabrication consists of an activated carbon/carbon nanotube/reduced graphene oxide hybrid composite. Here, the porous 3D framework is constructed from a 1D carbon nanotube, which is composed of 2D reduced graphene oxide by van der Waals attraction. The activated carbon/carbon nanotube-reduced graphene oxide composite device delivers an energy density of 30 Wh kg^{-1} at the operating voltage of ~ 3 V using the organic electrolyte. This synergistic effect of the hybrid nanostructure activated carbon/carbon nanotube/reduced graphene oxide composite enhances the device's performance [16].

The capacitive non-faradaic electrode materials, i.e., carbon-based materials like bio-derived activated carbon, multi-dimensional carbon materials such as single-walled and multi-walled carbon nanotubes, nanofibers, aerogels, foams, and graphene derivatives such as graphene, graphene oxide, and reduced graphene oxide, exhibit [17]:

- High surface area for electrolyte ions accessing
- High electronic conductivity
- High mechanical and thermal stability
- Eco-friendly and pollution-free
- Robust structure.

2.2.2 Capacitive Faradaic Charge Storage

Another type of hybrid nanostructure used in the supercapacitor application is pseudocapacitive-type charge storage mechanism-based materials. The prefix "pseudo" means "false," which implies that the charge storage mechanism is neither completely capacitive nor typical faradaic observed via cyclic voltammetry and galvanostatic charge–discharge studies. The pseudocapacitive combines both capacitive and faradaic mechanisms, where the reaction kinetics at the electrode surface are like the reactions occurring in capacitive materials (electric double-layer capacitors). In 1971, the first redox material, ruthenium oxide, exhibited a rectangular-type cyclic voltammogram that acts as a capacitor. Until 1990, ruthenium oxide was under the capacitor category. Similar electrochemical properties are observed in some transition metal oxides and conducting polymers. In the late 1990s, in addition to the electrostatic mechanism, some metal oxides exhibited faradaic reactions on the surface of the electrode, called the pseudocapacitive mechanism. There are two similarities between the capacitive and faradaic, i.e., the current's polarity is proportional to its scan rates, and the scan rate direction changes with the current polarity. These two similarities emphasize that it is pseudocapacitance [9, 10].

A typical faradaic-based material can exhibit extrinsic pseudocapacitive mechanisms only when the size of the material is reduced to the nanoscale level. This is due to the shortening of the solid-state diffusion path, which reduces the ions' intercalation, which induces successive surface redox reactions at the electrode. Hence, the electrochemical profiles of cyclic voltammetry and galvanic charge–discharge may appear close to the pseudocapacitor profile when the battery material thickness is at the nanoscale level [18].

In a pseudocapacitor, the charged electrodes are obtained due to the faradaic reaction, compensated by the counter ion from the electrolyte. The ruthenium oxide (RuO_2)-based capacitor has multiple redox sites that undergo redox reactions, leading to a change in sub-stoichiometric level [10].

$$\mathrm{RuO_2} + x\mathrm{H}^+ + x\mathrm{e}^- \rightleftharpoons \mathrm{RuOOH}_x + \mathrm{e}^-(0 \leq x \leq 2)$$

The transition metal oxides, such as MnO_2, Fe_2O_3, and V_2O_5, and conducting polymers, such as poly(3,4-ethylenedioxythiophene), polyaniline, and polypyrrole, are some of the pseudocapacitive materials that exhibit a similar redox charge storage mechanism on the surface.

Pseudocapacitive materials are limited in the range of operation potential and exhibit pseudocapacitive behavior at a certain potential range, and if there is an increase in potential, faradaic reactions occur. So, these materials are limited by a short range of operating potential. For example, polyaniline shows capacitive behavior between the potential range of 0.1 V–0.8 V. Below or over this range, pseudocapacitive materials follow the faradaic reaction, leading to an irreversible reaction. For the conducting polymers, at the reduction potential, they become insulators and irreversibly. At oxidation potential shows the current peaks and acts as a conductor. It is possible only when the electrolytic anions bind with the surface of the polymers and exhibit the capacitive behavior of the system [10].

The surface redox reactions take place by delocalized electrons, i.e., non-bonding electrons of the materials, at a distance of x << $(2Dt)^{1/2}$, where x is the thickness of the diffusion layer, D is the diffusion coefficient of the ions, and t is the time. The pseudocapacitive materials undergo three mechanisms: surface redox, ion intercalation pseudocapacitive, and underpotential deposition. At the surface, the redox pseudocapacitive mechanism, the electrolytic ions adsorb on the electrode surface and are involved in the reversible redox process commonly occurring in aqueous-based systems. In intercalation pseudocapacitive, ions of electrolyte diffuse into the pores and between the layers of the materials without any phase transformation of the materials undergo faradaic reactions. This behavior is observed mainly in non-aqueous organic-based electrolyte systems. The underpotential deposition happens only at overpotential, where the metal ions in the electrolyte tend to deposit on the material's surface [19] (Fig. 2.4).

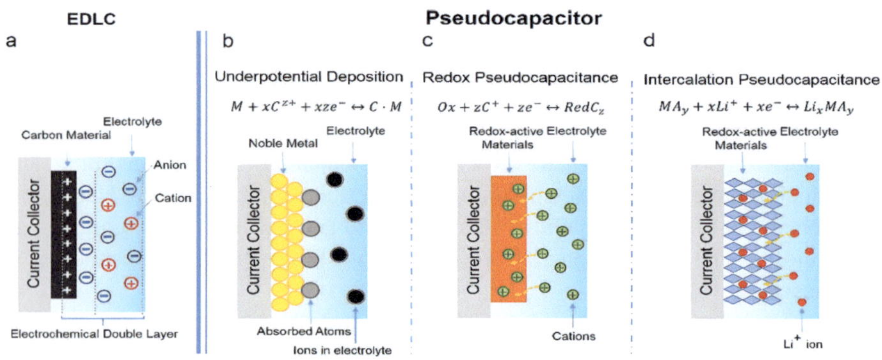

Fig. 2.4 Comparison of the electric double-layer capacitors and different types of pseudocapacitance charge storage mechanisms such as under potential deposition, redox pseudocapacitance and intercalation pseudocapacitance. Reprinted with permission of the American Chemical Society from ref. [20]

There are certain misunderstandings between pseudocapacitors and batteries in the research community. The justifications for some pseudocapacitor materials, which are considered battery materials, are as follows. (1) Both battery and pseudocapacitive materials exhibit faradaic reactions, but the battery gives intercalation redox reactions, and the pseudocapacitor gives surface redox at optimized operation potentials. (2) Even some pseudocapacitive materials exhibit charge intercalation storage with no phase change after long cycles, but in the battery, due to intercalation leads to phase transformation in the crystal lattice of the materials. (3) Variations in the current–voltage response due to the kinetics of ions by calculating with the normalized formula of $i_p = av^b$, where i_p is peak current obtained from cyclic voltammogram, v is sweep rate and a, b are the adjustable values in which value b is the slope of the $\log(i)$ versus $\log(v)$ from cyclic voltammogram. From this expression, capacitive and diffusion-controlled kinetics can easily be determined. If the b-slope is one or above, the peak current linearly increases for the capacitive materials showing surface capacitive non-faradaic process, but for pseudocapacitive material, the b value lies between 0.5 and 1, exhibiting both diffusion-controlled and capacitive process. If the b value lies below 0.5, typical battery-type behavior exists [9, 18, 19].

The charging mechanism of the pseudocapacitor comprises both non-diffusion-controlled capacitive behavior and diffusion-controlled faradaic behavior. The expression is as follows: $\boldsymbol{i = k_1 v + k_2 v^{1/2}}$, where $\boldsymbol{k_1 v}$ is capacitive behavior, where the current is proportional to the scan rate, and $\boldsymbol{k_2 v^{1/2}}$ is faradaic behavior, where the current is proportional to the square root of the scan rate [20]. The pore size of the active electrode materials influences the charge storage mechanism. Three types of porosity exist in the materials, i.e., macroporous, mesoporous, and microporous. The size of macropores is more than 500 Å, mesopores range from 20 to 500 Å, and micropores are less than 20 Å, respectively.

In capacitive-based devices, the electrolytic ions are solvated, which cannot diffuse through the microporous (< 20 Å), so diffusion occurs through the mesoporous and macroporous capacitive materials. But in a pseudocapacitor, the ions diffuse through the unsolvated mesopores, exhibiting the fast, reversible faradaic reactions at the surface. If the materials are microporous, the de-solvated ions diffuse into the micropores and utilize the complete structure of the materials by faradaic reaction inside the pores with low mass transport. The pseudocapacitor exhibits slightly capacitive, in addition to the fast-faradaic surface reactions. The various functional groups or active sites carry out this surface mechanism. These functional groups of the electrode materials are covalently bonded with electrolyte ions at the respective electrode terminals [19]. The main disadvantage of the pseudocapacitive transition metal oxides, conducting polymers, is their lower electrical conductivity and structural demolition during long-cycle life, which makes them unable to reach the theoretical capacitance. Therefore, carbon-based materials are composited with transition metal oxides and conducting polymers, which support the structural backbone for the respective pseudocapacitive materials and enhance the electrical conductivity along with long cyclic stability. Scientists are involved in synthesizing hybrid materials to overcome the demerits of pseudocapacitive materials.

Among the various transition metal oxides, iron oxide is one of the promising materials used in supercapacitors. The thumb-ring-like α-Fe_2O_3/reduced graphene oxide composite is prepared by the facile hydrothermal method. The α-Fe_2O_3/reduced graphene oxide composite electrode subjected to electrochemical studies exhibits efficient anode materials with wide operating windows between − 1 and 0 V. The specific capacitance of the α-Fe_2O_3/reduced graphene oxide composite electrode shows 255 F g^{-1} at a specific current of 1 A g^{-1}. The thumb-ring-like α-Fe_2O_3 gets wrapped with reduced graphene oxide sheets, which prevent the structural collapse of the α-Fe_2O_3 and enhance the cycle stability up to 11,000 cycles. Also, reduced graphene oxide prevents the agglomeration of the α-Fe_2O_3, which helps in the charge storage aspects. These properties made the α-Fe_2O_3/reduced graphene oxide composite a promising anode for the asymmetric type of supercapacitor [21].

The prepared reduced graphene oxide/carbon nanotube /MnO_2 nanocomposite via the chemical vapor deposition method, followed by the hydrothermal technique, shows the specific capacitance of 319 F g^{-1} at a specific current of 1 A g^{-1} in the neutral aqueous medium (aq. 1 M Na_2SO_4). The reduced graphene oxide/carbon nanotube/MnO_2 nanocomposite has a higher specific capacitance than MnO_2. Here, the reduced graphene oxide/carbon nanotube is a strong scaffold material to improve the electrical conductivity and enhance the cyclic stability of MnO_2. This nanocomposite has a porous structure, which makes it feasible for electrolytic ions to access. Then, the synergistic effects of reduced graphene oxide/carbon nanotube /MnO_2 nanocomposite exposed as a high-performing supercapacitor electrode [22].

The MnO_2-carbon nanotube composite was prepared by a facile precipitate method, and the electrode was fabricated as a freestanding paper-like film along with the carbon nanotubes. The MnO_2-carbon nanotube/carbon nanotube hybrid composite electrode material shows high areal capacitance and volumetric capacitance of 0.293 F cm^{-2} and 177.5 F cm^{-3} at the specific current of 0.2 A g^{-1}, respectively. The materials deliver 90% of capacitance retention even after 5000 cycles. The excellent performance of MnO_2-carbon nanotube/carbon nanotube electrode is due to the synergetic property of MnO_2 on the carbon nanotube scaffold, which also acts as a good adhesive for MnO_2-carbon nanotube with carbon nanotube, which enhances the electrical and ionic conductivity. Also, MnO_2-carbon nanotube/carbon nanotube hybrid electrode exhibits a 3D structure that facilitates high-charge storage [23]. The MnO_2-graphene/polyaniline hybrid structure is used as the electrode material in the supercapacitor. First, the MnO_2—graphene was synthesized by the simple precipitation method and stabilized by the Pickering emulsion technique. Then, stable MnO_2-graphene undergoes polymerization with aniline to MnO_2-graphene/polyaniline hybrid material. The MnO_2-graphene/polyaniline electrode exhibited a high specific capacitance of 743.8 F g^{-1} at a specific current of 0.5 A g^{-1} and showed 82.5% capacitance retention as the current moves from 0.5 to 10 A g^{-1} specific current.

The MnO_2-graphene/polyaniline electrodes' cyclic stability remains 93% even after 2000 cycles. The MnO_2-graphene/polyaniline electrode has a stable stack and hollow structure, which provides a facile pathway for the electrolytic ions to access the electrode surface and perform as an efficient electrode material [24]. The MnO_2/

$CuCo_2O_4$ hybrid structure was obtained as a MnO_2 nanosheet grown on the $CuCo_2O_4$ nanoflake via a step hydrothermal technique. The $MnO_2/CuCo_2O_4$ hybrid structure was fabricated as a symmetric device, which exhibits a high energy density of 64.1 Wh kg^{-1} at the power density of 1500 W kg^{-1}. The coulombic efficiency of the device was maintained even after 5000 cycles, and the capacitance retention was reduced to 80% from its initial value. The synergistic effect of MnO_2/ $CuCo_2O_4$ mounted on the carbon fiber support gives a strong charge storage ability and enhances the specific capacitance and stability [25] (Fig. 2.5).

The electrodeposition of $MoO_2/Cu(OH)_2$ composite on the nickel foam by a three-electrode configuration. The $MoO_2/Cu(OH)_2$ composite electrode exhibits a high specific capacitance of 800 F g^{-1} at a specific current of 2 A g^{-1}. The electrode materials are subjected to 5000 cycles, which deliver 97% capacitance retention at the end of the cycles. The high performance and robust $MoO_2/Cu(OH)_2$ composite supported on 3D nickel foam is due to synergistic effects on increasing ion diffusion, electron transfer across the electrode/electrolyte interface, and reduced structural deformation caused by the ion insertion/de-insertion mechanism [26]. The metal-based electrode materials exhibit high charge production by the faradaic reactions. So, the recent focus is on fabricating alloy-based materials as supercapacitor electrodes. Snowflake-like dendritic $CoNi_D$/reduced graphene oxide nanocomposite was synthesized using the co-precipitation method. The fabricated asymmetric supercapacitor consists of ($CoNi_D$)/reduced graphene oxide nanocomposite as cathode and reduced graphene oxide as anode with polyvinyl alcohol/3 M KOH/0.1 M $K_4[Fe(CN)_6]$ delivers the high energy density of 52.8 Wh g^{-1} at the power density of 2000 W kg^{-1} and exhibits the capacitance retention of 95% even after 4000 cycles. The high performance of the ($CoNi_D$)/reduced graphene oxide nanocomposite was that the snowflake-like dendritic morphology facilitates high ionic charge storage property, which enhances the electrochemical activity. Also, the high specific capacitance of the device was obtained by the addition of redox-active $K_4[Fe(CN)_6]$ electrolyte, which has $[Fe(CN)_6]^{4-}/[Fe(CN)_6]^{3-}$ redox couple, boosting the redox reaction. The redox-active $K_4[Fe(CN)_6]$ electrolyte acts as an electron buffer source occurring at the electrode/electrolyte interface in the electrochemical process [27].

Recently, MXene, the transition metal carbides, nitrides, and carbonitrides-based supercapacitors, have dominated performance in energy storage applications. This supercapacitor was developed in 2011 and consists of $M_{n+1}X_n$ stoichiometric materials, where M is a transition metal, X is carbon or nitrogen, and n is an integer. MXene exhibits the electronic conductivity of ~ 700 S cm^{-1}, and is hydrophilic, and has excellent mechanical strength. Gogotsi's group developed MXenes from MAX phase materials, where A is aluminum, by etching with hydrofluoric acid or lithium fluoride. The $Ti_3C_2T_x$ MXene supercapacitor exhibited 50–100 F g^{-1} in an aqueous medium, but the volumetric capacitance of the $Ti_3C_2T_x$ gave 900 F cm^{-3}, which is higher than that of $RuO_2.xH_2O$ supercapacitor [28]. When $Ti_3C_2T_x$ is composed with polypyrrole, it exhibits the specific capacitance of 416 F g^{-1} with outstanding cyclic stability of up to 25,000 cycles [29]. The MXene-carbon nanotube sandwich-type composite gave a high volumetric capacitance of 435 F cm^{-3}, which increased with an increase in scan rate. At a specific current of 10 A g^{-1}, the volumetric capacitance

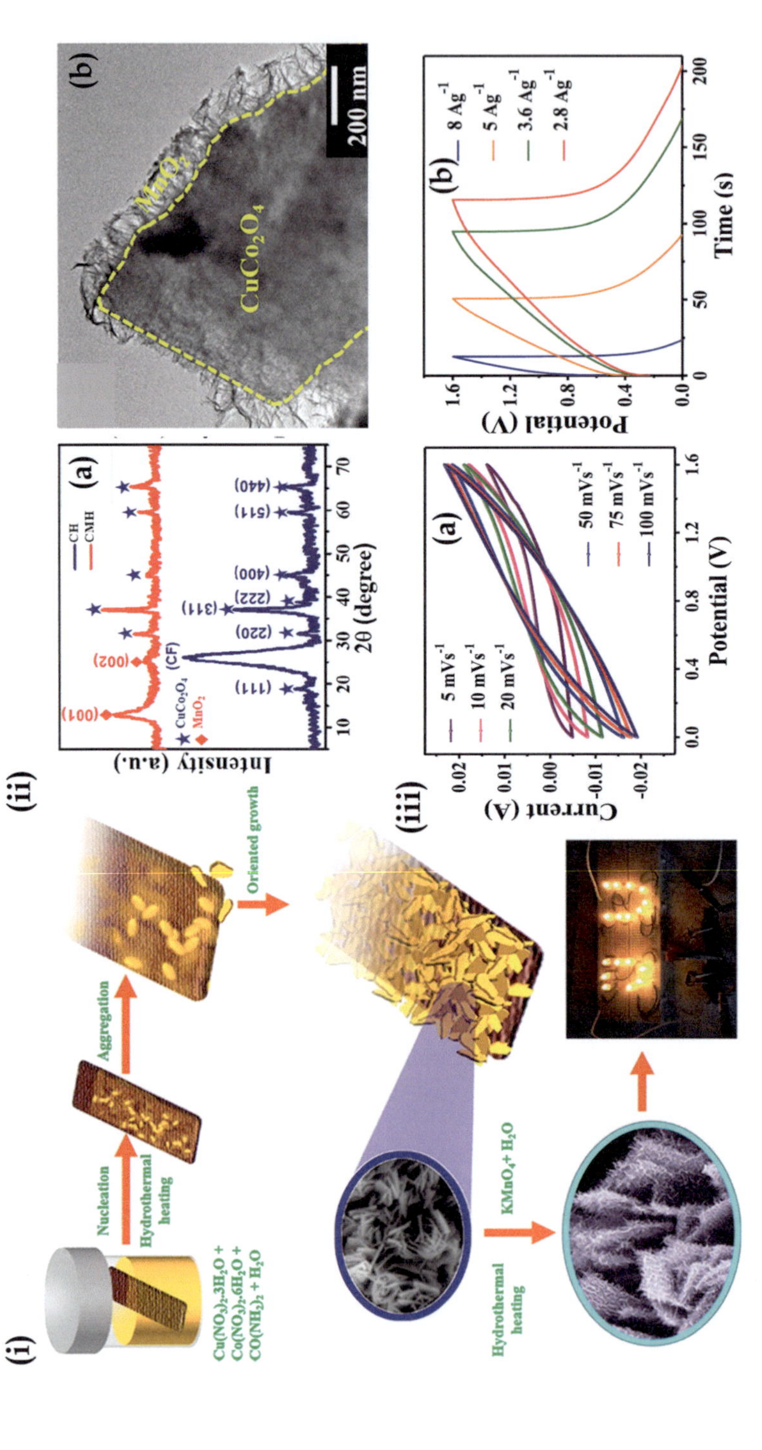

Fig. 2.5 (i) Schematic illustration of Hierarchically grown $CuCo_2O_4$–MnO_2 over carbon fiber; (ii) **a** X-ray diffraction pattern of $CuCo_2O_4$ and $CuCo_2O_4-MnO_2$ composite, and **b** high-resolution transmission electron microscope image of $CuCo_2O_4-MnO_2$ composite; (iii) **a** Cyclic voltammogram at various scan rates, and **b** charge–discharge profiles at different specific currents of the fabricated solid-state symmetric supercapacitor using $CuCo_2O_4-MnO_2$ composite as an electrodes. Reprinted with permission of the American Chemical Society from ref. [25]

of 300 F cm^{-3}, and after 20,000 cycles, the capacitance increases to 350 F cm^{-3}, respectively [30].

2.2.3 Non-capacitive Faradaic Charge Storage

Recently, scientists have come out with potential hybrid materials by combining typical faradic battery-type materials, which exhibit the synergistic effects of the charge storage mechanism. These materials have a wide bandgap of semiconductor, which delivers remarkable specific capacity (mAh g^{-1}) and energy density (Wh kg^{-1}). The faradaic-based materials are known for their crystallographic structure with high lattice sites for charge storage. These materials undergo a specific bulk faradaic charge storage mechanism at the material's lattice point serially by solid-state ion diffusion. These are highly involved in the diffusion-limited intercalation process, where the electrolytic ions stored occupy the interspace between the lattice points and undergo reactions at localized valence electrons of the materials. As the electrolytic ions are intercalated inside the structure of the materials, the possibility of the creation of strain in the material structure may cause the structure's instability. So, the material structure will automatically relax, which leads to changes or transformations in the phases of the electrode materials. This strain-relaxing phase transformation affects the kinetics and structural properties, which limits the cycle life of the battery-type materials [31]. Other limitations are low power density (W kg^{-1}), electrical conductivity, and mechanical stability of the materials. To overcome these drawbacks, these materials are made into a composite with carbon-based materials, which provide high electrical conductivity and cyclic stability. In this context, the battery-type materials of carbon nanotubes at Ni at $Ni_2(CO_3)(OH)_2$ nanocomposites were synthesized via the chemical reduction deposition method consisting of three layers. Here, the carbon nanotube was the inner layer that played a vital role in improving electrical conductivity and increasing cycle life stability. The superior capacity was delivered by the middle layer of nickel between the carbon nanotube and $Ni_2(CO_3)(OH)_2$ (outer layer). The outer layer $Ni_2(CO_3)(OH)_2$ involves the rapid redox kinetics in the reaction, which was mainly responsible for the capacity. Carbon nanotube at Ni at $Ni_2(CO_3)(OH)_2$ nanocomposites electrode exhibits a high specific capacity of 221 mAh g^{-1} at a specific current of 1 A g^{-1}. The carbon nanotube at Ni at $Ni_2(CO_3)(OH)_2$ nanocomposite electrode materials endure a capacity retention of 81% even after 9000 cycles and maintain a capacity retention of 76% up to 50 A g^{-1}. The carbon nanotube at Ni at $Ni_2(CO_3)(OH)_2$ composite electrode facilitates the short diffusion path of ions, high surface area accessibility, and good electrical conductivity [32].

The hybrid metal composite is highly used as an electrode material in supercapacitors. NiSe/ZnSe (NZSe) hybrid nanocomposites were fabricated using the facile electrodeposition technique. The NZSe hybrid composites showed a high specific capacity of 651.5 mAh g^{-1} compared to the counter materials of NiSe (267.5 mAh g^{-1}) and ZnSe (211.1 mAh g^{-1}) at a specific current of 1 A g^{-1}.

The capacitance retention of NZSe was 97.6% even after 2000 cycles. The high performance was obtained by the unique morphological and synergistic effect of NiSe and ZnSe, which increased redox activity, conductivity, and ionic diffusion [33]. Other than pure metal oxides and metal chalcogenides, other materials, such as nitrides, phosphides, and carbides, were efficiently used as electrodes in supercapacitors. The titanium oxynitride (TiON) is synthesized by nitrating the titanium oxide at various temperatures by passing ammonia gas. The sample at a temperature of 900 °C gives a high capacitance of 140 F g^{-1} compared to different temperatures of 800 °C (60 F g^{-1}) and 1000 °C (71 F g^{-1}). However, the TiON-900 shows very low capacitance compared with other works. So, the TiON-900 composite with binary metal spinel materials of $MnCo_2O_4$ boosted the specific capacitance from 140 to 515 F g^{-1}. This composite exhibits higher specific capacitance than the individual $MnCo_2O_4$. Cyclic stability was conducted for the three samples, where the TiON-900/$MnCo_2O_4$ showed a higher retention of 78% than TiON-900, and $MnCo_2O_4$ had a 71% and 68% retention, respectively. In the TiON-900 electrode material, the high specific capacity of the TiON-900/$MnCo_2O_4$ composite depends on the N/O ratio. If there is an excess of N content, leading to electrode corrosion, whereas slow diffusion kinetics are exhibited when the N content is low. Therefore, the optimal atomic percentage of N and O enhanced the specific capacity of the electrode materials. The electrochemical contributions of both TiON-900 and $MnCo_2O_4$ combine to deliver the synergistic performance of the electrode materials. Also, TiON-900/$MnCo_2O_4$ composite has a peculiar morphology that facilitates the high-charge storage ability of the material [34].

Recently, the scientist fabricated the hybrid composite nanostructure by combining the faradaic non-capacitive-based materials with capacitive faradaic-based materials by combining battery-type material ($NiCo_2O_4$) with pseudocapacitive material Ni-metal–organic framework via a two-step approach, the hydrothermal method. The Ni-metal-organic framework nanoarrays are made to self-generate $NiCo_2O_4$ nanowires, which are strong adhesives that induce synergistic effects to enhance the overall performance. The $NiCo_2O_4$/Ni-metal–organic framework electrode material exhibits a specific capacity of 208.8 mAh g^{-1} at an areal current of 2 mA cm^{-2}. The pristine metal–organic framework-based materials have a high surface area, porosity, and various redox-active sites. However, the metal–organic framework has its disadvantages, such as poor conductivity and cycle stability. To rectify these issues, the highly efficient nickel cobaltite ($NiCo_2O_4$) was used to fabricate the hybrid composite, which enhanced the stability and conductivity [35] (Fig. 2.6).

The NiSe were synthesized using two-step hydrothermal, and WO_3/polypyrrole hybrid nanostructure was synthesized using a polycondensation process. The NiSe shows a specific capacitance of 1274 F g^{-1} compared to the specific capacitance of 774 F g^{-1} for NiO at a specific current of 2 A g^{-1}. The specific capacitance of WO_3/polypyrrole is 586 F g^{-1}, which was high compared to individual counterparts, i.e., WO_3 (402 F g^{-1}) and polypyrrole (224 F g^{-1}). The asymmetric supercapacitor was constructed using WO_3/polypyrrole as anode and NiSe as cathode, delivering the energy density of 37.3 Wh kg^{-1} at the power density of 1249 W kg^{-1} at the specific

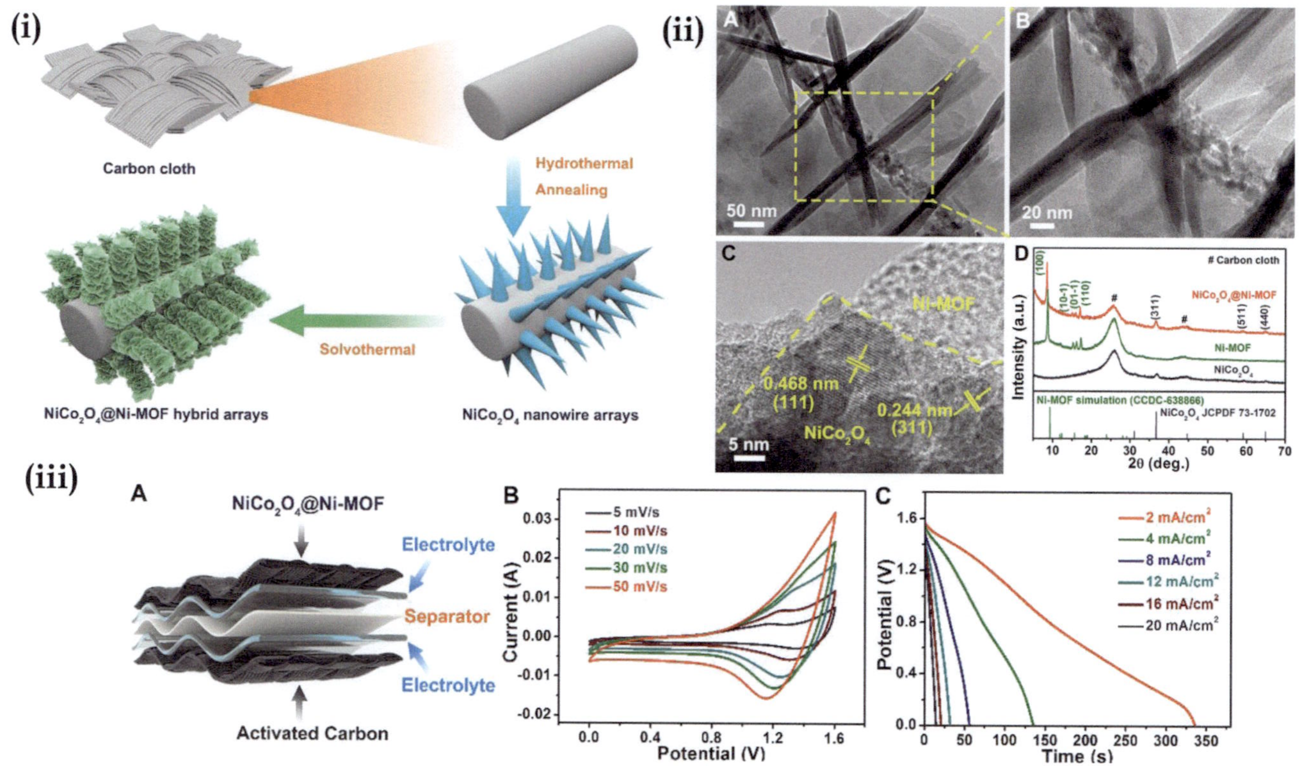

Fig. 2.6 (i) Synthetic approach for the $NiCo_2O_4$ at nickel-metal organic frameworks core/shell hybrid arrays on carbon cloth; (ii) **a, b** Typical transmission electron microscope of $NiCo_2O_4$ at nickel-metal-organic frameworks core/shell nanostructures, **c** High-resolution transmission electron microscope images of $NiCo_2O_4$ at nickel-metal–organic frameworks core/shell nanostructures, **d** X-ray diffraction patterns of the pristine $NiCo_2O_4$, nickel-metal–organic frameworks (36 h), and $NiCo_2O_4$—nickel-metal–organic frameworks (12 h) arrays on carbon cloth; (iii) **a** Fabricated hybrid supercapacitor device using activated carbon// $NiCo_2O_4$ nickel-metal-organic frameworks electrodes, **b** cyclic voltammogram of the hybrid supercapacitor device at different scan rates, **c** discharge curves at various specific currents. Reprinted with permission of the American Chemical Society from ref. [35]

current of 2 A g^{-1}. The porous nature of the NiSe structure allows the electrolyte ions to be driven by rapid kinetics between the interfaces. Polypyrrole became the strong backbone for the WO_3 structure, which prevents the anode electrode material degradation [36]. The $MnCo_2O_{4.5}$ at $Ni(OH)_2$ hybrid nanocomposite was synthesized using a facile hydrothermal technique. The $Ni(OH)_2$ is grown on a flower-like $MnCo_2O_{4.5}$ to form a core–shell structure, a highly efficient electrode material for the supercapacitor application. The $MnCo_2O_{4.5}$ at $Ni(OH)_2$ hybrid nanocomposite exhibits a high specific capacity of 318 mAh g^{-1} at a specific current of 3 A g^{-1}. The asymmetric supercapacitor device fabricated by $MnCo_2O_{4.5}$ at $Ni(OH)_2$ hybrid nanocomposite as cathode and hollow Fe_2O_3/reduced graphene oxide as an anode delivers the energy density of 56.33 Wh kg^{-1} with a power density of 1.9 kW kg^{-1} and retention of its capacitance of 90.4% even after 3000 cycles. The efficient performance given by the $MnCo_2O_{4.5}$ at $Ni(OH)_2$ hybrid nanocomposite is due to the synergistic effect of the core–shell nanostructure of $MnCo_2O_{4.5}$ combined with 2D $Ni(OH)_2$ nanosheet, which involves a high-charge storage mechanism and fast-faradaic reactions [37].

2.3 Role of Hybrid Nanostructures in Supercapacitor Applications

The supercapacitor performance is highly dependent on the structural morphologies which access the ionic storage efficiency to increase the energy density. Among them, the hybrid nanostructure contributes to the device's energy density.

The hybrid nanostructure exhibits the cumulative performance of the materials with varying properties. The morphological tuning is the main criterion for hybrid nanostructures. Here, $CoMn_2O_4$ mounted on 3D-reduced graphene oxide as hybrid nanostructures synthesized by hydrothermal method showed that 3D-reduced graphene oxide acts as a substrate, and $CoMn_2O_4$ was vertically aligned and randomly grown on 3D-reduced graphene oxide at the temperature of 50 °C for 10 h followed by further annealed at 600 °C for 5 h under argon atmosphere. The nanosheets were interconnected to form a flower-like shape. This morphology shows the interconnections between the nanosheets, providing a large electrode–electrolyte contact area and increasing the ion-conducting channels helps in supporting the volume change during the charge–discharge process of the material. The hybrid enhances the surface area and gives mechanical strength for structural stability. The formed heterostructure resulted in the increased potential window and high specific capacitance of 1028 F g^{-1} compared to 3D reduced graphene oxide and $CoMn_2O_4$. Also, 75% of its original capacitance is retained when the specific current slightly increases from 1 to 30 A g^{-1}. $CoMn_2O_4$/3D reduced graphene oxide assembles an asymmetric supercapacitor device as the 3D reduced graphene oxide as the negative electrode, and $CoMn_2O_4$/3D-reduced graphene oxide as the positive electrode. The potential working window increases to 1.8 V due to the synergetic effect of the hybrid. Asymmetric supercapacitor device shows a high specific capacitance of 103 F g^{-1} at

1 A g^{-1}, a good cyclability with 89.7% of the initial capacitance even after 5000 cycles. Due to the large surface area and increased ion conduction, a wide range of voltages can be reached, leading to increased specific capacitance and good cyclic stability, and hence it can facilitate high energy density supercapacitors [38].

The multi-dimensional nanostructure plays a vital role in energy applications. Particularly in a supercapacitor, the charge storage is enhanced by multi-dimensional nanostructures. The 3D cobalt oxide nanobeads attached to carbon nanotubes are decorated on a graphene nanostructure (Fig. 2.7).

The 3D cobalt oxide nanobead-CNT-Graphene hybrid nanostructure were obtained by microwave irradiation-assisted process at five different reaction times, showing variation in the morphology of nanostructures. Co_3O_4 nanobeads anchored on carbon nanotubes were grown on graphene nanosheets. The microwave irradiation for one minute showed no growth of carbon nanotubes. The Co_3O_4 nanoparticles were evenly distributed on the surface and edges with the help of functional groups such as epoxy, carboxyl, carbocyclic, and hydroxyl in the graphene over the graphene nanosheets surface. Increasing the reaction time of microwave irradiation by two minutes showed the initial growth of the carbon nanotubes. However, the microwave did not provide sufficient temperature for the complete development of nanobeads and longer carbon nanotubes. Graphene nanosheets acted as the carbon source for the formation of carbon nanotubes. Microwave irradiation for four minutes showed the formation of nanobeads in the form of necklaces on carbon nanotubes, and the length of carbon nanotubes also increased, leading to increased interfacial contact and subdued agglomeration of nanoparticles with enhanced electrochemical activity. Increased reaction time resulted in melting and agglomeration of Co_3O_4 nanobeads attached to carbon nanotubes due to higher irradiation. The size of the nanobead increased with unstable behavior, making it difficult for the carbon nanotubes to lift it from graphene nanosheets. Microwave irradiation for six minutes resulted in overweighed nanobeads leading to their agglomeration, which lay tangentially on the surface of graphene nanosheets.

In microwave irradiation for eight minutes, dense agglomeration of nanobeads covered the carbon nanotubes except for a few hangings on graphene nanosheets. Microwave irradiation for four minutes showed optimum growth conditions for forming the 3D hybrid nanostructure. These 3D-Co_3O_4 nanobeads on graphene electrodes show a specific capacitance of 600 F g^{-1} at 0.7 A g^{-1} in aqueous KOH electrolyte, which is 1.56 folds higher than the Co_3O_4-decorated graphene nanostructure. The capacitance retention of 94.5% from its initial capacitance, even after 5000 cycles, indicates the excellent cyclic stability of the fabricated electrode. Also, the Co_3O_4 nanobead on carbon nanotube forms a necklace-like structure that is responsible for high performance, such as stability, cyclic stability, and specific capacitance [39].

A simple chemical reduction technique was used to obtain W_2C and Mo_2C mixed with commercial MWCNT in a 1:1 ratio in ethanol, which was kept at 80 °C and stirred vigorously. Then the residues were centrifuged and dried in an oven overnight, which was further heated in a furnace at 850 °C for 3 h in an inert atmosphere. The Mo_2C nanoparticles accumulated in the shape of sphere-forming larger-sized grains.

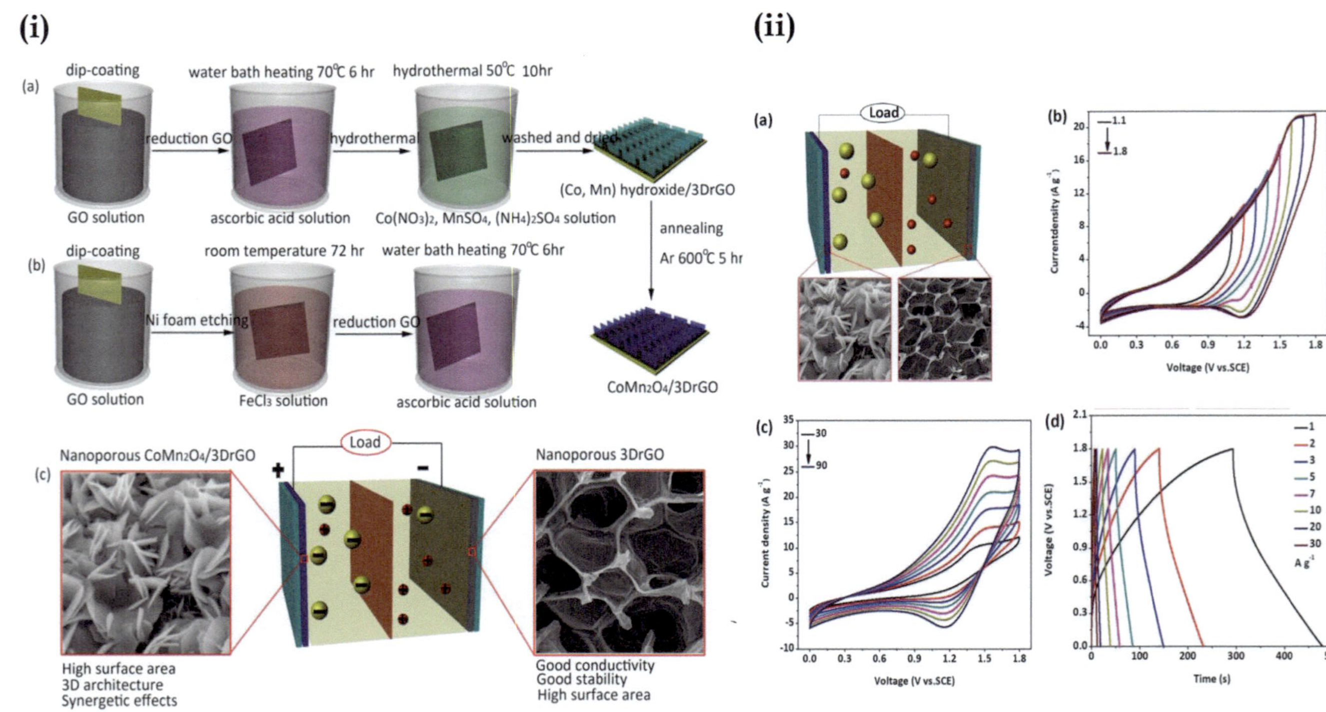

Fig. 2.7 (i) Fabrication of **a** the $CoMn_2O_4$/3D-reduced graphene oxide heterostructure on nickel foam, **b** 3D-reduced graphene oxide, and **c** all-solid-state flexible asymmetric supercapacitor; (ii) **a** Asymmetric supercapacitor device assembled by $CoMn_2O_4$/3D-reduced graphene oxide on nickel foam and a 3D-reduced graphene oxide on nickel foam, **b** Cyclic voltammogram of asymmetric supercapacitor at the scan rate of 60 mV s^{-1} with different voltage, **c** Cyclic voltammogram of asymmetric supercapacitor at various scan rates, **d** Charge–discharge profile of asymmetric supercapacitor at different current density. Reprinted with permission of the American Chemical Society from ref. [38]

Carbon nanotubes surround the Mo_2C grains with a sharp interface between the Mo_2C and carbon nanotubes. Mo_2C-carbon nanotube hybrid showed nanograin-like morphology containing both carbon nanotubes and Mo_2C. Pristine W_2C nanoparticles contain non-homogeneous nano-sized grains of different sizes. W_2C-carbon nanotubes showed W_2C particles covering the area between the nanotubes. The specific capacitance thus obtained for the Mo_2C-carbon nanotube, the W_2C-carbon nanotube, and the carbon nanotube hybrid electrodes was 367, 204, and 103 F g^{-1}, respectively, at a specific current of 1 A g^{-1}. The blending of carbon nanotubes with Mo_2C and W_2C facilitates fast electron movement. Carbon nanotube prevents aggregation by providing a large surface area for transition metal carbides interaction at the interface with porous structure, reducing electrolyte diffusion. The Mo_2C-carbon nanotube symmetric capacitor exhibits outstanding energy densities of 50.9, 46.1, 43.4, and 40.2 Wh kg^{-1} at power densities of 0.5, 1.0, 2.5, and 5.0 kW kg^{-1}, while comparatively lower energy densities of 28.3, 24.4, 20.8, and 18.0 Wh kg^{-1} are observed for the W_2C-carbon nanotube device, and 14.3, 12.5, 9.7, and 8.3 Wh kg^{-1} are observed for the carbon nanotube device, respectively. The transition metal carbide and carbon nanotube were combined without adding additives, giving better electrochemical properties of the two materials due to the synergistic effect, leading to flexible bonding and highly tuned conducting behavior [40]. The MoS_2 is a synthesized solution-based process that provides a flocculent nanostructure on 3D graphene foam. The 3D graphene sheets were interconnected to form a porous structure on which the MoS_2 deposits. The 3D graphene (graphene foam) skeleton is highly conductive, providing electric contact between the current collector and deposited MoS_2. MoS_2-3D graphene composite was taken as the anode, and nano-porous carbon as the cathode. The MoS_2-3D graphene composite showed uniform dispersion of flocculent MoS_2 nanosheets on graphene. MoS_2 without the 3D graphene skeleton resulted in thicker sheets. The specific capacitance of MoS_2-3D graphene /N-doped porous carbon is 88.3, 75.6, 68, 63.8, 60, and 55.7 F g^{-1} at the current densities of 0.1, 0.2, 0.5, 1, 2, and 4 A g^{-1}, respectively. MoS_2-3DG// N-doped porous carbon hybrid supercapacitor shows stable cycling life over 2000 cycles (from 97.2 to 75.4 Wh kg^{-1}, retention of 78% at 1 A g^{-1}). Porous composite MoS_2-3D graphene with uniformly incorporated MoS_2 flocculent nanostructure 3D graphene acts as an anode for high-performance hybrid supercapacitors. The device's high performance was due to the synergistic effects of the flocculent MoS_2 nanostructure and the interconnected porous structure in the composite and conductive graphene sheets, which show enhanced electrochemical properties [41] (Fig. 2.8).

Recently, conductive polymers have been catching interest among energy applications due to their structural flexibility and conductivity. 3D nanostructured polyaniline/MoS_2 quantum dot hydrogels synthesized by in-situ polymerization. Here, MoS_2 acts as a substrate, and N, N′-dibenzoyl-L-cystine acts as a dopant and gelling agent. MoS_2 quantum dots were crystalline on the inner and outer surfaces of polyaniline nanofibers. The D-doped polyaniline forms an interconnected nanofibers matrix in D-doped polyaniline gel, which shows a fibrillar network that increases the specific surface area and enhances the charge carrier transport properties. The larger surface area in 3D nanostructured polyaniline/MoS_2 quantum dot hydrogels may be due to

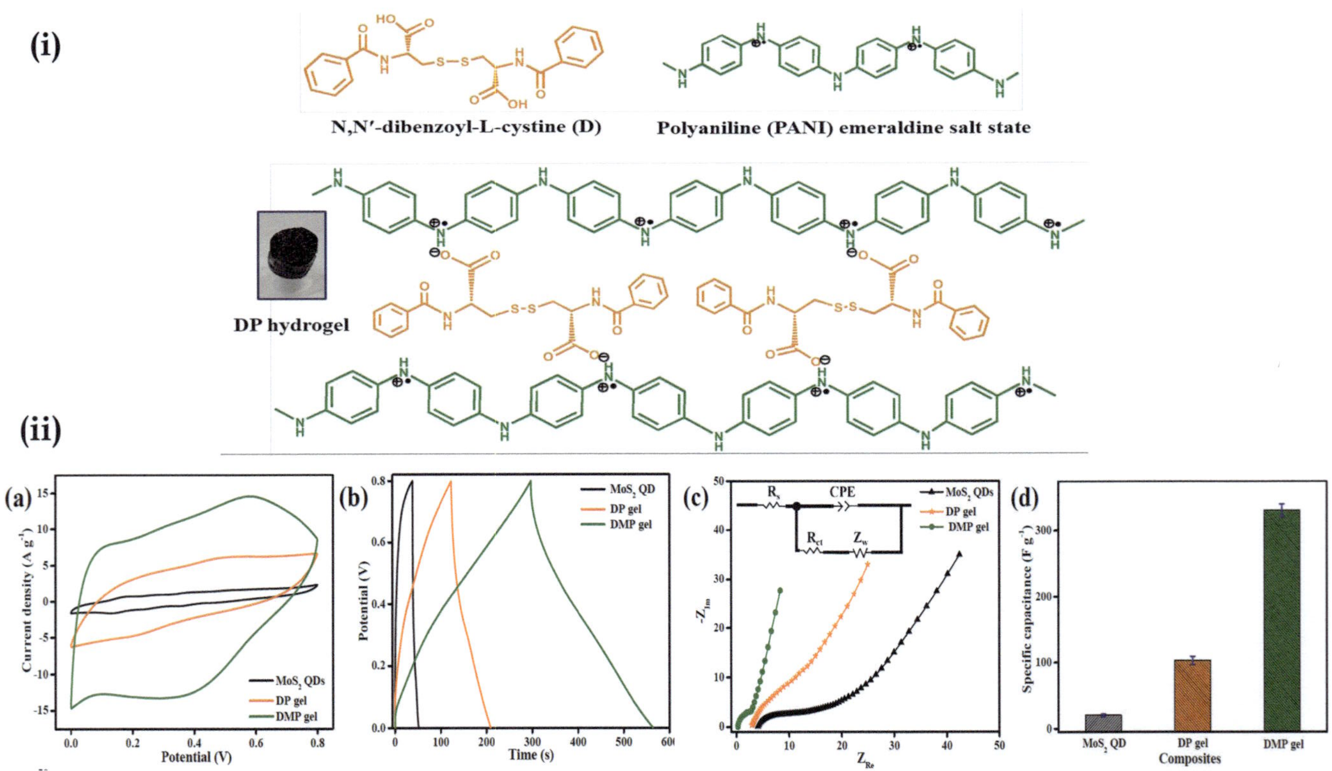

Fig. 2.8 (i) Structural representation of gelator N,N′-Dibenzoyl-L-cystine, polyaniline, and DP hydrogel; (ii) **a** Cyclic voltammogram of MoS_2 QD, DP, and DMP xerogels measured at a scan rate of 10 mV s^{-1}, **b** Galvanostatic charge−discharge curves of the MoS_2 QD, DP, and DMP xerogels measured at 1 A g^{-1}, **c** Nyquist plot in of MoS_2 QD, DP, and DMP xerogels based flexible device in the 1 to 10^5 Hz frequency range, **d** Bar diagram for specific capacitance of MoS_2 QD, DP, and DMP xerogels. Reprinted with permission of the American Chemical Society from ref. [42]. QD - quantum dot; DP - dibenzoyl-L-cystine-polyaniline; DMP - N,N′-dibenzoyl-L-cystine/MoS_2 quantum dot/PANI hydrogel

the synergistic interaction between polyaniline chains and MoS_2 quantum dots. A few MoS_2 quantum dots got entrapped into the gel fibers, preventing the MoS_2 quantum dots from self-aggregating. The amalgamation of MoS_2 quantum dots in the D-doped polyaniline gel increases pore volume and surface area. The interaction between the DP network and MoS_2 quantum dots inside the 3D nanostructured polyaniline /MoS_2 quantum dot hydrogel gel synergistically facilitates the movement of the charge from polyaniline fibers to MoS_2 quantum dots for their accumulation, improving its capacitance. The high specific surface area enhances the contact area between the electrode and electrolyte by reducing the diffusion ion path, resulting in better electrochemical performance. In a three-electrode system, the hybrid 3D nanostructured polyaniline/MoS_2 quantum dot hydrogel shows a high specific capacitance of 791 F g^{-1} at 1 A g^{-1}. The fabricated solid-state symmetric supercapacitor uses 3D nanostructured polyaniline/MoS_2 quantum dot hydrogels that exhibit enhanced specific capacitance of 331 F g^{-1} at a current density of 1 A g^{-1} [42].

Sandwich-structured 3D MnO_2 at graphite-like capsules at MnO_2 were obtained by decorating MnO_2 nanocrystals on 3D graphite-like capsules. The chemical vapor deposition technique synthesized graphite-like capsules using ZnO as a template and catalyst, which was finally acid-leached for removal, ZnO acts as a sacrificial template for the MnO_2 growth on the interior and exterior surface of the hollow graphite-like capsules and as a catalyst to catalyze the carbon precursor. Hybrid sandwich nanostructures obtained by the redox reaction between $KMnO_4$ and graphite-like capsules through a microwave-assisted approach. Here graphite-like capsules act as a reductant transporting electrons to the oxidant that converts MnO_4 to insoluble MnO_2 nanocrystals. Furthermore, the tough and rigid nature of graphite-like capsules provides essential structural stability, and the synergic effect between MnO_2 and graphite-like capsules leads to high electrochemical activity in neutral and alkaline electrolytes. This MnO_2 coated on the interior and exterior provides more surface area and redox sites, and graphite-like capsules act as conducting frameworks. A specific capacitance of 265.1 F g^{-1} at the specific current of 0.5 A g^{-1} in aqueous electrolytes (1 M Na_2SO_4) and 390 F g^{-1} at the specific current of 0.5 A g^{-1} in alkaline electrolytes (6 M KOH) is obtained. 3D MnO_2-graphite-like capsules-MnO_2 hybrid exhibited excellent cycling stability with 80.6% capacitance retention even after 6000 cycles at a specific current of 2 A g^{-1} [43]. The hybrid nickel cobalt nitride on carbon nano coils/nickel foam was synthesized by solvothermal reaction, followed by annealing under an ammonia atmosphere. The 3D carbon nano coils growth was uniform with high purity, as bundling of carbon nano coils was restricted, increasing the surface area for mass loading of active material. Carbon nano coils develop a network-like structure, forming cross-sections that help transmit current and electrolyte ions. The substrate carbon nano coils/nickel foam was synthesized using a chemical vapor deposition technique, then two-dimensional nickel–cobalt-layered double hydroxide nanosheets through the solvothermal method, and annealed at 700 °C for 3 h under an ammonia atmosphere. The morphology of nickel cobalt nitride/carbon nano coils/nickel foam varied depending on the annealing temperature. When the temperature was raised from 400–500 °C, the nickel cobalt nitride nanosheets were transformed into nanoparticles. Increasing the temperature from

500–600 °C resulted in a uniform distribution of nanoparticles. Vigorous etching occurred on increasing the temperature from 600–700 °C, resulting in decreased growth density of nickel cobalt nitride particles. Hence, the uniform distribution of nickel cobalt nitride nanoparticles with appropriate size obtained at 600 °C showed better electrochemical performance due to the synergistic effect of the carbon nano coils/nickel foam matrix. The carbon nano coils/nickel foam offers a high surface area for the growth of nickel cobalt nitride nanoparticles, which also increases active redox sites. The hybrid yields a specific capacitance of 523.5 F g^{-1} at 1 A g^{-1}, a rate capability of 86% at a specific current of 50 A g^{-1}, and 95.6% capacitance retention after 3000 cycles. The synergetic effect of nickel cobalt nitride and carbon nano coils/nickel foam helps achieve good redox reaction, electronic conductivity, and enhanced electrode performance [44].

The $NiCo_2O_4$/reduced graphene oxide hybrid nanostructures were grown on the nickel wires synthesized by combining solvothermal and hydrothermal decomposition methods. The etching of nickel wires resulted in a change in structure, surface morphology, and electrochemical performance. The etching process roughened the surface of nickel wire for 10 min which increased the interfacial contact area between the substrate and metal oxide, leading to increased film adhesion to the substrate. The surface roughness of nickel wire leads to the modification of surface morphology and the effective surface area of the $NiCo_2O_4$/reduced graphene oxide hybrid nanostructures. The rough surface of nickel wire resulted in the radial growth of numerous fine nanoneedles, leading to the substantial formation of microspheres with a hollow inner surface. The heterogeneous nucleation process is favored as the rough surface leads to a change in the contact angle between the substrate and the crystal, resulting in an energy change for the formation of nuclei on the surface. Compared to flat surfaces, many molecules can be absorbed/nucleation seeds on a rough surface, offering extra area for the accommodation of nuclei, resulting in a larger size and high density of the microsphere. The hybrid nanostructures provided more surface area and porosity. The electrode shows the highest volumetric capacitance (2.64 F cm^{-3}) and areal capacitance (25.3 mF cm^{-2}) at 200 μA g^{-1}. The 3D Ni/$NiCo_2O_4$/reduced graphene oxide hollow microspheres with large surface areas improved the electrolyte ion's contact area in the electrode–electrolyte interface with a reduced ion diffusion path, thus enabling faster redox reactions. The synergistic effect between the $NiCo_2O_4$ nanoparticles and reduced graphene oxide nanosheets improved electrochemical performance. Ni_{10}/$NiCo_2O_4$/reduced graphene oxide twisted wearable supercapacitor device winded for some laps on a metal rod with a diameter as small as 1 mm, and the wire-shaped device exhibited power densities of 10 μWh cm^{-3} (54 mWh kg^{-1}) and 4.95 mW cm^{-3} (27 W kg^{-1}) [45] (Fig. 2.9).

The reduced graphene oxide/$BiVO_4$ hybrid nanostructures were synthesized by the hydrothermal method. Reduced graphene oxide sheets were observed to be encapsulated in the especially dendritic fern-like structures of bismuth vanadate ($BiVO_4$). Due to the rotation of particles via Brownian motion or short-range interaction between the particles, branches (rod-like structures) are formed, which act as building blocks for fern development. The growth of these branches occurs via Ostwald ripening and always tries to adjust the structures to achieve minimal surface free

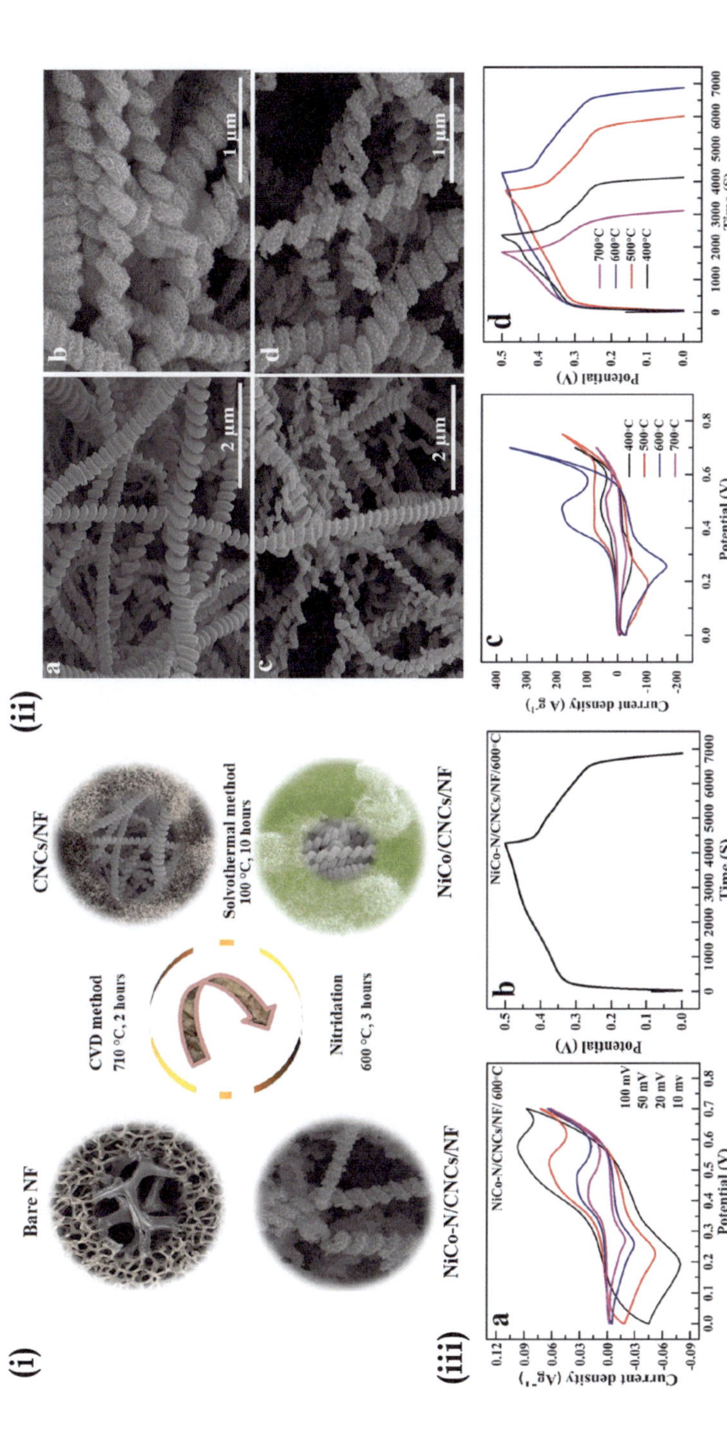

Fig. 2.9 (i) Synthesis process of nickel cobalt nitride/carbon nano coils on nickel foam; (ii) Field emission—scanning electron microscope images of **a** network growth of carbon nano coils on nickel foam, **b** Precursor growth of nickel cobalt layered double hydro-oxide/carbon nano coils/nickel foam, and **c, d** Low- and high-magnification images of nickel cobalt nitride/carbon nano coils/nickel foam; (iii) Electrochemical measurement of nickel cobalt nitride/carbon nano coils/nickel foam **a** Cyclic voltammogram of sample which annealing temperature at 600 °C with different scan rates, **b** Charge–discharge profile at a specific current of 1 A g^{-1}, **c** Cyclic voltammogram of samples at different annealing temperatures in an NH_3 atmosphere, **d** Charge–discharge profiles of samples at different annealing temperatures in an NH_3 atmosphere. Reprinted with permission of the American Chemical Society from ref. [44]

energy. A hybrid structure facilitates short ion diffusion and high surface area. The reduced graphene oxide /$BiVO_4$ symmetric cell exhibits a wider potential window of 1.6 V, leading to a significantly higher volumetric (gravimetric) energy density of 1.60 mWh cm^{-3} (33.7 Wh kg^{-1}) with a high volumetric (gravimetric) power density of 391 mW cm^{-3} (8.0 kW kg^{-1}). Improved cycling stability of 80% retention after 2000 cycles was achieved [46]. Ni_3S_2 nanosheets on 3D-reduced graphene oxide (Ni_3S_2/3D-reduced graphene oxide), supported by nickel foam, were synthesized by electrodeposition. Graphene oxide was deposited on the nickel foam by dripping and drying. Ni_3S_2 mesoporous nanosheets were vertically grown on the 3D-reduced graphene oxide scaffold without using conductive additives and polymeric binders. The obtained hybrid nanostructure showed enhanced electrical conductivity by providing more chemical reacting sites and shortening the ion and electron migration path. The specific capacitance of the Ni_3S_2/3D-reduced graphene oxide hybrid is higher than that of the reduced graphene oxide. Ni_3S_2/3D-reduced graphene oxide electrode electrodeposited with four cycles gives an areal capacitance of 11.31 F cm^{-2} (1886 F g^{-1}), while 3D-reduced graphene oxide only has 0.15 F cm^{-2} (136 F g^{-1}). When the number of cycles is increased, the mass loading of Ni_3S_2 increases, because of which the resistance increases, leading to poor transport of ions. Solid-state symmetric supercapacitor device fabricated by these hybrid nanostructures delivers a high energy density of 58.9 Wh kg^{-1} and a high power density of 3.7 kW kg^{-1}. The excellent cycling stability is of 92% capacitance retention even after 30,000 charge/discharge cycles [47].

From the above, the hybrid nanostructure formation involves two steps: forming individual metal oxides or chalcogenides and forming carbon scaffolds and combining them with metal oxides or chalcogenides. Combining metal oxides or chalcogenides can be grown, decorated, anchored, or embedded on carbon scaffolds. The advantage of creating the hybrid nanostructure in supercapacitors is that the potential window and capacitance of the device increase, which in turn, increases the overall performance of the supercapacitor device. Since the carbon-based material has electric double-layer capacitors behavior and metal oxides or chalcogenides have pseudocapacitive behavior, the hybrid of both helps to increase the potential window. The hybrid nanostructure offers new electrochemical properties that are different and highly efficient than their individual properties. The carbon-based material enhances the surface area and electrical conductivity because of the fast transportation of ions, which enhances the capacitance. So, the device fabricated from these hybrid materials will show improved energy density, which is listed in Table 2.1.

Table 2.1 List of hybrid nanostructured materials for supercapacitor applications

S. No.	Hybrid materials	Synthesis technique	Half-cell capacitance/ capacity	Device operation voltage	Energy density	Specific power density	Cyclic life (cycles)	References
1	Bi_2S_3 at g-C_3N_4	Hydrothermal	41.53 μAh cm^{-2}	1 V	3.17 μ Wh cm^{-2}	249 μ Wh cm^{-2}	7500	[48]
	Bi_2S_3		23.56 μAh cm^{-2}					
	g-C_3N_4		13.75 μAh cm^{-2}					
2	Co_8FeS_8/ Nitrogen-doped graphene	Hydrothermal	1,013 F g^{-1}	1.6 V	70.4 Wh kg^{-1}	598 Wh kg^{-1}	10,000	[49]
	Co_8FeS_8		861 F g^{-1}					
	Nitrogen-doped graphene		301 F g^{-1}					
3	$NiCo_2N$ at nitrogen-doped graphene	Pyrolysis	265 mAh g^{-1}	1.6 V	94.93 Wh kg^{-1}	790 W kg^{-1}	24,000	[50]
	$NiCo_2N$		166 mAh g^{-1}					
	N-doped graphene		48 mAh g^{-1}					
4	Sulfur-polyaniline	Polymerization	149 F g^{-1}	1.2 V	31.4 W h kg^{-1}	922.78 W kg^{-1}	5000	[51]
	Sulfur-polyaniline/ graphene		1107 F g^{-1}					

(continued)

Table 2.1 (continued)

S. No.	Hybrid materials	Synthesis technique	Half-cell capacitance/capacity	Device operation voltage	Energy density	Specific power density	Cyclic life (cycles)	References
5	Co–Metal–organic frameworks at $CoCr_2O_4$/Nickel foam	Hydrothermal	596.8 C g^{-1}	1.6 V	34.36 Wh kg^{-1}	201.03 W kg^{-1}	5000	[52]
	Co–Metal–organic frameworks/nickel foam		297.7 C g^{-1}					
6	$NiCo_2S_4$/nitrogen-graphene frameworks	Hydrothermal	1240 F g^{-1}	1.6 V	36.8 Wh kg^{-1}	375 W kg^{-1}	3000	[53]
	$NiCo_2S_4$		688.9 F g^{-1}					
	Nitrogen-graphene frameworks		111.11 F g^{-1}					
7	$CoFe_2O_4/Ti_3C_2T_x$	Co-precipitation, chemical etching, and ultrasonication method	1268.75 F g^{-1}	–	–	–	–	[54]
	$CoFe_2O_4$		594 F g^{-1}					
	$Ti_3C_2T_x$		1046.25 F g^{-1}					
8	$NiCo_2O_4$ at MnO_2	Electrodeposition	920 F g^{-1}	1.5 V	37.8 Wh kg^{-1}	187.5 W kg^{-1}	3000	[55]
	$NiCo_2O_4$		430 F g^{-1}					
	MnO_2		412 F g^{-1}					
9	VS_4/carbon nanotubes/reduced graphene oxide	Hydrothermal	490.7 F g^{-1}	2.25 V	72.07 Wh kg^{-1}	14.69 W kg^{-1}	5000	[56]

(continued)

Table 2.1 (continued)

S. No.	Hybrid materials	Synthesis technique	Half-cell capacitance/capacity	Device operation voltage	Energy density	Specific power density	Cyclic life (cycles)	References
	VS_4/carbon nanotubes/		271.6 F g^{-1}					
	VS_4/reduced graphene oxide		399.5 F g^{-1}					
	Carbon nanotubes/reduced graphene oxide		82.8 F g^{-1}					
10	α-MnS/N-reduced graphene oxide	Hydrothermal	933.6 F g^{-1}	1.6 V	27.7 Wh kg^{-1}	800 W kg^{-1}	2000	[57]
	α-MnS		189.4 F g^{-1}					
11	G-SnS_2	Ball milling	565 F g^{-1}	0.8 V	23.5 Wh kg^{-1}	880 W kg^{-1}	3000	[58]
	SnS_2		85.4 F g^{-1}					

2.4 The Behavior of Multi-dimensional Hybrid Nanostructures in Supercapacitor Applications

Carbon-based materials can store energy through the electrical double-layer formation, which is commercially relevant. These materials show high surface area and electrical conductivity, leading to increased power density, but lack energy density as electrochemical capacitors [59]. The proper redox centers should be incorporated with this carbon-based material to gain energy density. For the redox property with robust structures, the various spinel-type like $A + B_2 + X_4$, where A is a divalent cation (Zn, Ni, Mn, Cu), B is a trivalent cation (Co, Fe, Mn), and X is oxides and chalcogenides [60, 61], perovskites like ABX_3 structures [62], carbides, nitrides, and carbonitrides are used as electrode materials. Recently, spinel (AB_2X_4) structures have ruled over the energy storage applications rather than perovskite (ABX_3) structures due to their intrinsic conductivity, fast electron transport, high insertion/extraction of electrolytic ions, and structural stability over long cycles. Particularly, the electrocatalytic performance of these materials depends on features such as chemical composition, crystal structure, nanostructure, and cation substitution level, e.g., orbital filling or oxygen vacancies. Cobalt-based metal oxides are a promising electrode material among spinel-type metal oxides due to their excellent electrocatalytic properties. The specific capacitance of $CuCo_2O_4$, $ZnCo_2O_4$, and $NiCo_2O_4$ is 1846 F g^{-1}, 1983 F g^{-1}, and 2118 F g^{-1} at a scan rate of 2 mV s^{-1}, respectively [61]. While comparing the above cobalt-based spinel structures, the $NiCo_2O_4$ shows better pseudocapacitance behavior. However, these spinel-type materials still suffer from low electrical conductivity and stability over the long operational period, which hinders the storage performance of the device.

Meanwhile, carbon-based materials, e.g., carbon black, carbon nanotubes, activated carbon, and graphene, have excellent conductivity properties, high chemical stability, and tunable microstructure, and are abundant in nature. These two combined to mitigate these drawbacks and get high specific capacitance [63]. Carbonaceous materials rely only on their surface area and porosity [64, 65]. According to the porosity in nature, they are classified as porous structures comprised of macropores with a size range less than 50 nm, mesopores with a size range between 2 and 50 nm for transportation of the ion, and micropores with a size range less than 2 nm for charge storage. The hetero atoms, such as nitrogen and sulfur, doped with carbon materials, enhance energy storage due to the redox mechanism and surface reactivity [66]. So, for these reasons, the effect of various carbonaceous materials with $NiCo_2O_4$ was analyzed.

Carbonaceous materials are classified into dimensional structures such as 0D, 1D, 2D, and 3D, as shown in Fig. 2.10 [67]. Each dimension has its own merits and demerits. The one-dimensional materials have the advantages of having a large surface area, low mass density, and high tensile strength, and are suitable for weaving. However, 1D materials have increased risks of short-circuiting, lower energy density, and difficulty in functionalization. 2D materials like carbon nanosheets have the

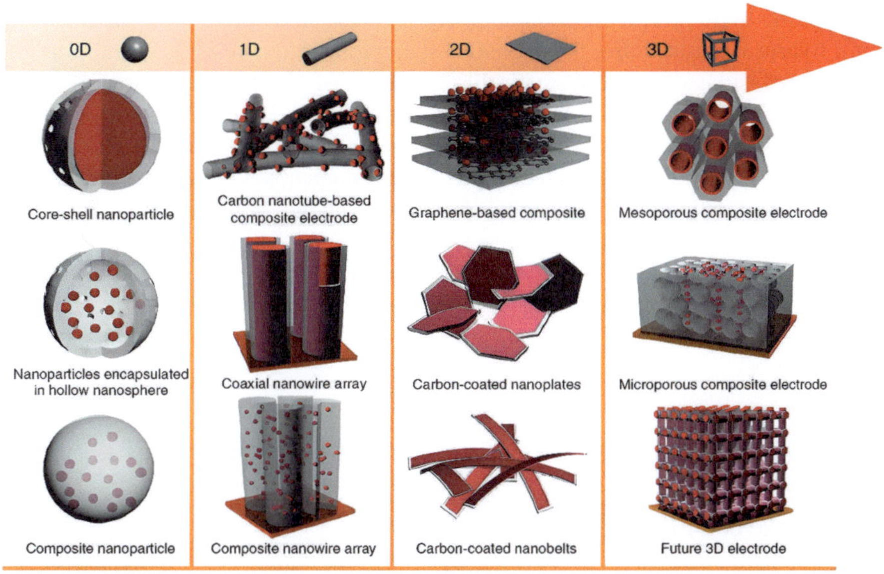

Fig. 2.10 Multi-dimensional nanostructured materials prepared by various physio and chemical techniques are used in a wide range of supercapacitor applications. Reprinted with permission of the Royal Society of Chemistry from ref. [67]

advantage of better physicochemical properties such as large surface area, excellent electrical conductivity, good thermal and chemical stabilities, a wide potential window, and cost-effective production, and are easy to functionalize too [68]. However, these are highly involved in restacking and reversible agglomeration, which is the main drawback of 2D materials. Compared to these, all the 3D porous carbons have excellent surface area. They have easy access to electrolyte diffusion and facile routes for charge movement because of the ease of loading other materials. These binder-free electrodes have reliable rate capability, strong cycling stability, and high specific capacitance. They are difficult to control and not cost-effective. The macropores in a large number affect the capacitance, power density, and gravimetric energy.

2.4.1 0D Carbon-Based Materials at $NiCo_2X_4$

The graphene dots and particles can be prepared by using electron beam lithography/electrochemical synthesis. The graphene dots can also be synthesized from graphene/photonic crystals by laser ablation due to their high surface area, good electrical conductivity, and fast charge transfer. Carbon dots and graphene quantum dots have the advantages of better surface grafting, nontoxicity, and good solubility, thus making them promising candidates for replacing inorganic quantum dots [69]. For a hybrid nanostructure, these conductive graphene quantum dots are homogenously encapsulated on the tremella-like structure of $NiCo_2O_4$. The tremella, like $NiCo_2O_4$, exhibits a capacitance of 790 F g^{-1} at 30 A g^{-1}. However, the hybrid structure of $NiCo_2O_4$-graphene quantum dots shows the specific capacitance of 1242 F g^{-1} at 30 A g^{-1} in 2 M KOH electrolyte. As in supercapacitor devices, these electrodes are assembled with activated carbon as the anode, which exhibits an energy density of 38 Wh kg^{-1} at a power density of 800 W kg^{-1} [70]. While the $NiCo_2O_4$ with carbon nanoparticle, these hybrids deliver the specific capacitance of 204.28 F g^{-1} for a specific current of 1 A g^{-1}, higher than pristine $NiCo_2O_4$ (178.78 F g^{-1}) and mesoporous carbon (25.06 F g^{-1}) [71].

2.4.2 1D Carbon-Based Materials at $NiCo_2X_4$

2.4.2.1 Carbon Fiber

The carbon fibers are microfibers manufactured into a highly conductive network with excellent mechanical flexibility and strength, containing ~ 90–100% carbon. This fabric includes carbon fiber clothes and carbon fiber papers. These have high strength, modulus, lightweight, excellent fatigue resistance, low density, and electrical conductivity of 106 S m^{-1} [72–74]. These significantly increase the composite stability because carbon fiber's thermal expansion coefficient is typically much lower than most matrices. The functionalization of carbon cloth is reported, and the highest specific capacitance of 88 mF cm^{-2} for 10 mV s^{-1} [75]. The composite $Ni_{0.25}Co_{0.25}O$/electrospun carbon nanofiber exhibits the specific capacitance of 697.5 F g^{-1} with 94% capacitance retention even after 2000 cycles at 10 A g^{-1} specific current [76]. The $NiCo_2O_4$ nano grass (NG) is synthesized hydrothermally at 130 °C for 5 h and grows on carbon fibers. This $NiCo_2O_4$ NG-CF acts as a better positive electrode to provide a pseudocapacitance, and the biowaste-derived carbon as a negative electrode provides the non-faradaic mechanism. The assembled device gives a specific capacitance of 17.5 F g^{-1} (25.03 mF cm^{-2}) and an energy density of 6.61 Wh kg^{-1} (9.46 μWh cm^{-2}) at a current of 1 mA [77].

The $NiCo_2O_4$ mesoporous nanowire arrays on carbon textiles with a highly conductive polypyrrole nanosphere shell layer enhance the storage and exhibit a specific capacitance of 2244 F g^{-1}, excellent rate capability, and cycling stability. The above positive electrode assembled with activated carbon as a negative electrode provides an energy density of 58.8 Wh kg^{-1} and the highest power density of 10.2 kW kg^{-1}. The capacitance retention is 89.2% even after the 5000 cycles [78].

2.4.2.2 Carbon Yarn

The carbon nanotubes have an average diameter of 3.2 nm and lengths between 1 and 20 μm. The carbon nanotube filaments are long, strong, and thin, with a diameter of usually 23 μm, but down to 10 μm is also possible, and are highly conductive. In a novel development, tens to hundreds of parallel carbon nanotube filaments form carbon nanotube yarn. The metal oxide and metal chalcogenides coated the carbon nanotube above and below the yarn surface, increasing the loading through the electrode deposition. A similar composite of $Ni(OH)_2$ at $NiCo_2O_4$/carbon nanotube fiber has already been reported and showed a specific capacitance of 373 F cm^{-2} at a specific current of 2 mA cm^{-2}. The fabricated devices exhibit an energy density of 0.13 mWh cm^{-2} and a specific power of 3.2 mW cm^{-2}. The device shows a capacitance retention of 80.3% even after 2000 cycles [79]. Also, $NiCo_2O_4$ was decorated on the Cu filament intertwined with carbon nanotube yarn hydrothermally. The high-charge transport efficiency of the copper as a current collector coated by $NiCo_2O_4$/carbon nanotube composite exhibits a specific areal capacitance of 277.3 mF cm^{-2} and energy density of 35.76 μWh cm^{-2} with a power density of 0.154 mW cm^{-2} [80].

2.4.2.3 Carbon Nanotube

Carbon nanotubes are one-dimensional nanostructured materials widely used and the most attractive candidates for application in energy storage. Carbon nanotubes have excellent electrical, thermal, and mechanical properties, as well as a high surface area and good storage capacity, which are highly suitable for supercapacitor electrodes. The surface area of pure carbon nanotubes varies from 120 to 500 m^2 g^{-1}, with specific capacitance ranging from 2 to 200 F g^{-1}. This shows a lower equivalent series resistance than activated carbon due to the more efficient diffusion within the porous network for the ions from the electrolyte. The single-walled carbon nanotubes (SWCNTs) as the electrode materials have a specific capacitance of 180 F g^{-1}, a power density of 20 kW kg^{-1}, and an energy density of up to 7 Wh kg^{-1} [81]. The surface area varies from 128 to 411 m^2 g^{-1} for the multi-walled carbon nanotubes (MWCNTs). Compared to the SWCNTs, the MWCNTs exhibited a better specific capacitance of 80 F g^{-1} in 6 M KOH electrolyte [82]. The $NiCo_2O_4$ is a group of self-assembled nano-sized particles with a particular hierarchical order, resistant to structural damage due to volume changes and internal connections, and is more efficient than 1D and 2D structures. However, the hierarchical 3D nano pins material provides 2168 F g^{-1} at a specific current of 5 A g^{-1} [83]. The conductive network has the property to accommodate the expansion of volume during the charge/discharge cycles and gives fast ion transfer. The core shell-like $NiCo_2O_4$ and MWCNTs exhibit the specific capacitance of 822 F g^{-1} at a scan rate of 5 mV s^{-1} [84]. Hierarchical $NiCo_2O_4$/polyaniline/carbon nanotubes hybrid composites exhibit the specific capacitance of 2250 F g^{-1} at a scan rate of 5 mV s^{-1} and 2000 F g^{-1} at a specific current of 1 A g^{-1} and exhibit good rate capability [85].

2.4.3 2D Carbon-Based Materials—$NiCo_2X_4$

2.4.3.1 Graphene

Compared to active carbon, freestanding carbon materials such as graphite foils and graphite sheets without current collectors are the emerging electrodes with lightweight, high electrical conductivity, high flexibility, and easy processability [86]. Graphene is a monolayer of carbon atoms that have undergone sp^2 hybridization and are closely arranged in a 2D honeycomb lattice. The graphene surface area per unit mass is 2.630 m^2 g^{-1}. Owing to the long-range conjugation, it has a wide range of intriguing properties, including outstanding electrical and thermal conductivity, excellent chemical stability, and strong mechanical strength. The semiconductor-perfect graphene has a zero-band gap and is highly chemically inert. Disordered graphene sheets containing topological defects, edge defects, functional groups, or heteroatom dopants have been thoroughly investigated to change the band gap and broaden their use in vast applications in electronic devices, electrochemical reactions, catalysis, luminescence [87, 88].

There are many methods for synthesizing graphene sheets. Most graphene oxides exhibit a capacitance of 550 F g^{-1} and a surface area of 2675 m^2 g^{-1}. The $NiCo_2O_4$/graphene composite is used as an electrode material and exhibits a specific capacitance of 1238 F g^{-1} at 1 A g^{-1}. These nanoparticles already have a hollow structure with a 10–20 nm void [89]. $NiCo_2O_4$ nanowires/nitrogen-doped graphene showed the maximum specific capacitance of 1273 F g^{-1} at a specific current of 0.5 A g^{-1} in 6 M KOH aqueous solution and exhibited excellent capacity retention without appreciable degradation after 3000 cycles at 4 A g^{-1} [90]. $NiCo_2S_4$ at N, S co-doped graphene exhibits structural stability and acts as a conductive network, which shows 1145 F g^{-1} at a specific current of 0.5 A g^{-1}. The $NiCo_2S_4$ provides numerous reaction sites and avoids the restacking of graphene sheets, showing an energy density of 33.8 Wh kg^{-1} and a power density of 799.8 W kg^{-1} [91].

2.4.3.2 Graphene Oxide

The oxygen functionalization in graphene has interesting properties that can differ from those of graphene. The graphene oxide has a conductivity of 10^{-6}–10^{-2} S cm^{-1} and a thickness of 200–500 nm. The pseudocapacitive nature of the graphene oxide delivers the specific capacitance of 189 F g^{-1}. Compared to graphene, graphene oxide has a higher conductivity. These make them one of the best materials for supercapacitors [92]. The metal–organic framework-based derivatives of novel $NiFe_2O_4$/$NiCo_2O_4$/graphene oxide composites exhibit an excellent specific capacitance of 1470 F g^{-1} at a specific current of 1 A g^{-1}. The $NiFe_2O_4$/$NiCo_2O_4$/graphene oxide//activated carbon delivers an energy density of 109 Wh kg^{-1} and a power density of 1051 W kg^{-1} with 90.7% capacitive retention after 1000 cycles [93].

2.4.3.3 Reduced Graphene Oxide

Reduced graphene oxide is formed by removing oxidized functional groups in graphene oxide. The reduced graphene oxide contains residual oxygen, other heteroatoms, and structural defects. Reduced graphene oxide has an electrical conductivity of 3112 S cm^{-1} and a specific surface area of 110.16 m^2 g^{-1} [94]. Reduced graphene oxide has been synthesized extensively by a facile hydrothermal method followed by calcination in one step. The porous reduced graphene oxide/ $NiCo_2O_4$ nano bundles give a higher specific capacitance of 1278 F g^{-1} at a specific current of 1 A g^{-1} and 719 F g^{-1} at a specific current of 20 A g^{-1}, showing remarkable capacitance retention. The efficient electrochemical performances could be ascribed to the unique structural feature with a larger area. The composite electrode material could be anticipated to gain potential candidates in supercapacitors and other devices because of its excellent characteristics of facile synthesis and controllable capacitance [95].

2.4.4 3D Carbon-Based Materials—$NiCo_2X_4$

2.4.4.1 3D Graphene

The surface area and other unique qualities of few-layer graphene are drastically reduced by the stable and high-performance use of graphene materials with a smart 3D layout of the graphene-based layers/units as opposed to a graphite-like ordering. Less than ten graphene layers should make up a 3D graphene material while retaining its basic characteristics [96, 97]. Graphene is classified into macroscopic (> 100 μm) and microscopic (< 100 μm) and is derived from graphene oxide, inorganic carbon compounds, and hydrocarbons. Some biomasses, such as cellulose, lignin, and cost-effective polymers with a small amount of graphene oxide, exhibit excellent surface area and conductivity. The porous 3D graphene-based materials are mostly made with pores > 1 nm but < 10 nm with a surface area of 3536 m^2 g^{-1}. The specific capacitance of 3D porous graphene is 231 F g^{-1}, and the energy density is 98 Wh kg^{-1} [98]. The 3D graphene, composed of the $Ni_xCo_{1-x}O$ nanoparticles, was well-dispersed on the graphene framework. The interconnected porous conductive framework enhances the connection between 3D graphene and $Ni_xCo_{1-x}O$, delivering a capacitance of 697.8 F g^{-1}. While assembling, the device gives an energy density of 27 Wh kg^{-1} and a power density of 725 W kg^{-1} with a cycle stability of 86% after 10,000 cycles at a specific current of 5 A g^{-1} [99].

2.4.4.2 Carbon Aerogel

The carbon aerogel from sugarcane, a 3D interconnected meso/microporous structure, gives a surface area of 2000 m^2 g^{-1} and a low bulk density of 0.10 g cm^{-3} and exhibits a specific capacitance of 385 F g^{-1} at a scan rate of 5 mV s^{-1}. The porous 3D interconnected structure of aerogel is beneficial for electrolyte ion diffusion ($D = 2.1 \times 10^{-11}$ cm^2 s^{-1}) and shortening pathways for electron transport. Hence, the rate performance of the carbon aerogel electrode enhances the specific capacity, and the hollow structure prohibits the volume changes during the charge–discharge process for significant improvement of cycling stability. While compositing with the spinel $NiCo_2O_4$ gives the ($D = 1.26 \times 10^{-11}$ cm^2 s^{-1}) specific capacitance of 473 F g^{-1} at 5 mV s^{-1} and power density of 6245 W kg^{-1} [100]. $NiCo_2O_4$ aerogel structure shows a high specific surface area (167.8 m^2 g^{-1}), high specific capacitance (903.2 F g^{-1} at a specific current of 1 A g^{-1}, excellent rate performance of 96.2% capacity retention from 1 to 10 A g^{-1}, and outstanding cycling stability nearly without capacitance loss after 3000 cycles at 10 A g^{-1} [101].

2.5 Conclusion

Electrochemical energy storage devices can rectify the future energy demand through supercapacitors. However, the main problem with supercapacitors is their low energy density. The research community focuses on developing new hybrid nanostructured materials or fabricating new hybrid devices for high energy density supercapacitors. Recently, many research works have been coming out with new hybrid nanostructured materials. Supercapacitor technologies should be focused on some factors given future perspectives.

- Proper understanding of the charge storage mechanism at electrode/electrolyte interfaces for each type of supercapacitor. The distinction between battery and intercalation pseudocapacitive-type materials should be adequately understood. Also, the kinetics of the diffuse ions inside the devices should be studied.
- Developing cost-effective and high-performance hybrid nanostructured materials that are abundant and eco-friendly. Also, the researchers should develop new synthetic techniques for the cost-efficient preparation of materials at the nanoscale level.
- Designing an alternative material that has high specific capacitance and energy density, which replaces carbon-based materials, as carbon materials exhibit low specific capacitance, which leads to low energy density.
- In pseudocapacitive-based and battery-based nanostructured materials, stability is the major factor. So, the researchers should focus on mechanical and electrical conductivity for long-cycle life.
- The selection of the electrode materials influences the device's performance. The proper awareness is required in selecting the current collectors, separators, and electrolytes for fabricating the devices. Every material has unique properties, so the positive electrode should have a high reduction potential, and the negative electrode should have a low reduction potential.

- In recent times, metal ion capacitors have been paid attention to improve the energy density factors. Significant works are concerned with the Li-ion and Na-ion capacitors, but the research should focus on the other monovalent ions like K^+ and Ca^{2+} and multivalent ions like Mg^{2+} and Al^{3+} which are in high abundance on the earth's crust.
- Integrating applications like a supercapacitor with solar cells, and thermoelectric devices will satisfy the future energy demand. Also, the development of wearable and flexible supercapacitors can be used in the fabrics and textiles industries.
- The major issue of the supercapacitor is self-discharge. So, the researchers should focus on this demerit and improve the capacity of the supercapacitor for longer periods.

This chapter discusses the essential needs of hybrid nanostructured materials and their factors for high energy density supercapacitors. Various synthetic methods for the development of efficient hybrid nanostructures have been discussed. They are sorting out the hybrid nanostructures based on their charge storage mechanism and structural response in the performance of the efficient high-energy supercapacitor. Therefore, the high influence of hybrid nanostructure in supercapacitor application, enhances energy density and leads to high electrochemical performance. These hybrid nanostructures based on high energy density devices will be commercialized and revolve for better future energy storage technologies.

Acknowledgements The author Karthick Sivalingam Nallathambi would like to thank RUSA 2.0BCTRC (Bharathiar Cancer Theranostic Research Centre) and UGC-BSR research start-up-grant (2018–2020) for the financial assistance and research facilities to undertake this research work. The authors Hemalatha Kuzhandaivel would like to thank the Technical Education and Quality Improvement Programme, Phase-III (TEQIP-III) for financial support.

References

1. K. Matuszek, M. Kar, J.M. Pringle, D.R. MacFarlane, Phase change materials for renewable energy storage at intermediate temperatures. Chem. Rev. **123**(1), 491–514 (2023). https://doi.org/10.1021/acs.chemrev.2c00407
2. M. Krichen, Y. Basheer, S.M. Qaisar, A. Waqar, A survey on energy storage: techniques and challenges. Energies **16**(5), 2271 (2023). https://doi.org/10.3390/en16052271
3. B. Kurc, M. Pigłowska, L. Rymaniak, P. Fuc, Modern nanocomposites and hybrids as electrode materials used in energy carriers. Nanomaterials **11**(2), 538 (2021). https://doi.org/10.3390/nano11020538
4. A.L.M. Reddy, S.R. Gowda, M.M. Shaijumon, P.M. Ajayan, Hybrid nanostructures for energy storage applications. Adv. Mater. **24**(37), 5045–5064 (2012). https://doi.org/10.1002/adma.201104502
5. S. Zhang, N. Pan, Supercapacitors performance evaluation. Adv. Energy Mater. **5**, 1401401 (2015). https://doi.org/10.1002/aenm.201401401
6. M. Inagaki, F. Kang, Chapter 3—Engineering and applications of carbon materials engineering and applications of carbon materials, in *Materials Science and Engineering of Carbon: Fundamentals*, 2nd edn., pp. 219–525 (2014)

7. E. Zhao, C. Qin, H.-R. Jung, G. Berdichevsky, A. Nese, S. Marder, G. Yushin, Lithium titanate confined in carbon nanopores for asymmetric supercapacitors. ACS Nano **10**, 3977–3984 (2016). https://doi.org/10.1021/acsnano.6b00479
8. Y. Gogotsi, R.M. Penner, Energy storage in nanomaterials—capacitive, pseudocapacitive or battery-like? ACS Nano **12**, 2081–2083 (2018). https://doi.org/10.1021/acsnano.8b01914
9. S. Balasubramaniam, A. Mohanty, S.K. Balasingam, S.J. Kim, A. Ramadoss, Comprehensive insight into the mechanism, material selection and performance evaluation of supercapatteries. Nano-Micro Lett. **12**, 85 (2020). https://doi.org/10.1007/2Fs40820-020-0413-7
10. G.Z. Chen, Supercapacitor and supercapattery as emerging electrochemical energy stores. Int. Mater. Rev. 173–202 (2016). https://doi.org/10.1080/09506608.2016.1240914
11. H. Yang, S. Ye, J. Zhou, T. Liang, Biomass-derived porous carbon materials for supercapacitor. Front. Chem. **7**, 274 (2019). https://doi.org/10.3389/fchem.2019.00274
12. Q. Cheng, J. Tang, J. Ma, H. Zhang, N. Shinya, L.-C. Qinc, Graphene and carbon nanotube composite electrodes for supercapacitors with ultra-high energy density. Phys. Chem. Chem. Phys. **13**, 17615–17624 (2011). https://doi.org/10.1039/C1CP21910C
13. J. Yan, T. Wei, B. Shao, F. Ma, Z. Fan, M. Zhang, C. Zheng, Y. Shang, W. Qian, F. Wei, Electrochemical properties of graphene nanosheet/carbon black composites as electrodes for supercapacitors. Carbon **48**, 1731–1737 (2010). https://doi.org/10.1016/j.carbon.2010.01.014
14. W. Lu, R. Hartman, L. Qu, L. Dai, Nanocomposite electrodes for high-performance supercapacitors. J. Phys. Chem. Lett. **2**, 655–660 (2011). https://doi.org/10.1021/jz200104n
15. C. Xie, S. Yang, X. Xu, J.-W. Shi, C. Niu, Core–shell structured carbon nanotubes/N-doped carbon layer nanocomposites for supercapacitor electrodes. J. Nanopart. Res. **22**, 25 (2020). https://doi.org/10.1007/s11051-019-4734-8
16. X. Li, Y. Tang, J. Song, W. Yang, M. Wang, C.Z.W. Zhao, J. Zheng, Y. Lin, Self-supporting activated carbon/carbon nanotube/reduced graphene oxide flexible electrode for high-performance supercapacitor. Carbon **129**, 236–244 (2018). https://doi.org/10.1016/j.carbon.2017.11.099
17. E. Pomerantseva, F. Bonaccorso, X. Feng, Y. Cui, Y. Gogotsi, Energy storage: the future enabled by nanomaterials. Science **366**, 969 (2019). https://doi.org/10.1126/science.aan8285
18. Y. Jiang, J. Liu, Definitions of pseudocapacitive materials: a brief review energy. Environ. Mater. **2**, 30 (2019). https://doi.org/10.1002/eem2.12028
19. J. Xie, P. Yang, Y. Wang, T. Qi, Y. Lei, C.M. Li, Puzzles and confusions in supercapacitor and battery: theory and solutions. J. Power Sour. **401**:213–223 (2018). https://doi.org/10.1016/j.jpowsour.2018.08.090
20. Y. Shao, M.F. El-Kady, J. Sun, Y. Li, Q. Zhang, M. Zhu, H. Wang, B. Dunn, R.B. Kaner, Design and mechanisms of asymmetric supercapacitors. Chem. Rev. **118**, 9233–9280 (2018). https://doi.org/10.1021/acs.chemrev.8b00252
21. Y. Zhu, S. Cheng, W. Zhou, J. Jia, L. Yang, M. Yao, M. Wang, J. Zhou, P. Wu, M. Liu, Construction and performance characterization of α-Fe2O3/rGO composite for long-cycling-life supercapacitor anode ACS sustainable. Chem. Eng. **5**, 5067–5074 (2017). https://doi.org/10.1021/acssuschemeng.7b00445
22. H. Jiang, Y. Dai, Y. Hu, W. Chen, C. Li, Nanostructured ternary nanocomposite of rGO/CNTs/MnO_2 for high-rate supercapacitors ACS sustainable. Chem. Eng. **2**, 70–74 (2014). https://doi.org/10.1021/sc400313y
23. P. Wu, S. Cheng, L. Yang, Z. Lin, X. Gui, X. Ou, J. Zhou, M. Yao, M. Wang, Y. Zhu, M. Liu, Synthesis and characterization of self-standing and highly flexible δ-MnO_2@CNTs/CNTs composite films for direct use of supercapacitor electrodes ACS. Appl. Mater. Interfaces **8**(36), 23721–23728 (2016). https://doi.org/10.1021/acsami.6b07161
24. Y. Wang, B. Hu, J. Luo, Y. Gu, X. Liu, Synthesis of polyaniline@MnO_2/graphene ternary hybrid hollow spheres via pickering emulsion polymerization for electrochemical supercapacitors ACS. Appl. Energy Mater. **4**(8), 7721–7730 (2021). https://doi.org/10.1021/acsaem.1c01087
25. K. Chanda, S. Maiti, S. Sarkar, P. Bairi, S. Thakur, K. Sardar, N. Besra, N.S. Das, K.K. Chattopadhyay, Hierarchical assembly of MnO_2 nanosheet on $CuCo_2O_4$ nanoflake over fabric

scaffold for symmetric supercapacitor ACS. Appl. Nano Mater. **4**, 1420–1433 (2021). https://doi.org/10.1021/acsanm.0c02958

26. K.M. Hercule, Q. Wei, A.M. Khan, Y. Zhao, X. Tian, L. Mai, Synergistic effect of hierarchical nanostructured $MoO_2/Co(OH)_2$ with largely enhanced pseudocapacitor cyclability. Nano Lett. **13**, 5685–5691 (2013). https://doi.org/10.1021/nl403372n
27. P. Makkar, N.N. Ghosh, Snowflake-like dendritic CoNi alloy-rGO nanocomposite as a cathode electrode material for an all-solid-state flexible asymmetric high-performance supercapacitor device. ACS Omega **5**, 10572–10580 (2020). https://doi.org/10.1021/acsomega.0c01053
28. M. Ghidiu, M.R. Lukatskaya, M.Q. Zhao, Y. Gogotsi, M.W. Barsoum, Conductive two-dimensional titanium carbide 'clay' with high volumetric capacitance. Nature **516**, 78–81 (2014). https://doi.org/10.1038/nature13970
29. M. Boota, B. Anasori, C. Voigt, M.Q. Zhao, M.W. Barsoum, Y. Gogotsi, Pseudocapacitive electrodes produced by oxidant-free polymerization of pyrrole between the layers of 2D titanium carbide (MXene). Adv. Mater. **28**, 1517–1522 (2016). https://doi.org/10.1002/adma.201504705
30. Q. Zhao, C.E. Ren, Z. Ling, M.R. Lukatskaya, C. Zhang, K.L. Van Aken, M.W. Barsoum, Y. Gogotsi, Flexible MXene/carbon nanotube composite paper with high volumetric capacitance. Adv. Mater. **27**, 339–345 (2015). https://doi.org/10.1002/adma.201404140
31. M.R. Lukatskaya, B. Dunn, Y. Gogotsi, Multidimensional materials and device architectures for future hybrid energy storage. Nat. Commun. **7**, 12647 (2016). https://doi.org/10.1038/ncomms12647
32. S. Chen, Q. Wu, M. Wen, C. Wang, Q. Wu, J. Wen, M. Zhu, Y. Wang, A tubular sandwich-structured $CNT@Ni@Ni_2(CO_3)(OH)_2$ with High stability and superior capacity as hybrid supercapacitor. J. Phys. Chem. C **121**(18), 9719–9728 (2017). https://doi.org/10.1021/acs.jpcc.7b01551
33. B. Ye, X. Cao, Q. Zhao, J. Wang, Coelectrodeposition of NiSe/ZnSe hybrid nanostructures as a battery-type electrode for an asymmetric supercapacitor. J. Phys. Chem. C **124**(39), 21242–21249 (2020). https://doi.org/10.1021/acs.jpcc.0c05125
34. J.S. Samdani, T.-H. Kang, B.-J. Lee, Y.H. Jang, J.-S. Yu, S. Shanmugam, Heterostructured titanium oxynitride-manganese cobalt oxide nanorods as high-performance electrode materials for supercapacitor devices. ACS Appl. Mater. Interfaces **12**(49), 54524–54536 (2020). https://doi.org/10.1021/acsami.0c13803
35. G. Li, H. Cai, X. Li, J. Zhang, D. Zhang, Y. Yang, J. Xiong, Construction of hierarchical $NiCo_2O_4@Ni$-MOF hybrid arrays on carbon cloth as superior battery-type electrodes for flexible solid-state hybrid supercapacitors. ACS Appl. Mater. Interfaces **11**, 37675–37684 (2019). https://doi.org/10.1021/acsami.9b11994
36. A.K. Das, S. Paria, A. Maitra, L. Halder, A. Bera, R. Bera, S.K. Si, A. De, S. Ojha, S. Bera, S.K. Karan, B.B. Khatua, Highly rate capable nanoflower-like NiSe and $WO_3@PPy$ composite electrode materials toward high energy density flexible all-solid-state asymmetric supercapacitor. ACS Appl. Electron. Mater. **1**, 977–990 (2019). https://doi.org/10.1021/acsaelm.9b00164
37. Y.-L. Liu, C. Yan, G.-G. Wang, H.-Y. Zhang, L.-Y. Dang, B.-W. Wu, Z.-Q. Lin, X.-S. An, J.-C. Han, Achieving ultrahigh capacity with self-assembled $Ni(OH)_2$ nanosheet-decorated hierarchical flower-like $MnCo_2O_{4.5}$ nanoneedles as advanced electrodes of battery-supercapacitor hybrid devices. ACS Appl. Mater. Interfaces **11**, 9984–9993 (2019). https://doi.org/10.1021/acsami.8b21803
38. C. Zhang, B. Zheng, C. Huang, Y. Li, J. Wang, S. Tang, M. Deng, Y. Du, Heterostructural three-dimensional reduced graphene Oxide/$CoMn_2O_4$ nanosheets toward a wide-potential window for high-performance supercapacitors. ACS Appl. Energy Mater. **2**(7), 5219–5230 (2019). https://doi.org/10.1021/acsaem.9b00904
39. R. Kumar, R.K. Singh, P.K. Dubey, D.P. Singh, R.M. Yadav, Self-assembled hierarchical formation of conjugated 3D cobalt oxide nanobead–CNT–graphene nanostructure using microwaves for high-performance supercapacitor electrode. ACS Appl. Mater. Interfaces **7**(27), 15042–15051 (2015). https://doi.org/10.1021/acsami.5b04336

40. S. Hussain, I. Rabani, D. Vikraman, A. Feroze, K Karuppasamy, Z. ul Haq, Y.-S. Seo, S.-H. Chun, H.-S. Kim, J. Jung, Hybrid design using carbon nanotubes decorated with Mo_2C and W_2C nanoparticles for supercapacitors and hydrogen evolution reactions. ACS Sustain. Chem. Eng. **8**, 12248–12259 (2020). https://doi.org/10.1021/acssuschemeng.0c04022
41. F. Zhang, Y. Tang, H. Liu, H. Ji, C. Jiang, J. Zhang, X. Zhang, C.-S. Lee, Uniform incorporation of flocculent molybdenum disulfide nanostructure into three-dimensional porous graphene as an anode for high-performance lithium ion batteries and hybrid supercapacitors. ACS Appl. Mater. Interfaces **8**(7), 4691–4699 (2016). https://doi.org/10.1021/acsami.5b11705
42. S. Das, R. Ghosh, D. Mandal, A.K. Nandi, Self-assembled nanostructured MoS_2 quantum dot polyaniline hybrid gels for high-performance solid state flexible supercapacitors. ACS Appl. Energy Mater **2**(9), 6642–6654 (2019). https://doi.org/10.1021/acsaem.9b01171
43. X. Jian, S. Liu, Y. Gao, W. Zhang, W. He, A. Mahmood, C.M. Subramaniyam, X. Wang, N. Mahmood, S.X. Dou, Facile synthesis of three-dimensional sandwiched MnO_2@GCs@MnO_2 hybrid nanostructured electrode for electrochemical capacitors. ACS Appl. Mater. Interfaces **9**(22), 18872–18882 (2017). https://doi.org/10.1021/acsami.7b04416
44. K.A. Sammed, L. Pan, A. Farid, M. Javid, S. Yang, M. Asif, Carbon nanocoil-supported three-dimensional structure of nickel–cobalt nitrides as the electrode material for supercapacitors. ACS Appl. Energy Mater **4**, 6678–6687 (2021). https://doi.org/10.1021/acsaem.1c00745
45. P.S. Shewale, K.-S. Yun, $NiCo_2O_4$/RGO hybrid nanostructures on surface-modified Ni core for flexible wire-shaped supercapacitor. Nanomaterials **11**(4), 852 (2021). https://doi.org/10.3390/nano11040852
46. S.S. Patil, D.P. Dubal, V.G. Deonikar, M.S. Tamboli, J.D. Ambekar, P. Gomez-Romero, S.S. Kolekar, B.B. Kale, D.R. Patil, Fern-like rGO/$BiVO_4$ hybrid nanostructures for high-energy symmetric supercapacitor. ACS Appl. Mater. Interfaces **8**(46), 31602–31610 (2016). https://doi.org/10.1021/acsami.6b08165
47. C. Zhang, Y. Huang, S. Tang, M. Deng, Y. Du, High-energy all-solid-state symmetric supercapacitor based on Ni_3S_2 mesoporous nanosheet-decorated three-dimensional reduced graphene oxide. ACS Energy Lett. **2**, 759–768 (2017). https://doi.org/10.1021/acsenergylett.7b00078
48. M. Karuppaiah, X. Benadict Joseph, S.F. Wang, B. Sriram, G. Antilen Jacob, G. Ravi, Engineering architecture of 3D urchin-like structure and 2D-nanosheets of Bi_2S_3@g-C_3N_4 as the electrode material for a solid-state symmetric supercapacitor. Energy Fuels **35**(15), 12569–12580 (2021). https://doi.org/10.1021/acs.energyfuels.1c01145
49. Y.A. Haj, J. Balamurugan, N.H. Kim, J.H. Lee, Nitrogen-doped graphene encapsulated cobalt iron sulfide as an advanced electrode for high-performance asymmetric supercapacitors. J. Mater. Chem. A **7**, 3941–3952 (2019). https://doi.org/10.1039/C8TA12396A
50. J. Balamurugan, T.T. Nguyen, V. Aravindan, N.H. Kim, J.H. Lee, Flexible solid-state asymmetric supercapacitors based on nitrogen-doped graphene encapsulated ternary metal-nitrides with ultralong cycle life. Adv. Funct. Mater. **28**, 1804663 (2018). https://doi.org/10.1002/adfm.201804663
51. P. Bandyopadhyay, T. Kuila, J. Balamurugan, T.T. Nguyen, N.H. Kim, J.H. Lee, Facile synthesis of novel sulfonated polyaniline functionalized graphene using m-aminobenzene sulfonic acid for asymmetric supercapacitor application Chem. Eng. J. **308**, 1174–1184 (2017). https://doi.org/10.1016/j.cej.2016.10.015
52. Q. Li, J. Zhou, S. Zhao, Y. Li, C. Chen, K. Tao, R. Liu, L. Han, Hollow and hierarchical cobalt-metal-organic framework@$CoCr_2O_4$ microplate array as a battery-type electrode for high-performance hybrid supercapacitors. ChemElectroChem. **7**, 437–444 (2020). https://doi.org/10.1002/celc.201902015
53. Y. Chen, T. Liu, L. Zhang, J. Yu, $NiCo_2S_4$ nanotubes anchored 3D nitrogen-doped graphene framework as electrode material with enhanced performance for asymmetric supercapacitors. ACS Sustain. Chem. Eng. **7**(13), 11157–11165 (2019). https://doi.org/10.1021/acssuschemeng.9b00284
54. I. Ayman, A. Rasheed, S. Ajmal, A. Rehman, A. Ali, I. Shakir, M.F. Warsi, $CoFe_2O_4$ nanoparticle-decorated 2D MXene: a novel hybrid material for supercapacitor. Appl. Energy Fuels **34**(6), 7622–7630 (2020). https://doi.org/10.1021/acs.energyfuels.0c00959

55. Y. Zhang, B. Wang, F. Liu, J. Cheng, X.W. Zhang, L. Zhang, Full synergistic contribution of electrodeposited three-dimensional $NiCo_2O_4$@MnO_2 nanosheet networks electrode for asymmetric supercapacitors. Nano Energy **27**, 627–637 (2016). https://doi.org/10.1016/j.nanoen.2016.08.013
56. X. Wang, Y. Zhang, J. Zheng, H. Jiang, X. Dong, X. Liu, C. Meng, Fabrication of vanadium sulfide (VS_4) wrapped with carbonaceous materials as an enhanced electrode for symmetric supercapacitors. J. Colloid Interface Sci. **574**, 312–323 (2020). https://doi.org/10.1016/j.jcis.2020.04.072
57. H.Q.B.C.D. Chen, X.S. Yanhe, X.S. Lei, One-pot synthesis of α-MnS/nitrogen-doped reduced graphene oxide hybrid for high-performance asymmetric supercapacitors. Electrochim. Acta **210**, 557–566 (2016). https://doi.org/10.1016/j.electacta.2016.05.031
58. S.P. Lonkar, V.V. Pillai, S.P. Patole, S.M. Alhassan, Scalable in situ synthesis of 2D–2D-type graphene-wrapped SnS_2 nanohybrids for enhanced supercapacitor and electrocatalytic applications. ACS Appl. Energy Mater. **3**, 4995–5005 (2020). https://doi.org/10.1021/acsaem.0c00519
59. H. Ji, X. Zhao, Z. Qiao, J. Jung, Y. Zhu, Y. Lu, L.L. Zhang, A.H. MacDonald, R.S. Ruoff, Capacitance of carbon-based electrical double-layer capacitors. Nat. Commun. **5**, 3317 (2014). https://doi.org/10.1038/ncomms4317
60. K. Malaie, M.R. Ganjali, Spinel nano-ferrites for aqueous supercapacitors; linking abundant resources and low-cost processes for sustainable energy storage. J. Energy Storage **33**, 102097 (2021). https://doi.org/10.1016/j.est.2020.102097
61. V. Kumar, C.R. Mariappan, R. Azmi, D. Moock, S. Indris, M. Bruns, H. Ehrenberg, G. Vijaya Prakash, Pseudocapacitance of mesoporous spinel-type MCo_2O_4 (M=Co, Zn, and Ni) rods fabricated by a facile solvothermal route. ACS Omega **2**, 6003–6013 (2017). https://doi.org/10.1021/acsomega.7b00709
62. Y. Cao, J. Liang, X. Li, L. Yue, Q. Liu, S. Lu, A.M. Asiri, J. Hu, Y. Luo, X. Sun, Recent advances in perovskite oxides as electrode materials for supercapacitors. Chem. Commun. **57**, 2343–2355 (2021). https://doi.org/10.1039/D0CC07970G
63. A. Rashti, B. Wang, E. Hassani, F. Feyzbar-Khalkhali-Nejad, X. Zhang, T.-S. Oh, Electrophoretic deposition of nickel cobaltite/polyaniline/rGO composite electrode for high-performance all-solid-state asymmetric supercapacitors. Energy Fuels **34**, 6448–6461 (2020). https://doi.org/10.1021/acs.energyfuels.0c00408
64. K. Xia, Q. Gao, J. Jiang, Hierarchical porous carbons with controlled micropores and mesopores for supercapacitor electrode materials. Carbon **46**, 1718–1726 (2008). https://doi.org/10.1016/j.carbon.2008.07.018
65. L. Zhang, X.S. Zhao, Carbon-based materials as supercapacitor electrodes. Chem. Soc. Rev. **38**, 2520–2531 (2009). https://doi.org/10.1039/B813846J
66. L. Wu, Y. Li, Z. Fu, B.-L. Su, Hierarchically structured porous materials: synthesis strategies and applications in energy storage. Nat. Sci. Rev. **7**, 1667–1701 (2020). https://doi.org/10.1093/nsr/nwaa183
67. R. Liu, J. Duaya, S.B. Lee, Heterogeneous nanostructured electrode materials for electrochemical energy storage. Chem. Commun. **47**, 1384–1404 (2011). https://doi.org/10.1039/C0CC03158E
68. H Chao, J Peng, Q Xia, X Wang, Recent progress on two-dimensional carbon materials for emerging post-lithium (Na^+, K^+, Zn^{2+}) hybrid supercapacitors. Polymers **13**, 2137 (2021). https://doi.org/10.3390/polym13132137
69. P. Tian, L. Tang, K.S. Teng, S.P. Lau, Graphene quantum dots from chemistry to applications. Mater. Today Chem. **10**, 221–258 (2018). https://doi.org/10.1016/j.mtchem.2018.09.007
70. J. Luo, J. Wang, S. Liu, W. Wu, T. Jia, Z. Yang, S. Mu, Y. Huang, Graphene quantum dots encapsulated tremella-like $NiCo_2O_4$ for advanced asymmetric supercapacitors. Carbon **146**, 1–8 (2019). https://doi.org/10.1016/j.carbon.2019.01.078
71. N.V. Nguyen, T.V. Tran, S.T. Luong, T.M. Pham, K.V. Nguyen, T.D. Vu, H.S. Nguyen, N.V. To, Facile synthesis of a $NiCo_2O_4$ nanoparticles mesoporous carbon composite as electrode materials for supercapacitor. Chem. Select **5**, 7060–7068 (2020). https://doi.org/10.1002/slct.202001410

72. S.S. Barton, J.E. Koresh, A study of the surface oxides on carbon cloth by electrical conductivity. Carbon **22**(6), 481–485 (1984). https://doi.org/10.1016/0008-6223(84)90079-4
73. F. Coeuret, E. Oliveira Vilar, E. Bezerra Cavalcanti, Carbon fiber cloth as an electrode material: electrical conductivity and mass transfer. J. Appl. Electrochem. **32**, 1175–1182 (2002). https://doi.org/10.1023/A:1021236210511
74. Q. Zhao, K. Zhang, S. Zhu, H. Xu, D. Cao, L. Zhao, R. Zhang, W. Yin, Review on the electrical resistance/conductivity of carbon fiber reinforced polymer. Appl. Sci. **9**, 2390 (2019). https://doi.org/10.3390/app9112390
75. G. Wang, H. Wang, X. Lu, Y. Ling, M. Yu, T. Zhai, Y. Tong, Y. Li, Solid-state supercapacitor based on activated carbon cloths exhibits excellent rate capability. Adv. Mater. **26**, 2676–2682 (2014). https://doi.org/10.1002/adma.201304756
76. A. Mohammadpour-Haratbar, P. Kiaeerad, S. Mazinani, A.M. Bazargan, F. Sharif, Bimetallic nickel–cobalt oxide nanoparticle/electrospun carbon nanofiber composites: preparation and application for supercapacitor electrode. Ceram. Int. **48**, 10015–10023 (2022). https://doi.org/10.1016/j.ceramint.2021.12.210
77. S.T. Senthilkumar, N. Fu, Y. Liu, L. Zhou, H. Huang, Flexible fiber hybrid supercapacitor with $NiCo_2O_4$ nanograss@carbon fiber and bio-waste derived high surface area porous carbon. Electrochimia Acta **211**, 411–419 (2016). https://doi.org/10.1016/j.electacta.2016.06.059
78. D. Kong, W. Ren, C. Cheng, Y. Wang, Z. Huang, H.Y. Yang, Three-dimensional $NiCo_2O_4$@polypyrrole coaxial nanowire arrays on carbon textiles for high-performance flexible asymmetric solid-state supercapacitor. ACS Appl. Mater. Interfaces **7**(38), 21334–21346 (2015). https://doi.org/10.1021/acsami.5b05908
79. H. He, X. Yang, L. Wang, X. Zhang, X. Li, W. Lü, Enhanced energy density of coaxial fiber asymmetric supercapacitor based on MoS_2@Fe_2O_3/carbon nanotube paper and $Ni(OH)_2$@$NiCo_2O_4$/carbon nanotube fiber electrodes. Chem. Eur. J. **26**, 17212–17221 (2020). https://doi.org/10.1002/chem.202004002
80. Q. Wang, D. Zhang, Y. Wu, T. Li, A. Zhang, M. Miao, Fabrication of supercapacitors from $NiCo_2O_4$ nanowire/carbon-nanotube yarn for ultraviolet photodetectors and portable electronics. Energy Technol. **5**, 1449–1456 (2017). https://doi.org/10.1002/ente.201600726
81. K.H. An, W.S. Kim, Y.S. Park, Y.C. Choi, S.M. Lee, D.C. Chung, D.J. Bae, S.C. Lim, Y.H. Lee, Supercapacitors using single-walled carbon nanotube electrodes. AIP Conf. Proc. **590**, 241 (2001). https://doi.org/10.1063/1.1420099
82. H. Ahmad, R.A. Khan, B.H. Kood, A. Alsalm, Systematic study of physicochemical and electrochemical properties of carbon nanomaterials. RSC Adv. **12**, 15593–15600 (2022). https://doi.org/10.1039/D2RA02533G
83. S. Hassanpoor, F. Aghely, Hierarchically self-assembled $NiCo_2O_4$ nano pins as a high-performance supercapacitor cathodic material: a morphology controlled study. RSC Adv. **10**, 35235–35244 (2020). https://doi.org/10.1039/D0RA07620A
84. T.H. Ko, S. Radhakrishnan, W.K. Choi, M.K. Seo, B.S. Kim, Core/shell-like $NiCo_2O_4$-decorated MWCNT hybrids prepared by a dry synthesis technique and its supercapacitor applications. Mater. Lett. **166**, 105–109 (2016). https://doi.org/10.1016/j.matlet.2015.12.053
85. G. Chaudhary, A.K. Sharma, P. Bhardwaj, K. Kant, I. Kaushal, A.K. Mishra, $NiCo_2O_4$ decorated PANI–CNTs composites as supercapacitor electrode materials. J. Energy Chem. **26**, 175–181 (2017). https://doi.org/10.1016/j.jechem.2016.09.013
86. B. Jang, H. Kim, S.W. Park, M. Lim, J. Lee, G.M. Go, Y.H. Choa, In situ exfoliation and modification of graphite foil in supercapacitor devices: a facile strategy to fabricate high-performance supercapacitors. RSC Adv. **11**, 4006–4010 (2021). https://doi.org/10.1039/D0RA10533C
87. A. Zuttel, P. Sudan, P. Mauron, P. Wenger, Model for the hydrogen adsorption on carbon nanostructures. Appl. Phys. A **78**, 941–946 (2004). https://doi.org/10.1007/s00339-003-2412-1
88. S. Gadipelli, Z.X. Guo, Graphene-based materials: synthesis and gas sorption, storage and separation. Progr. Mater. Sci. **69**, 1–60 (2015). https://doi.org/10.1016/j.pmatsci.2014.10.004

89. S. Sun, S. Li, S. Wang, Y. Li, L. Han, H. Kong, P. Wang, Fabrication of hollow $NiCo_2O_4$ nanoparticle/graphene composite for supercapacitor electrode. Mater. Lett. **182**, 23–26 (2016)
90. M. Yu, J. Chen, Y. Ma, J. Zhang, J. Liu, S. Li, J. An, Hydrothermal synthesis of $NiCo_2O_4$ nanowires/nitrogen-doped graphene for high-performance supercapacitor. Appl. Surf. Sci. **314**, 1000–1006 (2014). https://doi.org/10.1016/j.apsusc.2014.06.125
91. D. He, F. Li, Y. Xiao, S. Chen, Z. Zhu, H. Chen, X. Hu, W. Peng, S. Xin, Y. Bai, Electrostatic self-assembly assisted hydrothermal synthesis of bimetallic $NiCo_2S_4$@N, S co-doped graphene for high-performance asymmetric supercapacitors. Electrochim. Acta **404**, 139751 (2022). https://doi.org/10.1016/j.electacta.2021.139751
92. B. Xu, S. Yue, Z. Sui, X. Zhang, S. Hou, G. Cao, Y. Yang, What is the choice for supercapacitors: graphene or graphene oxide? Energy Environ. Sci. **4**, 2826–2830 (2011). https://doi.org/10.1039/C1EE01198G
93. J. Pan, S. Li, F. Li, T. Yu, Y. Liu, L. Zhang, L. Ma, M. Sun, X. Tian, The $NiFe_2O_4$/$NiCo_2O_4$/GO composites electrode material derived from dual-MOF for high performance solid-state hybrid supercapacitors. Colloids Surf. A Physicochem. Eng. Aspects **609**, 125650 (2021). https://doi.org/10.1016/j.colsurfa.2020.125650
94. Y. Chen, K. Fu, S. Zhu, W. Luo, Y. Wang, Y. Li, E. Hitz, Y. Yao, J. Dai, J. Wan, V.A. Danner, T. Li, L. Hu, Reduced graphene oxide films with ultrahigh conductivity as Li–ion battery current collectors. Nano Lett. **16**(6), 3616–3623 (2016). https://doi.org/10.1021/acs.nanolett.6b00743
95. Z. Wang, X. Zhang, Z. Zhang, N. Qiao, Y. Li, Z. Hao, Hybrids of $NiCo_2O_4$ nanorods and nano bundles with graphene as promising electrode materials for supercapacitors. J. Colloid Interface Sci. **460**, 303–309 (2015). https://doi.org/10.1016/j.jcis.2015.08.067
96. Q. Shi, Y. Cha, Y. Song, J.-I. Lee, C. Zhu, X. Li, M.-K. Song, D. Du, Y. Lin, 3D graphene-based hybrid materials: synthesis and applications in energy storage and conversion. Nanoscale **8**, 15414–15447 (2016). https://doi.org/10.1039/C6NR04770J
97. C. Xiong, B. Li, X. Lin, H. Liu, Y. Xu, J. Mao, C. Duan, T. Li, Y. Ni, The recent progress on three-dimensional porous graphene-based hybrid structure for supercapacitor. Compos. B **165**, 10–46 (2019). https://doi.org/10.1016/j.compositesb.2018.11.085
98. L. Zhang, F. Zhang, X. Yang, G. Long, Y.P. Wu, T. Zhang, K. Leng, Y. Huang, Y. Ma, A. Yu, Y. Chen, Porous 3D graphene-based bulk materials with exceptionally high surface area and excellent conductivity for supercapacitors. Sci. Rep. 1408 (2013). https://doi.org/10.1038/srep01408
99. Y. Zhou, L. Le Wen, K.Z. Ya, Y.B. Zhao, Three-dimensional porous graphene/nickel cobalt mixed oxide composites for high-performance hybrid supercapacitor. Ceram. Int. **44**(17), 21848–21854 (2018). https://doi.org/10.1016/j.ceramint.2018.08.292
100. R.B. Marichi, S. Goel, A. Kumar, T. Vikrant, S. Shubra, L.G.S. Raj, K. Sharma, Direct hydrothermal treatment of sugarcane juice for 3D oxygen-rich carbon Aerogel/$NiCo_2O_4$ based supercapacitor. Mater. Chem. Phys. **239**, 121957 (2020). https://doi.org/10.1016/j.matchemphys.2019.121957
101. J. Li, S. Chen, X. Zhu, X. She, T. Liu, H. Zhang, S. Komarneni, D. Yang, X. Yao, Toward Aerogel Electrodes of Superior Rate Performance in Supercapacitors through Engineered Hollow Nanoparticles of $NiCo_2O_4$. Adv. Sci. **4**(12), 1700345 (2017). https://doi.org/10.1002/advs.201700345

Chapter 3
Hybrid Nanostructures with Conducting Polymers for High-Performance Supercapacitors

Subramanian Arulmani, Sambandam Anandan, and Veeraswamy Davamani

Abstract With concern to global changes, energy has emerged as a considerable focus of the scientific community and world power needs. Lacking energy resources and environmental crises have driven us to explore cleaner energy storage devices like supercapacitors. Among various materials, conducting polymers are proposing unique chemistry in the supercapacitors that significantly enrich to design of multiple structures and functions. Further, scrutinizing the essential utilities by associating structure-function relationships in conducting polymer-based composites reveals the structural parameters and interfaces that govern the supercapacitors' achievements. This chapter aims to exploit the synthesis and synergism of the nanostructured conducting polymers with their support materials during their composite formation. It also emphasizes the progress and diverse support constructions for effective electrochemical exploitation of supercapacitor applications. And last, the future exploration methods and enduring tasks concerned with the rational strategy and development are also elaborated.

Keywords Conducting polymers · Nanostructures · Morphology · Nanohybrids · Nanocomposite · Supercapacitors · Energy density · Power density

S. Arulmani (✉)
Department of Chemistry, School of Advanced Science, Vellore Institute of Technology, Chennai, Tamil Nadu, India
e-mail: arulmaniyes@gmail.com; s.arulmani@vit.ac.in

S. Arulmani · S. Anandan
Nanomaterials and Solar Energy Conversion Lab, Department of Chemistry, Tiruchirappalli, Tamil Nadu, India

V. Davamani
Department of Environmental Sciences, Tamil Nadu Agricultural University, Coimbatore, Tamil Nadu, India

© The Author(s), under exclusive license to Springer Nature Switzerland AG 2026

K. Raju et al. (eds.), *Hybrid Nanostructures for Advanced Supercapacitors*, Modern Aspects of Electrochemistry 63, https://doi.org/10.1007/978-3-032-04593-5_3

3.1 Introduction

The story of polymer/plastic materials started at the beginning of the twentieth century, even though organic chemistry as a science engages back to the eighteenth century. Today, they are entered into manufacturing with several scientific, therapeutic, and technical products. These polymers have a high impact, and achievement is typically correlated with an extensive range of their characteristics. Properties like extraordinary mechanical strength, roughness, flexibility, plasticity, low density, and resistance to destructive chemicals such as solvents, acids, or bases are essential in various fields. Besides, the best feature is that it is easy to synthesize and fabricate, leading to instant pertinence to the industry. Due to better electrical resistivity, plastic is usually used as an insulating material for metallic conductors or photoresists.

In the meantime, metallic materials like silver, copper, and aluminum only exhibit an exceptionally high electrical conductivity of over 10^6 S/m. So, they have been used for numerous electrical/electronic applications. However, these materials are restricted due to their heavy-weight and ease of chemical corrosion. Compared to metals, polymeric materials reveal a very low density of about 1.0–1.3 g/cm^3 with more flexibility and corrosion resistive nature; however, they are insulators. Although polymers are insulators, some can potentially have conduct/semiconducting properties, confirmed by the ability to synthesize polymers with intrinsic conduction [1, 2]. Miraculously, the polymers' conducting nature was discovered in 1976 by Heeger, MacDiarmid, and Shirakawa, for which they were awarded the Nobel Prize in chemistry in the year 2000 [3].

After the above instance, the prominence of the field of conducting polymers has blossomed. Polymer plastics with alternating single and double bonds linked in their carbon atoms might conduct electricity. Opening from this innovation, the curiosity about this type of "synthetic metals"—polymeric materials has been exposed more in diverse fields. That is, the availability of an extended π-electron conjugation beside their backbone chain can be reflected together as beneficial in terms of providing intrinsic electron-conducting characteristics [4]. Conducting polymers exhibit high electrical conductivity, good electrochemical activity, unique optical properties, and biocompatibility. As a consequence of these remarkable properties, conducting polymers have been recognized special consideration as a promising aspirant in numerous areas like supercapacitors (SCs), flexible electronics, sensors, and electromagnetic interference (EMI) shielding [5].

Especially, conducting polymers are considered an essential material due to their pseudocapacitive property [6]. Further, their electrochemical properties can be improved by selecting monomers, oligomers, and selective chemistries for the polymer's fine-tuning. The meticulous construction of conducting polymers is responsible for structural and morphological variants, unveiling its physicochemical assets [5, 7].

This chapter is aimed to reveal the interactions of the diverse nanostructured conducting polymers as a sustainable material and their performance for next-generation energy devices. This chapter's succeeding section outlined the existing synthetic methods concerning different nanostructured conducting polymers, the significance of arrangements, predictable structural design, and elucidated their improved energy storage capabilities.

3.2 Conductivity Mechanism of Conducting Polymers

The polymers which are having saturated carbons in their structure are insulators. In those saturated polymers, all four valance electrons of carbon are used to form covalent bonds. Nevertheless, some polymers can exhibit different electronic configurations, such as conjugated polymers [8]. The available unpaired electrons, so-called π-electrons for each carbon atom, lead to electron delocalization adjacent to the polymers' backbone. This factor is most responsible for the charge mobility that enables semiconducting or even metallic properties in conducting polymers. The number and the nature of atoms within the polymers' repeating units may decide the polymeric chain's symmetry.

Further, the conductivity of the conducting polymers can be altered from insulating to highly conducting through fine-tuning the doping level [9]. The delocalized π-electron system is primarily available only in the doped state, and the conjugation in the backbone will be formed in the un-doped state. The tuning can be possible either during synthesis or by advanced dispersion techniques [10].

An insulating or semiconducting organic polymer which has low conductivity in its un-doped state (10^{-10}–10^{-5} S cm^{-1}) is transformed into a polymer through doping that displays its conductivity in a "metallic" regime (approximately 1–10^4 S cm^{-1}) [9]. Generally, the doping process is always reversible, and the polymer can return to its initial state without the main modifications in its basic assembly. The charge carriers (negative and positive) are used for the doping process, and the same is called "n" doping and "p" doping, respectively [2]. The following equations are the doping of conducting polymers.

$$\mathrm{CP} + n\ \mathrm{Red} \rightarrow \left(\mathrm{C}^{+}\right)_n \mathrm{CP}^{n-} + n\ \mathrm{Red}^{+} \tag{3.1}$$

$$\mathrm{CP} + n\ \mathrm{Oxi} \rightarrow \mathrm{CP}^{n+} \left(\mathrm{A}^{-}\right)_n + n\ \mathrm{Oxi}^{-} \tag{3.2}$$

where Eqs. (3.1) and (3.2) are the process of n-doping and p-doping, respectively; CP is conducting polymer, n is the number, (Oxi) and (Red) is the oxidant and reductant, (C^+) and (A^-) is the cation and anion, respectively [6].

The charge incorporation onto conjugated, semiconducting-macromolecular chains induces extensive research and applications in polymers that can be accomplished in many ways. Among them, (i) chemical doping by charge transfer (p-type doping, oxidation or n-type doping, reductions), (ii) electrochemical doping, (iii) photo doping (electron-hole pair creation and separation into "free" carriers), and (iv) charge injection at a metal-semiconducting polymer interface (hole injection into filled π-band or electron injection into an empty π*-band) is the primary process employed to enhance the activity of the conducting polymers [4]. The doping may significantly improve the properties of conducting polymers that enable their identity in solar cells, sensors, memory-based devices, electrochromic devices, and electrochemical supercapacitors. Thus, the conducting polymers based supercapacitor electrodes or devices might exhibit a range of benefits, for example, flexibility, better conductivity, and comparatively inexpensive through the most accessible way [7, 11]. Consequently, these conducting polymers have diverse redox states that shift the electrical conductivity from insulator to conductor range.

3.3 Classification of Conducting Polymers

The conducting polymers are classified majorly into three types as per their main chain components, i.e., the presence of double bonds, aromaticity, and a combination of both concepts. Poly(acetylene) (PAC) is the only conducting polymer comprising double bonds in its skeleton. The group belongs to the aromatic cycle and can be classified through heteroatoms in the polymer backbone. Poly(fluorene), polyphenylene, polypyrene, polyazulene, poly naphthalene are the best examples of conducting polymers containing aromatic cycles without heteroatoms. Likewise, poly(pyrrole) (PPY), polycarbazole, polyindole, polyazepine, and polyanilines (PANI) are the nitrogen-containing, and the poly(thiophene) (PTh), poly(3,4-ethylene dioxythiophene) (PEDOT), and poly(p-phenylene sulfide) (PPS) are sulfur-containing conducting polymers. Among the heteroatom group, polyanilines (PANI) and poly(p-phenylene sulfide) (PPS) are the exceptions in which the heteroatom is present outside the aromatic ring. Poly(p-phenylene vinylene) (PPV) is the only one coming under the category of containing both double bond and aromatic cycles [12]. The essential and widely used conducting polymers are listed as per the specified classifications (Fig. 3.1). The conductivity of the conducting polymers is due to the structure and orientation, which depend on their doping level [7]. The structure-based activity and doping are always related to synthetic approaches.

3.4 Electrochemistry of Conducting Polymers

Conducting polymers may store and release the charges over their redox processes, which comprise the π-conjugated polymer chains during the electrochemical doping-undoping process. The ensuing reaction showed the general mechanism of the conducting polymers.

$$[\mathrm{CP}] + n\mathrm{X}^- \leftrightarrow \left[(\mathrm{CP})^{n+}\, n\mathrm{X}^-\right] + n\mathrm{e} \tag{3.3}$$

where CP is conducting polymer, X^- is anion, e^- is electron, and n is numeric.

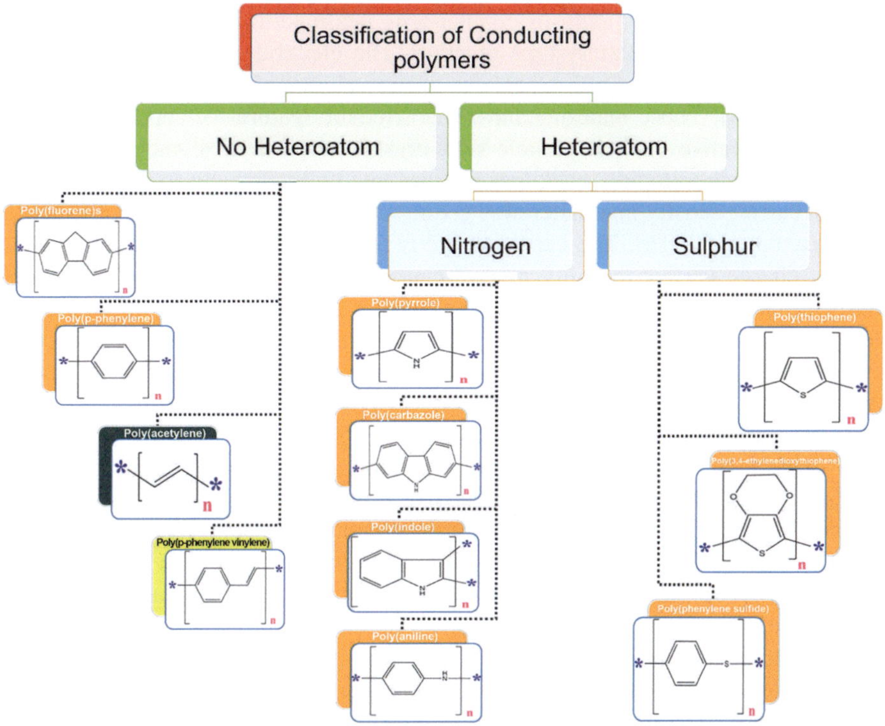

Fig. 3.1 Classification of conducting polymers based on their composition. (The orange, dark green, and yellow colors are depicted for the aromatic cycles, double bonds, and both in a colored version)

During the oxidization (p-doping) process, the anionic species X^- from the electrolyte is introduced into the polymer backbone and is re-released into the electrolyte upon the eduction process [13]. When the conducting polymer film was electrochemically grown, the reversible system's current peak was calculated using the following equation.

$$i = n^2F^2A\Gamma v\left[\frac{\exp\theta}{\mathrm{RT}(1+\exp\theta)}\right] \tag{3.4}$$

where $\theta = (n\mathrm{F/RT})(\mathrm{E}\text{-}\mathrm{E}^\circ)$, n is the number of electrons involved in the process, and A is the area of the electrode (cm^2). The current peak is proportionate to the surface coverage (Γ) and potential scan rate (), valid only for thin conducting polymer films with dopant ions with lesser diffusion coefficients. The study has conferred this occurrence on the electrochemical properties of electrochemically prepared PPy doped with a BF_4 molecule [14]. In such cases, cyclic voltammetry (CV) is obtained through entirely symmetrical redox peaks as one-electron redox processes [15].

The inserting and stripping of counter-ions facilitate an extraordinary specific capacitance through a battery-type reaction. The electrode material that delivers broad or nearly coincided redox peaks in cyclic voltammetry is considered a pseudocapacitive [4]. Those materials have characteristic features of capacitive charge storing, irrespective of their particle size, crystalline properties, and morphology. Meantime, a higher doping level does not turn into high conductivity. Nevertheless, the high carrier density in a material is getting more significant toward constructing of newer materials [16]. The carrier mobility (μ) is another chief feature because the low charge carrier mobility always causes insulating nature that suppresses charge-discharge cycles. In this case, the transfer of charges either through intra-chain or inter-chain in the conducting polymers backbone. The intra-chain mobility is endorsed through the conjugated sp^2 carbon available in the polymer that generates a resonance stabilized delocalized π system [17]. In this way, the charge carriers are spontaneously transferring to the entire chain.

Similarly, the conjugation also simplifies inter-chain charge mobility due to boosting π-π^* stacking between chains [6, 15]. However, conducting polymers suffer from volumetric changes in electrochemical reactions that display a deprived cycling performance due to the intrinsically brittle material's low mechanical properties that slow Faradic processes [18]. For improved supercapacitive performances and better cycling stability, a particular type of electrode architectures of conducting polymers should be prepared, either improved active sites with different morphologies or hybrid formation. Hence, care must be taken to identify the highly functional conducting polymers based electrode materials to enhance supercapacitors' energy storage capability [19].

3.5 Significance of Supercapacitors and their Mechanisms

Numerous efforts exist to harvest the energy, but the main barriers still are to store it in large-scale practice. So, the available energy storage devices should be preferred for better energy utilization [20]. In the last decade, the research on energy storage has sustained to advance and familiarize the necessities and developments. Generally, batteries and supercapacitors gained extreme consideration in recent years [21]. All energy storage devices mainly depend on their unique features like their own mechanisms, energy storage phenomenon, charging rate, power restrictions, and cycle life. The above features are also used to distinguish the batteries with supercapacitor (Table 3.1), which helps improve their performances [18, 22].

For example, batteries' charge/energy storage mainly depends on their chemical nature, but the supercapacitors depend on their physical nature. The intercalation/de-intercalation type of reactions is more favorable in batteries, but the surface-specified reactions are happening in the supercapacitors. Further, in comparison with lithium-ion batteries, the supercapacitors also reveal greater specific power. Likewise, it can store the electrical charge, particularly at the electrode's surface instead of the whole electrode. Also, it can be operated at higher rates and deliver more specific power

Table 3.1 Diverse features/parameters associated with battery and supercapacitor based on their performances

Parameters	Battery	Supercapacitor
Storage mechanism	Through reversible chemical reactions	Through the separation of charge at electrode/electrolyte interface
Energy storage	High-store in chemical form throughout the area	Limited-store electrostatically, enabling rapid charge and discharge cycles
Power limitation	Internal resistance, mass transport and chemical reaction rates	Based on the electrolyte conductivity
Charging rate	Depends on factors such as its chemistry, capacity, and charging circuitry, influencing the speed	Fast charging rate due to their ability to quickly adsorb and desorb ions at the electrode-electrolyte interface
Charge life limitation	Limited by factors such as chemical degradation, cycling, and operating conditions, impacting its ability to maintain capacity over time	Limited by factors such as self-discharge rates and electrode degradation, affecting their long-term energy storage capabilities

than batteries. Therefore, it has numerous benefits associated with batteries, such as extraordinary power density, rapid charge-discharge time with maximum efficiency, long life-cycle, a wide range of functioning temperatures, eco-friendly, and safety issues [23, 24].

Besides, supercapacitors can store charge electrostatically on the electrode's surface, contrasting to the batteries where a phase change occurs. In these devices, the charge gets accumulated at the electrode/electrolyte interface, so the material's surface area is always playing a significant role. The electrochemical process is fast and too reversible as a consequence of the physical charge storage. The electrodes in a cell with some charges (either positive or negative) consume their opposite electrolyte charges when connected with an external source. As a result, a double-layer of charge is created at each electrode called an electric double-layer capacitor (EDLC) [20]. Combining the double-layers phenomenon with high surface area electrodes will decrease the distance among the electrodes, which permits EDLC to realize higher energy densities. In the EDLCs, there is no charge transfer among electrode and electrolyte and no chemicals or composition variations related to non-Faradaic processes (Fig. 3.2). Hence, the charge storage in EDLC is too reversible, which is evident as mirror image voltammograms [25, 26]. The charge storage mechanism in EDLC is established through double-layer theory and stated by three models: the Helmholtz model, Gouy–Chapman model, and Stern model [27, 28]. Carbon-based materials such as graphite, activated carbons, fullerene, carbon nanotubes, graphene, carbons derived from various sources are widely keeping their place in search of EDLC-based electrode materials.

In contrast to EDLC, the charge storage on pseudocapacitors can be attained through the Faradaic process, which leads to the charge transfer among electrode and electrolyte [18]. The charge storage in pseudocapacitive materials involves electron transfer through oxidation or reduction changes (Fig. 3.2) [26]. Also, three types of

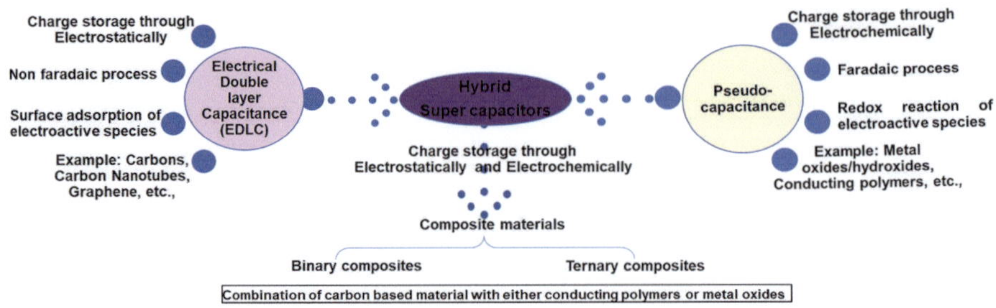

Fig. 3.2 Schematic illustration of materials and mechanisms involved in various supercapacitors (EDLC, pseudocapacitors and hybrid supercapacitors)

charge storage are skilled in pseudocapacitors, such as sorption pseudocapacitance, redox pseudocapacitance, and intercalation pseudocapacitance. The sorption pseudocapacitance can be formed during the adsorption/desorption of ions (electrolytes) on the electrode material's surface. The second, redox pseudocapacitance, initiates from fast revocable Faradaic reactions due to electro-absorption or intercalation of electrolyte ions on the electrode's surface. In the third, the ions may intercalate on the Van der Waals gaps or redox-active material's lattice [28]. These Faradaic processes can permit pseudocapacitors to realize superior capacitances and higher energy densities than EDLC. The better redox property of the electrode materials like conducting polymers (such as polyaniline, polypyrrole, polythiophene, etc.) and metal oxides (RuO_2, NiO, Co_3O_4, MnO_2, etc.) or hydroxides ($Co(OH)_2$ and $Ni(OH)_2$) delivers the enhanced pseudocapacitance performance.

3.6 Role of Nanostructured Materials in (Pseudo) Supercapacitors

Expansions in the field of nanoscience and nanotechnology are creating a more consistent way through its construction concern to energy storage. An extensive history is available about a systematic study of the material dimension in chemistry and physics [29]. Nanostructured materials are more beneficial by offering a large surface-to-volume ratio, promising carriage properties, different physical properties, and confinement effects ensuing as of the nanoscale magnitudes [30]. The improvement in specific surface area and related surface energy also depicts approximately identical or significant applications [31].

The fine-tuning of material construction, design, and performance on the nanoscale for electrochemical supercapacitors was fascinating, and prolonged consideration has been given to them over the past few years. Understanding the structure-based functionality at nanoscales is significant to reveal a rational design of efficient hybrid/composite systems with better activity and durability for supercapacitors [30]. The fascinating primary scientific concerns with superior execution are essential to determine the response for the arrangement of atoms or another building block of elements into their suggested constructions with a specified size of lengths and diameters and those materials intrinsic properties [32].

Further, inadequate, and slow diffusion kinetics and inaccessibility of better active bulk materials lead to new and advanced electrode materials. The field of supercapacitors has appreciated a revival in the last two decades as its implication to the ever-expanding domination of nanoscience [21]. The electrochemical reaction occurs on a flat surface, and the nanoarchitectured surface is different from each other. In the flat electrode, the response takes place consistently on the entire electrode surface. On the other hand, it is recognizable on the nanoarchitectured electrode with a scattered electrode reaction, and its rate is influenced by the nature/morphology of the material [33].

Recently numerous advanced materials with different nanostructures were designed and revealed. Their properties are correlated to electron transfer, ion diffusion that can be investigated through electrochemical techniques and utilized in diverse fields [34]. The inter-relation between interfacial electrochemistry and nanoscience produces innovative opportunities for the designing of chemical surfaces. The surface structure at the molecular level leads to innovative material and charge transfer extents [35].

The nanostructured electrode materials may be required to better utilize electrochemical benefits at the nanoscale level with higher active sites. Such integrated electrodes can minimize the loss of interfacial resistance and further improve the reaction kinetics. Emerging novel nanoarchitectures by adjusting the morphology, assembly, size of nanomaterials, and material processing considerations are frequently anticipated by the influence of boundaries on electrons' transport [36]. Nanostructures enclosed with high-energy surfaces and well-defined atomic arrangement might offer a promising platform to explore the surface structure-functionality at nanoscales [30].

3.7 Synthesis and Tailoring of Nanostructured Conducting Polymers

The capacitance and charge storage of supercapacitors familiarly governed by the electrode materials used. Consequently, the most significant to overcome these tasks are further emerging new materials with extraordinary capacitance and enriched performance compared with current electrode materials. The following methods can be utilized to prepare the conducting polymers and their composites/hybrids.

3.7.1 Chemical Polymerization

Chemical synthesis is used while an enormous amount of polymer is required. In this way, the conducting polymers are chemically produced via the oxidation of monomers. This method's foremost benefit is mass production at a low cost, easy handling, providing different promising directions for synthesizing diverse metal/ metalloid hybrid conducting polymers. Besides, using an electrode is not essential in this kind of synthesis [37]. The chemical method is further categorized into addition polymerization (chain-growth) and condensation polymerization (via step-growth). The various investigation has been made to enhance the quality and yield of the polymers or their composites. In this way, numerous transition metal compounds (Fe (III) and Cu (II) complexes) were used as both dopants and as well as initiators for the conversion of pyrrole to PPY, which leads to electrically conducting [38]. Recently, cross-linked polyaniline nanorods (CPANI) were synthesized *through* the oxidative chemical copolymerization of aniline using *p*-phenylenediamine (PPDA) and triphenylamine (TPA). PPDA and TPA usage improve the electrochemical properties of CPANI by improving electrical conductivity at 33.3 S cm^{-1}, and specific capacitance is 455.1 F g^{-1} than PANI [39]. The in-situ oxidative polymerization has been meritoriously engaged for the synthesis of various nanostructures. It deals with a simple, quick, and practical approach to preparing conducting polymers and their hybrids to avoid difficulties in solubility [40].

3.7.2 Sol-Gel Polymerization

Sol-gel is an adaptable practice that discovers favorable applications in preparing hybrid conducting polymers and nanocomposite materials [40]. This practice has main benefits: lower reaction temperature simplifies the straightforward synthesis of hybrid composites, superior integration of inorganic materials into the organic moieties, and vice versa. For example, PEDOT: SiO_x film was constructed with a silicate linkage through oxidative polymerization followed by *TEOS's* in-situ sol-gel process. The synthetic procedure permits more excellent link and bonding strength among the conducting polymer and glass substrate. The as-prepared material is an anode for the ITO-free solar cells with excellent transparency and minimum surface resistance [41]. A novel conducting polymer hydrogel (CPH) was recently prepared by a sol-gel method and promised to assemble a flexible, solid-state supercapacitor. The prepared device displayed a better area-normalized capacitance of 2.2 F cm^{-2} and a gravimetric capacitance of 1573.6 F g^{-1} with a maximum energy density of 0.18 mWh cm^{-2} (or 128.7 Wh Kg^{-1}). This study represents a new flexible, solid-state supercapacitor with future importance [42].

3.7.3 Emulsion Polymerization

It is the unique oxidative polymerization method for conducting polymer hybrids. Generally, emulsions may be formed by anyone of the following, i.e., oil in water emulsion (oil droplets dispersed in an aqueous medium) and water in oil emulsion (small water drops distributed in an oily medium). In this method, the surfactant, monomer, and initiator are mixed with the solvent, usually water, causing micelles' construction. The polymerization starts within the micelles and leads to stable colloidal dispersions termed emulsions [43]. Aniline is taken as a monomer and coated on the MWCNT surface through in-situ emulsion polymerization to synthesize PANI-MWCNT. Here, sodium dodecyl sulfate (SDS) acts as an emulsifier. The PANI corals have grown homogenously on the MWCNT and are used to construct symmetric supercapacitors. An exciting specific capacitance (360 F g^{-1}) was gained for the PANI-MWCNT nanocomposites at 0.4 A g^{-1} in 1 M H_2SO_4 with good capacitance retention (98.3%) after 5000 cycles [44].

Further, a reverse microemulsion system assisted with a block copolymer was applied to synthesize superfine shuttle shaped PANI nanoflowers with radiated whiskers on the flowers' corner. In this process, the copolymer poly(ethylene oxide)-block-poly(propylene oxide)-block-poly(ethylene oxide) (P-123) assisted as reverse micelles for the ordered arrangement of flower-like nanostructures having the size of 40–150 nm in widths and 80–180 nm in lengths. The exclusive structural design exhibited a high specific capacitance of 622 F g^{-1} at 0.5 A g^{-1} and good durability (76% after 1000 cycles) because of V-typed networks [45].

3.7.4 Electrochemical Polymerization

It is a flexible practice used to synthesize conducting polymers and their hybrids with diverse structures. It is executed through the three-electrode systems: working, counter, reference electrodes, monomers in a suitable solvent, or electrolyte (dopant) [46]. The monomers with low anodic-oxidation potential subject to electrophilic switch are mainly used for electro-polymerization. The solvent or electrolyte (conductive medium) may be chosen according to their monomer's oxidation potential strength. The electro-polymerization can be done through galvanostatic (using constant current), potentiostatic (using constant potential), and potential scanning/cycling or sweeping methods [47]. For instance, the flexible graphene/PANI hybrid (GPCP) nanocomposite was fabricated through electro-polymerization. The PANI was electrochemically deposited via a constant potential (0.75 V) on the graphene paper with various time intervals (60, 300, and 900 s). The maximum gravimetric capacitances of GPCP-900 s show 233 F g^{-1} and volumetric of 135 F cm^{-3} caused by a higher PANI ratio in the porous bucky paper assists greater lodging polyaniline [48].

Likewise, PANI was vertically grown on the interconnected porous carbon (ICPC), forming nanostructured fibrous PANI on ICPC through in-situ electro-polymerization. The PANI was created with different aniline molar ratios (0.05, 0.1, 0.15 M) in 0.5 M of H_2SO_4 with 0.75 V for a period of 300 s. The exploitation of unique 3D cross-linked porous carbon construction, and PANI's perpendicular deposition provides an easy ion diffusion pathway. The extreme specific capacitance was perceived for PANI/ICPC is 861 F g^{-1} compared to their individuals. The as-prepared hybrid shows 73% of capacitance retention after 100 cycles, and up to 300 cycles, it maintains its initial capacitance (100%). The maximum capacitance and improved cycle stability are attained due to the synergism of PANI incorporation at ICPC [49].

A fast and straightforward electro-polymerization technique effectively organizes the aligned array PANI nano-fibers on carbon paper (PANI-NF/CP). The PANI-NF/CP was fabricated without using any polymer binders and conducting additives with a galvanostatic approach. The aligned array PANI nano-fibers are homogeneously coated at 60 mA cm^{-2} in 3 min on carbon paper's surface. The electrochemical results of PANI-NF/CP composites display the highest specific capacitance of 531 F g^{-1} at 3 mA cm^{-2} with 94% discharge/charge efficacy [50].

3.7.5 Miscellaneous Practices

It is renowned as a significant one-step, economical process toward the preparation of conductive polymer films. They are noticeably more even, smoother, prominently adhesive, and pinhole-free films. The chains arbitrarily diverge and smaller, with a higher-level cross-linking nature. The PANI/CNT composite was synthesized using a microwave plasma polymerization procedure. In this process, the aniline dispersed with MWCNTs and MWCNT-COOH was inserted into the plasma prompted through the microwave discharge that showed spherical, granular, massive spheres uniform distribution on the titanium substrate. The calculated Cs of PANI/MWCNT-COOH nanocomposite film was 112 F g^{-1} for the 1.5 wt% loading of MWCNT [51].

A low-temperature modified vapor-phase synthesis favors depositing polypyrrole (PPY) onto 3D fibrous substrates with a high aspect ratio, more conductive and capacitive 1D nanostructure. Nano-fibrillar structured PPY similar to nano-brushes deposited at low temperatures (50 °C) without rigid templates. This PPY deposited on hard carbon paper shows a high specific capacitance (144.7 F g^{-1}). Simultaneously, fabricated two-electrode supercapacitors reveal specific energy of about 13 Wh kg^{-1}. Besides high cycling stability, the devices can go through 200,000 cycles with 70% capacitance retention when charged to 0.6 V. The self-assembly of nano-fibrillar reduces the need for challenging and costly templates removal, denotes an advance in the outstanding formation of nanostructures from the vapor phase [52].

Furthermore, the structure-specific materials for special applications can be prepared either template-based or template-free methods. For example, 1D conducting polymers nanostructures such as nanowires, nanotubes, nanorods, nanoribbons, nanocones, etc., are usually organized through this template/template-free approach [53]. 1D nanostructure, well-ordered conducting polymer nanowire arrays are produced using the template and template-free methods. Martin and Parthasarathy explored this approach. It was used to create well-aligned arrays of micro and nano-fibers and tubes of conducting polymers, owing to its ease, adaptability, and controllability [54]. In this concern, aluminum oxide (AAO), particle track-etched membranes (PTM), porous di-block copolymer films are mainly used as templates. For example, PEDOT nanotubes and nanowires were electrochemically produced in porous alumina by applying different potentials through a potentiostatic approach. The nanotubes were made at 1.6 V in acetonitrile containing 20 mM EDOT, whereas the nanowires were prepared at 1.4 V in 100 mM EDOT. Symmetric supercapacitor based on PEDOT nanotube electrodes succeeds in an extraordinary power density (25 kW kg^{-1}), preserving 80% of energy density (5.6 Wh kg^{-1}). The hollow nanotubular assemblies permit counter-ions to enter into the polymer and contact their internal surfaces voluntarily. In contrast, the thin wall affords a short diffusion space to simplify ion conveyance [55].

Single-step, template-free method is used to create enormous arrays of vertically aligned polyaniline (PANI) nanowires on many conducting substrates through a Galvano-static approach. From the SEM study, the PANI layer was self-possessed of aligned nanowires (average of 50 nm diameter and 0.4 μm of length) and is perpendicular to the substrate when the sample was tilted by 30°. The PANI nanowire arrays possess a maximum specific capacitance of about 950 F g^{-1} at 1 A g^{-1}. The importance of vertically aligned nanowires helps ion diffusion from an electrolyte to the surface of PANI nanowires. PANI nanowire arrays' capacitance performance was explored in some electrolytes such as Perchloric acid ($HClO_4$), Lithium bis(trifluoromethanesulfonyl)imide (LiTFSI) aqueous solution, and non-solvent electrolyte EMITFSI ionic liquids [56].

Some of the essential synthetic approaches for conducting polymers and their hybrids/composites with their process, advantages and disadvantages were summarized in Table. 3.2. The above-said methods are most helpful in preparing diverse nanostructures/nanoarchitectures of conducting polymers.

However, every procedure has some limitations. For example, the chemical polymerization will give homogeneous mixing, powdery materials at a low cost instead of free-standing/flexible hybrid thin films. While electro-polymerization can offer hybrid thin films with enhanced properties like conductivity and constancy, it doesn't help prepare massive amounts. Also, electrochemical methods need specific electrode substrates. The best properties of the conducting polymers can be simply realized through both chemical and electro-polymerization practices. It has also been witnessed that the morphology of the hybrid performs an imperative character toward essential properties. Also, the morphology of the combination is consecutively governed by the synthetic approach [57].

Table 3.2 Synthetic strategies for the fabrication of nanostructured conducting polymer and its hybrid composites

Synthetic strategy	Description	Advantages	Disadvantages
Chemical polymerization	Polymerization of monomers through chemical reactions to form conducting polymers	• Ability to scale production • Precision in controlling polymer characteristics • Compatibility with a variety of monomers	• Severe reaction requirements • Restricted ability to manipulate nanostructure
Electrochemical deposition	Deposition of conducting polymers onto substrates via electrochemical methods	• Accurate regulation of film shape and thickness • Ability to work with diverse substrate materials	• Necessitates specialized machinery • Constrained by specific geometric configurations
Template-assisted synthesis	Fabrication of nanostructured conducting polymers using templates to control size, shape, and orientation	• Exact management of nanostructure • High surface area and porosity	• Challenges in removing templates • Restricted scalability
In-situ polymerization on nanostructured substrates	Polymerization directly onto nanostructured substrates, promoting enhanced interfacial interactions	• Synthesis directly onto substrates • Improved interfacial interactions • Customized functionalities	• Dependent on compatible substrates • Limited to certain polymer-substrate combinations
Hybridization with nanomaterials	Incorporation of nanomaterials (e.g., carbon nanotubes, graphene) into conducting polymer matrices	• Combined properties lead to synergy • Increased conductivity, mechanical strength, and stability	• Difficulties in dispersing nanomaterials • Risk of aggregation • Concerns regarding compatibility
Layer-by-layer assembly	Alternating deposition of conducting polymers and other materials layer-by-layer to form nanostructured composites	• Accurate adjustment of film thickness and composition • Properties can be tailored and adjusted	• Time-consuming • Limited to certain materials compatibility

3.8 Strategies to Fabricate Advanced Functional Nanohybrids/Composites

Recently, conducting polymers based nanocomposites have been established with size, shape, aspect ratio, dispersion, alignment within the matrix, controlled morphology, and tunable electrical properties [58]. Their superior properties and performances fascinated the energy storage field, and that trust the quality and morphology of individual components and significantly be contingent on the fabrication approach. The insertion of various nanomaterials tends to revise the electrical and electronic properties of hybrid conducting polymers significantly. So the composite electrodes with conducting polymers have a progressive impression on supercapacitor electrodes' performance compared to individuals [19]. Further, it is best to motivate present supercapacitors toward forthcoming innovative energy storage requests due to their significant redox-active and integral elastic/polymeric nature. Besides, structural control for conducting polymer composites has been reflected as a proficient methodology for expanding the cycling stability of the conducting polymers. The development of nanohybrids of conducting polymers is more critical when the conducting polymers are integrated with distinct categories of materials like metals, carbonaceous materials, and inorganic compounds. The strong interactions like chemical bonding among the two phases and the weak interactions (hydrogen bonding, van der Waals, and electrostatic interaction, are the most responsible for this kind of hybrid formation [57].

3.8.1 Conducting Polymers with Metals (Doping)

The construction of conducting polymer-based nano-hybrids with metal nanoparticles is getting advantages by combining organic and inorganic components that have the individuals' collective properties. This hybrid formation offers an intelligent method to strengthen the electronic interaction between the two members through morphological alteration and structural renovations. The outcome of this increases the electrochemical properties compared to pure conducting polymers. Designing diverse morphologies/architecture involves dissimilar initial resources and fabrication methods. While, the conducting polymer/metal nano-hybrids with several designs such as 3D network, core-shell, and multi-layered heterostructure have been effectively established through simple techniques like oxidative chemical polymerization, electrochemical deposition, and electrospinning practices. Further, in-situ oxidative polymerization is often used to blend metal particles with monomers with different molar ratios. The electrochemical approach encompasses the direct deposition of metal nanoparticles, whereas the depositing of conducting polymers can be done simultaneously.

Besides, the doping of conducting polymers with metal ions has been extensively explored by several works. It is well believed that doping increases the charge delocalization in the polymer backbone, and hence the conductivity and capacitance also increase. Many metals like Au, Ag, Ni, Co, Mn, etc., are in use. For example, the core-shell of nano-Au@PANI nanocomposite was prepared through in-situ polymerization on the nano-gold surface. The average diameter of nano-Au@PANI is 22.17 nm, whereas nano-Au is 18.52. The highly conductive core part of nano-Au is combined with the nano-PANI layer's thin shell, enhancing the specific capacitance of the hybrid material (285 F g^{-1}) compared to the pure PANI (115 F g^{-1}). An asymmetric supercapacitor device is assembled, and that delivers a high energy density (7.01 Wh kg^{-1}) and power density (4.2 kW kg^{-1}) [59].

The specific capacitance of PANI (285 F g^{-1}) was increased easily through doping of Ag (0.9 wt%) in PANI (Ag/PANI) (512 F g^{-1}) by dip-coating method. The Ag/PANI exposed a nano-fiber-like morphology for PANI and the nano-fibers with Ag particles' bright spots for the Ag/PANI films. The change in the morphology and capacitance was due to the synergism between PANI and silver nanoparticles. Ag nanoparticles' existence on PANI nano-fibers delivers a minimum resistance pathway to electrons. It increases the electron movement concerning the current collectors and the active materials that boost the current density and, therefore, the specific capacitance [60].

A remarkably transparent, stretchable, and flexible supercapacitor was constructed by evolving Ag/Au/Polypyrrole core-shell nanowire network as an electrode through a thin layer of gold coating surface of Ag nanowires. A solution process synthesizes the Ag/Au core-shell nanowires, and the polypyrrole is deposited on that through the electrochemical polymerization process. The Ag/Au/Ppy revealed higher mechanical strength in several bending and stretching conditions. Additionally, the anticipated supercapacitor exhibited sufficient photosensitive conduction composed with five-fold better areal capacitance associated with pristine Ag/Au core-shell nanowire mesh-based supercapacitor [61].

The doping of nickel into PANI is getting more advantages toward the supercapacitor applications. Many methodologies are used to synthesis Ni^{2+} doped PANI and analyze their electrochemical performances. In this regard, Ni^{2+} doped PANI (PANI/Ni^{2+}) was produced through cyclic voltammetry on stainless steel (SS) substrates with nickel sulfate ($NiSO_4$). Nanorods with a coarse surface and porous structured (PANI/Ni^{2+}) film offer a maximum specific capacitance of 658.3 F g^{-1} and lower resistance [62]. Likewise, Ni^{2+}/PANI was synthesized over simple oxidative chemical polymerization using $NiCl_2$ in the presence of ultrasonic irradiations. The ultrasonic waves encourage the nano-fibrillar formation and avoid agglomerations, whereas the absence of ultrasound displays an irregular granular-like morphology. The better specific capacitance of 370 F g^{-1} is achieved by supporting ultrasonic irradiation, which is 1.3 times greater than the usual stirring process. The unified nano-fibrillar arrangement boosts the electron transferral rate with ion diffusion [63].

3.8.2 Conducting Polymers with Carbonaceous Materials

The carbon materials usually use the electrode materials for EDLC due to the large surface area and conductivity. Moreover, the easy accessibility, higher stability in acidic and basic media, and eco-friendly are the extra unique compensations. The combination of conducting polymers with carbon through tailored nanostructures will always be responsible for superior capacitive properties. Various carbon morphologies have been obtained, including spheres, hollow spheres, core-shells, onion-like carbons, nanotubes, and graphene were utilized as adequate supports. Besides, the carbons' key features are possessed different sizes, shapes, and structures that modify the catalytic activity through the synergism among the conducting polymers and carbon supports [64]. Usually, the nanostructured carbon materials retain an excellent surface area and enormous permeability that can offer more embellishment provisions to endorse the nanostructured conducting polymers. Consequently, the utilization of different nanostructured carbon materials (one-dimensional (1D), two-dimensional (2D), and three-dimensional (3D)) may boost up the electrochemical performance not only by way of providing EDLC but also by providing a backbone for the long life during its electrochemical reactions [21].

3.8.3 Conducting Polymers with One-Dimensional Carbon

The hybrids 1D carbon with conducting polymers are considered due to many benefits like porosity, better conductivity, excellent chemical strength, and extensive working temperature. Carbon nanofibers (CNFs), carbon nanotubes (CNTs), and graphene nanoribbons (GNRs) have been widely considered as 1D carbon for the accommodation of conducting polymers. The electrochemical enhancements are observed via a synergistic effect resulting from the electrostatic attractions (ions) in the EDL and surplus faradaic reactions commencing the conducting polymers. Usually, two significant factors, such as the CNTs dispersion and the interfacial contact concerning CNTs and conducting polymers, affect the performance of the hybrid materials. Several studies focused on refining the distribution of MWCNTs in the polymer over the surface chemical alteration of CNTs, realized over the oxidation of MWCNTs. 3D architecture PANI/MWCNT nanocomposite films were arranged using Xanthosoma sagittifolium leaves as a template via the nano casting method. The biomimetic surface morphology containing multiscale papilla-like (nanoscale texture) PANI/MWCNT nanocomposite demonstrates a specific capacitance of 535 F g^{-1} at specific MWCNT loading (5 wt%). The nanocomposite's specific capacitance slightly declined up to 90%, only after 1000 cycles [65].

Similarly, Triblock copolymer poly(ethylene glycol)-block-poly(propylene glycol)-block-poly(ethylene glycol) (P-123) assisted PANI embraced CNTs (PANI@CNT) nano-hybrid with outcropping whiskers at the outer surface were synthesized through soft template chemical polymerization method. It has shown a specific capacitance of 348.9 F g^{-1} and better cycling stability (78.6%) over 1000 cycles. The improved electrochemical activity because of porous morphology and outcroppings whiskers which may facilitate the ion movement. PANI's unique structure with one-dimensional CNTs deals a maximum contact concerning the electrode/electrolyte interfaces. CNTs may be turned into a hybrid backbone for easy electron relocation, and the porous PANI executes the capacitance boosting [66].

A new synthetic route was recognized concerning polyaniline (PANI) grafted carbon nano-fiber (CNF) composite (PANI-g-TCNF) contrasting to conventional ways through van der Waals forces. In this work, initially, toluene diisocyanate was functionalized to carboxylated CNF through amidation and then reacted with surplus aniline to form urea derivative and residual aniline polymerized and grafted with urea derivative. The composite, which contains TCNF (6 wt%), shows 557 F g^{-1} of specific capacitance with 86% retention after 2000 continuous cycles, ensuring rapid ion diffusion and fast charge transfer via chemical grafting synergistic influence. The performance development is due to the phenyl substituted amide connection's significance in expanding the π-conjugated assembly [67].

In connection to the 1D graphene nanoribbons (GNRs), a simplistic and low-cost method is used to manufacture well-arranged, vertically ranged PANI nanorods on GNRs direct polymerization. The resultant two-electrode supercapacitor has an excellent specific capacitance (340 F g^{-1}) with a better cycling performance of 90% retention over 4200 cycles. It is due to the synergistic arrangement of electrically conductive GNRs and extremely capacitive PANI [68].

3.8.4 Conducting Polymers with Two-Dimensional Carbon

Graphite nanoflakes, graphene oxide (GO), and reduced graphene oxide (rGO) are the main components of 2D carbon materials. They are receiving more interest in supercapacitors because of their theoretical and practical benefits, such as outstanding electrical conductivity, excellent thermal stability, large surface area, wrinkled structures, and large inter-layer spaces for the electrolyte's stress-free movement. These features create 2D carbon materials as a fascinating substrate for the preparation of combinations with conducting polymers.

Graphene oxide/polyaniline (GO/PANI) nanocomposites were synthesized on flexible substrates via single-step interfacial electro-polymerization. During this process, PANI nanorods were deposited on the GO's surface and designed as a 3D nanostructure at the water/chloroform interface. A well adhesive and homogeneously dark-green film confirm the formation of GO/PANI nanocomposites. The prepared PANI nanorods existed in an excessively doped state and disclosed a sturdy π-π electronic contact with GO, which improved conjugation. Besides, the flexible all-solid-state supercapacitor constructed by this composite shows maximum specific capacitance (1095 F g^{-1} at 1 A g^{-1}) with cycling performance (91.1% of initial capacitance next 1000 cycles), along with high energy density (24.3 Wh kg^{-1})

and the maximum power density (28.1 kW kg^{-1}). The favorable electrochemical properties are observed because of GO and PANI's synergistic influences [69].

Apart from the distinct synthetic ways, the three different conducting polymers, PEDOT, PANI, and PPy, were coated on reduced graphene oxide (RGO) sheets through a simple in-situ polymerization. The usage of ethanol played the leading part in realizing a uniform covering of the conducting polymers on RGO sheets. RGO sheets' surface is coated with many bumps ascribed to the PEDOT coating, and the RGO-PANI reveals a worm-like structure. The RGO-PANI composite demonstrates a high specific capacitance of 361 F g^{-1} compared to RGO-PPy (248 F g^{-1}) and RGO-PEDOT (108 F g^{-1}) at 0.3 A g^{-1}. This contribution to the specific capacitance is due to the synergism [70].

3.8.5 Conducting Polymers with Three-Dimensional Carbon

3D architectured carbon-based materials deliver enormous ion-accessible surface areas, rich pores, and interconnected architecture compared to 1D and 2D nanostructures. These are gradually promoted to the deposition of conducting polymers. Their extraordinary surface area can efficiently lessen the current density and polarization in bulk electrodes, increasing the electrodes' charge storing kinetics. Carbon aerogels, fibers, cloths, and other porous carbons are always used to insert conducting polymers in its skeleton. Graphene aerogels are receiving importance toward various energy storage uses due to tailoring pores' structure and composition in the walls. The formation of a highly porous structure in GO sheets' leads to self-assembly and interconnected skeletons (pore walls) through the freeze-casting procedure. In this regard, free-standing porous graphene beads holding radially oriented over networks from the sphere's center to the bead surface were fabricated. The rGO beads were filled with pseudo polymer (PANI) to make them protected, available composite beads with an exclusive pore morphology through the electrodeposition technique. The polymerization is done on both sides, the pore walls and the junctions among adjacent pores ensuing in identical polymer-graphene-polymer sandwiched structures all over the bead. The composite beads exhibited high specific capacitance (669 F g^{-1}) and outstanding cycling strength (72.6% after 3000 cycles) signifying the future use as porous electrodes for energy storage devices [71].

A well-arranged polyaniline nanocone arrays were produced on a 3D graphene linkage by template-free electrodeposition procedure. In this regard, the 3D graphene was organized by chemical vapor deposition (CVD), and vertical aligned PANI nanocone arrays were electrodeposited. As per the morphological analysis, the PANI nanocones are well separated from each other. Each nanocone's size has 50 nm, and the length increases when the deposition time increases (137, 207, 303, and 355 nm for GPN1-4). Interestingly, the specific capacitance of GPN1-4 is 751.3, 561.7, 536.6, and 472.6 F g^{-1}, respectively, at 1 A g^{-1}. The inversional relationship of specific capacitance with the nanocones' size shows that the lengthier nanocones obstruct PANI's electron conduction. Therefore, the nanocone's outer part becomes less active than the inner [72].

3.8.6 Conducting Polymers with Inorganic Materials

Generally, the composites of conducting polymers with a mixture of inorganic material like metal oxides, hydroxides, and sulfides can buffer the volume transformation throughout the continual redox cycles and simultaneously increase the mechanical strength and specific capacitance of the conducting polymers. The conducting polymers in these composites can inhibit the aggregation and restack of inorganic materials over space steric interference and electrostatic influence. This phenomenon permits the inorganic particles for homogeneous dispersion into the conducting polymer matrix.

The transition metal oxides (TMO) have been considered to construct the batteries and pseudo capacitor materials for several years, attributable to their exceptional strength and outstanding charge storage features. The arrangement of conducting polymers and metal oxides would afford some advantages, particularly in supercapacitors such as a flexible buffering layer with increasing electrodes' conductivity, lifetime and improving rate performance [73]. Several transition metal oxides like ruthenium (RuO_2), manganese (MnO_2), vanadium (V_2O_5), cobalt (CoO_x), and iron (Fe_3O_4) have been discovered as stabilizer electroactive materials for conducting polymers due to their diverse transition oxidation states.

For example, a redox exchange made heterogeneous RuO_2-conducting polymer nanowires were prepared by merely soaking the PPY and PEDOT nanowire arrays in potassium perruthenate ($KRuO_4$) solution. PPY and PEDOT nanowires were produced potentiostatically using anodic aluminum oxide (AAO) as a template. Each polymer nanowire array was individually exposed to KRuO4 for about 6 min to develop the heterogeneous RuO_2 decorated conducting polymer (PEDOT or PPY nanowires) nanocomposites. An extraordinary specific capacitance of 371 F g^{-1} for PEDOT-based and 500 F g^{-1} PPY-based composite nanowires. The RuO_2 nanoparticles were distributed all over the PEDOT nanowires via redox response among the perruthenate ion and the sulfur atom present in the PEDOT thiophene assembly. Later it was oxidized to sulfone. Simultaneously, the RuO_2 thin layer was coated on the exterior part of the PPY nanowires by reaction with PPY (PPY was over-oxidized) because of the variations in super capacitance values between them composites [73].

Metal sulfides are exciting materials with unique physicochemical possessions, for example, electrical conductivity, thermal and mechanical constancy. The hybrids of conducting polymer with metal sulfides will provide rich redox chemistry, which is most useful in supercapacitor applications. For example, a "pizza-like" MoS_2/PPy/PANI ternary nanostructure was synthesized through two consecutive polymerizations of PPy and PANI on a few-layer MoS_2 (f-MoS_2) templates. Remarkably, such a ternary nanocomposite's rational strategy facilitates significantly enhanced specific capacitance up to 1273 F g-1 and 83% retention after 3000 cycles. The outstanding capacitance and its retention for the MoS_2/PPy/PANI electrode recognized to the availability of MoS_2 nanosheets offers a 2D nanostructure for the proficient holding of two different conducting polymers. The exclusive "pizza-like" PPy-PANI nanostructures on the MoS_2 thus suspend the wrinkling/accumulation of conducting polymers during cycles [74, 75].

3.9 Conducting Polymers-Based Ternary Nanocomposites

Recently, conducting polymers based ternary nanocomposites (a combination of three components including conducting polymer) with inorganic materials (metal, metal oxides, etc.) and carbon nanomaterials have been fascinated due to their admirable performance over binary arrangements. In ternary architecture design, numerous synergistic properties seemed in their claim as the electrode material for supercapacitors. A new stretchable, flexible, linear supercapacitor was constructed based on polyaniline@Au@carbon nanotube (PANI@Au@CNT) with extreme axial electron conveyance. The ternary composite was synthesized through Au nanograin's decoration on aligned MWCNT sheets, subsequently incorporating polyaniline (PANI). The device was invented by twisting two PANI@Au@CNT yarns that display an overall volumetric capacitance of ~ 6 F cm^3. The enhanced electrochemical activity is due to the linear electrodes' fast radial ion dissemination and axial electron passage. The stretchable supercapacitors expose a significant volumetric capacitance of about 0.2 F cm^3 at a scan rate of 1 V s^1, even at a strain of 400% [76].

Likewise, a coaxial ternary nanohybrid was prepared with amorphous $Ni(OH)_2$ deposited on MWCNTs wrapped with PEDOT: PSS (poly (3,4-ethylene dioxythiophene)-poly(styrene sulfonate)). Coordinative etching and the precipitating process were mainly used to synthesize amorphous $Ni(OH)_2$ thin film. The ensuing ternary (MWCNT/amorNi$(OH)_2$ /PEDOT: PSS) composite shows a too high specific capacitance (3262 F g^{-1} at 5 mV s^{-1}) and outstanding rate ability (71.9%). Further, asymmetric supercapacitors were fabricated using positive and negative electrodes of MWCNT/amorNi$(OH)_2$ /PEDOT: PSS and rGO/CNT hybrid. It provides an excellent energy density (58.5 Wh kg^{-1}) and power density (780 W kg^{-1}) along with long cycle life (86% retention holding after 30,000 cycles). Further, two series-connected supercapacitors can light up two LED lights simultaneously and drive a motor that validates the unique possibilities for real-world use [77].

The formation of hybrids/composite can be done by one of the above-said combinations. Compared with their synthetic approaches, nanostructured conducting polymers and their hybrid/nanocomposites are utilized in supercapacitors, nanostructure formation, and proper applications (Table 3.3). However, research has been directed on the production and use of nanostructured conducting polymers. Innovative theoretical and experimental investigations are pretty required to support investigators to attain a more in-depth understanding. The necessary knowledge regarding the unique properties of nanostructures on charge transfer and surface chemistry. Advanced in-situ and ex-situ practices are too essential to revise the fundamental electrochemistry of conducting polymers and electrochemical dynamics of interfaces. Integrative research through chemistry, physics, nanotechnology, and materials science engineering will meet electrifying future progresses.

3.10 Conclusion

To date, a massive number of methodologies have been established to form nanostructured conducting polymers and their hybrids/composites due to advanced applications in supercapacitors. This chapter demonstrates the critical utilization of various materials to construct conducting polymer-based combinations with diverse nanoarchitectures concerning the supercapacitors applications. Different types of nanohybrid electrodes have been developed to harvest the maximum electrochemical performance from the component materials.

Recently, there has been growing global demand to develop nanostructured materials for various applications, particularly energy fields. Making a new nanoarchitecture with a large internal surface area to facilitate electrolyte ion penetration cannot solve the present problem. Nevertheless, low-rate capability, less electrical conductivity, and short cycling stability are still significant problems, improving through diverse nanoarchitectured materials. The integration of materials and mechanisms will help achieve better performance, which is possible through individual components. The question arises of how the materials and their functions should be promoted through their architecture. Efforts are needed to improve the nanoarchitectured materials' ability through interdisciplinary energy conversion and storage applications. The targeted strategy to design and synthesize tailored conducting polymer-based nanohybrids has the budding to cultivate the next-generation materials with more significant action to commercialize energy storage devices.

Based on the above scrutiny, one can do future research in the following directions:

1. The interconnected porous assembly should be intended to boost electron transport and avoid the deterioration of conducting polymers through supreme structural intention.
2. The preparation of conducting polymer nanohybrids with better dispersion through acceptable investigational practices avoids stacking, thus refining electrochemical performance and cycle stability.
3. Micro/nano-structural control of the conducting polymer hybrid fabrication through different tactics like anchoring, wrapping, and sandwiches of either conducting polymer or its other substrates and vice versa.
4. Unraveling the problem of low strength (shrinkage and expansion) of conducting polymers and improving the stability during consecutive cycles and at high current densities in supercapacitors.
5. Reliable, extensive fabrication system needs to be progressive in the future to promise the dependability of the nanocomposite properties.

Table 3.3 Various nanostructured conducting polymers and their hybrid/nanocomposites are utilized in supercapacitors

Nanostructure	Material/hybrid	Synthetic method	Configuration	Specific capacitance	Capacitance retention	Energy density at a power density	Reference
Pure conducting polymers							
Cross-linked nanorods	PANI	Chemical oxidative copolymerization	Three electrodes	297 F g^{-1} at 1 A g^{-1}	65% (1300 cycles)	–	[39]
Hydrogels	PANI- sodium alginate	In-situ polymerization	Three electrodes	252 F g^{-1} at 0.5 A g^{-1}	71% (1000 cycles)	–	[78]
Nanowire	PEDOT	The micellar soft-template method subsequently vacuum-assisted filtration	Three electrodes	667.5 mF cm^{-2} at 1 mA cm^{-2}	94.3% (10,000 cycles)	–	[79]
			Symmetric-flexible solid-state supercapacitor	413.5 mF cm^{-2} at 1 mA cm^{-2}	–	48.3 mWh cm^{-2} at 0.22 mW cm^{-2}	
Superfine shuttle-shaped nano flowers	PANI	Reverse micelle emulsion polymerization	Three electrodes	622 F g^{-1} at 0.5 A g^{-1}	76% (1000 cycles)	–	[45]
Conducting polymers with metals							
Nanofiber	PANI-Ni	Electrochemical polymerization with hydrothermal	Three electrodes	622 F g^{-1} at 0.5 A g^{-1}	93% (1000 cycles)	–	[80]
Linked virgulate particles	Zn^{2+} and H^{+} doped PANI	Chemical polymerization -stirring	Three electrodes	369 F g^{-1} at 0.1 A g^{-1}	90% (1000 cycles)	–	[81]
Nanoparticles	Transparent PEDOT: PSS (4 layers)/ag grids	Inject printed	Translucent flexible solid-state supercapacitor	4.72 mF cm^{-2}	–	0.59 mWh cm^{-2} at 0.036 W cm^{-2}	[82]

(continued)

Table 3.3 (continued)

Nanostructure	Material/hybrid	Synthetic method	Configuration	Specific capacitance	Capacitance retention	Energy density at a power density	Reference
Coral-like, nanorods	$PANI/Co^{2+}$	Electrochemical deposition	Three electrodes	736 F g^{-1} at 3 mA cm^{-2}	58% (250 cycles)	–	[83]
Conducting polymers with carbons							
Smooth fibers	PEDOT / cellulose paper	Vapor phase polymerization	Flexible solid-state supercapacitors	13.7 Fcm^{-3}	–	0.76 mWh cm^{-3} at 0.01 W cm^{-3}	[84]
Nanowire Array	fCC-PANI array-rGO	Chemical polymerization	Three electrodes	471 mF cm^{-2} at 0.5 mA cm^{-2}	75.5% (10,000 cycles)	–	[85]
			Flexible solid-state supercapacitors	197 mF cm^{-2} at 0.1 mA cm^{-2}	91.3% (7000 cycles)	0.22 mWh cm^{-3} at 0.50 mW cm^{-3}	
Aerogel	3D rGO/PEDOT: PSS	Hydrothermal followed by acid treatment	Three electrodes	471 F g^{-1} at 0.2 A g^{-1}	98.71% (20,000 cycles)	–	[86]
			Symmetric supercapacitor	–	–	48.18 Wh kg^{-1} and 1.4 kW kg^{-1}	
3D hybrid nanoarchitecture	Graphene/ particulate PPy (GPPy)	Physical synthesis route	Flexible solid-state supercapacitor	–	–	8.4 Wh kg^{-1} at 3.2 kW kg^{-1}	[87]
Conducting polymers with inorganics							
Well-aligned nanowire	CoO@PPY	CoO grown on 3D nickel foam with PPy	Three electrodes	2223 F g^{-1} at 1 mA cm^{-1}	99.8% (2000 cycles)	–	[88]
			Asymmetric supercapacitors	–	91.5% (20,000 cycles)	43.5 Wh kg^{-1} at 87.5 W kg^{-1}	
3-D urchin shaped coaxial	MnO_2@PANI	Hydrothermal route followed by oxidation	symmetric all-solid-state flexible	129.2 F g^{-1} at 0.5 A g^{-1}	–	17.9 Wh kg^{-1} at 261 W kg^{-1}	[89]

(continued)

Table 3.3 (continued)

Nanostructure	Material/hybrid	Synthetic method	Configuration	Specific capacitance	Capacitance retention	Energy density at a power density	Reference
			solid-state asymmetric supercapacitor	–	89% (5000 cycles)	37 Wh kg^{-1} at 386 W kg^{-1}	
3-D tubular nanowire array	MoS_2/PANI	In situ oxidative polymerization	Three electrodes	552 F g^{-1} at 0.5 A g^{-1}	88% (1000 cycles)	–	[90]
			Symmetric supercapacitor	124 F g^{-1} at 1 A g^{-1}	79% (6000 cycles)	–	
Conducting polymers-based ternary							
Hierarchical core-shell nanotube arrays	$NiCo_2S_4$@$Ni(OH)_2$@PPy	Hydrothermal method accompanies the ion-exchange process	Three electrodes	9.1125 F cm^{-2} at 5 mA cm^{-2}	–	–	[91]
			Asymmetric supercapacitor	–	98.87% (30,000 cycles)	34.67 Wh kg^{-1} at 120.127 W kg^{-1}	
Thin film/ nanoparticles	CNT-Fe_3O_4-PANI	Hydrothermal and electrodeposition	Flexible solid-state supercapacitor	201 F g^{-1}	96.4% (10,000 cycles)	28 Wh kg^{-1} at 5.3 kW kg^{-1}	[92]
Nanofibers, vertically aligned	CNF/CNT/PANI	In situ chemical polymerization	Three electrodes	315 F g^{-1} at 1 A g^{-1}	–	–	[93]
			Flexible solid-state supercapacitor	–	92% (10,000 cycles)	5.1 Wh kg^{-1} at 10.1 kW kg^{-1}	
Coaxial hierarchical core-shell, nanowire	Ag@$Ni(OH)_2$@PPy	Chemical bath deposition (CBD)	Three electrodes	3103.5 F g^{-1} at 2.6 A g^{-1}	92.2% (20,000 cycles)	–	[94]
			Solid-state asymmetric supercapacitor	–	97.4% (10,000 cycles)	121 Wh kg^{-1} at 613 W kg^{-1}	

Thus, the challenge is still open to emerging nanoarchitectured materials with maximum efficiency and stability and proper mechanistic insights to successfully commercialize sustainable clean energy techniques.

References

1. H. Naarmann, The development of electrically conducting polymers. Adv. Mater. **2**, 345–348 (1990). https://doi.org/10.1002/adma.19900020802
2. H. Shirakawa, E.J. Louis, A.G. MacDiarmid, C.K. Chiang, A.J. Heeger, Synthesis of electrically conducting organic polymers: halogen derivatives of polyacetylene, (CH) x. J. Chem. Soc. Chem. Commun., 578 (1977). https://doi.org/10.1039/c39770000578
3. H. Shirakawa, The discovery of Polyacetylene film: The dawning of an era of conducting polymers (Nobel lecture). Angew. Chem. Int. Ed. **40**, 2574–2580 (2001). https://doi.org/10.1002/1521-3773(20010716)40:14<2574::AID-ANIE2574>3.0.CO;2-N
4. A.J. Heeger, Semiconducting and metallic polymers: The fourth generation of polymeric materials. J. Phys. Chem. B **105**, 8475–8491 (2001). https://doi.org/10.1021/jp011611w
5. D. Nguyen, H. Yoon, Recent advances in nanostructured conducting polymers: from synthesis to practical applications. Polym. (Basel) . **8**, 118 (2016). https://doi.org/10.3390/polym8040118
6. A.M. Bryan, L.M. Santino, Y. Lu, S. Acharya, J.M. D'Arcy, Conducting polymers for Pseudocapacitive energy storage. Chem. Mater. **28**, 5989–5998 (2016). https://doi.org/10.1021/acs.chemmater.6b01762
7. M.E. Abdelhamid, A.P. O'Mullane, G.A. Snook, Storing energy in plastics: A review on conducting polymers & their role in electrochemical energy storage. RSC Adv. **5**, 11611–11626 (2015). https://doi.org/10.1039/c4ra15947k
8. P.J. Nigrey, A.G. MacDiarmid, A.J. Heeger, Electrochemistry of polyacetylene, (CH) x : electrochemical doping of (CH) x films to the metallic state. J. Chem. Soc. Chem. Commun., 594 (1979). https://doi.org/10.1039/c39790000594
9. C.C. Han, R.L. Elsenbaumer, Protonic acids: Generally applicable dopants for conducting polymers. Synth. Met. **30**, 123–131 (1989). https://doi.org/10.1016/0379-6779(89)90648-6
10. A.G. MacDiarmid, "Synthetic metals": A novel role for organic polymers (Nobel lecture). Angew. Chem. Int. Ed. **40**, 2581–2590 (2001). https://doi.org/10.1002/1521-3773(20010716)40:14<2581::AID-ANIE2581>3.0.CO;2-2
11. J. Yang, Y. Liu, S. Liu, L. Li, C. Zhang, T. Liu, Conducting polymer composites: Material synthesis and applications in electrochemical capacitive energy storage. Mater. Chem. Front. **1**, 251–268 (2017). https://doi.org/10.1039/c6qm00150e
12. H. Naarmann, Polymers, Electrically conducting, in *Ullmann's Encycl. Ind. Chem.*, (Wiley-VCH Verlag GmbH & Co. KGaA, Weinheim, Germany, 2000), pp. 295–311. https://doi.org/10.1002/14356007.a21_429
13. A.K. Bakhshi, G. Bhalla, Electrically conducting polymers: Materials of the twentyfirst century. J. Sci. Ind. Res. **63**, 715–728 (2004)
14. A. Diaz, Electrochemistry of conducting polypyrrole films. J. Electroanal. Chem. **129**, 115–132 (1981). https://doi.org/10.1016/0368-1874(81)87387-X
15. T.-H. Le, Y. Kim, H. Yoon, Electrical and electrochemical properties of conducting polymers. Polym. (Basel) . **9**, 150 (2017). https://doi.org/10.3390/polym9040150
16. G.a. Snook, P. Kao, A.S. Best, Conducting-polymer-based supercapacitor devices and electrodes. J. Power Sources **196**, 1–12 (2011). https://doi.org/10.1016/j.jpowsour.2010.06.084
17. K. Lee, S. Cho, S. Heum Park, A.J. Heeger, C.-W. Lee, S.-H. Lee, Metallic transport in polyaniline. Nature **441**, 65–68 (2006). https://doi.org/10.1038/nature04705
18. J. Liu, J. Wang, C. Xu, H. Jiang, C. Li, L. Zhang, J. Lin, Z.X. Shen, Advanced energy storage devices: basic principles, analytical methods, and rational materials design. Adv. Sci. **5** (2018). https://doi.org/10.1002/advs.201700322

19. Y. Shi, L. Peng, Y. Ding, Y. Zhao, G. Yu, Nanostructured conductive polymers for advanced energy storage. Chem. Soc. Rev. **44**, 6684–6696 (2015). https://doi.org/10.1039/c5cs00362h
20. M. Winter, R.J. Brodd, What are batteries, fuel cells, and supercapacitors? Chem. Rev. **104**, 4245–4270 (2004). https://doi.org/10.1021/cr020730k
21. X. Zhang, X. Cheng, Q. Zhang, Nanostructured energy materials for electrochemical energy conversion and storage: A review. J. Energy Chem. **25**, 967–984 (2016). https://doi.org/10.1016/j.jechem.2016.11.003
22. W. Raza, F. Ali, N. Raza, Y. Luo, K.-H. Kim, J. Yang, S. Kumar, A. Mehmood, E.E. Kwon, Recent advancements in supercapacitor technology. Nano Energy **52**, 441–473 (2018). https://doi.org/10.1016/j.nanoen.2018.08.013
23. S.P.S. Badwal, S.S. Giddey, C. Munnings, A.I. Bhatt, A.F. Hollenkamp, Emerging electrochemical energy conversion and storage technologies. Front. Chem. **2** (2014). https://doi.org/10.3389/fchem.2014.00079
24. I. Hadjipaschalis, A. Poullikkas, V. Efthimiou, Overview of current and future energy storage technologies for electric power applications. Renew. Sust. Energ. Rev. **13**, 1513–1522 (2009). https://doi.org/10.1016/j.rser.2008.09.028
25. P. Simon, Y. Gogotsi, Materials for electrochemical capacitors. Nat. Mater. **7**, 845–854 (2008). https://doi.org/10.1038/nmat2297
26. L. Guan, L. Yu, G.Z. Chen, Capacitive and non-capacitive faradaic charge storage. Electrochim. Acta **206**, 464–478 (2016). https://doi.org/10.1016/j.electacta.2016.01.213
27. Y. Zhang, H. Feng, X. Wu, L. Wang, A. Zhang, T. Xia, H. Dong, X. Li, L. Zhang, Progress of electrochemical capacitor electrode materials: A review. Int. J. Hydrog. Energy **34**, 4889–4899 (2009). https://doi.org/10.1016/j.ijhydene.2009.04.005
28. K.D. Verma, P. Sinha, S. Banerjee, K.K. Kar, Characteristics of electrode materials for supercapacitors, in *Handb. Nanocomposite Supercapacitor Mater*, ed. by K.K. Kar, (I Charact., Springer, 2020), pp. 269–285. https://doi.org/10.1007/978-3-030-43009-2_9
29. S. Arulmani, S. Anandan, M. Ashokkumar, Introduction to Advanced Nanomaterials, in *Nanomater. Green Energy*, (Elsevier, 2018), pp. 1–53. https://doi.org/10.1016/B978-0-12-813731-4.00001-1
30. T. Sheng, Y.-F. Xu, Y.-X. Jiang, L. Huang, N. Tian, Z.-Y. Zhou, I. Broadwell, S.-G. Sun, Structure design and performance tuning of nanomaterials for electrochemical energy conversion and storage. Acc. Chem. Res. **49**, 2569–2577 (2016). https://doi.org/10.1021/acs.accounts.6b00485
31. Q. Zhang, E. Uchaker, S.L. Candelaria, G. Cao, Nanomaterials for energy conversion and storage. Chem. Soc. Rev. **42**, 3127–3171 (2013). https://doi.org/10.1039/c3cs00009e
32. J. Hu, T.W. Odom, C.M. Lieber, Chemistry and physics in one dimension: Synthesis and properties of nanowires and nanotubes. Acc. Chem. Res. **32**, 435–445 (1999). https://doi.org/10.1021/ar9700365
33. R. Liu, J. Duay, S.B. Lee, Heterogeneous nanostructured electrode materials for electrochemical energy storage. Chem. Commun. (Camb.) **47**, 1384–1404 (2011). https://doi.org/10.1039/c0cc03158e
34. P. Balaya, Size effects and nanostructured materials for energy applications. Energy Environ. Sci. **1**, 645–654 (2008). https://doi.org/10.1039/b809078p
35. H. Niroumand, M.F.M. Zain, The role of nanomaterials in Nanoarchitecture, Procedia - Soc. Behav. Sci. **89**, 27–30 (2013). https://doi.org/10.1016/j.sbspro.2013.08.804
36. A.L.M. Reddy, S.R. Gowda, M.M. Shaijumon, P.M. Ajayan, Hybrid nanostructures for energy storage applications. Adv. Mater. **24**, 5045–5064 (2012). https://doi.org/10.1002/adma.201104502
37. C.I. Awuzie, Conducting polymers. Mater Today Proc **4**, 5721–5726 (2017). https://doi.org/10.1016/j.matpr.2017.06.036
38. R.E. Myers, Chemical oxidative polymerization as a synthetic route to electrically conducting polypyrroles. J. Electron. Mater. **15**, 61–69 (1986). https://doi.org/10.1007/BF02649904
39. X. Wang, J. Deng, X. Duan, D. Liu, J. Guo, P. Liu, Crosslinked polyaniline nanorods with improved electrochemical performance as electrode material for supercapacitors. J Mater Chem A **2**, 12323–12329 (2014). https://doi.org/10.1039/C4TA02231A

40. N.Y. Abu-Thabit, Chemical oxidative polymerization of polyaniline: A practical approach for preparation of smart conductive textiles. J. Chem. Educ. **93**, 1606–1611 (2016). https://doi.org/10.1021/acs.jchemed.6b00060
41. Y. Soo Kim, S. Bin Oh, J. Hyeok Park, M. Suk Cho, Y. Lee, Highly conductive PEDOT/silicate hybrid anode for ITO-free polymer solar cells. Sol. Energy Mater. Sol. Cells **94**, 471–477 (2010). https://doi.org/10.1016/j.solmat.2009.11.009
42. G. Guo, Y. Sun, Q. Fu, Y. Ma, Y. Zhou, Z. Xiong, Y. Liu, Sol-gel synthesis of ternary conducting polymer hydrogel for application in all-solid-state flexible supercapacitor. Int. J. Hydrog. Energy **44**, 6103–6115 (2019). https://doi.org/10.1016/j.ijhydene.2019.01.080
43. H. Berber, Emulsion polymerization: effects of polymerization variables on the properties of vinyl acetate based emulsion polymers, in *Polym. Sci.*, ed. by F. Yılmaz, (InTech, 2013). https://doi.org/10.5772/51498
44. B. Sydulu Singu, P. Srinivasan, K.R. Yoon, Emulsion polymerization method for polyaniline-multiwalled carbon nanotube nanocomposites as supercapacitor materials. J. Solid State Electrochem. **20**, 3447–3457 (2016). https://doi.org/10.1007/s10008-016-3309-1
45. H. Mi, J. Zhou, Z. Zhao, C. Yu, X. Wang, J. Qiu, Block copolymer-guided fabrication of shuttle-like polyaniline nanoflowers with radiating whiskers for application in supercapacitors. RSC Adv. **5**, 1016–1023 (2015). https://doi.org/10.1039/c4ra10273h
46. N.K. Guimard, N. Gomez, C.E. Schmidt, Conducting polymers in biomedical engineering. Prog. Polym. Sci. **32**, 876–921 (2007). https://doi.org/10.1016/j.progpolymsci.2007.05.012
47. C. Li, H. Bai, G. Shi, Conducting polymer nanomaterials: electrosynthesis and applications. Chem. Soc. Rev. **38**, 2397 (2009). https://doi.org/10.1039/b816681c
48. D. Wang, F. Li, J. Zhao, W. Ren, Z. Chen, J. Tan, Z. Wu, I. Gentle, G.Q. Lu, H. Cheng, Fabrication of graphene/polyaniline composite paper via in situ anodic Electropolymerization for high-performance flexible electrode. ACS Nano **3**, 1745–1752 (2009). https://doi.org/10.1021/nn900297m
49. S. Arulmani, S. Michelraj, C. Suresh, C. Sivakumar, Interconnected porous carbon: Unique versatile matrix for the growth of nanostructured polyaniline and its enhanced performance towards electrochemical energy storage. Chem. Sel. **2**, 2197–2204 (2017). https://doi.org/10.1002/slct.201602064
50. G. Wang, Y. Zhang, F. Zhou, Z. Sun, F. Huang, Y. Yu, L. Chen, M. Pan, Simple and fast synthesis of polyaniline nanofibers/carbon paper composites as supercapacitor electrodes. J Energy Storage **7**, 99–103 (2016). https://doi.org/10.1016/j.est.2016.05.011
51. N.I. Ibrahim, A.S. Wasfi, Electrochemical evaluation of polyaniline/multi-walled carbon nanotube composite synthesized by microwave plasma polymerization as a supercapacitor electrode. IOP Conf. Ser. Mater. Sci. Eng. **757**, 012036 (2020). https://doi.org/10.1088/1757-899X/757/1/012036
52. L.M. Santino, S. Acharya, J.M. D'Arcy, Low-temperature vapour phase polymerized polypyrrole nanobrushes for supercapacitors. J Mater Chem A **5**, 11772–11780 (2017). https://doi.org/10.1039/C7TA00369B
53. K. Wang, H. Wu, Y. Meng, Z. Wei, Conducting polymer nanowire arrays for high performance supercapacitors. Small **10**, 14–31 (2014). https://doi.org/10.1002/smll.201301991
54. C.R. Martin, R. Parthasarathy, V. Menon, Template synthesis of electronically conductive polymers—preparation of thin films. Electrochim. Acta **39**, 1309–1313 (1994). https://doi.org/10.1016/0013-4686(94)E0052-2
55. R. Liu, S. Il Cho, S.B. Lee, Poly(3,4-ethylenedioxythiophene) nanotubes as electrode materials for a high-powered supercapacitor. Nanotechnology **19**, 215710 (2008). https://doi.org/10.1088/0957-4484/19/21/215710
56. K. Wang, J. Huang, Z. Wei, Conducting polyaniline nanowire arrays for high performance supercapacitors. J. Phys. Chem. C **114**, 8062–8067 (2010). https://doi.org/10.1021/jp9113255
57. S. Iqbal, S. Ahmad, Recent development in hybrid conducting polymers: Synthesis, applications and future prospects. J. Ind. Eng. Chem. **60**, 53–84 (2018). https://doi.org/10.1016/j.jiec.2017.09.038

58. Q. Meng, K. Cai, Y. Chen, L. Chen, Research progress on conducting polymer based supercapacitor electrode materials. Nano Energy **36**, 268–285 (2017). https://doi.org/10.1016/j.nanoen.2017.04.040
59. Y. Tan, Y. Zhang, L. Kong, L. Kang, F. Ran, Nano-au@PANI core-shell nanoparticles via in-situ polymerization as electrode for supercapacitor. J. Alloys Compd. **722**, 1–7 (2017). https://doi.org/10.1016/j.jallcom.2017.06.068
60. D.S. Patil, J.S. Shaikh, S.A. Pawar, R.S. Devan, Y.R. Ma, A.V. Moholkar, J.H. Kim, R.S. Kalubarme, C.J. Park, P.S. Patil, Investigations on silver/polyaniline electrodes for electrochemical supercapacitors. Phys. Chem. Chem. Phys. **14**, 11886–11895 (2012). https://doi.org/10.1039/c2cp41757j
61. H. Moon, H. Lee, J. Kwon, Y.D. Suh, D.K. Kim, I. Ha, J. Yeo, S. Hong, S.H. Ko, Ag/au/Polypyrrole Core-shell nanowire network for transparent, stretchable and flexible supercapacitor in wearable energy devices. Sci. Rep. **7**, 41981 (2017). https://doi.org/10.1038/srep41981
62. H. Xu, J. Li, Z. Peng, J. Zhuang, J. Zhang, Investigation of polyaniline films doped with Ni2+ as the electrode material for electrochemical supercapacitors. Electrochim. Acta **90**, 393–399 (2013). https://doi.org/10.1016/j.electacta.2012.12.047
63. S. Arulmani, J.J. Wu, S. Anandan, Ultrasound promoted transition metal doped polyaniline nanofibers: Enhanced electrode material for electrochemical energy storage applications. Ultrason. Sonochem. **51**, 469–477 (2019). https://doi.org/10.1016/j.ultsonch.2018.07.006
64. G. Yu, X. Xie, L. Pan, Z. Bao, Y. Cui, Hybrid nanostructured materials for high-performance electrochemical capacitors. Nano Energy **2**, 213–234 (2013). https://doi.org/10.1016/j.nanoen.2012.10.006
65. C.-M. Chang, C.-J. Weng, C.-M. Chien, T.-L. Chuang, T.-Y. Lee, J.-M. Yeh, Y. Wei, Polyaniline/carbon nanotube nanocomposite electrodes with biomimetic hierarchical structure for supercapacitors. J Mater Chem A **1**, 14719 (2013). https://doi.org/10.1039/c3ta13758a
66. S. Arulmani, J.J. Wu, S. Anandan, Amphiphilic triblock copolymer guided polyaniline embraced CNT nanohybrid with outcropping whiskers as an energy storage electrode. Electrochim. Acta **246**, 737–747 (2017). https://doi.org/10.1016/j.electacta.2017.06.101
67. M. Kotal, A.K. Thakur, A.K. Bhowmick, Polyaniline–Carbon Nanofiber Composite by a Chemical Grafting Approach and Its Supercapacitor Application. ACS Appl. Mater. Interfaces **5**, 8374–8386 (2013). https://doi.org/10.1021/am4014049
68. L. Li, A.-R.O. Raji, H. Fei, Y. Yang, E.L.G. Samuel, J.M. Tour, Nanocomposite of polyaniline Nanorods grown on graphene nanoribbons for highly capacitive Pseudocapacitors. ACS Appl. Mater. Interfaces **5**, 6622–6627 (2013). https://doi.org/10.1021/am4013165
69. D. Li, Y. Li, Y. Feng, W. Hu, W. Feng, Hierarchical graphene oxide/polyaniline nanocomposites prepared by interfacial electrochemical polymerization for flexible solid-state supercapacitors. J Mater Chem A **3**, 2135–2143 (2015). https://doi.org/10.1039/C4TA05643D
70. J. Zhang, X.S. Zhao, Conducting polymers directly coated on reduced graphene oxide sheets as high-performance supercapacitor electrodes. J. Phys. Chem. C **116**, 5420–5426 (2012). https://doi.org/10.1021/jp211474e
71. A. Ouyang, A. Cao, S. Hu, Y. Li, R. Xu, J. Wei, H. Zhu, D. Wu, Polymer-coated graphene aerogel beads and supercapacitor application. ACS Appl. Mater. Interfaces **8**, 11179–11187 (2016). https://doi.org/10.1021/acsami.6b01965
72. M. Yu, Y. Ma, J. Liu, S. Li, Polyaniline nanocone arrays synthesized on three-dimensional graphene network by electrodeposition for supercapacitor electrodes. Carbon N. Y. **87**, 98–105 (2015). https://doi.org/10.1016/j.carbon.2015.02.017
73. C.D. Lokhande, D.P. Dubal, O.-S. Joo, Metal oxide thin film based supercapacitors. Curr. Appl. Phys. **11**, 255–270 (2011). https://doi.org/10.1016/j.cap.2010.12.001
74. K. Wang, L. Li, Y. Liu, C. Zhang, T. Liu, Constructing a "pizza-like" MoS2/Polypyrrole/polyaniline ternary architecture with high energy density and superior cycling stability for supercapacitors. Adv. Mater. Interfaces **3**, 1600665 (2016). https://doi.org/10.1002/admi.201600665

75. J. Yang, Y. Liu, S. Liu, L. Li, C. Zhang, T. Liu, Conducting polymer composites: material synthesis and applications in electrochemical capacitive energy storage. Mater. Chem. Front. **1**, 251–268 (2017). https://doi.org/10.1039/C6QM00150E
76. J. Xu, J. Ding, X. Zhou, Y. Zhang, W. Zhu, Z. Liu, S. Ge, N. Yuan, S. Fang, R.H. Baughman, Enhanced rate performance of flexible and stretchable linear supercapacitors based on polyaniline@au@carbon nanotube with ultrafast axial electron transport. J. Power Sources **340**, 302–308 (2017). https://doi.org/10.1016/j.jpowsour.2016.11.085
77. W. Jiang, D. Yu, Q. Zhang, K. Goh, L. Wei, Y. Yong, R. Jiang, J. Wei, Y. Chen, Ternary hybrids of amorphous nickel hydroxide-carbon nanotube-conducting polymer for supercapacitors with high energy density, excellent rate capability, and long cycle life. Adv. Funct. Mater. **25**, 1063–1073 (2015). https://doi.org/10.1002/adfm.201403354
78. H. Huang, X. Zeng, W. Li, H. Wang, Q. Wang, Y. Yang, Reinforced conducting hydrogels prepared from the in situ polymerization of aniline in an aqueous solution of sodium alginate. J. Mater. Chem. A **2**, 16516–16522 (2014). https://doi.org/10.1039/C4TA03332A
79. D. Ni, Y. Chen, H. Song, C. Liu, X. Yang, K. Cai, Free-standing and highly conductive PEDOT nanowire films for high-performance all-solid-state supercapacitors. J. Mater. Chem. A **7**, 1323–1333 (2019). https://doi.org/10.1039/c8ta08814d
80. Y. Xie, X. Sha, Electrochemical cycling stability of nickel (II) coordinated polyaniline. Synth. Met. **237**, 29–39 (2018). https://doi.org/10.1016/j.synthmet.2018.01.011
81. J. Li, M. Cui, Y. Lai, Z. Zhang, H. Lu, J. Fang, Y. Liu, Investigation of polyaniline co-doped with Zn^{2+} and H^{+} as the electrode material for electrochemical supercapacitors. Synth. Met. **160**, 1228–1233 (2010). https://doi.org/10.1016/j.synthmet.2010.03.014
82. T. Cheng, Y.-Z. Zhang, J.-P. Yi, L. Yang, J.-D. Zhang, W.-Y. Lai, W. Huang, Inkjet-printed flexible, transparent and aesthetic energy storage devices based on PEDOT: PSS/ag grid electrodes. J. Mater. Chem. A **4**, 13754–13763 (2016). https://doi.org/10.1039/C6TA05319J
83. H. Xu, J.x. Wu, C.l. Li, J.l. Zhang, X.x. Wang, Investigation of polyaniline films doped with Co^{2+} as the electrode material for electrochemical supercapacitors. Ion.S (Kiel) . **21**, 1163–1170 (2015). https://doi.org/10.1007/s11581-014-1267-0
84. B. Li, H. Lopez-Beltran, C. Siu, K.H. Skorenko, H. Zhou, W.E. Bernier, M.S. Whittingham, W.E. Jones, Vaper phase polymerized PEDOT/cellulose paper composite for flexible solid-state supercapacitor. ACS Appl. Energy Mater. **3**, 1559–1568 (2020). https://doi.org/10.1021/acsaem.9b02044
85. P. Du, Y. Dong, H. Kang, X. Yang, Q. Wang, J. Niu, S. Wang, P. Liu, Graphene-wrapped polyaniline nanowire Array modified functionalized of carbon cloth for high-performance flexible solid-state supercapacitor. ACS Sustain. Chem. Eng. **6**, 14723–14733 (2018). https://doi.org/10.1021/acssuschemeng.8b03278
86. H. Huang, L. Xia, Y. Zhao, H. Zhang, T. Cong, J. Wang, N. Wen, S. Yang, Z. Fan, L. Pan, Three-dimensional porous reduced graphene oxide/PEDOT: PSS aerogel: facile preparation and high performance for supercapacitor electrodes. Electrochim. Acta, 137297 (2020). https://doi.org/10.1016/j.electacta.2020.137297
87. Y. Lee, H. Choi, M.S. Kim, S. Noh, K.J. Ahn, K. Im, O.S. Kwon, H. Yoon, Nanoparticle-mediated physical exfoliation of aqueous-phase graphene for fabrication of three-dimensionally structured hybrid electrodes. Sci. Rep. **6**, 1–10 (2016). https://doi.org/10.1038/srep19761
88. C. Zhou, Y. Zhang, Y. Li, J. Liu, Construction of high-capacitance 3D CoO@Polypyrrole nanowire Array electrode for aqueous asymmetric supercapacitor. Nano Lett. **13**, 2078–2085 (2013). https://doi.org/10.1021/nl400378j
89. K. Ghosh, C.Y. Yue, M.M. Sk, R.K. Jena, Development of 3D urchin-shaped coaxial manganese dioxide@polyaniline (MnO2@PANI) composite and self-assembled 3D pillared graphene foam for asymmetric all-solid-state flexible supercapacitor application. ACS Appl. Mater. Interfaces **9**, 15350–15363 (2017). https://doi.org/10.1021/acsami.6b16406
90. L. Ren, G. Zhang, Z. Yan, L. Kang, H. Xu, F. Shi, Z. Lei, Z.-H. Liu, Three-dimensional tubular MoS 2/PANI hybrid electrode for high rate performance supercapacitor. ACS Appl. Mater. Interfaces **7**, 28294–28302 (2015). https://doi.org/10.1021/acsami.5b08474

91. M. Liang, M. Zhao, H. Wang, J. Shen, X. Song, Enhanced cycling stability of hierarchical NiCo 2 S 4 @Ni(OH) 2 @PPy core–shell nanotube arrays for aqueous asymmetric supercapacitors. J Mater Chem A **6**, 2482–2493 (2018). https://doi.org/10.1039/C7TA10413H
92. J. Li, W. Lu, Y. Yan, T.W. Chou, High performance solid-state flexible supercapacitor based on Fe3O4/carbon nanotube/polyaniline ternary films. J Mater Chem A **5**, 11271–11277 (2017). https://doi.org/10.1039/c7ta02008b
93. F. Miao, C. Shao, X. Li, K. Wang, N. Lu, Y. Liu, Electrospun carbon nanofibers/carbon nanotubes/polyaniline ternary composites with enhanced electrochemical performance for flexible solid-state supercapacitors. ACS Sustain. Chem. Eng. **4**, 1689–1696 (2016). https://doi.org/10.1021/acssuschemeng.5b01631
94. Y. Mao, J. Xie, H. Liu, W. Hu, Hierarchical core-shell ag@Ni(OH)2@PPy nanowire electrode for ultrahigh energy density asymmetric supercapacitor. Chem. Eng. J. **405**, 126984 (2021). https://doi.org/10.1016/j.cej.2020.126984

Chapter 4
Graphene-Metal Oxide Hybrid Nanostructures for High-Performance Supercapacitors

P. Mathan Kumar, Rajagopal Peri, T. Mohana Selvi, and B. Muthuraaman

Abstract Supercapacitors received a lot of attention among energy storage systems due to their high power, fast charge and discharge rate, long cyclic stability, and environmental-friendly. Till now researchers have been focusing on developing novel materials to enhance the supercapacitor performance. Transition metal oxides are one among the many electrode materials showing excellent supercapacitor performance due to their unique properties such as changeable valence and intercalation of electrons and ions into the lattice of the metallic compounds. However, the metal oxide-based electrodes lag in electronic conductivity at the electrode–electrolyte interface, which impedes its applications. Over a decade, the hybridization of metal oxides with graphene became of great interest due to its peculiar properties of graphene. Moreover, graphene enhances conductivity and induces synergistic effects that result in promising supercapacitor performances. In this section, a comprehensive review of various metal oxide-graphene hybrids and their supercapacitor performances was discussed. To summarize the discussions, the specific capacitance, current density, preparation methods, and cyclic stability of metal oxide-graphene hybrids were represented in the respective tables.

Keywords Supercapacitors · Metal oxides · Graphene · Reduced graphene oxide · Nanostructures · Electrolytes · Electrodes · Carbon derivatives

4.1 Introduction

Energy is a magical thing that makes other things happen, we may not see it but it's everywhere around us. Our energy needs, demand, consumption, and shortage are increasing day by day so issues like environmental pollution, the greenhouse effect, and fossil fuel depletion. Efficient and clean energy storage technologies

P. Mathan Kumar · R. Peri · T. Mohana Selvi · B. Muthuraaman (✉)
Electrochemical Energy Devices Laboratory, Department of Energy, University of Madras, Guindy Campus, Chennai, India
e-mail: muthuraaman@unom.ac.in

© The Author(s), under exclusive license to Springer Nature Switzerland AG 2026
K. Raju et al. (eds.), *Hybrid Nanostructures for Advanced Supercapacitors*, Modern Aspects of Electrochemistry 63, https://doi.org/10.1007/978-3-032-04593-5_4

are necessary to satisfy future energy requirements for renewable energy. Various renewable energy storage appliances were developed such as solar and wind energies, fuel cells, conventional capacitors, lithium-ion batteries, and supercapacitors. Among these, each has some drawbacks like fuel cells have a larger size and low power densities. Batteries have limited power densities and cycle lives, but supercapacitors also called ultracapacitors or electrochemical capacitors are promising energy storage devices which provide higher energy density than conventional capacitors and higher power density than batteries, hence they have drawn a lot of attention due to their unique characteristics, such as high power density, excellent columbic efficiency, fast charging and discharging rate and large cycling stability. Moreover, they do not depend on any complex chemical reaction to store energy and can be configured into flexible energy storage devices with simple architecture at low cost, they also have low maintenance and environment-friendly nature with the potential to replace traditional batteries partly or fully. However substandard energy density has restricted its practical applications [68] and an imperative to find an operative way to improve its energy density while preserving power density in the contemporary encounter as none of the single energy sources reported so far has fulfilled both power and energy densities at the same time. This is a prerequisite for insistence breakthrough that would complement existing supercapacitor structures via maximizing geometric and material parameters with enhanced energy storage capacity, density, and power that induces rechargeable batteries.

The typical supercapacitor device is a sandwich of two identical electrodes (anode and cathode) that contains the electrolyte and separator. The energy storage mechanism involves charging and discharging whereas, in charging, charge separation at the interface between the electrode and the electrolyte will happen and in discharging combination of separated charges in the electrolyte happens, overall, the device delivers the stored energy into an external load. Out of the two key components electrolyte can be a diluted acid, base and salt solution, or an ionic liquid ascribed to an ion-conducting substance. Electrode the main component, more specifically electrode material that is the same can act as both cathode and anode which plays a pivotal role in charge transfer and requires that the electrode materials must withstand high electric fields and maintain a high dielectric constant, to contribute to the capacitance, power density, energy density, and charge–discharge rates.

Supercapacitors are categorized into three categories based on their charge storage phenomena such as (i) electric double-layer capacitors (EDLCs), (ii) pseudocapacitors, and (iii) hybrid supercapacitors. EDLCs store physical charges (physisorption) via the formation of an electric double layer, due to the adsorption of negative ions from the electrolyte on the positively charged electrode and vice versa associated with the charge separation and accumulation at the interface between electrode and electrolyte. The capacitance stored in the electric double layer is proportional to the specific surface area of the electrode. Nanostructured carbon materials, such as mesoporous carbon, activated carbon, carbon nanotubes, carbon nanofibers, carbon aerogels, graphene, graphene oxide, and reduced graphene oxide, have been widely studied as the electrode materials of EDLCs. Pseudocapacitors can store charge not only through EDL formation but also from reversible oxidation–reduction, i.e., reversible faradaic redox reactions. Their capacitance is derived from the fast and reversible faradaic redox occurring on the surface or near the surface of the electrode materials. The redox process involved here is different from the reaction process of the battery because pseudocapacitors are not limited by the diffusion process, but batteries are controlled by ion diffusion within the crystalline framework of active material. The characteristic of pseudocapacitors is closer to capacitors which possess highly dynamic reversibility. The electrode materials of pseudocapacitors mainly include transition metal oxides and conductive polymers. Hybrid supercapacitors have been trending in recent years due to their high specific capacitance compared to both EDLCs and pseudocapacitors. Generally, hybrid supercapacitors are formed by the combination of EDLC and pseudocapacitor (asymmetric approach) which is the main reason for its enhanced capacitance value.

Among the three categories, EDLCs are mature technologically and being used in commercial applications. Currently, researchers across the world are trying to develop electrode materials with high capacitance and electrolytes of lower electrochemical potential or both. The most used electrode materials are carbon materials, metal oxides, and conductive polymers. The carbon materials exhibit EDLC behavior, their production process is very simple and cheap, also show good cyclic performance but the drawback is they have low specific capacitance. The metal oxides and the conductive polymers have pseudocapacitive behavior, showing higher specific capacity but costly and cyclic stability is low. Moreover, researchers have come up with a finding that a composite of different kinds of electrode materials can overcome their drawbacks and attain the advantages of high capacity, excellent cyclic ability, and rate performance. Out of all the composites reported graphene-based metal oxide composites got great attention. The objective of this article is to discuss the recent progress of graphene-metal oxide hybrid nanostructured electrode materials for high-performance supercapacitors. In this chapter, we focus on the various nano-architectures of graphene and different metal oxides and their performances as electrocatalytic electrodes for supercapacitor devices. In the sight of discussing the composites, better to have a glimpse of the individual compounds their properties, performance, and drawbacks.

Starting with graphene, which exists as a two-dimensional sp^2 bonded carbon material with ideal single atomic layer thickness, with high theoretical specific surface area (2630 m^2/g), stability, electrical and thermal conductivity, superior charge transfer properties, i.e., the unique electronic structure and large surface enhance the charge transfer, ultra-high mechanical strength, i.e., due to Van der Waals forces it easily agglomerates to sheets, with reduced specific surface area and volume, and excellent electrochemical activity which has created astonishing research interests which in turn make it a promising electrode material. Graphene has enhanced the performance of not only supercapacitors but also batteries, fuel cells, photovoltaics, photocatalysis, hydrogen generations, water splitting, sensors, and many more. However, the presence of delocalized electrons in the single-layer graphene deteriorates its electrocatalytic activity. In order to overcome this limit and improve the performance, electrochemical properties and to further enhance the faster electron transfer processes, incorporating a variety of metal oxides in carbon nanomaterials has been done for the development of next-generation supercapacitors. In this regard, graphene nanohybrid materials have opened diverse new frontiers by offering new concepts of novel nano-architectures, performance enhancement, and improvements in the reliability of supercapacitor devices.

Some of the electroactive metal oxides such as RuO_2, MnO_2, NiO, SnO_2, Fe_3O_4, V_2O_5, CuO and $LiFePO_4$ etc. are explored as electrodes for supercapacitors anticipating larger specific capacitance than carbon materials due to their multiple oxidation states, fast and reversible redox reactions and desirable pseudocapacitive properties. They possess high energy density but low power density and cycle life. Among all the metal oxides, ruthenium oxide (RuO_2) has been extensively researched as it exhibits high specific capacitance and good electrical conductivity [116]. Acute toxicity, high cost, and rareness of ruthenium elements have hampered its wide range of applications. After that naturally abundant metal oxides like NiO, Co_3O_4, MnO_2, and V_2O_5 were widely investigated which are not only lucrative but also environmentally benign [113]. However, relatively low specific capacitances have laden their practical application. To overcome this, researchers have come up with mixed metal oxides such as $NiCo_2O_4$, $NiCoO_2$, ZnV_2O_4, etc., the coexistence of two distinct metal species in a single crystal structure has catalyzed diverse redox reactions. The synergistic effect between the metals has helped develop more electroactive sites and improved the working potential which leads to better results compared to single metal oxides [27, 112]. However, they suffer from poor conductivity, inferior cycling stability, and prevalence of accretion during the continuous charge–discharge process because of their sensitivity toward the atmosphere and the high and low pH of the electrolyte [129]. The cost and scarcity of high pure metal oxides have further limited their commercial usage. Besides, researchers have observed the capacitive performance of supercapacitor banks on their morphology, porosity, specific surface area, and process parameters such as calcination temperature and mass loading. Nevertheless, of material, nanotechnology has enhanced its performance. Preparing various nanomaterials, nanostructures, nano-architectures, and nanohybrids as electrode materials has become a trend. Enduring the trend researchers have attempted various structures and morphologies, such as thin films, nanowires, nanorods, nanotubes, nanosheets, nanoflowers, and hollow nanospheres, which were fabricated through different synthesis routes. Gradually, due to their unique structure, large specific surface area, abundant accessible electroactive sites, and short ion diffusion pathway hierarchically nanostructured have become an enduring research hotspot in supercapacitor application.

The hybridization of metal oxides with carbonaceous material, especially graphene, accommodates volume expansion and engenders more electroactive sites which enhance the kinetics of ions and electron diffusion at the electrode–electrolyte interface. Additionally, the synergistic effect aids in increasing electrical conductivity and favors Faradic redox reactions. Furthermore, metal oxides entrenched between the layers of graphene can effectively block restacking which helps in delivering high charge capacity. [26] overtly demonstrated the starring role of transition metal salts in the process of exfoliation of graphite. The starting material, i.e., metal oxide regulates the graphene structure and improves the electrochemical properties of the composites.

The structure of the substrate and the number of surface functional groups achieve similar effects [12, 119]. This chapter extensively discusses the mechanism of development of electrochemical capacitance, the synthetic methods of graphene-metal oxide hybrid nanostructures, and the application of composites in supercapacitors. The number of reports, articles per compound, and year-wise representation have been plotted and shown in Fig. 4.1.

Recently, various researchers were discussing batteries and supercapacitor electrode materials in terms of their performance metrics such as specific capacitance, rate capability, operating voltage, etc. The problem arises when metal oxides are prepared in nanoscale resulting in an increased power density due to the short transport paths for ions and electrons. However, this increase in power density does not necessarily transform nanoscale materials into oxide supercapacitors since its Faradic peaks and galvanostatic profile remain battery-like. Therefore, Patrice Simon et al. stated that testing a battery material using a symmetric cell or hybrid device at low rates and claiming that as high energy density supercapacitors is misleading [98].

Later, the issue raised with the performance metrics of supercapacitor electrode materials was widely discussed by many researchers in the following articles [6, 10, 56].

In this chapter, the supercapacitor performances of various metal oxides such as binary and ternary metal compounds in combination with graphene, reduced graphene oxide, and graphene oxide have been discussed. This chapter includes many articles which come under the issue of high specific capacitance values for metal oxide nanohybrids. In addition, an insight into the various reported preparation methods was also presented. This chapter also extensively discusses the mechanism of development of electrochemical capacitance, the synthetic methods of graphene-metal oxide hybrid structures, and the application of their respective composite in supercapacitors.

4.2 Binary Metal Oxides—Graphene Hybrids

4.2.1 Manganese Oxide-Reduced Graphene Oxide

Manganese oxide (MnO_2) is an excellent Faradic material which has a theoretical specific capacitance of 1370 F/g and works at extended operating potentials. MnO_2 is widely explored as a supercapacitor electrode material due to its price, availability, high theoretical specific capacity, and non-toxicity. However, poor ionic (10^{-13} S/cm) and electrical (10^{-5}–10^{-6} S/cm) conductivity and easy dissolution of MnO_2 limits its practical applications. Many attempts have been developed to overcome these conductivity issues such as modifying its structures, doping with other materials, and graphene incorporation. Graphene-supported MnO_2 leads to enhanced supercapacitor performance; the lack of conductivity can be encountered with the addition of graphene [13].

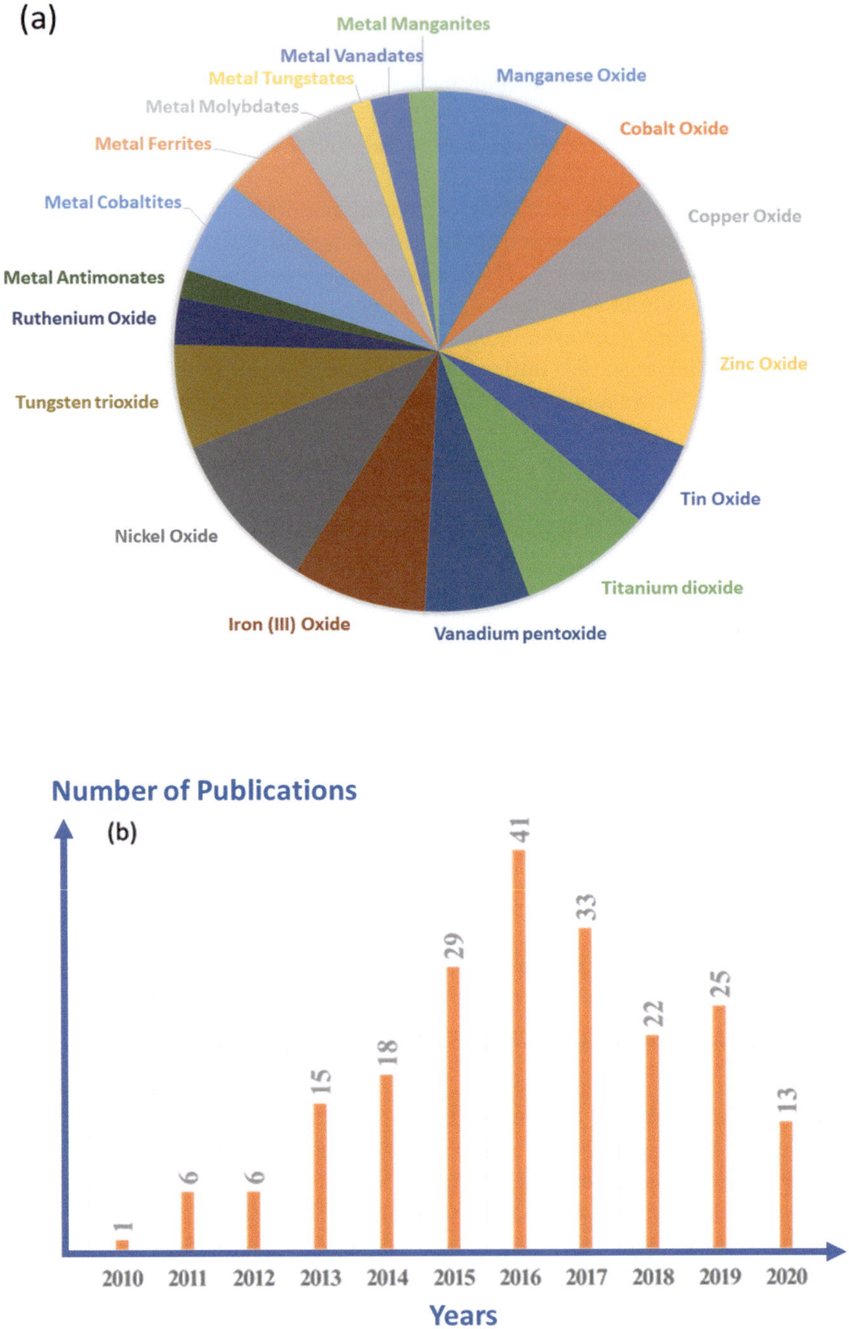

Fig. 4.1 **a** Pie chart representing papers per metal, metal oxide compounds and graphene nanohybrids. **b** Number of publications per year on the graphene-metal oxide and graphene-metal hybrid structures used as electrode materials for supercapacitor applications

Chen et al. [13] prepared needle-like MnO_2 supported by graphene oxide composite with different mass ratios using a simple soft chemical route and studied its supercapacitor behavior in 1 M Na_2SO_4 solution on the potential range of 0–1 V. They observed that the specific capacitance increases with the increase in the graphene oxide: MnO_2 ratios from 0.5 to 10, further increases do not show any improvement in specific capacitance values. The electrochemical stability for the prepared composite retained about 84.1% (165.9 F/g) of initial capacitance after 1000 cycles, while that of the as-prepared MnO_2 retained only about 69.0% (145.7 F/g). The authors suggest that by increasing the feeding ratio of MnO_2 - graphene oxide from 3 to 15, more MnO_2 anchors on graphene oxide sheets which causes a synergistic effect that leads to the formation of chemical bonds between MnO_2 and graphene oxide.

Dong et al. [21] deposited 3D graphene on nickel foam (1 mm thick) via the chemical vapor deposition method. Later, MnO_2 were in situ synthesized on 3D graphene-Ni foam using a simple hydrothermal approach. The specific capacitance for the prepared MnO_2-reduced graphene oxide was 560 F/g at 0.2 A/g which is higher than the as-prepared one. The loading of MnO_2 content on the 3D graphene-nickel foam leads to a synergetic effect in which the MnO_2 nanostructures give high electrochemical capacity, the three dimensionally multiplexed and highly conductive and the large specific surface area of 3D graphene ensures a large loading capacity for MnO_2 nanostructures and a large active surface area. Gao et al. [29] prepared graphene hydrogel with 3D interconnected pores using reduced graphene oxide by adding L-glutathione into graphene oxide dispersion followed by heating it to 95 °C. The prepared graphene hydrogel was used as a positive electrode.

On the other hand, vertically aligned MnO_2 nanoplates were electrodeposited on nickel foam used as negative electrodes. An asymmetric supercapacitor was fabricated using configuration graphene hydrogel-MnO_2 electrodes. Consequently, symmetric capacitors were also fabricated using graphene hydrogel-graphene hydrogel and MnO_2//MnO_2 for comparison. They achieved an energy density of 23.2 Wh/kg with a power density of 1.0 kW/kg. In contrast, the energy densities of the symmetric supercapacitor were 5.5 Wh/kg and 6.7 Wh/kg for graphene hydrogel and MnO_2-nickel foam respectively. Moreover, the graphene hydrogel-MnO_2 supercapacitor also presents stable cycling performance with 83.4% capacitance retention after 5000 cycles. This enhanced performance of asymmetric graphene hydrogel-MnO_2 supercapacitor is due to its desired porous structure, high specific capacitance, rate capability, elimination of polymer binders, and conductive additives [29]. Ye et al. [132] reported a facile electrochemical method to prepare MnO_2-reduced graphene oxide on carbon cloth by electrodeposition method and achieved a specific capacitance of 31.8 F/g at 2 mV/s with a rate capability reached 63.3% from 2 to 200 mV/s, indicating the good performance of the as-fabricated device. Also, the fabricated device exhibits excellent long-term stability, which retained about 100% retention of the initial capacitance after 5000 cycles.

Kim et al. [51] investigated the effect of different chemical reducing agents on preparing the graphene oxide-MnO_2 nanocomposite. They synthesized graphene oxide-MnO_2 by chemical reduction method using different reducing agents such as hydrazine hydrate and sodium borohydrate. The electrochemical behavior for both electrodes was studied using cyclic voltammetry and galvanostatic charge and discharge tests and electrochemical impedance spectroscopy in a three-electrode experimental setup using a 1 M Na_2SO_4 aqueous solution as the electrolyte. Hydrazine-reduced graphene oxide-MnO_2 shows superior conductivity and delivers a specific capacitance of 327.5 F/g at 10 mV/s, which is higher than that of the sodium borohydrate-reduced graphene oxide-MnO_2 (278.6 F/g). The cyclic stability for both the hydrazine and sodium borohydride-reduced graphene oxide-MnO_2 retained 88.26% and 88.19% of the initial specific capacitance over 1000 cycles. They concluded that both the hydrazine and sodium borohydride reduce the oxygen content which leads to the improvement in electrical conductivity and thus the supercapacitive behavior.

Sawangphruk et al. [96] prepared the MnO_2-reduced graphene oxide nanocomposite electrode by spray coating technique on flexible carbon foil paper. The MnO_2-reduced graphene oxide-coated carbon foil paper delivered an ultra-high specific capacitance of 393 F/g and retained 98.5% of the original capacitance after 2000 cycles. The authors concluded that the improved performance was owing to its high conductivity of reduced graphene oxide, high active surface area of tiny MnO_2, and high porosity between each MnO_2-reduced graphene oxide nanosheet coated on porous carbon foil paper. [31] developed a facile dip and dry process to fabricate hierarchical graphene-MnO_2 nanostructured sponges as high-performance supercapacitor electrode material and attained the highest specific capacitance of 450 F/g at a scan rate of 2 mV/s. The synergetic combination of EDLC active material reduced graphene nanosheets and pseudocapacitive material MnO_2 on skeletons of sponge delivered a maximum energy and power density of 2.08 Wh/kg and 94 kW/kg respectively at an operating voltage of 0.8 V. The cyclic stability of the prepared sponge-reduced-graphene oxide-MnO_2 retained 90% of its capacitance after 10,000 cycles under a scan rate of 10 mV/s.

Chan et al. [9] prepared MnO_2-reduced graphene oxide via facile electrodeposition method on stainless steel substrate with various deposition times such as 13, 15, and 18 min. The electrode which was deposited for 15 min achieved a specific capacitance of 378 F/g at a scan rate of 1 mV/s which retained 83% of its initial specific capacitance value after 1000 cycles than the later ones. [140] fabricated an asymmetric supercapacitor using MnO_2-graphene oxide as positive and a carbon material derived from an Artemia cyst shell with a genetic 3D hierarchical porous structure as the negative electrode. The maximum specific capacitance values can reach 308 F/g and 238 F/g for MnO_2—graphene oxide composites and hierarchical porous electrodes, respectively. Moreover, all the curves are symmetrical at different current densities from 0.5 to 10 A/g. The fabricated device delivered an energy density of 46.7 Wh/kg and a power density of 100 W/kg respectively. Moreover, the fabricated device showed superior cyclic stability with ~100% capacitance retention after 1000 cycles and ~93% after 4000 cycles. These superior electrochemical behaviors were due to P123 micelles that enhance the dispersion in-between the graphene oxide sheets via hydrophobic effect which unlocks the pathways for electrolyte ions, provides a high contact area between MnO_2 and graphene oxide, and its unique hierarchical porous structure as a negative electrode which complements the continuous electron and ion transport. [104] prepared nanostructured porous MnO_2-reduced graphene oxide-nickel via a simple chemical method; exhibited a pseudocapacitance behavior with a specific capacitance of 479.0 F/g at 1.0 A/g, with a good rate capability of 313.0 F/g at 10.0 A/g. Furthermore, the MnO_2-reduced graphene oxide-Ni electrode retained 83.5% of its initial capacitance value after 1000 cycles.

Miniach et al. [77] reported thermally reduced graphene oxide where the graphite oxide is thermally reduced under an inert atmosphere followed by hydrothermal synthesis to prepare MnO_2-reduced graphene oxide composite. An asymmetric supercapacitor was fabricated which consists of MnO_2-reduced graphene oxide (3:1) and activated carbon was used as positive and negative electrodes with an operating voltage of 2.1 V in Na_2SO_4 solution. The device delivered an energy density of 21 and 24 Wh/kg for the devices charged up to 2.0 and 2.1 V, respectively. Cyclic stability was carried out at different operating voltages such as 2.0 V and 2.1 V followed by surface morphology studies in order to study the morphological conditions after cyclic tests. The prepared composite shows 100% retention and 92% retention of its initial values after 5000 cycles with an operating cell voltage of 2.0 V and 2.1 V respectively. The small decrease in capacitance (8%) after 5000 cycles of charge/discharge for the asymmetric system with an operating cell voltage of 2.1 V was the result of MnO_2 morphological transformation (Table 4.1).

Ghasemi et al. [33] described needle-like MnO_2 nanostructures and MnO_2-reduced graphene oxide composite using sonochemical and electrostatic pulse co-precipitation methods. The prepared MnO_2-reduced graphene oxide hybrid delivered a specific capacitance of 375 F/g at a current density of 1 A/g and good cycle stability with 93% capacitance retention after 500 cycles at a scan rate of 200 mV/s. Zhang et al. [133] prepared uniform MnO_2-reduced graphene oxide films via the traditional electrodeposition method on nickel foam and found enhanced supercapacitive behavior. The electrodeposited MnO_2-reduced graphene oxide shows a specific capacitance of 467 and 307 F/g whereas the as-prepared MnO_2 shows specific capacitance increasing from 153 and 103 F/g to 270 and 190 F/g at the current densities of 1 and 5 A/g, respectively. Moreover, the specific capacitance of MnO_2 and MnO_2-reduced graphene oxide electrodes keep 83.9% and 93.1% retention after 2500 cycles at the large current density of 5 A/g, respectively. [141] prepared a novel three-dimensional reduced graphene oxide aerogel and MnO_2-reduced graphene oxide via mild and freeze-drying method followed by electrodeposition on nickel foam achieving a specific capacitance of 288 F/g at 0.5 A/g. The fabricated device delivers a maximum power density of 8.61 kW/g with 94.7% of initial capacitance retained over 1000 cycles. Hence, the above reports suggest that the coordination of graphene-MnO_2 nanohybrids can be used as effective electrode materials for supercapacitors.

Table 4.1 List of various reported manganese oxide-graphene composites and their supercapacitance performance

Reported compound	Method	Electrolyte	Specific capacitance	Current density/scan rate	Cyclic stability	References
Graphene oxide/MnO_2	Soft chemical route	1.0 M Na_2SO_4	197.7 F/g	200 mA	84.1% (1000 cycles)	[13]
MnO_2 on 3D graphene-nickel foam	Hydrothermal growth	1.0 M Na_2SO_4	560.0 F/g	0.2 A/g	79.0% (1000 cycles)	[21]
Graphene Hydrogel: MnO_2/Nickel foam	Electrodeposition	0.5 M Na_2SO_4	41.7 F/g	1.0 A/g	83.4% (5000 cycles)*	[29]
MnO_2/rGO	Electrodeposition	0.5 M Na_2SO_4	31.8 F/g	2 mV/s	100% (5000 cycles)	[132]
GO/MnO_2	Chemical reduction	1.0 M Na_2SO_4	327.5 F/g	10 mV/s	88.26% (1000 cycles)	[51]
MnO_2/rGO on CFP	Spray coating	0.5 M Na_2SO_4	393.0 F/g	0.1 A/g	98.5% (2000 cycles)	[96]
Sponge@rGO@MnO_2	Dip coating	1.0 M Na_2SO_4	450.0 F/g	2 mV/s	90% (10,000 cycles)	[31]
RGO-wrapped MnO_2	Electrodeposition	1.0 M Na_2SO_4	378.0 F/g	1 mV/s	83% (1000 cycles)	[9]
MnO_2/GO nanoflakes	Chemical method	1.0 M Na_2SO_4	308.0 F/g	0.5 A/g	87% (4000 cycles)	[140]
MnO_2/rGO/Ni	Simple chemical method	6.0 M KOH	479.0 F/g	1.0 A/g	83.5% (1000 cycles)	[104]
MnO_2/rGO	Hydrothermal growth	1.0 M Na_2SO_4	145.0 F/g	0.5 A/g	92% (5000 cycles)	[77]
Needle-like MnO_2/rGO	Sonochemical and electrostatic pulse co-precipitation	0.5 M Na2SO4	375.0 F/g	1.0 A/g	93% (500 cycles)	[33]
MnO_2/rGO @ nickel foam	Electrodeposition	1.0 M Na_2SO_4	467.0 F/g	1.0 A/g	93.1% (2500 cycles)	[133]
rGO/MnO_2/Ni	Electrodeposition	1.0 M Na_2SO_4	288.0 F/g	0.5 A/g	94.7% (1000 cycles)*	[141]

(*represents the cyclic stability of the fabricated device, rGO—reduced graphene oxide, CFP—carbon foil paper)

4.2.2 Cobalt Oxide—Reduced Graphene Oxide

Cobalt oxide (Co_3O_4) is one of the promising alternative electrode materials with pseudocapacitive behavior, excellent redox properties, and a high theoretical specific capacitance of 3560 F/g. On the other hand, cobalt monoxide (CoO) has a higher electrochemical property with a theoretical capacitance larger than 4000 F/g [106]. However, there are crucial factors affecting its electrochemical performances such as ion diffusion and poor conductivity, morphology dependent, and electronic states of the metal Co^{2+} and Co^{3+} [78]. Several attempts have been made to enhance its electrochemical behavior. [76] synthesized ultra-layered Co_3O_4 high porous structure by hydrothermal method and achieved a remarkable specific capacitance of 548 F/g at a current density of 8 A/g retaining 66% of initial capacitance. The cyclic stability measurements showed an excellent retention capability of around 98.5% after 2000 cycles at a current density of 16 A/g. [106] reported CoO nanowall structures via a combination of solvothermal and annealing processes and obtained nano-sized pores of 50 nm with a high specific area of 65 m^2/g. The prepared CoO nanoparticles delivered a specific capacitance of 352 F/g with a capacitance retention of 72.7% and 92.9% at a current density of 1 and 20 A/g depicting its good cyclic stability. This enhanced performance is due to unique porous nanowall structures which lead to a large surface and improve the charge transfer electrode–electrolyte interfaces.

Song et al. [99] prepared Co_3O_4 and various Co_3O_4-reduced graphene oxide nanosheets (rGONS) via a facile hydrothermal method, followed by calcination such as Co_3O_4/rGONS-I, II, III, IV (prepared with 2, 3, 4 and 5 mmol of $CoCl_2{\cdot}6H_2O$) and obtained a flower-like Co_3O_4 fine nanoparticles on reduced graphene surface. The electrochemical behavior of the prepared material was investigated using cyclic voltammetry and charge/discharge in 1 M KOH aqueous solution. The prepared Co_3O_4/rGONS-II delivered a specific capacitance of 445 F/g compared to the as-prepared and other composites as 278, 406, 284, and 236 F/g for Co_3O_4, Co_3O_4/rGONS-I, III, and IV, respectively. The charge and discharge characteristics of Co_3O_4/rGONS-II composite were further investigated with different current densities and obtained 445, 440, 438, 402, 314, and 262 F/g at the current density of 0.5, 0.8, 1.0, 2.0, 4.0, and 5.0 A/g, respectively. However, the specific capacitance was decreased by increasing current densities in the range of 2.0 to 5.0 A/g. The cyclic stability was tested in 1 M KOH for 1500 consecutive cycles and retained a specific capacitance of 244 F/g. The enhanced specific capacitance and good cyclic stability are due to the nanosheet structure and porous nature of the Co_3O_4/rGONS composite which enhances the contact between the Co_3O_4 and reduced GO in the inter-molecule level and results in a synergistic effect.

Deng et al. [17] deposited CoO nanowalls on 3D graphene foams via a simple hydrothermal process and subsequent thermal treatment, where the 3D graphene foam was prepared using chemical vapor deposition. The hybrid structure in which the CoO nanowalls were separated by graphene foams with robust adhesion can directly serve as electrode materials without binder and conductive agents. The CoO nanowall at graphene foam delivers a specific capacitance of 231.87 F/g at

1 A/g current density and 98% retention in capacitance after 1000 cycles at 7 A/g current. The enhancement in specific capacitance and cycle stability is due to excellent conductivity, lightweight, and rational architectures that provide fast electron pathways and low ion diffusion resistance.

Ahamed et al. [2] reported a facile method to prepare CoO-reduced graphene oxide nanocomposite via a simple co-precipitation method and obtained CoO nanoparticles that were 10–20 nm in size and encapsulated by reduced graphene oxide shells. The specific capacitance for the prepared CoO-reduced graphene oxide nanocomposite was calculated to be 175 F/g at 10 mV/s and 91.8 F/g at a high scan rate of 50 mV/s. This enhancement in specific capacitance is due to the synergistic effect of reduced graphene oxide-CoO, in which graphene oxide acts as an electric double-layer capacitance for fast charging and discharging. The composite also acts as a stable layer structured block that restricts the change in CoO volume resulting in more electrochemically active sites. Numan et al. [81] prepared Co_3O_4-reduced graphene oxide via a facile hydrothermal method with variant molar concentrations of cobalt precursors such as C1, C2, C3, and C4 intercalated into reduced graphene oxide. The preliminary studies confirm the strong interaction between Co_3O_4 and reduced graphene oxide matrix. Morphological studies showed cubical Co_3O_4 particles were decorated on a reduced graphene oxide matrix with an average size of 45 nm. The prepared Co_3O_4-reduced graphene oxide composite, namely C2, electrode obtained a specific capacitance of 278 F/g at a current density of 200 mA/g with good rate capability and retained 91.6% of its initial capacitance after 2000 cycles. The authors concluded that this enhanced performance is due to the effective ion transfer, less particle aggregation of Co_3O_4 on the reduced graphene oxide matrix than in the other electrodes, and uniform intercalation of CoO over reduced graphene oxide.

Zhou et al. [144] reported a facile direct flame approach to preparing CoO nanoparticles with an approximate size of 20 nm grown on reduced graphene oxide layers which are subjected to study its electrochemical behavior. The prepared CoO-reduced graphene oxide nanocomposite delivered a specific capacitance of 1615.0 F/g and 737.5 F/g at a current density of 1 A/g and 50 A/g respectively with 88.12% retention over 15,000 cycles. The sheet resistance (R_s) values for the CoO-reduced graphene oxide electrode before and after 15,000 cycles were 0.39 and 0.81 Ω, respectively, showed good conductivity of the electrolyte and low internal resistance of the electrode, whereas the charge transfer resistance (R_{ct}) values for the CoO-reduced graphene oxide electrode before and after 15,000 cycles were 3.9 and 7.1 Ω, respectively, which resulted in better electrode properties. An asymmetric supercapacitor was fabricated using CoO-reduced graphene oxide and activated graphene as positive and negative electrodes, respectively. The fabricated device delivers a maximum power and energy density of 62.46 Wh/kg and 1600 W/kg with an operating voltage of 1.6 V.

Xiang et al. [123] synthesized 20 nm Co_3O_4 nanoparticles, grown in situ on reduced graphene oxide sheets and had their electrochemical behaviors studied in a 2 M KOH aqueous electrolyte solution. The Co_3O_4-reduced graphene oxide showed a pseudocapacitive nature with a specific capacitance of 472 F/g at a scan rate of 2 mV/s with 82.6% retention of its initial capacitance on increasing the scan rate to

Table 4.2 List of various reported cobalt oxide-graphene composites and their supercapacitance performance

Reported compound	Method	Electrolyte	Specific capacitance	Current density/ scan rate	Cyclic stability	References
Co_3O_4/ rGO nanosheets	Hydrothermal growth	1.0 M KOH	445.0 F/g	0.5 A/g	45.2% (1500 cycles)	[99]
CoO nanowall @3D graphene	Hydrothermal growth	3.0 M NaOH	231.87 F/g	1.0 A/g	98% (1000 cycles)	[17]
rGO-CoO	Co-precipitation method	1.0 M KOH	175.0 F/g	10.0 mV/s	–	[2]
Cubical Co_3O_4/ rGO	Hydrothermal growth	1.0 M KOH	278.0 F/g	200 mA/g	91.6% (2000 cycles)	[81]
CoO/rGO	Direct flame approach	2.0 M KOH	1615.0 F/g	1.0 A/g	88.12% (15,000 cycles)	[144]
GO/Co_3O_4	Hydrothermal growth	2.0 M KOH	472.0 F/g	2.0 mV/s	95.6% (1000 cycles)	[123]

(GO—graphene oxide, rGO—reduced graphene oxide)

100 mV/s. At a current density of 2.0 A/g the reduced graphene oxide-Co_3O_4 composite electrode showed long-term stability with 95.6% retention of initial capacitance after 1000 continuous cycles. The electrochemical impedance spectroscopic data shows a low charge transfer and series resistance due to the combination of the Co_3O_4 nanoparticles with the reduced graphene oxide sheets. The Co_3O_4-reduced graphene oxide nanocomposite exhibits high energy density which reaches 39.0 Wh/kg at a power density of 8.3 kW/kg. The authors suggested that these superior performances were due to the synergistic effect of the small size and excellent redox activity of the Co_3O_4 particles combined with the high electronic conductivity of the reduced graphene oxide sheet (Table 4.2).

4.2.3 Copper Oxide-Reduced Graphene Oxide

Cupric oxide is attractive due to its non-toxicity, abundance, low cost, and desirable optical, electrical, and electrochemical properties, and it is highly utilized as electrodes for batteries and supercapacitor electrodes. However, there are a few factors like low electrical conductivity and structural damage during intercalation and de-intercalation which limit its electrochemical performance [86]. These issues can be overcome by tuning the CuO morphologies which provide accessibility for the electrolyte ions and the addition of carbon-based materials like graphene, carbon nanotubes etc., can improve the electronic conductivity.

A simple approach to the synthesis of CuO-reduced graphene oxide composite via hydrothermal method was reported by [86] in their work a symmetric supercapacitor was fabricated composed of Cu_2O-reduced graphene oxide positive and negative electrodes separated by polypropylene and their electrochemical behaviors were analyzed in 0.5 M K_2SO_4 aqueous electrolyte. The electrochemical impedance data showed a very low 2.47 Ω charge transfer resistance favors the charge–discharge capability. The prepared hybrid delivered a high specific capacitance of 326 F/g at a current density of 0.5 A/g, along with a high energy density of 65.7 Wh/kg at a power density of 302 W/kg and 97 F/g with a power density of 72 W/kg at 0.2 A/g. Furthermore, the CuO-reduced graphene oxide material did not show any loss of specific capacity after consecutive 1500 charge–discharge cycles. The authors have concluded that these superior electrochemical performances were due to the synergy between CuO and reduced graphene oxide in the storage of charge.

Dong et al. [20] prepared Cu_2O-reduced graphene oxide composite films grown directly on the copper foil substrates via the hydrothermal method and the electrochemical behavior was studied in 1 M KOH aqueous solution. Morphological analysis shows that the reduced graphene oxide is reduced to reduced graphene oxide and the Cu_2O nanoparticles are well-covered by reduced graphene oxide sheets. The Cu_2O-reduced graphene oxide composite delivers a specific capacitance of 98.5 F/g at a current density of 1 A/g and good cyclic stability. The excellent electrochemical behavior is due to the unique structure where the entangled graphene sheets interact with the Cu_2O layer to form an open pore system, where electrolyte ions can easily gain access to the surface graphene to form electrical double layers. Li et al. [63] investigated nitrogen-doped-reduced graphene oxide-CuO nanocomposite as a promising electrode material for supercapacitors. The composite was synthesized by refluxing the ammonia-added precursor solution for 12 h at 90 °C in an oil bath and the resulting black product was annealed at a low temperature of 200 °C. The CuO nanoparticles of 5 nm diameter were anchored homogeneously on the nitrogen-doped-reduced graphene oxide nanosheets and were confirmed by inductively coupled plasma mass spectroscopy (ICP-MS) measurements. The electrochemical behavior for the prepared composite was analyzed using cyclic voltammetry, galvanostatic charge, and discharge and electrochemical impedance spectroscopy in 6 mol/dm^3 potassium hydroxide (KOH) aqueous electrolyte solution. Electrochemical impedance spectroscopy data showed that reduced charge transfer resistance leads to excellent conductivity at the electrode–electrolyte interface. The prepared nitrogen-doped-reduced graphene oxide-CuO showed a high specific capacitance of 340 F/g at 0.5 A/g in the potential window of 1.4 V with retention of 80% of initial capacitance after 500 charge–discharge cycles. These enhanced performances can be attributed to the contributions of both electrical conductivity and large effective surface area of nitrogen-doped-reduced graphene oxide, and the pseudocapacitance of tiny CuO nanoparticles.

Wang et al. [114] investigated Cu_2O-CuO-reduced graphene oxide nanocomposite as an alternative electrode material for supercapacitor application. The Cu_2O-CuO-reduced graphene oxide was prepared by a simple hydrothermal-assisted redox reaction of elemental Cu and graphene oxide without any reagents. Morphological studies confirmed the uniform CuO spread over the reduced graphene oxide sheets and high-resolution transmission electron microscope images confirm the bi-component Cu_2O and CuO. Moreover, the morphology reveals that the uniform Cu_2O-CuO nanoparticles with double-sized scales 25 and 5 nm were anchored on reduced graphene oxide sheets. The prepared composite shows a specific capacitance from 173.4 F/g to 136.3 F/g at a current density of 1 A/g to 10 A/g respectively with 21.4% loss in capacitance whereas the capacity remains unchanged over 100,000 cycles with 98.2% retention in capacitance at a current density of 10 A/g. The low charge transfer resistance shown by electrochemical impedance spectroscopic plots further proves the excellent charge–discharge ability of the prepared composite. This enhanced cyclic stability is due to the redox property of Cu_2O-CuO and its unique nanostructure.

Dubal et al. [24] investigated the CuO addition into reduced graphene oxide sponge on the improvement of supercapacitor performance. CuO-reduced graphene oxide-sponge hybrid was prepared by depositing CuO nanostructures directly on the reduced graphene oxide sponge by a chemical bath deposition method. An asymmetric supercapacitor was fabricated with CuO-reduced graphene oxide sponge and reduced graphene oxide sponge as positive and negative electrodes respectively. The prepared electrode showed a maximum specific capacitance of 519 F/g at a high scan rate of 200 mV/s, whereas the fabricated device showed 77.84 F/g and a high energy density of 31.24 Wh/kg at an operating voltage of 1.7 V. Furthermore, the fabricated device retained 83% of its initial capacitance after 2000 consecutive cycles. The authors concluded that this enhanced supercapacitor performance was due to reduced graphene oxide addition on both electrodes which not only supports the CuO nanosheets but also maintains the mechanical integrity and high conductivity. The enhancement in power density is due to the macroporous conducting framework which supports the electrolyte ions and provides a large surface area for the charge transfer reactions.

Saraf et al. [94] prepared CuO-reduced graphene oxide via a facile and effective ultrasonication method in which the reduced graphene oxide is chemically synthesized separately. The prepared composite showed a specific capacitance of 244 F/g at a current density of 0.125 A/g which is six times higher than the prepared reduced graphene oxide and CuO nanoparticles. The maximum energy and power density of the prepared composite obtained 8.47 Wh/g and 73.49 W/kg, respectively. Moreover, the prepared material exhibits a high-rate performance, i.e., 82.78% of its initial capacitance was retained and exhibited excellent long cycle stability with 90% retention over 1000 cycles. The superior electrochemical performance is due to the synergistic effect between CuO and reduced graphene oxide that provides mechanical strength and short diffusion path charge transfer that leads to excellent conductivity at the electrode/electrolyte interface. Liu et al. [65] reported a CuO-reduced graphene oxide nanocomposite, where the reduced graphene oxide is wrapped with dandelion-like CuO microspheres; and prepared via the hydrothermal method. The dandelions like CuO nanospheres with the size of 2.0 μm were well-wrapped by ultrathin-reduced graphene oxide nanosheets. The prepared electrode delivers a specific capacitance of 296 F/g at 1.0 A/g and a retention of 96.1% after 2000 cycles at 2.0 A/g. The authors concluded that this excellent supercapacitor performance was due to the wrapping effect of reduced graphene oxide which effectively improved the electronic conductivity and buffered the volume changes (Table 4.3).

Sudhakar et al. [102] prepared CuO/reduced graphene oxide nanocomposite via ultrasonication with reduced graphene oxide solution, where the reduced graphene oxide was prepared by reducing graphene oxide using piperine as a green reducing agent. Transmission electron microscopic images revealed CuO nanoparticles trapped on reduced graphene oxide networks. The specific capacitance of the prepared CuO-reduced graphene oxide material shows 137 F/g at a scan rate of 5 mV/s. The prepared electrodes were further studied under different electrolyte systems such as Na_2SO_4, H_3PO_4, H_2SO_4 and (9:1) H_2SO_4:H_3PO_4 and found to be 151, 153,

Table 4.3 Copper oxide-graphene and their supercapacitance performance

Reported compound	Method	Electrolyte	Specific capacitance	Current density/ scan rate	Cyclic stability	References
Cu_2O/rGO	Hydrothermal growth	0.5 M K_2SO_4	326.0 F/g	0.5 A/g	100% (1500 cycles)	[86]
Cu_2O/rGO films on Cu foil	Hydrothermal growth	1.0 M KOH	98.5 F/g	1.0 A/g	54.3% (1000 cycles)	[20]
N-rGO/CuO	Chemical method	6.0 M KOH	340.0 F/g	0.5 A/g	80% (500 cycles)	[63]
Cu_2O/CuO/rGO	Hydrothermal growth	6.0 M KOH	173.4 F/g	1.0 A/g	98.2% (100,000 cycles)	[114]
CuO@rGO@sponge	Chemical bath deposition	1.0 M Na_2SO_4	519.0 F/g	200.0 mV/s	83% (2000 cycles)	[24]
CuO/rGO	Chemical method	1.0 M Na_2SO_4	244.0 F/g	0.125 A/g	90% (1000 cycles)	[94]
Dandelion-like CuO/rGO	Hydrothermal growth	2.0 M KOH	296.0 F/g	1.0 A/g	96.1% (2000 cycles)	[65]
rGO-CuO	Sonochemical	0.1 N H_2SO_4: H_3PO_4 (9:1)	224.0 F/g	5.0 mV/s	–	[102]
CuO/rGO nanosheets	Hydrothermal growth	1.0 M Na_2SO_4	78.7 F/g	10.0 mV/s	–	[7]

(rGO—reduced graphene oxide, CuO—Copper oxide, KOH—Potassium hydroxide)

185, and 224 F/g, respectively. The supercapacitor performance of the nanocomposite electrode was attributed to the synergistic effect of double-layer capacitance of reduced graphene oxide and redox capacitance of CuO nanoparticles. The prepared CuO-reduced graphene oxide nanocomposite via a simple hydrothermal method and compared to the as-prepared CuO nanoparticles. Morphological analysis revealed well-dispersed CuO nanoparticles on the surface of reduced graphene oxide nanosheets. The prepared CuO-reduced graphene oxide composite material shows a specific capacitance of 78.72 F/g which is higher than that for CuO (13.68 F/g). The enhanced supercapacitor performance was attributed to improved electrical contact between CuO nanoparticles and reduced graphene oxide nanosheets. Additionally, the CuO nanoparticle hooks on the reduced graphene oxide nanosheets effectively prevent the CuO from agglomerating and reduce graphene oxide from restacking, thus improving ion transport [7].

4.2.4 Zinc Oxide—Reduced Graphene Oxide

Zinc oxide (ZnO) is one of the promising alternative electrode materials in view of its simple synthesis and various morphologies. Interestingly, ZnO possesses excellent electrochemical behavior with a high energy density of 650 Ah/g that makes its way into supercapacitor applications [125]. Moreover, ZnO has a few advantages such as low cost, availability, and environment-friendly which makes it a step further toward commercial applications [34]. However, poor conductivity is still a challenge for the ZnO-based electrodes to achieve their maximum performance, and this can be tuned by the addition of graphene contents [64]. Initial works on performance enhancement were carried out by [37] and reported reduced graphene oxide-ZnO nanorods composite as a substitute electrode for supercapacitor applications. Graphene was deposited on copper foil via chemical vapor deposition and transferred into polyethene terephthalate (PET) substrate, whereas reduced graphene oxide was prepared by modified hummers' method and spin-coated on graphene polyethene terephthalate substrate. Finally, ZnO nanorods were grown on reduced graphene oxide-graphene-polyethylene terephthalate substrate and covered with reduced graphene oxide film. Morphological studies have shown that the ZnO nanorods (~1 μm) in length with diameters ranging from 50 to 100 nm, in the electrodes prevent the aggregation of reduced graphene oxide sheets and developed porous structure. The electrochemical studies were carried out in 1 M KCl aqueous solution and the specific capacitance of the prepared material is 51.6 F/g at the scan rate of 10 mV/s [37].

Chen et al. [14] investigated the reduced graphene oxide-ZnO nanocomposite varying the mass ratios of ZnO to reduced graphene oxide. The ZnO-reduced graphene oxide was prepared as KOH reacts in the aqueous graphene oxide dispersion, subsequently, $Zn(NO_3)_2$ precursors were dissolved in that solution above to form $Zn(NO_3)_2$-graphene oxide aqueous precursor followed by a thermal treatment in air. The results found that the ZnO to reduced graphene oxide mass ratio of 93.3:6.7

exhibits a specific capacitance as high as 308 F/g at the specific current of 1 A/g which is 128% higher than the 135 and 157 F/g for pure ZnO and bare reduced graphene oxide respectively, and found that the reduced graphene oxide dispersion within the composite is the key to achieving an excellent capacitive performance.

Li et al. [64] reported a 3D ZnO-reduced graphene oxide-ZnO sandwich structure where the ZnO nanorod arrays with sizes of 20–40 nm were grown directly on the reduced graphene oxide sheets via the chemical vapor deposition method. The electrochemical measurements were carried out in 1 M Na_2SO_4 aqueous solution. The prepared sandwich 3D ZnO-reduced graphene oxide-ZnO delivered a specific capacitance of 275 F/g at a scan rate of 5 mV/s, which is much larger than 98 F/g for reduced graphene oxide. The specific capacitance at the current density of 1.0 A/g is 270 F/g and remains 125 F/g when the current density increases to 15 A/g which is still larger than 24 F/g of reduced graphene oxide. On the other hand, the prepared nanohybrid delivered an energy and power density of 37.5 Wh/kg and 26.9 kW/kg, respectively, showing excellent rate capability. Also, the nanohybrid shows only a 2% loss from its initial capacitance over 2000 cycles whereas the reduced graphene oxide retained only 30%.

Li et al. [61] reported a new strategy to fabricate functional 3D graphene and ZnO composite materials with a specific capacitance of 554.23 F/g at a scan rate of 5 mV/s. They fabricated a porous graphene network with ZnO nanorods composite via chemical vapor deposition and hydrothermal method. In their preparation method, the nucleation step could influence the ZnO nanorod's growth directly on the graphene surface and enhance the affinity between ZnO and graphene. The electrochemical stability on the other hand retained 94.4% of the initial capacitance after 2300 cycles showing excellent cyclic stability. Zhang et al. [137] prepared ZnO-graphene nanosheets via one-step electrodeposition method, which showed a specific capacitance of 291 F/g at a current density of 3 A/g whereas, the pure ZnO displayed 118.8 F/g. On the other hand, [88] reported a scalable chemical reduction method to prepare ZnO-reduced graphene oxide. The prepared ZnO-reduced graphene oxide achieved a specific capacitance of 280 F/g, whereas the pure ZnO and reduced graphene oxide showed 125 F/g and 137 F/g respectively.

Saranya et al. [95] reported well-crystalline ZnO nanoparticles with a size range of 30–70 nm uniformly deposited on graphene sheets by a facile solvothermal approach, and the electrochemical properties were carried out in 6 M KOH aqueous solution. The prepared composite delivered a specific capacitance of 122.4 F/g, whereas the as-prepared graphene oxide and reduced graphene oxide had 2.13 F/g and 102.5 F/g at a 5 mV/s scan rate, respectively. Samuel et al. [93] investigated a binder-free nanocomposite consisting of ZnO nanoparticles which were grown directly on graphene sheets by electrospray deposition method with various concentrations of graphene and ZnO. The electrochemical studies were analyzed in 1 M Na_2SO_4 aqueous solution. The fabricated symmetric supercapacitor delivered an energy and power density of 67 mWh/cm^3 and 6000 mW/cm^3 retaining 90% of its capacitance after 1000 galvanostatic charge/discharge cycles. Lee et al. [57] reported a hierarchical graphene oxide-ZnO composite prepared via one-step precipitation method. The electrochemical studies were carried out at 1 M Na_2SO_4 aqueous electrolyte and the specific capacitance obtained for the prepared electrode is 97 F/g at a current density of 0.5 A/g with 90.8% retention of the specific capacitance for 5000 consecutive cycles.

Jayachandiran et al. [47] prepared a ZnO-reduced graphene oxide nanocomposite via an ultrasonic-assisted solution process and the electrochemical studies were analyzed in a 1 M Na_2SO_4 aqueous solution. The morphological studies confirm the embellishment of ZnO nanoparticles on the reduced graphene oxide sheets. The prepared materials offered a specific capacitance of 312 F/g compared to pure ZnO 200 F/g at a lower scan rate of 5 mV/s, with better cycling stability up to 1000 cycles. The authors concluded that this superior electrochemical performance is attributed toward its morphological variations as the ZnO homogenously intercalated on a reduced graphene oxide surface which facilitates the electrolyte ion transport, higher surface area, and electrical conductivity.

Du et al. [23] reported a facile filtration self-assembly and thermal reduction process to prepare ZnO nanoflakes and reduced graphene oxide where the ZnO nanoflakes were inserted into graphene sheets to obtain a unique layer-to-layer structure. The maximum specific capacitance obtained for the prepared electrode is 167.2 F/g at a scan rate of 5 mV/s, which is higher than the prepared reduced graphene oxide and ZnO nanoflakes. The cyclic stability studies show a 92.7% retention of capacitance at a scan rate of 1 V/s which suggests its excellent cycling stability. Furthermore, electrochemical impedance spectroscopic analysis showed low electrical resistance that improved the ion diffusion capacity and surface charge transfer performance of the working electrode. The authors concluded that this enhancement in specific capacitance was owing to the synergistic effect between the electroactive ZnO nanoflake and high-conducting graphene sheet networks.

Subramani and Sathish [101] demonstrated a facile and scalable synthesis of crystalline hexagonal ZnO nanoflowers-reduced graphene oxide nanocomposite via direct chemical decomposition of zinc hexacyanoferrate (ZnHCF) over reduced graphene oxide nanosheets. Morphological studies confirmed the hierarchical ZnO nanoflowers which consist of 2D nanosheets wrapped with reduced graphene oxide nanosheets. The electrochemical measurements were carried out in 3.5 M KOH solution and showed a specific capacitance of 203 F/g at the current density of 1 A/g along with excellent electrochemical stability (~98%) at a current density of 20 A/g for 10,000 cycles. [125] prepared ZnO-nitrogen-doped graphene (ZnO@NG) via a convenient hydrothermal method in which ethylene glycol is the reducing agent for the reaction of Zn precursor and graphene oxide dispersion. The electrochemical measurements were made in a 6 M KOH aqueous solution. The prepared composite showed a high specific capacitance of 237.18 F/g at a current density of 0.5 A/g retaining 90.8% over 2000 cycles at a current density of 0.5 A/g (Table 4.4).

Table 4.4 List of various reported zinc oxide-graphene and their supercapacitance performance

Reported compound	Method	Electrolyte	Specific capacitance	Current density/ scan rate	Cyclic stability	References
rGO/ZnO nanorods/rGO	Hydrothermal growth	1.0 M KCl	51.6 F/g	10 mV/s	–	[37]
rGO/ZnO	Solvothermal growth	1.0 M Na_2SO_4	308.0 F/g	10.0 A/g	93.5% (1000 cycles)	[14]
3D ZnO/rGO/ZnO	Chemical vapor deposition	1.0 M Na_2SO_4	270.0 F/g	1.0 A/g	98% (2000 cycles)*	Li et al. 2014b
3D graphene/ZnO	Chemical vapor deposition	1.0 M KOH	554.23 F/g	5.0 mV/s	94.4% (2300 cycles)	[61]
ZnO/graphene nanosheets	Electrodeposition	1.0 M KOH	291.0 F/g	3.0 A/g	–	[137]
ZnO/rGO	Hydrothermal	1.0 M Na_2SO_4	280.0 F/g	1.0 A/g	–	[88]
Graphene/ZnO	Hydrothermal	6.0 M KOH	122.4 F/g	5.0 mV/s	–	[95]
ZnO@graphene	Electrospinning	1.0 M Na_2SO_4	89.0 mF/cm^2	1.5 mA/cm^2	–	[93]
GO/ZnO	One-step precipitation	1.0 M Na_2SO_4	91.0 F/g	0.5 A/g	90.8% (5000 cycles)	[57]
rGO/ZnO	Solution route	1.0 M Na_2SO_4	312.0 F/g	5.0 mV/s	–	[47]
ZnO/rGO	Self-filtration and thermal reduction	1.0 M KCl	167.2 F/g	5.0 mV/s	90% (2000 cycles)	[23]
Hexagonal ZnO nanoflowers/rGO	Chemical decomposition	3.5 M KOH	203.0 F/g	1.0 A/g	98% (10,000 cycles)	[101]
ZnO/ Nitrogen-graphene	Hydrothermal growth	6.0 M KOH	237.18 F/g	0.5 A/g	90.8% (2000 cycles)	[125]

(*represents the cyclic stability of the fabricated device, rGO—reduced graphene oxide, ZnO—zinc oxide, KCl—potassium chloride)

4.2.5 *Tin Oxide—Reduced Graphene Oxide*

Tin oxide (SnO_2) is a semiconductor material which was used in electrodes for lithium-ion batteries and supercapacitors due to its low cost and high power density [111]. An interesting investigation was carried out by [32] and reported SnO_2

quantum dots via microwave-assisted method and achieved a tetragonal rutile structure with an average particle size diameter of 4–6 nm. The prepared SnO_2 quantum dots achieved a maximum of 315 F/g at a current density of 1 A/g. On the other hand, a symmetric capacitor was fabricated with SnO_2 quantum dots that exhibited an excellent energy density of 62 Wh/kg at a power density of 1 kW/kg. To improve the conductivity and enhance the overall supercapacitor performance, [11] introduced the addition of graphene content into the SnO_2 and reported a novel method to prepare SnO_2-reduced graphene oxide droplet aerogel with complex morphologies. In their method, graphene oxide suspension was added dropwise into the tin chloride with hydrochloric acid precursor solution, then the mixture was freeze-dried to obtain the SnO_2-reduced graphene oxide aerogels. Morphological studies confirm that the prepared aerogel has egg tart and mushroom shapes. The prepared aerogels showed a specific capacitance of 310 F/g at a scan rate of 5 mV/s. The energy and power densities for the prepared material were 30 Wh/kg and 8.3 kW/kg, respectively and retained 90% of its initial capacitance over 1000 cycles indicating its good cyclic stability.

Haldorai et al. [38] achieved a specific capacitance of 396 F/g at a current density of 4.5 A/g using flower-like SnO_2 architecture. The SnO_2-reduced graphene oxide composite was prepared using a simple one-pot hydrothermal method, the transmission electron microscopic image confirmed the flower-like SnO_2 was homogenously dispersed onto the reduced graphene oxide surface. Furthermore, the prepared hybrid material retained 92.6% of its initial capacitance even after 10,000 continuous charge–discharge cycles. The enhanced performance is due to the synergistic effect between flowers like SnO_2 and reduced graphene oxide sheets. [111] prepared SnO_2-graphene via a simple wet chemical route and achieved tetragonal SnO_2 nanostructures uniformly distributed on graphene sheets with a particle size of 50 nm. The electrochemical behavior was analyzed in a 6 M KOH electrolyte and achieved a maximum specific capacitance of 818.6 F/g at a scan rate of 1 mV/s, with 95.1% of capacitance retention after 1000 cycles. [43] prepared a SnO_2-reduced graphene oxide nanocomposite via combined sol–gel and thermal reduction method and fabricated a solid-state supercapacitor. The device achieved a specific capacitance of 229.1 F/g at a scan rate of 2 mV/s retaining 80% of its initial capacitance after 1000 cycles at 1 A/g.

Liu et al. [67] reported ultrathin SnO_2 nanorods decorated on reduced graphene oxide nanosheets and achieved an excellent supercapacitor performance. Morphological studies confirmed that the prepared SnO_2 nanorods were 20 and 100 nm in diameter and length, respectively. The electrochemical behavior was analyzed in 1 M Na_2SO_4 aqueous electrolyte and obtained a specific capacitance of 184.6 F/g at a current density of 100 mA/g, which is higher than the as-prepared SnO_2 which has 62.4 F/g. The composite exhibits superior cyclic stability and retained 98% after 6000 cycles when the scan rate was 50 mV/s. This excellent electrochemical performance is due to the synergistic effect of both reduced graphene oxide and ultrathin SnO_2 nanorods. Recently, Zhang et al. [135] synthesized SnO_2 nanorods- reduced graphene oxide hybrid via a two-step hydrothermal method and achieved a specific

Table 4.5 List of various reported tin oxide-graphene composites and their supercapacitance performance

Reported compound	Method	Electrolyte	Specific capacitance	Current density/ scan rate	Cyclic stability	References
SnO_2/graphene	Solution processing and freeze drying	2.0 M KOH	310.0 F/g	5.0 mV/s	90% (1000 cycles)*	[11]
Horn-like SnO_2/rGO	Hydrothermal	1.0 M H_2SO_4	396.0 F/g	4.5 A/g	92.6% (10,000 cycles)	[38]
Tetragonal SnO_2/rGO	Wet chemical route	6.0 M KOH	818.0 F/g	1.0 mV/s	95% (1000 cycles)	[111]
SnO_2/rGO	Sol–gel and thermal reduction	PVA/KOH	229.1 F/g	2.0 mV/s	80% (1000 cycles)	[43]
SnO_2 nanorods /rGO	Hydrothermal	1.0 M Na_2SO_4	184.6 F/g	0.1 A/g	98% (6000 cycles)	[67]
SnO_2 nanorods /rGO	Hydrothermal	1.0 M Na_2SO_4	262.2 F/g	0.1 A/g	96.1% (6000 cycles)	[135]

(*represents the cyclic stability of the fabricated device, rGO—reduced graphene oxide, SnO_2—tin oxide, PVA—polyvinyl alcohol)

capacitance of 262.2 F/g at the current density of 100 mA/g that retained 96.1% of its initial capacitance after 6000 cycles (Table 4.5).

4.2.6 *Titanium (II) Oxide—Reduced Graphene Oxide*

Initial investigations were made by [124] who reported both TiO_2 nanobelt and nanoparticles along with reduced graphene oxide to obtain TiO_2-reduced graphene oxide nanocomposites. They investigated the optimum ratio of TiO_2 and reduced graphene oxide for an enhanced supercapacitor performance. The group found that the mass ratio of 7:3 of reduced graphene oxide-TiO_2 delivered excellent specific capacitances of 225 F/g and 62.8 F/g for reduced graphene oxide-TiO_2 nanobelt and nanoparticles at a current density of 0.125 A/g, respectively. The prepared nanomaterials retained 91.1% and 86.5% of the initial specific capacitance over 200 and 2000 cycles, respectively, at a current density of 0.625 A/g, showing excellent long-term stability. Furthermore, the nanobelt TiO_2-reduced graphene oxide composite had an energy density of 20.1 Wh/kg at a power density of 170 W/kg. The authors concluded that the nanobelt structures resulted in enhanced supercapacitor performance than the nanoparticles which is due to the unique nanobelt structures that facilitate the charge transport property and large contact area with the reduced graphene oxide nanosheet.

Ramadoss et al. [89] reported a TiO_2 nanorods-2D reduced graphene oxide as an efficient electrode material for supercapacitors. The prepared hybrid material showed a maximum specific capacitance of this reduced graphene oxide-TiO_2 electrode to be 114.5 F/g at a scan rate of 5 mV/s. On the other hand, the prepared electrode retained 85% of its initial capacitance after 4000 cycles. Electrochemical impedance spectroscopic analysis showed a low charge transfer resistance which is attributed to the arrays of nanorods which were in contact with reduced graphene oxide nanosheets reduced the interfacial resistance; the nanorod prevents the aggregation of graphene nanosheets. The enhanced supercapacitor performance is due to the combined pseudocapacitive and EDLC behavior of TiO_2 nanorods and graphene, respectively, which facilitated the electron transport at the electrode–electrolyte interface [89]. The same group reported another method to prepare TiO_2—graphene hybrid structure via the microwave-assisted method and achieved a specific capacitance of 165 F/g at a scan rate of 5 mV/s with 90% retention of capacitance after 5000 cycles [90].

Kim et al. [50] synthesized TiO_2-reduced graphene oxide via a simple chemical route and achieved an excellent supercapacitor performance. The electrochemical studies were analyzed in non-aqueous 1 M lithium hexafluorophosphate in a 1:1 mixture of ethylene carbonate and dimethyl carbonate in a glove box. The specific capacitance for the prepared material was studied under various scan rates from 2, 5, 10, and 20 mV/s and showed values of 89, 75, 63, and 53 F/g respectively. On the other hand, varying current densities from 0.4, 0.8, 1.2, 1.6, 2.0, and 4.0 A/g showed 150, 90, 70, 55, 50, and 30 F/g respectively. The cyclic stability studies displayed an 80% capacitance retention after 100 cycles at 0.4 A/g and an energy density of 42 Wh/kg at a power of 800 W/kg. [18] prepared a 3D-reduced graphene oxide-TiO_2 using cellulose as structure structure-directing agent and intermediate agent which reduces the negative effects of graphene oxide on the growth of TiO_2 crystals. Transition electron microscopic analysis revealed the effect of cellulose on the morphology of TiO_2-reduced graphene oxide hybrid. The electrochemical behavior was analyzed in 6 M KOH aqueous solution and obtained a specific capacitance of 397 F/g with an energy density and power density of 55.7 Wh/kg and 1327 W/kg respectively. The cyclic stability studies displayed a 75% retention of its initial capacitance after 5000 cycles.

Kim and Park [52] reported interesting, reduced graphene oxide-thorn-like TiO_2 nanofibers aerogels via a combined electrospinning, silica etching, and hydrothermal method and investigated the effect of TiO_2 and graphene content ratios and found that 20% of TiO_2 and 80% reduced graphene oxide exhibited an excellent electrochemical performance. The specific capacitance for the prepared material showed a specific capacitance of 178 F/g at 1 A/g with 90% capacitance retention obtained over 3000 cycles. The authors concluded that the enhanced cyclic stability and specific capacitance is due to the unique thorn-like structure which enhances the chemical interactions between the reduced graphene oxide and TiO_2. Zhang et al.

[136] obtained elongated TiO_2 nanotubes along with reduced graphene oxide via a facile hydrothermal method in which commercial TiO_2 was used as starting material. The prepared compound showed an excellent specific capacitance of 338 F/g at 5 mV/s with 93.3% capacitance retention at 2.5 A/g over 3000 cycles displayed its long-term stability.

Agharezaei et al. [1] reported a sandwich-type TiO_2-reduced graphene oxide-TiO_2 prepared via a facile low-cost solution process. The electrochemical studies were carried out in 1 M KOH aqueous electrolyte and the sandwich-type TiO_2/ reduced graphene oxide /TiO_2 electrode material displayed a specific capacitance of 83.7 F/g at a scan rate of 5 mV/s. It was found that at different scan rates of 5, 10, 20, and 50 mV/s the specific capacitance starts to degrade slowly as 83.7, 69.8, 43.28, and 29 F/g. For the first time, [107] reported mixed-phase TiO_2 nanofibers-hydrogen exfoliated graphene; prepared by combined electrospinning and chemical methods. The prepared electrode material delivered a specific capacitance of 210.5 F/g at the current density of 1 A/g and retained 97% of its initial capacitance after 1000 cycles. The enhancement in supercapacitor behavior is due to the synergy between the mixed-phase TiO_2 nanofibers and reduced graphene oxide. The large specific surface area of mixed-phase TiO_2 increases the active sites for the adsorption of ions onto reduced graphene oxide-wrapped TiO_2, whereas the reduced graphene oxide improves the conductivity of charge carriers across the electrode–electrolyte interfaces.

Reddy et al. (2018) prepared a TiO_2-reduced graphene oxide nanocomposite via the hydrothermal method and investigated various mass ratios of reduced graphene oxide and TiO_2. The electrochemical behavior was carried out in 1 M H_2SO_4. The prepared composite with 5-reduced graphene oxide-TiO_2 exhibited an excellent specific capacitance of 160 F/g at 0.2 A/g retaining 99% of initial capacitance. The enhanced capacitance was due to the synergistic effect of both pseudo and double-layer capacitance behavior of both reduced graphene oxide and TiO_2. Sundriyal et al. [105] modified the electrolyte by adding 0.2 M $K_3[Fe(CN)_6]$ into the Na_2SO_4 electrolyte to analyze the electrochemical performance. They have synthesized TiO_2-reduced graphene oxide via a facile one-step hydrothermal method and achieved a remarkably high specific capacitance of 1565 F/g at a current density of 3 A/g. An asymmetric TiO_2-reduced graphene oxide-TiO_2-reduced graphene oxide supercapacitor was fabricated and delivered a specific capacitance of 204.5 F/g at 1.5 A/g. Furthermore, the fabricated device delivered an energy density and power density of 15.5 Wh/kg and 1.1 kW/kg, respectively. Moreover, the device exhibits excellent cyclic stability with 87% capacitance retention over 1000 consecutive cycles. The authors concluded that the synergistic effect of both porous TiO_2 and reduced graphene oxide nanosheets with redox electrolyte greatly enhanced the supercapacitor performance (Table 4.6).

Table 4.6 Titanium dioxide-graphene composites and their supercapacitance performance

Reported compound	Method	Electrolyte	Specific capacitance	Current density/scan rate	Cyclic stability	References
rGO: TiO_2	Hydrothermal growth	1.0 M Na_2SO_4	225.0 F/g	0.1 A/g	86.5% (2000 cycles)	[124]
TiO_2 Nanorods/2D rGO	Hydrothermal growth	1.0 M Na_2SO_4	114.5 F/g	5.0 mV/s	85% (4000 Cycles)	[89]
rGO/TiO_2	Microwave-assisted	1.0 M Na_2SO_4	165.0 F/g	5.0 mV/s	90% (5000 cycles)	[90]
rGO/TiO_2	Chemical route	1.0 M $LiPF_6$	150.0 F/g	0.4 A/g	80% (100 cycles)	Kim et al. 2013a
3D rGO/TiO_2	Hydrothermal growth	6.0 M KOH	397.0 F/g	0.36 A/g	75% (5000 cycles)	[18]
rGO/ hornlike TiO_2	Electrospinning	1.0 M Na_2SO_4	178.0 F/g	1.0 A/g	90% (3000 cycles)	[52]
rGO/TiO_2 Nanotubes	Hydrothermal	1.0 M Na_2SO_4	338.0 F/g	5.0 mV/s	93.3% (3000 cycles)	[136]
TiO_2/rGO/TiO_2	Low-cost solution route	1.0 KOH	83.5 F/g	5.0 mV/s	–	[1]
Hydrogen exfoliated graphene/TiO_2	Electrospinning	1.0 M H_2SO_4	210.5 F/g	1.0 A/g	97% (1000 cycles)	[107]
rGO/TiO_2	Hydrothermal growth	1.0 M H_2SO_4	160.0 F/g	0.2 A/g	99% (2000 cycles)	[85]
rGO/TiO_2	Hydrothermal growth	0.2 M $K_3[Fe(CN)_6]$	1565.0 F/g	3.0 A/g	87% (1000 cycles)*	[105]

(*represents the cyclic stability of the fabricated device, rGO—reduced graphene oxide, TiO_2—titanium oxide)

4.2.7 Vanadium Pentoxide—Reduced Graphene Oxide

A 3D-reduced graphene oxide coated on a V_2O_5 nanoribbon scaffold via a simple hydrothermal method was first reported by [131]. To investigate the performance of the hybrid, 3D VO_2-reduced graphene oxide and V_2O_5 powder were also prepared. A symmetric cell was fabricated with 1 M Na_2SO_4 aqueous electrolyte, and its electrochemical behavior was analyzed using various techniques. The prepared device exhibited a specific capacitance of 437 F/g at an operating voltage of 0–1 V and a scan rate of 50 mV/s. This was much higher than the 3D VO_2-reduced graphene oxide and V_2O_5 powder, which exhibited 262 F/g and 112 F/g, respectively. Electrochemical impedance analysis confirmed the low charge transfer resistance for the prepared hybrid material. The cyclic stability shows a 96% retention in initial capacitance over 10,000 cycles at 10 A/g.

Nagaraju et al. [79] reported a 2D V_2O_5 nanosheets-reduced graphene oxide nanocomposite via hydrothermal method and fabricated an asymmetric supercapacitor. The electrochemical analysis was studied in 1 M KCl electrolyte, and the prepared device exhibits a specific capacitance of 253 F/g and 635 F/g for 2D V_2O_5 nanosheets and 2D V_2O_5-reduced graphene oxide at 1 A/g, whereas the energy density of 39 Wh/kg and 79.5 Wh/kg were achieved for the same two electrodes at a power density of 900 W/kg. The cyclic stability studies displayed a 94% capacitance retention for the 2D V_2O_5-reduced graphene oxide electrodes over 3000 cycles.

Kiruthiga et al. [53] synthesized V_2O_5-reduced graphene oxide nanocomposite via a sol gel method. They used porous carbon with a large surface area of 1554 m^2/g obtained from honey as cathode and the prepared nanocomposite used as an anode for the fabrication of sodium ion capacitor. The electrochemical studies were analyzed in 0.75 M $NaPF_6$ in ethylene carbonate (EC)–diethyl carbonate (DEC) (1:1) electrolyte. The prepared nanocomposite exhibited a specific capacitance of 289 F/g at 0.01 A/g vs $Na/Na^{+,}$ whereas the obtained porous carbon showed 224 F/g. The cyclic stability studies showed 85% capacitance retention over 1000 cycles at 0.06 A/g for the V_2O_5-reduced graphene oxide nanocomposite. On the other hand, the fabricated sodium ion capacitor delivered an energy density of 65 Wh/kg and power density of 72 W/kg at 0.03 A/g with 74% retention of its initial capacitance over 1000 cycles at 0.06 A/g.

Van Hoa et al. [109] reported a hierarchical flower-like V_2O_5 grown on graphene and nickel foam prepared using chemical vapor deposition followed by hydrothermal method. The prepared hybrid displayed a high specific capacitance of 1235 F/g at 2 A/g and retained 92% of its initial capacitance after 5000 cycles. The superior electrochemical performance is due to the advantages of the high conductivity of graphene, flower-like V_2O_5 nanostructure with a very high surface area that increases the electroactive sites and open framework of 3D nickel foam which in turn enhances the charge carrier transport across the electrode/electrolyte interface. Hu et al. [41] synthesized V_2O_5-reduced graphene oxide via simple solvothermal method and an asymmetric supercapacitor was fabricated using V_2O_5 and reduced graphene oxide as positive and negative electrode, respectively. The prepared electrode material exhibits a discharge specific capacity of 574.5 C/g at a current density of 1 A/g with good stability and rate capability. However, the device showed a specific capacity of 236.8 C/g and an energy density of 46.05 Wh/Kg with 82.7% capacitance retention after 5000 cycles. It is found that the enhanced performance was due to the unique hierarchical structure, more redox active sites and good conductivity.

A thin film supercapacitor was first reported by [110] where the electrode materials were prepared by a thermal evaporation technique. Moreover, they fabricated a symmetric device with a configuration of V_2O_5/PVA-KOH/V_2O_5 with both as-prepared and annealed thin films separately. The electrochemical studies exhibited an areal capacitance of 9.7 mF/cm^2 at 10 mV/s and the fabricated device displayed an excellent cyclic stability retention of 95% from its initial capacitance over 30,000 cycles along with 99% coulombic efficiency. [66] investigated the addition of graphene content on the performance enhancement of supercapacitors. They prepared V_2O_5—reduced graphene oxide via simple solvothermal method; the electrochemical behavior was analyzed in 1 M $LiNO_3$ aqueous electrolyte. Morphological studies confirmed the porous V_2O_5 nanorods dispersed on reduced graphene oxide nanosheets. The prepared hybrid delivered a maximum specific capacitance of 450.5 F/g at 0.5 A/g and retained 73.3% at 10 A/g. A symmetric capacitor was fabricated using the prepared material coated on nickel foam, with mass loading of 1 mg, which delivered an energy density of 33.5 Wh/Kg at a power density of 425.6 W/Kg. However, the cyclic stability showed a decreasing trend from 66% during 100 cycles and dropped to 34.5% after 1000 cycles, this was due to the dissolution of V_2O_5 in the aqueous electrolyte (Table 4.7).

Table 4.7 Vanadium pentoxide-graphene composites and their supercapacitance performance

Reported compound	Method	Electrolyte	Specific capacitance	Current density/ scan rate	Cyclic stability	References
3D rGO/V_2O_5	Hydrothermal growth	1.0 M Na_2SO_4	437.0 F/g	1.0 A/g	96% (1000 cycles)	[131]
2D V_2O_5 Nanosheets/ rGO	Hydrothermal growth	1.0 M KCl	253.0 F/g	1.0 A/g	94.1% (3000 cycles)*	[79]
V_2O_5@rGO	Sol gel	0.75 M $NaPF_6$	289.0 F/g	0.01 A/ g	74% (1000 cycles)*	[53]
V_2O_5@graphene@nickel foam	Chemical vapor deposition	1.0 M Na_2SO_4	1235.0 F/g	2.0 A/g	92% (5000 cycles)	[109]
V_2O_5/rGO	Solvothermal growth	1.0 M LiCl	574.5 C/g	1.0 A/g	82% (5000 cycles)*	[41]
V_2O_5/rGO	Solvothermal growth	1.0 M $LiNO_3$	450.5 F/g	0.5 A/g	66% (100 cycles)*	[66]

(*represents the cyclic stability of the fabricated device, rGO—reduced graphene oxide, V_2O_5—vanadium pentoxide, KCl—potassium Chloride, $LiNO_3$—lithium nitrate)

4.2.8 Iron (II, III) Oxide—Reduced Graphene Oxide

Wang et al. [118] reported a Fe_2O_3–reduced graphene oxide nanocomposite prepared using urea in a single-step hydrothermal method that is simple and green. In the formed composite, Fe_2O_3 nanoparticles were well-grown on the surface of graphene, the formation of nanoparticles hindered the aggregation of graphene. The advantages of the EDLC from reduced graphene oxide, and the pseudocapacitance of Fe_2O_3 to make it a stable and suitable material for supercapacitor and demonstrated the capacitance of 226 F/g at a current density of 1 A/g in 1 M Na_2SO_4 electrolyte. The report has simply proposed Fe_2O_3-reduced graphene oxide nanocomposites as a promising electrode material for supercapacitors without any further discussion about its retention capacity.

Taking a step further, [75] reported a facile one-step hydrothermal method to prepare Fe_2O_3 supported on nitrogen-doped graphene hydrogel with homo-dispersed Fe_2O_3 nanoparticles with various sizes between 20 and 60 nm. The prepared composite showed almost three-fold improvement in specific capacitance than the previous one, Fe_2O_3-reduced graphene oxide which is as high as 618 F/g at a current density of 0.5 A/g in 1 M KOH, and showed a specific capacitance of 350 F/g at a current density of 10 A/g. The presence of nitrogen and Fe_2O_3 inside the 3D graphene network displayed a positive synergistic effect which helped the Fe_2O_3/N-reduced graphene oxide composite in gaining more capacitance and durability with improved electrical conductivity. The composite showed a promising potential as an advanced electrode material for supercapacitors with a retention of 56.7% at a current density of 4 A/g after 5000 cycles.

After that [130] has reported porous α-Fe_2O_3-graphene composites by hydrothermal method followed by a slow annealing route. Due to the porous α-Fe_2O_3 nanoparticles and crumbled sheets of graphene, the prepared nanomaterial delivered a specific capacitance of 343.7 F/g at current density of 3 A/g with, a capacitance retention of 95.8% even after 50,000 charge-discharge cycles at a current density of 10 A/g. Following this, [87] has reported two-dimensional α-Fe_2O_3-reduced graphene oxide nanocomposites using solvothermal method. The mesoporous two-dimensional nanostructure composes α-Fe_2O_3 nanoplates with a diameter of around 300 nm and a thickness of approximately 30 nm homogeneously encapsulated in reduced graphene oxide nanosheets. The prepared composite showed a specific capacitance of 903 F/g at 1 A/g current density but showed less cyclic stability with only 70% retention at 5 A/g after 1000 cycles.

Using the same solvothermal method, [30] reported a quasi-hexagonal α-Fe_2O_3 nanoplates of size 20–80 nm dispersed over reduced graphene nanosheets and showed a specific capacitance of 1083 F/g at 2 A/g current density and 517 F/g at 10 A/g which is still as high and exhibited a retention of 75% after 1000 cycles. The combination of Fe_2O_3 and graphene had been developed further by varying the structure of Fe_2O_3, such as Fe_3O_4 nanorods [16], Fe_2O_3 nanoparticle clusters [44], flower-like nanoparticles [117], and Zhao et al. reported Fe_2O_3/reduced graphene oxide/Fe_3O_4 nanocomposite on iron foil [139]. Furthermore, the researchers advanced into many more shapes and structures like thumb-ring-like [146], pyramid-like [149], spindle-like [22], mulberry-shaped [70], and hollow nanospheres α-Fe_2O_3 [46]. Among the reports rod-shaped Fe_3O_4 nanostructures anchored on reduced graphene oxide by the spontaneous in situ deposition of the rod-like FeOOH nanostructure onto the self-reduced graphene oxide surface followed by a thermal annealing process, having an average size of 150 nm and specific surface area of 186 m^2/g has showed a specific capacitance of 315 C/g at 5 A/g current density with a retention of 95% after 2000 cycles [16].

Hu et al. [44] reported a highly flexible asymmetric supercapacitors using hybrid Fe_2O_3 nanoparticle clusters of size around ~30 nm coupled with reduced graphene oxide sheets prepared by gel formation followed by hydrothermal growth, vacuum filtration, and electrochemical reduction methods. This hybrid nanoparticle cluster had been used as a negative electrode in flexible solid-state asymmetric supercapacitor, providing better energy and power densities. Thus, providing new insights into the future design of hybrid-type electrode materials for energy storage. The Fe_2O_3-reduced graphene oxide-Fe_3O_4 nanocomposite on iron foil reported by [139] through one-step hydrothermal growth process, the composite exhibited the specific capacitance of 337.5 mF/cm^2 at 20 mA/cm^2 in 2 M KOH with only a 2.3% capacity loss from 600 to 2000th cycle showed a prominent way for energy storage.

Out of the various shapes and structures, the thumb-ring-like α-Fe_2O_3-reduced graphene oxide reported by Zhu et al. [146], the composite was synthesized via a simple hydrothermal method accompanied by surface potential tuning has shown a stable specific capacitance value of 255 F/g at 0.5 A/g current density with almost 90% retention after 11,000 charge and discharge cycles, exhibited only 10% decay in the first 1000 cycles and then barely any drop during the following 10,000 cycles. During cycling, the ring-like hollow structures (nanorings) collapsed into nanoparticles and gradually wrapped by reduced graphene oxide which prevented the mass loss of the composite and supported long-term cycling making it a promising material as negative electrode in asymmetric aqueous supercapacitor. The mulberry-shaped α-Fe_2O_3-reduced graphene oxide nanoparticles of around 100 nm size reported by Liu et al. [70] via single-step hydrothermal reaction showed a remarkable specific capacitance of 1051 F/g at 1 A/g current density and excellent rate capability of 548 F/g at 30 A/g and showed a retention of 85.4% after 1000 cycles at 4 A/g current density. Pyramid shape reported by [149] has used a new method, 20-s flame treatment. The pyramid shape shows a high specific capacitance of 965.7 F/g at 5 mV/s and stable 80% retention at 5 A/g even after 2000 cycles. The authors also observed no significant loss in specific capacitance over 1400 cycles at 5 A/g. Along with the electrochemical behavior, the composite also possesses photo-degradation capacity which revealed a great potential in applications of energy storage and water treatment (Table 4.8).

The spindle-like α-Fe_2O_3-reduced graphene oxide reported by [22] via simple hydrothermal method and the composite showed a specific capacitance of 455 F/g at 1 A/g and good cycling life of about 87.5% reserved after 10,000 cycles. The composite electrode was used as anode in solid state asymmetric supercapacitor which exhibited an outstanding capacitance and energy density of 73 W/kg at the power density

Table 4.8 Iron (II, III) oxide-graphene composites and their supercapacitance performance

Reported compound	Method	Electrolyte	Specific capacitance	Current density/ scan rate	Cyclic stability	References
Fe_2O_3/rGO	Hydrothermal growth	1.0 M Na_2SO_4	226.0 F/g	1.0 A/g	–	[118]
Fe_2O_3/N-rGO hydrogel	Hydrothermal growth	1.0 M KOH	350.0 F/g	10.0 A/g	56% (5000 cycles)	[75]
α-Fe_2O_3 /graphene	Hydrothermal growth	1.0 M Na_2SO_4	343.7 F/g	3.0 A/g	95.8% (50,000 cycles)	[130]
α-Fe_2O_3 nanoplates/ rGO	Solvothermal growth	1.0 M KOH	903.0 F/g	1.0 A/g	70% (1000 cycles)	[87]
α-Fe_2O_3 nanoplates/ rGO	Solvothermal growth	1.0 M KOH	1083.0 F/g	2.0 A/g	75% (1000 cycles)	[30]
Fe_2O_3 nanorods/ rGO	Thermal annealing	1.0 M KOH	315 C/g	5.0 A/g	95% (2000 cycles)	[16]
Fe_2O_3/rGO	Electrochemical reduction	3.0 M KOH	178.3 F/ cm^3	1 mV/s	83.1% (10,000 cycles)	[44]
Flower-like Fe_2O_3/ rGO	Hydrothermal growth	6.0 M KOH	318.9 F/g	0.1 A/g	81.9% (2000 cycles)	[115]
Fe_2O_3/rGO/Fe_2O_3	Hydrothermal growth	2.0 M KOH	337.5 mF/ Cm^2	20.0 mA/ cm^2	97.6% (2000 cycles)	[139]
Thumb-like Fe_2O_3/ rGO	Hydrothermal growth	1.0 M Na_2SO_4	255.0 F/g	0.5 A/g	90% (11,000 cycles)	[146]
Pyramid-shaped Fe_2O_3/rGO	Flame treatment	2.0 M KOH	965.7 F/g	5.0 mV/s	80% (2,000 cycles)	[149]
Spindle-like α-Fe_2O_3/rGO	Hydrothermal growth	2.0 M KOH	455.0 F/g	1.0 A/g	87.5% (10,000 cycles)	[22]
Mulberry-shaped α-Fe_2O_3/rGO	Hydrothermal growth	1.0 M KOH	1051.0 F/g	1.0 A/g	85.4% (1000 cycles)	[70]
Fe_2O_3/rGO	Chemical reduction	0.5 M H_2SO_4	50.0 F/g	0.1 V/s	–	[5]

(rGo—reduced graphene oxide, Fe_2O_3—iron (III) oxide, KOH—potassium hydroxide) of 1357 W/kg, and more than 219% capacitance retention after 10,000 cycles. [5] reported a cost-effective chemical approach to provide an alternative pathway for the large-scale production, but the electrochemical results were not interesting. Recently, a freestanding flexible cellulose-reduced graphene oxide-silver-Fe_2O_3 hybrid film synthesized in multiple steps using in situ reduction, vacuum filtration, and SILAR deposition method with a maximum thickness of 620 μm. The as-prepared electrodes can be directly used for construction of high-performance flexible solid-state supercapacitor. Single electrode in three-electrode systems show maximum areal and volumetric capacitance of 2044 mF/cm^2 and 75.7 F/cm^3, and the parallel device offers an areal and volumetric capacitance of 1132 mF/cm^2 and 18.2 F/cm^3, respectively. The electrode can also work in bent conditions and finds applications in advanced wearable electronics [139].

4.2.9 Nickel Oxide—Reduced Graphene Oxide

A nickel foam-supported graphene sheet/porous NiO hybrid film prepared using graphene sheets as substrates by combining electrophoretic deposition and chemical bath deposition was reported by [122] and obtained a graphene sheet film of about nineteen layers, which helped in forming a highly porous NiO film with a thickness of 10–20 nm. The composite exhibited a superior pseudocapacitive performance outsmarting porous NiO and showed 400 F/g at 2 A/g and 324 F/g at 40 A/g, maintaining 94% of capacitance retention over 2000 cycles at a 2 A/g current density.

Later [138] reported a unique two-dimension graphene-NiO nanosheets for that graphene oxide synthesized from the modified Hummer method, was used as starting material following a multiple step synthesis route which results in NiO nanoparticles of size 5–7 nm uniformly dispersed on the surface of graphene with a surface area of 134.5 m^2/g. The nanocomposite showed a specific capacitance of 525 F/g at 0.2 A/g with a retention of 95.4% after 1000 cycles. [145] reported reduced graphene oxide-NiO nanosheet-based microsphere composite by a simple approach using homogeneous co-precipitation and subsequent annealing. The nanocomposite showed a specific capacitance of ~700 F/g at a current density of 0.2 A/g and still as high as 455 F/g at 5 A/g with coulombic efficiency of more than 96% for each cycle during 1000 cycles at 2 A/g current density. This method not only provided an efficient material for energy storage applications but also provided an approach to distribute any metal oxide nanoparticles onto the surface of reduced graphene oxide sheets.

Wu et al. [121] reported NiO-graphene oxide using electrophoretic deposition in which graphene oxide nanosheets were deposited on stainless steel substrate using electrophoretic deposition and converting adsorbed nickel ions to nickel nanoparticles through electrochemical reduction followed by annealing at 300 °C to form NiO nanoparticles. The specific capacitance reached as high as 569 F/g which is 40 times higher than that of the bare graphene oxide electrode (13 F/g) at a discharge current density of 5 A/g. Later, without any complicated procedure and controlling agent Su et al. reported petal-like NiO sheets dispersed on the surfaces of graphene nanosheets-NiO through a microwave-assisted method. The graphene nanosheets-NiO composite exhibits a specific capacitance of 799 F/g at a current density of 0.3 A/g with 90% retention after 1000 cycles at 0.6 A/g. Significantly, the method can be used for the preparation of any other graphene nanosheets/metal oxide composite [100]. A one-pot green method was developed by [60] to prepare 3D flower-like hierarchical reduced graphene oxide-NiO composites using hydrogen as a reducing agent. Also, in this method, NiO acts as an additive and catalyst to facilitate the reduction. The composite showed a maximum specific capacitance of 428 F/g at a current density of 0.38 A/g maintaining 90% retention after 5000 cycles. Thus, producing high-performance hybrid composites at low cost.

Lee et al. [55] had studied the influence of the NiO composition and morphology on the electrochemical properties of NiO-reduced graphene oxide-based supercapacitor electrodes prepared by solvothermal methods. The reduced graphene oxide coating on electrode improves the electrical conductivity and mechanical stability of the electrodes. Moreover, morphology controls like particle size and thickness can be done simply by adding glucose during synthesis. Zhou et al. [143] also reported a green approach using glucose and graphene oxide as starting materials via hydrothermal method and synthesized NiO–graphene nanosheets composite with layered sandwich nanostructure which showed a specific capacitance of 1495 F/g at 0.5 A/g with 90% retention after 1000 cycles at 1 A/g.

Wu et al. [120] have developed a new 3D NiO-ultrathin-derived graphene hybrid on nickel foam (NiO/UDG/NF) as a binder-free supercapacitor electrode through a two-step approach, in which the 3D UDG-nickel foam scaffold was prepared via a simple nano casting process consisting of hydrothermal reaction and subsequent thermal transformation from an amorphous carbon source and NiO nanoflakes were grown (in situ) by a chemical bath deposition technique and showed a capacitance of 425 F/g at 2 A/g and 300 F/g at 10 A/g current density with retention of 79% after 2000 cycles. The chemical bath deposition method affords low-temperature large-scale preparation of 3D graphene hybrid composite materials. After that [147] reported a self-assembled hierarchical porous Ni/NiO/reduced graphene oxide nanohybrid, synthesized through a green sol–gel process method combined with subsequent in situ treatment at 400 °C. The prepared complex showed a specific capacitance of 1027.27 F/g at 2 A/g and 720 F/g at 20 A/g with retention of almost 93% after 1000 cycles at 5 A/g current density. The synergistically active sites of Ni–NiO complexes and graphene showed improved electronic conductivity and favors Faradic reaction.

Again, [108] reported an effective approach to synthesize 3D porous graphene-NiO nanoparticle composites. 3D porous graphene framework was produced by the self-assembly of poly (methyl methacrylate) (PMMA) particles and graphene oxide, followed by adsorption of Ni precursor on graphene oxide with simultaneous high-temperature treatment. The as-synthesized composite exhibits a specific capacitance of 1328 F/g at 1 A/g maintaining 87% retention after 2000 cycles at 2 A/g current density. Cao et al. [8] have come up with one more low-cost method to synthesize mesoporous NiO-reduced graphene oxide composites through a facile hydrothermal method which results from a specific capacitance of 1016.6 F/g at 1 A/g and 402.6 F/

g at 20 A/g, with almost 95% retention after 5000 cycles. Even after vast exploration of the applications of graphene-based metal oxides and their synthesis methods, researchers are coming up with new methods to make it easier with high capacitance but low material cost.

Lamiel et al. [54] reported reduced graphene oxide-NiO nanocomposites prepared using a simple ultrasonication method in non-aqueous solvent (tert-butanol) followed by calcination. Ultrasmall nickel oxide nanoparticles of 8.0 nm in size were deposited on the surface of the graphene sheets simultaneously with the partial reduction of graphene oxide and showed a specific capacitance of 689 F/g at 0.8 A/g with retention of 86.34% after 1500 cycles at 4 A/g current density. A facile, green, and effective method which can be readily applicable for the preparation of metal oxide-reduced graphene oxide composites was reported by Li et al. In their prepared hybrid architecture, graphene materials can be incorporated into hexagonal nanostructures to form high surface area, conductive porous networks. Using this method, they have synthesized NiO-reduced graphene oxide composite with well-controlled loose multi-layered structure by heterogeneous assembly approach with subsequent thermal treatment. The composite exhibited a specific capacitance of 782 F/g at 0.5 A/g with an excellent cycling stability of 94.1% retention after 3000 cycles at 2 A/g current density. An asymmetric supercapacitor was fabricated using NiO-reduced graphene oxide as cathode and activated carbon as anode which was able to light up a LED bulb. The fabricated device exhibited a maximum energy density of 32.5 Wh/kg at 375 W/kg power density and 19.78 Wh/kg at 7500 W/kg. Furthermore, the fabricated device showed high cycling stability with 92.7% retention after 3000 cycles at 2 A/g current density [59]. Another one-step electrochemical co-deposition method to produce binder-free composite electrodes reported by Shahrokhian et al. prepared a 3D NiO-reduced graphene oxide composite on a nickel foam substrate which showed a specific capacitance of 1715 F/g at 2 A/g and as high as 1066 F/g at 40 A/g current density, and retains more than 62% of its initial value [97]. Kahimbi et al. [49] also reported a one-step process for a high-performance NiO-reduced graphene oxide nanocomposites using ball milling method by taking graphene oxide and nickel powder as starting materials in a prescribed ratio. The as-prepared composite showed a specific capacitance of 590 F/g at 1 A/g and 522 F/g at 15 A/g with 100% retention after 1000 cycles at 1 A/g current density (Table 4.9).

4.2.9.1 Tungsten Trioxide—Reduced Graphene Oxide

Ma et al. [74] reported a flower-like $WO_3.H_2O$-reduced graphene oxide composite via the hydrothermal method. The preliminary studies confirmed that the prepared composite is of orthorhombic phase and morphological studies revealed the flower-like architecture which was evenly distributed over the reduced graphene oxide sheets. The hybrid delivered specific capacitances of 244, 226, 173, and 78 F/g at current densities of 1, 2, 4, and 10 A/g, respectively, with a 20% loss in capacitance over 900 cycles. [83] used poly (ionic liquid) as linker which literally enhances the contact between WO_3 and graphene. Morphological studies confirm the uniform

Table 4.9 Nickel oxide-graphene and their supercapacitance performance

Reported compound	Method	Electrolyte	Specific capacitance	Current density/ scan rate	Cyclic stability	References
NiO/ graphene films	Electrophoretic/ chemical bath deposition	1.0 M KOH	400.0 F/g	2.0 A/g	94%(2000 cycles)	[122]
2D graphene/ NiO	Pyrolysis	6.0 M KOH	525.0 F/g	0.2 A/g	95.4% (1000 cycles)	[138]
rGO/NiO	Co-precipitation	6.0 M KOH	700.0 F/g	0.2 A/g	96% (1000 cycles)	[145]
rGO/NiO	Electrophoretic	0.5 M KOH	564.0 F/g	5.0 A/g	–	[121]
Petal-like NiO/ graphene	Microwave-assisted	6.0 M KOH	799.0 F/g	0.3 A/g	90% (1000 cycles)	[100]
3D flower-like rGO/NiO	Solution route	6.0 M KOH	428.0 F/g	0.38 A/ g	90% (5000 cycles)	[60]
CP/NiO/ rGO	Solvothermal growth	6.0 M KOH	600.0 F/g	1.0 A/g	99% (1000 cycles)	[55]
NiO/ graphene nanosheets	Hydrothermal growth	6.0 M KOH	1495.0 F/g	0.5 A/g	90% (1000 cycles)	[143]
3D NiO/ ultrathin graphene	Simple nano casting and chemical bath deposition	5.0 M KOH	300.0 F/g	10.0 A/ g	79% (2000 cycles)	[120]
Porous Ni/ NiO/rGO	Sol–gel	6.0 M KOH	1027.0 F/g	2.0 A/g	92.95% (1000 cycles)	[147]
3D porous graphene/ NiO	High-temperature treatment	2.0 M KOH	1328.0 F/g	1.0 A/g	87% (2000 cycles)	[108]
rGO/NiO	Hydrothermal growth	6.0 M KOH	402.6 F/g	20.0 A/ g	95% (5000 cycles)	[8]
rGO/NiO	Ultrasonication	6.0 M KOH	689.0 F/g	0.8 A/g	86.34 (1500 cycles)	[54]
rGO/NiO	Chemical route	6.0 M KOH	782.0 F/g	0.5 A/g	94.1% (3000 cycles)	[59]

(continued)

Table 4.9 (continued)

Reported compound	Method	Electrolyte	Specific capacitance	Current density/ scan rate	Cyclic stability	References
3D NiO/ rGO@nickel foam	Electrochemical deposition	1.0 M KOH	1715.0 F/g	2.0 A/g	98.8% (2000 cycles)	[97]
rGO/NiO	Ball milling	1.0 M KOH	590.0 F/g	1.0 A/g	100% (1000 cycles)	[49]

(rGO—reduced graphene oxide, NiO—nickel oxide, KOH—potassium hydroxide)

distribution of WO_3 nanoparticles on reduced graphene oxide sheets. The prepared hybrid delivers a specific capacitance of 316 F/g at 1 A/g with 84.3% retention of capacitance over 8000 cycles. The enhanced performance of the prepared material due to the poly(ionic liquid) that acts as tight contact of both reduced graphene oxide and WO_3 layers which improves the electrical conductivity.

Chu et al. [15] improved the specific capacitance of WO_3-reduced graphene oxide nanohybrids to 495 F/g at 1 A/g with 87.5% retention after 1000 cycles. The hybrid was prepared by electrostatic attraction between a positive charge and negative charge modified WO_3 and reduced graphene oxide sheets, respectively, and found that the enhancement of the supercapacitor performance was due to the excellent interfacial contact and synergistic effect of both WO_3 and reduced graphene oxide. Guan et al. [36] also used a similar method to prepare a WO_3-reduced graphene oxide hybrid however, this hybrid achieved a specific capacitance of 343 F/g at a current density of 0.2 A/g for 1% incorporation of graphene content to the main material. Nayak et al. [80] fabricated an asymmetric supercapacitor with a WO_3-reduced graphene oxide nanohybrid achieving an energy density of 26.7 Wh/kg at a power density of 6 kW/kg. The group prepared the nanohybrid via the facile solvothermal method and studied its electrochemical behavior in 0.1 M H_2SO_4 aqueous electrolyte. The WO_3-reduced graphene oxide electrode delivered a specific capacitance of 465 F/g at 1 A/g and retained 97.7% of its initial capacitance over 2000 cycles. Moreover, the fabricated device retained 25 Wh/kg and 6 kWh/kg of energy and power density after 4000 cycles.

Samal et al. [92] improved the specific capacitance of WO_3-reduced graphene oxide further to 801 F/g with 75.7% capacitance retention. This nanohybrid was prepared via a one-pot hydrothermal method and investigated optimum graphene content to improve the electrochemical performance. The prepared hybrid exhibited that 80 mg of reduced graphene oxide content in the composite improved the supercapacitor performance, whereas the individual component demonstrated the specific capacitances of 71.11 F/g and 94.22 F/g only for reduced graphene oxide and WO_3, respectively. Pieretti et al. [84] reported a fast (10 min) preparation of WO_3-reduced graphene oxide nanohybrids based on nascent hydrogen reduction. The prepared

Table 4.10 Tungsten trioxide-graphene nanohybrid composites and their supercapacitance performance

Reported compound	Method	Electrolyte	Specific capacitance	Current density/ scan rate	Cyclic stability	References
$WO_3.H_2O$ / rGO	Hydrothermal growth	1.0 M H_2SO_4	244.0 F/g	1.0 A/g	80% (900 cycles)	[74]
WO_3/rGO	Hydrothermal growth	0.5 M H_2SO_4	316.0 F/g	1.0 A/g	84.3% (8000 cycles)	[83]
WO_3/rGO	Electrostatic attraction	0.5 M H_2SO_4	495.0 F/g	1.0 A/g	87.5% (1000 cycles)	[15]
WO_3/rGO	Hydrothermal growth	0.5 M H_2SO_4	343.0 F/g	0.2 A/g	–	[36]
WO_3/rGO	Solvothermal	0.1 M H_2SO_4	465.0 F/g	1.0 A/g	97.7% (2000 cycles)	[80]
WO_3/rGO	Hydrothermal growth	3.0 M KOH	801.0 F/g	4.0 A/g	75.7%	[92]
WO_3/rGO	Hydrogen reduction	1.0 M H_2SO_4	287.0 F/g	0.3 A/g	70% (3000 cycles)	[84]

(rGO—reduced graphene oxide, WO_3—tungsten oxide, H_2SO_4—sulfuric acid)

hybrid delivered a specific capacitance of 287 F/g at 0.3 A/g with 70% capacitance retention over 3000 cycles (Table 4.10).

4.2.9.2 Ruthenium Oxide—Reduced Graphene Oxide

Amir et al. [3] reported a specific capacitance of 418 F/g for the RuO_2-reduced graphene oxide nanohybrid with capacitance retention of 88.7% over 10,000 cycles. The RuO_2-reduced graphene oxide nanohybrid was synthesized using a sol gel method followed by electrophoretic deposition. Morphological studies showed a uniform anchoring of ultra-small RuO_2 nanoparticles on reduced graphene oxide nanosheets. Moreover, the rate capability of this nanohybrid supercapacitor displayed a capacitance retention of 85% on increasing the current density to over 20 A/g. Furthermore, the device was subjected to bend studies, which displayed a 4.9% decay of capacitance performance while bent to 180°.

Hassan et al. [39] reported a one-pot hydrothermal synthesis method to prepare Ru-nanoparticles-reduced graphene oxide and studied its electrochemical performance in 1 M $NaNO_3$ electrolyte. The prepared nanohybrid delivered a specific capacitance of 270 F/g with an energy density of 15 Wh/kg and a power density of 76 kWh/kg.

Table 4.11 Ruthenium oxide-graphene composites and their supercapacitance performance

Reported compound	Method	Electrolyte	Specific capacitance	Current density/ scan rate	Cyclic stability	References
RuO_2/rGO	Sol–gel and electrophoretic	0.5 M H_2SO_4/PVA	418.0 F/g	1.0 A/g	88.7% (1000 cycles)	[3]
Ru NP / rGO	Hydrothermal growth	1.0 M Na_2NO_3	270.0 F/g	5 mV/s	92.5% (5000 cycles)	[39]
PVC (RuO_2/ rGO)	Liquid phase reduction	1.0 M H_2SO_4	2698.0 F/g	2 mV/s	–	[4]
RuO_2/ rGO/CNT	Dip coating	0.5 M H_2SO_4	973.0 F/g	25 mV/s	60.5%	[42]
RuO_2/ CNT/RuO_2	Precipitation method	1.0 M H_2SO_4	1846.3 F/g	10 mV/s	98.37% (500 cycles)	[40]

(rGO—reduced graphene oxide, CNT—carbon nanotubes, RuO_2—ruthenium oxide, PVC—polyvinyl chloride, H_2SO_4—sulfuric acid)

Moreover, the prepared nanohybrid showed better cyclic stability with 92.5% capacitance retention over 5000 cycles at a current density of 24 A/g. On the other hand, RuO_2-reduced graphene oxide with other materials enormously improves electrochemical performance. Ates et al. [4] prepared RuO_2-reduced graphene oxide with the addition of polyvinyl carbazole and obtained a specific capacitance of 2698 F/g at 2 mV/s with an energy density and power density of 17.711 Wh/kg and 2684.36 W/kg, respectively. A RuO_2-reduced graphene oxide/carbon nanotubes composite was reported by [42] and achieved a specific capacitance of 973 F/g at 25 mV/s retaining, 60.5% of its initial capacitance when varying scan rate from 5 to 500 mV/s [42]. The further addition of MWCNT into the RuO_2-reduced graphene oxide composite exhibited a huge improvement in electrochemical performance with a specific capacitance of 1846.3 F/g at 10 mV/s with 98.3% capacitance retention over 500 cycles at 100 A/g [40] (Table 4.11).

4.3 Ternary Metal Compounds—Reduced Graphene Oxide Hybrids

Among all the transition metals an impressive amount of work has been done with nickel cobaltites compared to other transition metals. Preliminarily, Ma et al. [73] synthesized a composite with porous $NiCo_2O_4$ hexagonal nanoplates with an average side length of 61 nm and thickness 9.5 nm on reduced graphene oxide sheets through a simple hydrothermal method that was followed by annealing. The composite exhibited a specific capacitance of 947.4 F/g at 0.5 A/g and 725.4 F/g at 10 A/g, with a retention of 97.9% after 3000 cycles at a 10 A/g current density. Sun et al. [103] reported a porous $NiCo_2O_4$ nanograss grown in situ on nickel foam achieved a remarkable enhancement in electrochemical performance through facile hydrogenation for 3 h at 300 °C, and concluded that the hydrogenation process increased the specific capacitance of pristine compound (i.e., non-hydrogenated) by 240% and stability retention to 71% from 54%.

Zhu et al. [148] reported a simple hydrothermal method without any binders to grow porous flower-like $NiCoO_2$-reduced graphene oxide composite nanoarrays on nickel foam. The composite showed a specific capacity of 1286 C/g at 0.5 A/g and 1071.5 C/g at 12 A/g, respectively. An asymmetric supercapacitor assembled with $NiCoO_2$-reduced graphene oxide and reduced graphene oxide-nickel foam working electrodes showed an energy density of 38.5 Wh/kg at a 288 W/kg power density and cycle stability of 86.9% capacitance retention after 2000 cycles. [45] reported a nearly two-dimensional serrated structure which can be optimized by controlling the time of $NiCoO_2$ on nickel foam using a hydrothermal method and heat treatment. The composite showed a specific capacitance of 778.5 C/g at 0.5 A/g and 603.5 C/g at 40 A/g with 93% retention after 2000 cycles at a 5 A/g current density. To address the poor intrinsic conductivity and inevitable agglomeration issues, [58] synthesized a well-controlled layered $NiCo_2O_4$-reduced graphene oxide nanocomposite through a layer-by-layer assembly of exfoliated Co–Ni layered double hydroxide and graphene oxide nanosheets with opposite charges, followed by freeze-drying and annealing treatment. The composite showed a specific capacitance of 1388 F/g at 0.5 A/g and 840 F/g at 30 A/g, with a retention of 90.2% after 20,000 cycles at 5 A/g. An asymmetric supercapacitor assembled using prepared composite and activated carbon as working electrodes achieved a high energy density of 57 Wh/kg at a power density of 375 W/kg.

Among the nickel cobaltites reports, [72] reported a hydroxide composite instead of an oxide hybrid nanolayer. The Ni, Co–OH-reduced graphene oxide nanocomposite with an average thickness of 1.37 nm were synthesized using a one-pot hydrothermal method. Using the advantage of a face-to-face contact model between Ni − Co hydroxide and reduced graphene oxide sheets, an asymmetric supercapacitor was developed using the composite and three-dimensional (3D) hierarchical porous carbon. The device exhibited an energy density of 56.1 Wh/kg, with 80% retention after 17,000 cycles.

Xu et al. [128] synthesized two types of porous cobalt manganese oxide nanowires such as $MnCo_2O_4$ and $CoMn_2O_4$ with different structures through the thermal decomposition of organometallic compounds. Nitrilotriacetic acid was used as a chelating agent to coordinate Co(II) and Mn(II) ions in various molar ratios, in hydrothermal conditions. $CoMn_2O_4$ showed specific capacitance of 2108 F/g at 1 A/g and 1191 F/g at 20 A/g, whereas $MnCo_2O_4$ showed 1342 F/g at 1 A/g and 988 F/g at 20 A/g. [91] reported ternary $ZnCo_2O_4$-reduced graphene oxide-NiO nanowire arrays grown on nickel foam using a hydrothermally assisted thermal annealing process without nickel precursor salt and used as a binder-free supercapacitor electrode. The resulting

3D ternary composite possessed an ultra-high-specific capacitance of 1256 F/g at 3 A/g with 80% capacitance retention after 3000 cycles. The device showed an energy density of 62.8 Wh/kg at a power density of 7492.5 W/kg, and a low equivalent series resistance of 0.58 Ω.

Metal molybdates are the most investigated ternary transition metal compounds next to the metal cobaltites. Liu et al. [71] fabricated an asymmetric supercapacitor with $NiMoO_4$-reduced graphene oxide nanohybrid which was synthesized via microwave-assisted solvothermal method. Mesoporous morphology of $NiMoO_4$ and uniformly anchoring of nanoparticles onto graphene sheets were revealed by scanning electron micrograph. Electrochemical studies were carried out in 6 M KOH aqueous electrolyte solution. The prepared $NiMoO_4$-reduced graphene oxide and as-prepared $NiMoO_4$ showed a specific capacitance of 1274 F/g and 800 F/g at a current density of 1 A/g, respectively. Moreover, the $NiMoO_4$-reduced graphene oxide nanocomposite retained 81.7% of its initial capacitance over 1000 cycles. The fabricated device showed an energy density of 30.3 Wh/kg at a power density of 187 W/kg. The authors concluded that these enhanced performance was due to the mesoporous structure with large surface area greatly improved the supercapacitor performance.

Li et al. [62] achieved an improved specific capacitance of $NiMoO_4$-reduced graphene oxide further with the influence of sheet-on-sheet nanostructures. The sheets were interconnected together as 3D network results in mesoporous with large surface area. The prepared composite delivered a specific capacitance of 1888 F/g at 1 A/g which remained 1271 F/g at a high current density of 10 A/g. Moreover, the prepared hybrid shows only 5.1% loss from its initial capacitance over 2300 discharge cycles. This excellent performance is due to its "sheet-on-sheet" nanostructures which enhance the rapid electron and ions transportation across electrode–electrolyte interface.

A honeycomb-like $NiMoO_4$ nanosheets on reduced graphene oxide were reported by [142] the material was directly grown on nickel foam by simple hydrothermal method. The prepared nanohybrid showed an areal capacitance of 2165 mF/cm^2 at 1 mA/cm^2 with 93.8% capacitance retention over 1000 charge–discharge cycles. An asymmetric supercapacitor was fabricated with the prepared nanohybrid on nickel-reduced graphene oxide and nickel-reduced graphene oxide-$NiMoO_4$ as negative and positive electrodes, respectively, and achieved an energy density of 0.579 mWh/cm^2, with 93.1% retention over 5000 cycles. The authors concluded that the nickel foam served as skeleton which supported reduced graphene oxide that directly grown on Ni foam unlocks more pathways for the carriers, whereas the $NiMoO_4$ acted as active substance results in high areal capacitance.

On the other hand, [48] reported a $CoMoO_4$ flower-like-reduced graphene oxide was grown directly on nickel foam via one-step hydrothermal method. The prepared nanohybrid exhibited a specific capacitance of 856.2 F/g with a loss of 5.5% from its initial capacitance over 2000 cycles at a current density of 1 A/g. The interfacial charge transfer resistance was improved with the addition of reduced graphene oxide and $CoMoO_4$, which resulted in improved conductivity at the electrode–electrolyte interface. The authors concluded that the enhanced supercapacitance performance attributed to the large specific surface area, synergistic effect and flower-like structure of $CoMoO_4$, increased the electrode active sites for electrochemical reactions.

Metal ferrites have also attracted applications as supercapacitor electrode material, Zhang et al. [134] reported a facile approach to preparing copper ferrite nanoparticles approximately 100 nm diameter attached to a graphene nanosheet, $CuFe_2O_4$-graphene nanosheet using an efficient, scalable, green, and controllable single-step solvothermal method. Graphene in the composites played a considerable role in obtaining uniform $CuFe_2O_4$ nanoparticles and enhancing the electrochemical properties compared to pure $CuFe_2O_4$ nanoparticles. The composite showed an electrochemical capacitance of 576.6 F/g at 1 A/g, with a cyclic stability of 85% capacitance retention after 1000 cycles. The versatile $CuFe_2O_4$-graphene nanosheet composites are very promising candidates for supercapacitor materials. They can find application in a wide range of electrochemical fields because of the remarkable synergistic effects between $CuFe_2O_4$ nanoparticles and graphene.

Dong et al. [19] reported perpendicular ultrathin cobalt ferrite nanosheets assembled on the surface of reduced graphene oxide, $CoFe_2O_4$ nanosheets-reduced graphene oxide via adjusting the volume ratio of ethanol and deionized water followed by calcination treatment. The composite exhibited a specific capacitance of 2238.3 F/g at 1 A/g and 1225 F/g at 20 A/g current density, with a capacitance retention of 71% after 3000 cycles. In addition to that, it also finds an application as anode material in lithium-ion batteries.

For the first time [69] recycled waste steel oil sludge containing high amounts of iron via a facile hydrothermal method and used it as a ferric ions source for the preparation of $CoFe_2O_4$-graphene nanocomposite electrode. The complex exhibited a rough morphology and porous nanostructure with a high specific surface area of 191.5 m^2/g. It also exhibited a specific capacity of 166.5 C/g at 0.5 A/g and 127 C/g at 10 A/g current density, with a capacitance retention of 79.3% after 5000 cycles. The first approach of metal tungstates as electrode material for supercapacitors was attempted by Xu et al. [127] in their consecutive works, they prepared $CoWO_4$-reduced graphene oxide nanospheres and $NiWO_4$-reduced graphene oxide nanohybrid using the simple hydrothermal method and the resulting supercapacitor performance was compared with the as-prepared one $CoWO_4$ and $NiWO_4$, respectively. The morphological studies showed that the prepared $CoWO_4$ and $NiWO_4$ nanoparticles are well-anchored on reduced graphene oxide sheets. The electrochemical behavior was analyzed using a 2 M KOH aqueous electrolyte and obtained a specific capacitance of 159 F/g and 60.6 F/g for the prepared $CoWO_4$-reduced graphene oxide and as-prepared $CoWO_4$ at a scan rate of 5 mV/s, respectively. Moreover, the prepared hybrid showed a 94.7% capacitance retention over 1000 cycles. In the case of $NiWO_4$-graphene nanocomposite, they have obtained a specific capacitance of 1031.3 F/g at a current density of 0.5 A/g with no significant loss from its initial capacitance over 5000 cycles [126].

Metal vanadates as supercapacitor electrode material were first reported by [82] which prepared $BiVO_4$-reduced graphene oxide nanocomposites via a simple hydrothermal method. Morphological studies revealed that the prepared $BiVO_4$-reduced graphene oxide showed a unique dendrite-like structure. The prepared nanohybrid delivered a specific capacitance of 151 F/g at the current density of 0.15 mA/cm^2. A symmetric cell was fabricated with the prepared $BiVO_4$-reduced graphene oxide nanohybrid and it delivered an energy density of 33.7 Wh/kg with a power density of 8.0 kW/kg. Dutta et al. [25] fabricated a symmetric cell based on $BiVO_4$-reduced graphene oxide prepared via hydrothermal reaction and studied its electrochemical behavior in both polyvinyl alcohol-H_2SO_4 and aqueous Na_2SO_4 electrolytes. The prepared nanohybrid showed a specific capacitance of 400 F/g at a scan rate of 5 mV/s, and an energy density of 35.37 Wh/kg with 98% retention from its initial capacitance over 1000 cycles.

Metal manganites have also been reported as supercapacitor electrode materials. Giri et al. [35] developed $NiMnO_3$-nitrogen-doped graphene nanohybrid via a simple hydrothermal method and achieved a specific capacitance of 750 F/g at a scan rate of 1 mV/s with 88.5% retention of its initial capacitance over 1000 cycles. The authors proposed that the incorporation of nitrogen into graphene sheets decreases the decomposition of electrolytes. On the other hand, it also produces pyridine-N and pyrrole-N which reduces the internal resistance and thus increase the electrical conductivity of graphene. Wang et al. 2016a prepared Ni_6MnO_8-reduced graphene oxide via a template route in which the as-prepared Ni_6MnO_8 was loaded uniformly on the reduced graphene oxide nanosheets. The prepared hybrid obtained a specific capacitance of 862.5 F/g which is higher than that of the as-prepared Ni_6MnO_8 (498.3 F/g) at a current density of 0.5 A/g with only 10% loss from its initial capacitance over 9000 cycles. An ultrathin $CoMn_2O_4$ nanosheets-reduced graphene oxide nanohybrid was reported by [28] and achieved a specific capacitance of 876.1 F/g over 5000 cycles with 6.4% capacitance loss. However, the initial capacitance of $CoMn_2O_4$-reduced graphene oxide was remarkably high at the first cycle, i.e., 1028 F/g and decreased to 973.8 F/g at 4 A/g, even after 5000 cycles. These enhanced electrochemical performances of the prepared nanohybrids are due to the novel hierarchical hybrid nanostructure which facilitates the electrolyte penetration and electron transport between the electrode and electrolyte interface (Table 4.12).

4.4 Conclusions

In this chapter, various binary and ternary metal oxide-graphene nanohybrids and their electrochemical behavior were presented and listed in tables. The incorporation of graphene, reduced graphene oxide and graphene oxide surpasses the poor conductivity of metal oxides and leads to an enhanced electronic conductivity at the electrode–electrolyte interface, whereas the metal oxides prevent restacking of 2D graphene sheets to deliver the full functionality of graphene. In addition to that, metal oxides with various morphologies, large surface area, and porosity increased the electrochemical active sites shorten the ion diffusion pathway for the electrolyte ions and improved the charge storage capability, rate capability, energy density, and

Table 4.12 Ternary metal oxides-reduced graphene oxide and their supercapacitance performance

Reported compound	Method	Electrolyte	Specific capacitance	Current density/ scan rate	Cyclic stability	References
Hexagonal $NiCo_2O_4$/rGO	Hydrothermal growth	3.0 M KOH	947.4 F/g	0.5 A/g	97.9% (3000 cycles)	[73]
$NiCoO_2$/rGO/NF	Hydrothermal growth	6.0 M KOH	1286.0 F/g#	0.5 A/g	86.9% (2000 cycles)*	[148]
Layered $NiCoO_2$/rGO	Co-precipitation method	6.0 M KOH	1388.0 F/g	0.5 A/g	90.2% (20,000 cycles)*	[58]
Ni, Co–OH/rGO	Hydrothermal growth	6.0 M KOH	1691.0 F/g	0.5 A/g	78.5% (17,000 cycles)	[72]
$ZnCo_2O_4$/rGO/NiO	Hydrothermal growth	6.0 M KOH	1256.0 F/g	3.0 A/g	80% (3000 cycles)	[91]
Mesoporous $NiMoO_4$/rGO	Solvothermal growth	6.0 M KOH	1274.0 F/g	1.0 A/g	81.7% (1000 cycles)	[71]
$NiMoO_4$/rGO	Hydrothermal growth	2.0 M KOH	1888.0 F/g	1.0 A/g	94.9% (2300 cycles)	[62]
Honeycomb-like $NiMoO_4$@rGO@NF	Hydrothermal growth	3.0 M KOH	2165.0 mF/cm^2	1.0 mA/cm^2	93.8% (1000 cycles)	[142]
Flower-like $CoMoO_4$/rGO	Hydrothermal growth	2.0 M KOH	856.2 F/g	1.0 A/g	94.5% (2000 cycles)	[48]
$CuFe_2O_4$-GN	Hydrothermal growth	3.0 M KOH	576.6 F/g	1.0 A/g	85% (1000 cycles)	[134]
$CoFe_2O_4$ NSs@rGO	Simple chemical route	2.0 M KOH	2238.3 F/gs	1.0 A/g	71% (3000 cycles)	[19]
$CoFe_2O_4$/graphene	Hydrothermal growth	5.0 M KOH	166.5 F/g	0.5 A/g	79.3% (5000 cycles)	[69]
$CoWO_4$/rGO	Hydrothermal growth	2.0 M KOH	159.0 F/g	5.0 mV/s	94.7% (1000 cycles)	[127]
$NiWO_4$/rGO	Hydrothermal growth	2.0 M KOH	1031.3 F/g	0.5 A/g	100% (5000 cycles)	[126]

(continued)

Table 4.12 (continued)

Reported compound	Method	Electrolyte	Specific capacitance	Current density/ scan rate	Cyclic stability	References
Dendrite-like $BiVO_4$/rGO	Hydrothermal growth	6.0 M KOH	151.0 F/g	0.15 mA/ cm^2	80.3% (2000 cycles)*	[82]
$BiVO_4$/rGO	Hydrothermal growth	1.0 M Na_2SO_4	400.0 F/g	5.0 mV/s	98% (1000 cycles)*	[25]
$NiMnO_3$N-doped graphene	Hydrothermal growth	1.0 M Na_2SO_4	750.2 F/g	1.0 mV/s	88.5% (1000 cycles)	[35]
Murdochite-type Ni_6MnO_8/rGO	Chemical Route	6.0 M KOH	862.5 F/g	0.5 A/g	90% (9000 cycles)	[115]
$CoMn_2O_4$ NS/GO	Chemical Route	6.0 M KOH	1089.3 F/g	1.0 A/g	94.7% (5000 cycles)	[28]

(*represents the cyclic stability of the fabricated device, rGO—reduced graphene oxide, GO—graphene oxide, NS—nanosheet, $BiVO_4$—bismuth vanadate, KOH—potassium hydroxide)

cyclic stability. The combination of metal oxides with graphene results in a synergistic effect of both electrochemical double layer and pseudocapacitive behavior which greatly improves the supercapacitor performance.

In the material preparation aspect, the hydrothermal method has been widely reported as a common method, due to its low cost and simplicity, to prepare metal oxide-graphene nanohybrids of various morphologies resulted in better electrochemical performance in which the reduced graphene oxide and graphene oxide were synthesized using modified Hummer's method, whereas graphene sheets were prepared using chemical vapor deposition. However, the optimization of graphene content in the metal oxide matrix, morphology tuning, uniform distribution of active material on graphene sheets, choice of metal oxides to graphene and electrolyte, and preparation methods are the crucial factors that limit these nanohybrids from large-scale production. The design and fabrication of supercapacitor devices should possess an operating temperature range, long lifetime, self-discharge rate, packaging etc., for commercial applications. Further investigations on understanding the mechanism of electrochemical interactions should be undertaken with the aid of theoretical as well as experimental routes required to develop novel designs and development of various metal oxide-graphene hybrids-based supercapacitor devices.

Acknowledgements The authors acknowledge the financial assistance received from CSIR 80 (0093)/20/EMR-II, and MHRD-RUSA (C5/RUSA 1.0 /2018/1452 & RUSA 2.0/T2/P3 PF8/2021/ 035) in the form of research projects.

References

1. P. Agharezaei, H. Abdizadeh, M.R. Golobostanfard, Flexible supercapacitor electrodes based on TiO_2/rGO/TiO_2 sandwich type hybrids. Ceram. Int. **44**, 4132–4141 (2018). https://doi.org/10.1016/j.ceramint.2017.11.214
2. A.J. Ahamed, P.V. Kumar, G. Srikesh, Low-temperature synthesis and characterization of rGO-CoO nanocomposite with efficient electrochemical properties. J. Environ. Nanotechnol. **4**, 01–08 (2015). https://doi.org/10.13074/jent.2015.06.152147
3. F.Z. Amir, V.H. Pham, D.W. Mullinax, J.H. Dickerson, Enhanced performance of HRGO-RuO2 solid-state flexible supercapacitors fabricated by electrophoretic deposition. Carbon N. Y. **107**, 338–343 (2016). https://doi.org/10.1016/j.carbon.2016.06.013
4. M. Ates, M. Yildirim, O. Kuzgun, H. Ozkan, The synthesis of rGO, rGO/RuO2 and rGO/RuO2/PVK nanocomposites, and their supercapacitors. J. Alloys Compd. **787**, 851–864 (2019). https://doi.org/10.1016/j.jallcom.2019.02.126
5. R. Bhujel, S. Rai, U. Deka, B.P. Swain, Electrochemical, bonding network and electrical properties of reduced graphene oxide-Fe_2O_3 nanocomposite for supercapacitor electrodes applications. J. Alloys Compd. **792**, 250–259 (2019). https://doi.org/10.1016/j.jallcom.2019.04.004
6. T. Brousse, D. Bélanger, J.W. Long, To be or not to be pseudocapacitive? J. Electrochem. Soc. **162**, A5185–A5189 (2015). https://doi.org/10.1149/2.0201505jes
7. I.Y.Y. Bu, R. Huang, Fabrication of CuO-decorated reduced graphene oxide nanosheets for supercapacitor applications. Ceram. Int. **43**, 45–50 (2017). https://doi.org/10.1016/j.ceramint.2016.08.136
8. P. Cao, L. Wang, Y. Xu et al., Facile hydrothermal synthesis of mesoporous nickel oxide/reduced graphene oxide composites for high performance electrochemical supercapacitor. Electrochim. Acta **157**, 359–368 (2015). https://doi.org/10.1016/j.electacta.2014.12.107
9. P.Y. Chan, M.S.R. Rusi, RGO-wrapped MnO_2 composite electrode for supercapacitor application. Solid State Ionics **262**, 226–229 (2014). https://doi.org/10.1016/j.ssi.2013.10.005
10. G.Z. Chen, Supercapacitor and supercapattery as emerging electrochemical energy stores. Int. Mater. Rev. **62**, 173–202 (2017). https://doi.org/10.1080/09506608.2016.1240914
11. M. Chen, H. Wang, L. Li et al., Novel and facile method, dynamic self-assemble, to prepare SnO_2/rGO droplet aerogel with complex morphologies and their application in supercapacitors. ACS Appl. Mater. Interfaces **6**, 14327–14337 (2014). https://doi.org/10.1021/am5036169
12. S. Chen, W. Xing, J. Duan et al., Nanostructured morphology control for efficient supercapacitor electrodes. J. Mater. Chem. A **1**, 2941–2954 (2013). https://doi.org/10.1039/c2ta00627h
13. S. Chen, J. Zhu, X. Wu et al., Graphene oxide-Mno_2 nanocomposites for supercapacitors. ACS Nano **4**, 2822–2830 (2010). https://doi.org/10.1021/nn901311t
14. Y.L. Chen, Z.A. Hu, Y.Q. Chang et al., Zinc oxide/reduced graphene oxide composites and electrochemical capacitance enhanced by homogeneous incorporation of reduced graphene oxide sheets in zinc oxide matrix. J. Phys. Chem. C **115**, 2563–2571 (2011). https://doi.org/10.1021/jp109597n
15. J. Chu, D. Lu, X. Wang et al., WO3 nanoflower coated with graphene nanosheet: synergetic energy storage composite electrode for supercapacitor application. J. Alloys Compd. **702**, 568–572 (2017). https://doi.org/10.1016/j.jallcom.2017.01.226
16. A.K. Das, S. Sahoo, P. Arunachalam et al., Facile synthesis of Fe_3O_4 nanorod decorated reduced graphene oxide (RGO) for supercapacitor application. RSC Adv. **6**, 107057–107064 (2016). https://doi.org/10.1039/c6ra23665k
17. W. Deng, W. Lan, Y. Sun et al., Porous CoO nanostructures grown on three-dimension graphene foams for supercapacitors electrodes. Appl. Surf. Sci. **305**, 433–438 (2014). https://doi.org/10.1016/j.apsusc.2014.03.107

18. Y. Ding, W. Bai, J. Sun et al., Cellulose tailored anatase TiO_2 nanospindles in three-dimensional graphene composites for high-performance supercapacitors. ACS Appl. Mater. Interfaces **8**, 12165–12175 (2016). https://doi.org/10.1021/acsami.6b02164
19. B. Dong, M. Li, C. Xiao et al., Tunable growth of perpendicular cobalt ferrite nanosheets on reduced graphene oxide for energy storage. Nanotechnology **28** (2017). https://doi.org/10.1088/1361-6528/28/5/055401
20. X. Dong, K. Wang, C. Zhao et al., Direct synthesis of RGO/Cu_2O composite films on Cu foil for supercapacitors. J. Alloys Compd. **586**, 745–753 (2014). https://doi.org/10.1016/j.jallcom.2013.10.078
21. X. Dong, X. Wang, J. Wang et al., Synthesis of a MnO_2-graphene foam hybrid with controlled MnO_2 particle shape and its use as a supercapacitor electrode. Carbon N. Y. **50**, 4865–4870 (2012). https://doi.org/10.1016/j.carbon.2012.06.014
22. Y. Dong, L. Xing, F. Hu et al., α-Fe_2O_3/rGO nanospindles as electrode materials for supercapacitors with long cycle life. Mater. Res. Bull. **107**, 391–396 (2018). https://doi.org/10.1016/j.materresbull.2018.07.038
23. X. Du, S. Wang, Y. Liu et al., Self-assembly of free-standing hybrid film based on graphene and zinc oxide nanoflakes for high-performance supercapacitors. J. Solid State Chem. **277**, 441–447 (2019). https://doi.org/10.1016/j.jssc.2019.06.003
24. D.P. Dubal, N.R. Chodankar, G.S. Gund et al., Asymmetric supercapacitors based on hybrid CuO@reduced graphene oxide@sponge versus reduced graphene oxide@sponge electrodes. Energy Technol. **3**, 168–176 (2015). https://doi.org/10.1002/ente.201402170
25. S. Dutta, S. Pal, S. De, Hydrothermally synthesized BiVO4-reduced graphene oxide nanocomposite as a high performance supercapacitor electrode with excellent cycle stability. New J. Chem. **42**, 10161–10166 (2018). https://doi.org/10.1039/c8nj00859k
26. A. Ejigu, K. Fujisawa, B.F. Spencer et al., On the role of transition metal salts during electrochemical exfoliation of graphite: antioxidants or metal oxide decorators for energy storage applications. Adv. Funct. Mater. **28**, 1–12 (2018). https://doi.org/10.1002/adfm.201804357
27. E.R. Ezeigwe, P.S. Khiew, C.W. Siong, M.T.T. Tan, Solvothermal synthesis of $NiCo_2O_4$ nanocomposites on liquid-phase exfoliated graphene as an electrode material for electrochemical capacitors. J. Alloys Compd. **693**, 1133–1142 (2017). https://doi.org/10.1016/j.jallcom.2016.09.244
28. G. Gao, S. Lu, Y. Xiang et al., Free-standing ultrathin $CoMn_2O_4$ nanosheets anchored on reduced graphene oxide for high-performance supercapacitors. Dalt Trans. **44**, 18737–18742 (2015). https://doi.org/10.1039/c5dt03696h
29. H. Gao, F. Xiao, C.B. Ching, H. Duan, High-performance asymmetric supercapacitor based on graphene hydrogel and nanostructured MnO_2. ACS Appl. Mater. Interfaces **4**, 2801–2810 (2012). https://doi.org/10.1021/am300455d
30. Y. Gao, D. Wu, T. Wang et al., One-step solvothermal synthesis of quasi-hexagonal Fe_2O_3 nanoplates/graphene composite as high performance electrode material for supercapacitor. Electrochim. Acta **191**, 275–283 (2016). https://doi.org/10.1016/j.electacta.2016.01.072
31. J. Ge, Y.H. Bin, W. Hu et al., Facile dip coating processed graphene/MnO_2 nanostructured sponges as high performance supercapacitor electrodes. Nano Energy **2**, 505–513 (2013). https://doi.org/10.1016/j.nanoen.2012.12.002
32. J. Geng, C. Ma, D. Zhang, X. Ning, Facile and fast synthesis of SnO_2 quantum dots for high performance solid-state asymmetric supercapacitor. J. Alloys Compd. **825**, 153850 (2020). https://doi.org/10.1016/j.jallcom.2020.153850
33. S. Ghasemi, S.R. Hosseini, O. Boore-talari, Sonochemical assisted synthesis MnO_2/RGO nanohybrid as effective electrode material for supercapacitor. Ultrason. Sonochem. **40**, 675–685 (2018). https://doi.org/10.1016/j.ultsonch.2017.08.013
34. M. Ghorbani, M.R. Golobostanfard, H. Abdizadeh, Flexible freestanding sandwich type ZnO/rGO/ZnO electrode for wearable supercapacitor. Appl. Surf. Sci. **419**, 277–285 (2017). https://doi.org/10.1016/j.apsusc.2017.05.060
35. S. Giri, D. Ghosh, C.K. Das, One pot synthesis of ilmenite-type NiMnO3-"nitrogen-doped" graphene nanocomposite as next generation supercapacitors. Dalt Trans. **42**, 14361–14364 (2013). https://doi.org/10.1039/c3dt51807h

36. X.H. Guan, Z.W. Zhang, L. Yang, G.S. Wang, One-pot hydrothermal synthesis of hexagonal WO3 nanorods/graphene composites as high-performance electrodes for supercapacitors. ChemPlusChem **82**, 1174–1181 (2017). https://doi.org/10.1002/cplu.201700288
37. G. Guo, L. Huang, Q. Chang et al., Sandwiched nanoarchitecture of reduced graphene oxide/ZnO nanorods/reduced graphene oxide on flexible PET substrate for supercapacitor. Appl. Phys. Lett. **99**, 98–101 (2011). https://doi.org/10.1063/1.3629789
38. Y. Haldorai, Y.S. Huh, Y.K. Han, Surfactant-assisted hydrothermal synthesis of flower-like tin oxide/graphene composites for high-performance supercapacitors. New J. Chem. **39**, 8505–8512 (2015). https://doi.org/10.1039/c5nj01442e
39. H.K. Hassan, N.F. Atta, M.M. Hamed et al., Ruthenium nanoparticles-modified reduced graphene prepared by a green method for high-performance supercapacitor application in neutral electrolyte. RSC Adv. **7**, 11286–11296 (2017). https://doi.org/10.1039/c6ra27415c
40. M.G. Hosseini, E. Shahryari, Anchoring RuO2 nanoparticles on reduced graphene oxide-multi-walled carbon nanotubes as a high-performance supercapacitor. Ionics (Kiel) **25**, 2383–2391 (2019). https://doi.org/10.1007/s11581-018-2668-2
41. B. Hu, C. Guo, C. Xu et al., Rational construction of V2O5@rGO with enhanced pseudocapacitive storage for high-performance flexible energy storage device. ChemElectroChem **6**, 5845–5855 (2019). https://doi.org/10.1002/celc.201901680
42. C.C. Hu, C.W. Wang, K.H. Chang, M.G. Chen, Anodic composite deposition of RuO2/reduced graphene oxide/carbon nanotube for advanced supercapacitors. Nanotechnology **26** (2015). https://doi.org/10.1088/0957-4484/26/27/274004
43. R. Hu, J. Zhao, J. Zheng, Synthesis of SnO_2/rGO hybrid materials by sol-gel/thermal reduction method and its application in electrochemical capacitors. Mater. Lett. **197**, 59–62 (2017). https://doi.org/10.1016/j.matlet.2017.03.082
44. Y. Hu, C. Guan, Q. Ke et al., Hybrid Fe_2O_3 nanoparticle clusters/rGO paper as an effective negative electrode for flexible supercapacitors. Chem. Mater. **28**, 7296–7303 (2016). https://doi.org/10.1021/acs.chemmater.6b02585
45. H. Huang, H. Zhang, Y. Fan et al., Serrated-like $NiCoO_2$ nanoarrays on Ni foam for high-performance supercapacitors. Appl. Surf. Sci. **481**, 1220–1227 (2019). https://doi.org/10.1016/j.apsusc.2019.03.260
46. M.S. Javed, A.J. Khan, S. Batool et al., Ultra-fast one-step synthesis and surface engineering of mesoporous 3D α-Fe_2O_3 hollow nanospheres for high-performance and stable negative electrode for supercapacitors. Appl. Surf. Sci. **526**, 146634 (2020). https://doi.org/10.1016/j.apsusc.2020.146634
47. J. Jayachandiran, J. Yesuraj, M. Arivanandhan et al., Synthesis and electrochemical studies of rGO/ZnO nanocomposite for supercapacitor application. J. Inorg. Organomet. Polym. Mater. **28**, 2046–2055 (2018). https://doi.org/10.1007/s10904-018-0873-0
48. L. Jinlong, Y. Meng, K. Suzuki, H. Miura, Synthesis of CoMoO4@RGO nanocomposites as high-performance supercapacitor electrodes. Microporous Mesoporous Mater. **242**, 264–270 (2017). https://doi.org/10.1016/j.micromeso.2017.01.034
49. H. Kahimbi, S.B. Hong, M.H. Yang, B.G. Choi, Simultaneous synthesis of NiO/reduced graphene oxide composites by ball milling using bulk Ni and graphite oxide for supercapacitor applications. J. Electroanal. Chem. **786**, 14–19 (2017). https://doi.org/10.1016/j.jelechem.2017.01.013
50. H. Kim, M.Y. Cho, M.H. Kim et al., A novel high-energy hybrid supercapacitor with an anatase tio 2-reduced graphene oxide anode and an activated carbon cathode. Adv. Energy Mater. **3**, 1500–1506 (2013). https://doi.org/10.1002/aenm.201300467
51. M. Kim, Y. Hwang, J. Kim, Graphene/MnO_2-based composites reduced via different chemical agents for supercapacitors. J. Power. Sources **239**, 225–233 (2013). https://doi.org/10.1016/j.jpowsour.2013.03.146
52. T.W. Kim, S.J. Park, Synthesis of reduced graphene oxide/thorn-like titanium dioxide nanofiber aerogels with enhanced electrochemical performance for supercapacitor. J. Colloid Interface Sci. **486**, 287–295 (2017). https://doi.org/10.1016/j.jcis.2016.10.007

53. R. Kiruthiga, C. Nithya, R. Karvembu, B. Venkata Rami Reddy, Reduced graphene oxide embedded V2O5 nanorods and porous honey carbon as high performance electrodes for hybrid sodium-ion supercapacitors. Electrochim. Acta **256**, 221–231 (2017). https://doi.org/10.1016/j.electacta.2017.10.049
54. C. Lamiel, V.H. Nguyen, D. Tuma, J.J. Shim, Non-aqueous synthesis of ultrasmall NiO nanoparticle-intercalated graphene composite as active electrode material for supercapacitors. Mater. Res. Bull. **83**, 275–283 (2016). https://doi.org/10.1016/j.materresbull.2016.06.005
55. G. Lee, Y. Cheng, C.V. Varanasi, J. Liu, Influence of the nickel oxide nanostructure morphology on the effectiveness of reduced graphene oxide coating in supercapacitor electrodes. J. Phys. Chem. C **118**, 2281–2286 (2014). https://doi.org/10.1021/jp4094904
56. J. Lee, P. Srimuk, S. Fleischmann et al., Redox-electrolytes for non-flow electrochemical energy storage: a critical review and best practice. Prog. Mater. Sci. **101**, 46–89 (2019). https://doi.org/10.1016/j.pmatsci.2018.10.005
57. K.S. Lee, C.W. Park, S.J. Lee, J.D. Kim, Hierarchical zinc oxide/graphene oxide composites for energy storage devices. J. Alloys Compd. **739**, 522–528 (2018). https://doi.org/10.1016/j.jallcom.2017.12.248
58. Q. Li, C. Lu, C. Chen et al., Layered $NiCo_2O_4$/reduced graphene oxide composite as an advanced electrode for supercapacitor. Energy Storage Mater **8**, 59–67 (2017). https://doi.org/10.1016/j.ensm.2017.04.002
59. Q. Li, Q. Wei, L. Xie et al., Layered NiO/reduced graphene oxide composites by heterogeneous assembly with enhanced performance as high-performance asymmetric supercapacitor cathode. RSC Adv. **6**, 46548–46557 (2016). https://doi.org/10.1039/c6ra04998b
60. W. Li, Y. Bu, H. Jin et al., The preparation of hierarchical flowerlike NiO/reduced graphene oxide composites for high performance supercapacitor applications. Energy Fuels **27**, 6304–6310 (2013). https://doi.org/10.1021/ef401190b
61. X. Li, Z. Wang, Y. Qiu et al., 3D graphene/ZnO nanorods composite networks as supercapacitor electrodes. J. Alloys Compd. **620**, 31–37 (2015). https://doi.org/10.1016/j.jallcom.2014.09.105
62. Y. Li, J. Jian, L. Xiao et al., Synthesis of $NiMoO_4$ nanosheets on graphene sheets as advanced supercapacitor electrode materials. Mater. Lett. **184**, 21–24 (2016). https://doi.org/10.1016/j.matlet.2016.08.012
63. Y. Li, K. Ye, K. Cheng et al., Anchoring CuO nanoparticles on nitrogen-doped reduced graphene oxide nanosheets as electrode material for supercapacitors. J. Electroanal. Chem. **727**, 154–162 (2014). https://doi.org/10.1016/j.jelechem.2014.05.009
64. Z. Li, P. Liu, G. Yun et al., 3D (Three-dimensional) sandwich-structured of ZnO (zinc oxide)/rGO (reduced graphene oxide)/ZnO for high performance supercapacitors. Energy **69**, 266–271 (2014). https://doi.org/10.1016/j.energy.2014.03.003
65. A. Liu, Y. Bai, Y. Liu et al., Hierarchical dandelion-like copper oxide wrapped by reduced graphene oxide: hydrothermal synthesis and their application in supercapacitors. Mater. Res. Bull. **84**, 85–92 (2016). https://doi.org/10.1016/j.materresbull.2016.07.034
66. H. Liu, W. Zhu, D. Long et al., Porous V2O5 nanorods/reduced graphene oxide composites for high performance symmetric supercapacitors. Appl. Surf. Sci. **478**, 383–392 (2019). https://doi.org/10.1016/j.apsusc.2019.01.273
67. M. Liu, S. Sun, L. Yang, S. Yin, Ultrathin SnO_2 nanorod/reduced grapheme oxide nanosheet composites for electrochemical supercapacitor applications with excellent cyclic stability. Int. J. Mater. Res. **109**, 743–750 (2018). https://doi.org/10.3139/146.111662
68. M.C. Liu, K.L. Bin, L. Kang et al., Synthesis and characterization of M3V2O8 (M = Ni or Co) based nanostructures: a new family of high performance pseudocapacitive materials. J Mater Chem A **2**, 4919–4926 (2014). https://doi.org/10.1039/c4ta00582a
69. P. Liu, H. Chen, X. Chang et al., Novel method of preparing $CoFe_2O_4$/graphene by using steel rolling sludge for supercapacitor. Electrochim. Acta **231**, 565–574 (2017). https://doi.org/10.1016/j.electacta.2017.02.088
70. P. Liu, D. Wu, Y. Gao et al., Reduced graphene oxide-coated mulberry-shaped α-Fe_2O_3 nanoparticles composite as high performance electrode material for supercapacitors. J. Alloys Compd. **738**, 89–96 (2018). https://doi.org/10.1016/j.jallcom.2017.12.131

71. T. Liu, H. Chai, D. Jia et al., Rapid microwave-assisted synthesis of mesoporous $NiMoO_4$ nanorod/reduced graphene oxide composites for high-performance supercapacitors. Electrochim. Acta **180**, 998–1006 (2015). https://doi.org/10.1016/j.electacta.2015.07.175
72. H. Ma, J. He, D.B. Xiong et al., Nickel Cobalt hydroxide @reduced graphene oxide hybrid nanolayers for high performance asymmetric supercapacitors with remarkable cycling stability. ACS Appl. Mater. Interfaces **8**, 1992–2000 (2016). https://doi.org/10.1021/acsami.5b10280
73. L. Ma, X. Shen, H. Zhou et al., High performance supercapacitor electrode materials based on porous $NiCo_2O_4$ hexagonal nanoplates/reduced graphene oxide composites. Chem. Eng. J. **262**, 980–988 (2015). https://doi.org/10.1016/j.cej.2014.10.079
74. L. Ma, X. Zhou, L. Xu et al., Hydrothermal preparation and supercapacitive performance of flower-like WO3·H_2O/reduced graphene oxide composite. Colloids Surf. A Physicochem. Eng. Asp. **481**, 609–615 (2015). https://doi.org/10.1016/j.colsurfa.2015.06.040
75. Z. Ma, X. Huang, S. Dou et al., One-pot synthesis of Fe_2O_3 nanoparticles on nitrogen-doped graphene as advanced supercapacitor electrode materials. J. Phys. Chem. C **118**, 17231–17239 (2014). https://doi.org/10.1021/jp502226j
76. S.K. Meher, G.R. Rao, Ultralayered Co_3O_4 for high-performance supercapacitor applications. J. Phys. Chem. C **115**, 15646–15654 (2011). https://doi.org/10.1021/jp201200e
77. E. Miniach, A. Śliwak, A. Moyseowicz et al., MnO_2/thermally reduced graphene oxide composites for high-voltage asymmetric supercapacitors. Electrochim. Acta **240**, 53–62 (2017). https://doi.org/10.1016/j.electacta.2017.04.056
78. M.A.A. Mohd Abdah, N.H.N. Azman, S. Kulandaivalu, Y. Sulaiman, Review of the use of transition-metal-oxide and conducting polymer-based fibres for high-performance supercapacitors. Mater. Des. **186**, 108199 (2020). https://doi.org/10.1016/j.matdes.2019.108199
79. D.H. Nagaraju, Q. Wang, P. Beaujuge, H.N. Alshareef, Two-dimensional heterostructures of V2O5 and reduced graphene oxide as electrodes for high energy density asymmetric supercapacitors. J. Mater. Chem. A **2**, 17146–17152 (2014). https://doi.org/10.1039/c4ta03731f
80. A.K. Nayak, A.K. Das, D. Pradhan, High performance solid-state asymmetric supercapacitor using green synthesized graphene-WO3 nanowires nanocomposite. ACS Sustain. Chem. Eng. **5**, 10128–10138 (2017). https://doi.org/10.1021/acssuschemeng.7b02135
81. A. Numan, N. Duraisamy, F. Saiha Omar et al., Enhanced electrochemical performance of cobalt oxide nanocube intercalated reduced graphene oxide for supercapacitor application. RSC Adv. **6**, 34894–34902 (2016). https://doi.org/10.1039/c6ra00160b
82. S.S. Patil, D.P. Dubal, V.G. Deonikar et al., Fern-like rGO/BiVO4 hybrid nanostructures for high-energy symmetric supercapacitor. ACS Appl. Mater. Interfaces **8**, 31602–31610 (2016). https://doi.org/10.1021/acsami.6b08165
83. C. Peng, Y. Yang, C. Li et al., A novel WO3/graphene composite using poly(ionic liquid) as a linker for enhanced supercapacitive performance. Nanotechnology **31** (2020). https://doi.org/10.1088/1361-6528/ab814a
84. J.C. Pieretti, T.B. Trevisan, M.M.M. de Moraes et al., High capacitive rGO/WO3 nanocomposite: the simplest and fastest route of preparing it. Appl. Nanosci. **10**, 165–175 (2020). https://doi.org/10.1007/s13204-019-01089-z
85. A.K.R. Police, M. Chennaiahgari, R. Boddula et al., Single-step hydrothermal synthesis of wrinkled graphene wrapped TiO_2 nanotubes for photocatalytic hydrogen production and supercapacitor applications. Mater. Res. Bull. **98**, 314–321 (2018). https://doi.org/10.1016/j.materresbull.2017.10.034
86. K.K. Purushothaman, B. Saravanakumar, I.M. Babu et al., Nanostructured CuO/reduced graphene oxide composite for hybrid supercapacitors. RSC Adv. **4**, 23485–23491 (2014). https://doi.org/10.1039/c4ra02107j
87. H. Quan, B. Cheng, Y. Xiao, S. Lei, One-pot synthesis of α-Fe_2O_3 nanoplates-reduced graphene oxide composites for supercapacitor application. Chem. Eng. J. **286**, 165–173 (2016). https://doi.org/10.1016/j.cej.2015.10.068

88. M. Raja, A.B.V.K. Kumar, N. Arora, J. Subha, Studies on electrochemical properties of ZnO/rGO nanocomposites as electrode materials for supercapacitors. Fullerenes Nanotub. Carbon Nanostructures **23**, 691–694 (2015). https://doi.org/10.1080/1536383X.2014.971117
89. A. Ramadoss, G.S. Kim, S.J. Kim, Fabrication of reduced graphene oxide/TiO_2 nanorod/reduced graphene oxide hybrid nanostructures as electrode materials for supercapacitor applications. CrystEngComm **15**, 10222–10229 (2013). https://doi.org/10.1039/c3ce41517a
90. A. Ramadoss, S.J. Kim, Improved activity of a graphene-TiO_2 hybrid electrode in an electrochemical supercapacitor. Carbon N. Y. **63**, 434–445 (2013). https://doi.org/10.1016/j.carbon.2013.07.006
91. S. Sahoo, J.J. Shim, Facile synthesis of three-dimensional ternary $ZnCo_2O_4$/reduced graphene oxide/NiO composite film on nickel foam for next generation supercapacitor electrodes. ACS Sustain. Chem. Eng. **5**, 241–251 (2017). https://doi.org/10.1021/acssuschemeng.6b01367
92. R. Samal, B. Chakraborty, M. Saxena et al., Facile production of mesoporous WO3-rGO hybrids for high-performance supercapacitor electrodes: an experimental and computational study. ACS Sustain. Chem. Eng. **7**, 2350–2359 (2019). https://doi.org/10.1021/acssuschemeng.8b05132
93. E. Samuel, P.U. Londhe, B. Joshi et al., Electrosprayed graphene decorated with ZnO nanoparticles for supercapacitors. J. Alloys Compd. **741**, 781–791 (2018). https://doi.org/10.1016/j.jallcom.2017.12.320
94. M. Saraf, R.A. Dar, K. Natarajan et al., A binder-free hybrid of CuO-microspheres and rGO nanosheets as an alternative material for next generation energy storage application. ChemistrySelect **1**, 2826–2833 (2016). https://doi.org/10.1002/slct.201600481
95. M. Saranya, R. Ramachandran, F. Wang, Graphene-zinc oxide (G-ZnO) nanocomposite for electrochemical supercapacitor applications. J. Sci. Adv. Mater. Devices **1**, 454–460 (2016). https://doi.org/10.1016/j.jsamd.2016.10.001
96. M. Sawangphruk, P. Srimuk, P. Chiochan et al., High-performance supercapacitor of manganese oxide/reduced graphene oxide nanocomposite coated on flexible carbon fiber paper. Carbon N. Y. **60**, 109–116 (2013). https://doi.org/10.1016/j.carbon.2013.03.062
97. S. Shahrokhian, R. Mohammadi, E. Asadian, One-step fabrication of electrochemically reduced graphene oxide/nickel oxide composite for binder-free supercapacitors. Int. J. Hydrogen Energy **41**, 17496–17505 (2016). https://doi.org/10.1016/j.ijhydene.2016.07.087
98. P. Simon, Y. Gogotsi, B. Dunn, Where do batteries end and supercapacitors begin? Science (80-) **343**, 1210–1211 (2014). https://doi.org/10.1126/science.1249625
99. Z. Song, Y. Zhang, W. Liu et al., Hydrothermal synthesis and electrochemical performance of Co_3O_4/reduced graphene oxide nanosheet composites for supercapacitors. Electrochim. Acta **112**, 120–126 (2013). https://doi.org/10.1016/j.electacta.2013.08.155
100. X. Su, H. Chai, D. Jia et al., Effective microwave-assisted synthesis of graphene nanosheets/NiO composite for high-performance supercapacitors. New J. Chem. **37**, 439–443 (2013). https://doi.org/10.1039/c2nj40785j
101. K. Subramani, M. Sathish, Facile synthesis of ZnO nanoflowers/reduced graphene oxide nanocomposite using zinc hexacyanoferrate for supercapacitor applications. Mater. Lett. **236**, 424–427 (2019). https://doi.org/10.1016/j.matlet.2018.10.111
102. Y.N. Sudhakar, H. Hemant, S.S. Nitinkumar et al., Green synthesis and electrochemical characterization of rGO–CuO nanocomposites for supercapacitor applications. Ionics (Kiel) **23**, 1267–1276 (2017). https://doi.org/10.1007/s11581-016-1923-7
103. D.S. Sun, Y.H. Li, Z.Y. Wang et al., Understanding the mechanism of hydrogenated $NiCo_2O_4$ nanograss supported on Ni foam for enhanced-performance supercapacitors. J. Mater. Chem. A **4**, 5198–5204 (2016). https://doi.org/10.1039/c6ta00928j
104. Y. Sun, W. Zhang, D. Li et al., Facile synthesis of MnO_2/rGO/Ni composite foam with excellent pseudocapacitive behavior for supercapacitors. J. Alloys Compd. **649**, 579–584 (2015). https://doi.org/10.1016/j.jallcom.2015.07.212
105. S. Sundriyal, V. Shrivastav, M. Sharma et al., Significantly enhanced performance of rGO/TiO_2 nanosheet composite electrodes based 1.8 V symmetrical supercapacitor with use of redox additive electrolyte. J. Alloys Compd. **790**, 377–387 (2019). https://doi.org/10.1016/j.jallcom.2019.03.150

106. N. Tang, W. Wang, H. You, et al.: Morphology tuning of porous CoO nanowall towards enhanced electrochemical performance as supercapacitors electrodes. Catal. Today 240–245 (2019). https://doi.org/10.1016/j.cattod.2018.03.024
107. L. Thirugnanam, R. Sundara, Few layer graphene wrapped mixed phase TiO_2 nanofiber as a potential electrode material for high performance supercapacitor applications. Appl. Surf. Sci. **444**, 414–422 (2018). https://doi.org/10.1016/j.apsusc.2018.02.286
108. N.B. Trung, T.T. Van, D.K. Dang et al., Facile synthesis of three-dimensional graphene/nickel oxide nanoparticles composites for high performance supercapacitor electrodes. Chem. Eng. J. **264**, 603–609 (2015). https://doi.org/10.1016/j.cej.2014.11.140
109. N. Van Hoa, T.T.H. Quyen, N.H. Nghia et al., In situ growth of flower-like V2O5 arrays on graphene@nickel foam as high-performance electrode for supercapacitors. J. Alloys Compd. **702**, 693–699 (2017). https://doi.org/10.1016/j.jallcom.2017.01.241
110. R. Velmurugan, J. Premkumar, R. Pitchai et al., Robust, flexible, and binder free highly crystalline V2O5 thin film electrodes and their superior supercapacitor performance. ACS Sustain. Chem. Eng. **7**, 13115–13126 (2019). https://doi.org/10.1021/acssuschemeng.9b02302
111. V. Velmurugan, U. Srinivasarao, R. Ramachandran et al., Synthesis of tin oxide/graphene (SnO_2/G) nanocomposite and its electrochemical properties for supercapacitor applications. Mater. Res. Bull. **84**, 145–151 (2016). https://doi.org/10.1016/j.materresbull.2016.07.015
112. S. Vijayakumar, S.H. Lee, K.S. Ryu, Synthesis of Zn3V2O8 nanoplatelets for lithium-ion battery and supercapacitor applications. RSC Adv. **5**, 91822–91828 (2015). https://doi.org/10.1039/c5ra13904j
113. H. Wang, H. Dai, Strongly coupled inorganic–nano-carbon hybrid materials for energy storage. Chem. Soc. Rev. **42**, 3088–3113 (2013). https://doi.org/10.1039/c2cs35307e
114. K. Wang, X. Dong, C. Zhao et al., Facile synthesis of Cu_2O/CuO/RGO nanocomposite and its superior cyclability in supercapacitor. Electrochim. Acta **152**, 433–442 (2015). https://doi.org/10.1016/j.electacta.2014.11.171
115. L. Wang, G. Duan, J. Zhu et al., High capacity supercapacitor material based on reduced graphene oxide loading mesoporpus murdochite-type Ni6MnO8 nanospheres. Electrochim. Acta **219**, 284–294 (2016). https://doi.org/10.1016/j.electacta.2016.09.118
116. Y. Wang, J. Guo, T. Wang et al., Mesoporous transition metal oxides for supercapacitors. Nanomaterials **5**, 1667–1689 (2015). https://doi.org/10.3390/nano5041667
117. Y. Wang, P. Mao, F. Yan et al., Flower-like Fe_2O_3/reduced graphene oxide composite for electrochemical energy storage. Synth. Met. **222**, 198–204 (2016). https://doi.org/10.1016/j.synthmet.2016.09.020
118. Z. Wang, C. Ma, H. Wang et al., Facilely synthesized Fe_2O_3-graphene nanocomposite as novel electrode materials for supercapacitors with high performance. J. Alloys Compd. **552**, 486–491 (2013). https://doi.org/10.1016/j.jallcom.2012.11.071
119. W. Wei, X. Cui, W. Chen, D.G. Ivey, Manganese oxide-based materials as electrochemical supercapacitor electrodes. Chem. Soc. Rev. **40**, 1697–1721 (2011). https://doi.org/10.1039/c0cs00127a
120. C. Wu, S. Deng, H. Wang et al., Preparation of novel three-dimensional NiO/ultrathin derived graphene hybrid for supercapacitor applications. ACS Appl. Mater. Interfaces **6**, 1106–1112 (2014). https://doi.org/10.1021/am404691w
121. M.S. Wu, Y.P. Lin, C.H. Lin, J.T. Lee, Formation of nano-scaled crevices and spacers in NiO-attached graphene oxide nanosheets for supercapacitors. J. Mater. Chem. **22**, 2442–2448 (2012). https://doi.org/10.1039/c1jm13818a
122. X. Xia, J. Tu, Y. Mai et al., Graphene sheet/porous NiO hybrid film for supercapacitor applications. Chem. A Eur. J. **17**, 10898–10905 (2011). https://doi.org/10.1002/chem.201100727
123. C. Xiang, M. Li, M. Zhi et al., A reduced graphene oxide/Co_3O_4 composite for supercapacitor electrode. J. Power. Sources **226**, 65–70 (2013). https://doi.org/10.1016/j.jpowsour.2012.10.064
124. C. Xiang, M. Li, M. Zhi et al., Reduced graphene oxide/titanium dioxide composites for supercapacitor electrodes: shape and coupling effects. J. Mater. Chem. **22**, 19161–19167 (2012). https://doi.org/10.1039/c2jm33177b

125. J. Xu, J. Huang, M. Zini et al., A facile synthesis of ZnO nanorods on nitrogen-doped graphene sheets for supercapacitor applications. Int. J. Electrochem. Sci. **15**, 765–773 (2020). https://doi.org/10.20964/2020.01.65
126. X. Xu, L. Pei, Y. Yang et al., Facile synthesis of NiWO4/reduced graphene oxide nanocomposite with excellent capacitive performance for supercapacitors. J. Alloys Compd. **654**, 23–31 (2016). https://doi.org/10.1016/j.jallcom.2015.09.108
127. X. Xu, J. Shen, N. Li, M. Ye, Facile synthesis of reduced graphene oxide/CoWO4 nanocomposites with enhanced electrochemical performances for supercapacitors. Electrochim. Acta **150**, 23–34 (2014). https://doi.org/10.1016/j.electacta.2014.10.139
128. Y. Xu, X. Wang, C. An et al., Facile synthesis route of porous $MnCo_2O_4$ and $CoMn_2O_4$ nanowires and their excellent electrochemical properties in supercapacitors. J. Mater. Chem. A **2**, 16480–16488 (2014). https://doi.org/10.1039/c4ta03123g
129. Y. Xu, J. Wei, L. Tan et al., A facile approach to $NiCoO_2$ intimately standing on nitrogen doped graphene sheets by one-step hydrothermal synthesis for supercapacitors. J. Mater. Chem. A **3**, 7121–7131 (2015). https://doi.org/10.1039/c5ta00298b
130. S. Yang, X. Song, P. Zhang, L. Gao, Heating-rate-induced porous α-Fe_2O_3 with controllable pore size and crystallinity grown on graphene for supercapacitors. ACS Appl. Mater. Interfaces **7**, 75–79 (2015). https://doi.org/10.1021/am507910f
131. G. Ye, Y. Gong, K. Keyshar et al., 3D reduced graphene oxide coated V2O5 nanoribbon scaffolds for high-capacity supercapacitor electrodes. Part. Part. Syst. Charact. **32**, 817–821 (2015). https://doi.org/10.1002/ppsc.201500045
132. K.H. Ye, Z.Q. Liu, C.W. Xu et al., MnO_2/reduced graphene oxide composite as high-performance electrode for flexible supercapacitors. Inorg. Chem. Commun. **30**, 1–4 (2013). https://doi.org/10.1016/j.inoche.2012.12.035
133. M. Zhang, D. Yang, J. Li, Effective improvement of electrochemical performance of electrodeposited MnO_2 and MnO_2/reduced graphene oxide supercapacitor materials by alcohol pretreatment. J. Energy Storage **30** (2020a). https://doi.org/10.1016/j.est.2020.101511
134. W. Zhang, B. Quan, C. Lee et al., One-step facile solvothermal synthesis of copper ferrite-graphene composite as a high-performance supercapacitor material. ACS Appl. Mater. Interfaces **7**, 2404–2414 (2015). https://doi.org/10.1021/am507014w
135. Y. Zhang, M. Liu, S. Sun, L. Yang, The preparation and characterization of SnO_2/rGO nanocomposites electrode materials for supercapacitor. Adv. Compos. Lett. **29**, 2633366X2090983 (2020b). https://doi.org/10.1177/2633366x20909839
136. Y. Zhang, F. Wang, H. Zhu et al., Elongated TiO_2 nanotubes directly grown on graphene nanosheets as an efficient material for supercapacitors and absorbents. Compos. Part A Appl. Sci. Manuf. **101**, 297–305 (2017). https://doi.org/10.1016/j.compositesa.2017.06.026
137. Z. Zhang, L. Ren, W. Han et al., One-pot electrodeposition synthesis of ZnO/graphene composite and its use as binder-free electrode for supercapacitor. Ceram. Int. **41**, 4374–4380 (2015). https://doi.org/10.1016/j.ceramint.2014.11.127
138. B. Zhao, J. Song, P. Liu et al., Monolayer graphene/NiO nanosheets with two-dimension structure for supercapacitors. J. Mater. Chem. **21**, 18792–18798 (2011). https://doi.org/10.1039/c1jm13016a
139. C. Zhao, X. Shao, Y. Zhang, X. Qian, Fe_2O_3/reduced graphene oxide/Fe_3O_4 composite in situ grown on Fe foil for high-performance supercapacitors. ACS Appl. Mater. Interfaces **8**, 30133–30142 (2016). https://doi.org/10.1021/acsami.6b09594
140. Y. Zhao, W. Ran, J. He et al., High-performance asymmetric supercapacitors based on multi-layer MnO_2/graphene oxide nanoflakes and hierarchical porous carbon with enhanced cycling stability. Small **11**, 1310–1319 (2015). https://doi.org/10.1002/smll.201401922
141. Z. Zhao, T. Shen, Z. Liu et al., Facile fabrication of binder-free reduced graphene oxide/MnO_2/Ni foam hybrid electrode for high-performance supercapacitors. J. Alloys Compd. **812**, 152124 (2020). https://doi.org/10.1016/j.jallcom.2019.152124
142. D. Zhou, P. Cheng, J. Luo et al., Facile synthesis of graphene@NiMoO4 nanosheet arrays on Ni foam for a high-performance asymmetric supercapacitor. J. Mater. Sci. **52**, 13909–13919 (2017). https://doi.org/10.1007/s10853-017-1467-x

143. M. Zhou, H. Chai, D. Jia, W. Zhou, The glucose-assisted synthesis of a graphene nanosheet-NiO composite for high-performance supercapacitors. New J. Chem. **38**, 2320–2326 (2014). https://doi.org/10.1039/c3nj01351k
144. Y. Zhou, X. Zou, Z. Zhao et al., CoO/rGO composite prepared by a facile direct-flame approach for high-power supercapacitors. Ceram. Int. **44**, 16900–16907 (2018). https://doi.org/10.1016/j.ceramint.2018.06.128
145. X. Zhu, H. Dai, J. Hu et al., Reduced graphene oxide-nickel oxide composite as high performance electrode materials for supercapacitors. J. Power. Sources **203**, 243–249 (2012). https://doi.org/10.1016/j.jpowsour.2011.11.055
146. Y. Zhu, S. Cheng, W. Zhou et al., Construction and performance characterization of α-Fe_2O_3/rGO composite for long-cycling-life supercapacitor anode. ACS Sustain. Chem. Eng. **5**, 5067–5074 (2017). https://doi.org/10.1021/acssuschemeng.7b00445
147. Y. Zhu, W. Chu, N. Wang et al., Self-assembled Ni/NiO/RGO heterostructures for high-performance supercapacitors. RSC Adv. **5**, 77958–77964 (2015). https://doi.org/10.1039/c5ra14790e
148. Y. Zhu, H. Huang, G. Li et al., Graphene-anchored $NiCoO_2$ nanoarrays as supercapacitor electrode for enhanced electrochemical performance. Electrochim. Acta **248**, 562–569 (2017). https://doi.org/10.1016/j.electacta.2017.07.158
149. X. Zou, Y. Zhou, Z. Wang et al., Facile synthesis of α-Fe_2O_3 pyramid on reduced graphene oxide for supercapacitor and photo-degradation. J. Alloys Compd. **744**, 412–420 (2018). https://doi.org/10.1016/j.jallcom.2018.02.126

Chapter 5
Nanostructures in Hybrid Lithium and Sodium-Ion Capacitors

Uchenna C. Obi, Emmanuel C. Obetta, Yusuf A. Olanrewaju, Omolara V. Oyelade, Sharafadeen A. Adeniji, Odette F. Ngasoh, and Abdulhakeem Bello

Abstract Hybrid ion capacitors are promising energy storage devices with the potential to bridge the gap between commercial supercapacitors and batteries. This innovative technology conserves energy while providing supercapacitor-like power and cyclability. In recent years, hybrid ion capacitors have received interest due to their increased energy density without compromising power density, and because of their hybrid combination, they can also achieve greater capacitance. Moreover, hybrid

---

O. V. Oyelade · S. A. Adeniji · A. Bello (✉)
Department of Theoretical and Applied Physics, African University of Science and Technology (AUST), Abuja, Nigeria
e-mail: abello@aust.edu.ng

O. V. Oyelade
e-mail: oyeladeov@binghamuni.edu.ng

S. A. Adeniji
e-mail: sadeniji@aust.edu.ng

U. C. Obi · E. C. Obetta · Y. A. Olanrewaju · O. F. Ngasoh · A. Bello
Department of Materials Science and Engineering, African University of Science and Technology (AUST), Abuja, Nigeria
e-mail: uobi@aust.edu.ng

E. C. Obetta
e-mail: eobetta@aust.edu.ng

Y. A. Olanrewaju
e-mail: yolanrewaju@aust.edu.ng

O. F. Ngasoh
e-mail: fngasoh@aust.edu.ng

O. V. Oyelade
Department of Physics, Bingham University, Karu, Nasarawa State, Nigeria

S. A. Adeniji
Department of Physics, Nile University of Nigeria, Jabi, Abuja, Federal Capital Territory, Nigeria

A. Bello
Centre for Cyber Physical Food, Energy and Water System (CCP-FEWS), Electrical and Electronic Engineering Science, University of Johannesburg, Johannesburg, South Africa

© The Author(s), under exclusive license to Springer Nature Switzerland AG 2026

K. Raju et al. (eds.), *Hybrid Nanostructures for Advanced Supercapacitors*, Modern Aspects of Electrochemistry 63, https://doi.org/10.1007/978-3-032-04593-5_5

ion capacitors are a fast-expanding energy storage technology that provides residential and transportation applications with low-cost, high-efficiency storage and power-intensity solutions. To maximize the energy and power densities of next-generation storage devices, many multifunctional hybrid nanostructured materials have been studied. This article provides a comprehensive overview of recent developments in anode and cathode materials for hybrid ion capacitors. We discussed the origins and evolution of hybrid ion capacitors. This chapter discusses the historical development of lithium-ion batteries, characterized by their high energy density and low power density, supercapacitors, characterized by their high-power density and low energy density, and a hybrid system that combines the high energy and power densities of lithium-ion batteries and supercapacitors, respectively. For hybrid lithium-ion capacitor applications, this chapter also examines the use of other charge carriers, such as sodium and potassium, as well as many electrode materials, including activated carbon, graphene, carbon for lithium and sodium electrodes, and titanium-oxide-based compounds. The electrode materials' specific capacities and current densities are summarized and tabulated. We also addressed the necessity to provide high energy and high power in a single energy storage system by utilizing hybrid ion capacitors due to their configurable and adaptable performance characteristics, especially in energy and power densities. The combination of a capacitor-type cathode and a battery-type anode to form a hybrid supercapacitor, such as a sodium-ion hybrid capacitor, a lithium-ion hybrid capacitor, and a potassium-ion hybrid capacitor, was considered for charge storage with both capacitor and battery advantages. Relevant mechanisms are demonstrated for nanostructured electrode materials, pseudocapacitive oxides V_2O_5, MXenes, conversion compounds, and battery-related intercalation ceramics are demonstrated.

Keywords Supercapacitors · Lithium-ion batteries · Sodium ion capacitors · Nanostructures · Carbon · Transition metal oxide

5.1 Introduction

The need for energy storage devices that facilitate the shift from fossil-based fuel, which releases hazardous gases into the atmosphere, to electric power has inspired several research efforts to develop electrodes for energy storage technologies. Energy storage systems, such as batteries, supercapacitors, and dielectric capacitors, are advanced types of electrochemical energy storage technologies currently being developed (Fig. 5.1) and are driven electrochemically by adsorption i.e., electro sorption and insertion of ions into solid materials, respectively [1]. Lithium-ion batteries store energy by inserting Li-ion into the anode and cathode hosts [2, 3] and converting the chemical energy contained in electrode active materials into electrical energy through some electrochemical reactions. On the other hand, supercapacitors store energy by ion electrosorption onto high surface area carbon electrodes [4]. The energy density,

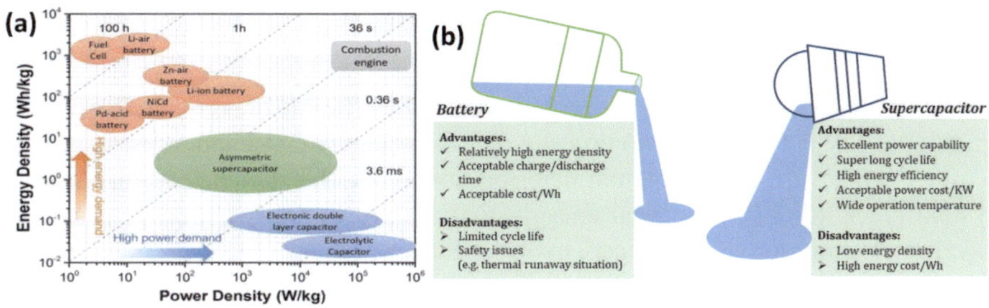

Fig. 5.1 **a** The Ragone plot showing the energy density versus power density of several energy storage technologies that are commercial and **b** shows the significant differences between Li-ion batteries and electrochemical capacitors in key fundamental characteristics. Reprinted with permission of the American Chemical Society from Shao et al. [5]. Reprinted with permission of Elsevier from Zhao and Burke [6]

i.e., how much energy is stored and power density i.e., how rapidly the energy can be delivered are critical aspects of these storage systems, as seen in Fig. 5.1.

Lithium-ion batteries have dominated and become an integral part of our daily life and have facilitated the development of portable digital electronics, such as cellphones and laptops, since their introduction in 1991 by Sony Corporation for digital camcorders [7], Goodenough and Park, [8, 9]. This class of batteries is on the verge of entering the utility market for grid-energy storage and transportation sectors such as electric cars, buses, and bikes. In addition, they are anticipated to play a major role in enabling the replacement of fossil-fuel-based technology with renewable energy sources, such as solar and wind, thereby providing a cleaner and more sustainable environment. The Nobel Prize in Chemistry in 2019 was awarded to John Goodenough, Stanley Whittingham, and Akira Yoshino [10–14],this reaffirms the importance of lithium-ion battery research. Li-ion batteries have high energy densities. However, they are limited by their power densities [4]. Special and superior energy-storage devices with energy density like that of lithium-ion batteries, power density, and long cycle life like that of electrochemical capacitors are desired. Scientists have proposed the mechanism of combining Li-ion batteries and electrochemical capacitors to simultaneously inherit their advantages by adapting their complementary charge-storage processes.

This chapter focuses on reviewing the recent advances in anode and cathode materials for hybrid ion capacitors which have improved the stability, power, and energy densities of the devices. Various designs of hybrid supercapacitors for efficient storage devices are also reviewed, including the use of nano-structured materials for next-generation devices. Furthermore, the relevance of carbon and non-carbon-based materials as electrodes is elucidated through their design configurations and working principles. In other words, this chapter provides insights into the materials, designs, and working principles of high-performance hybrid electrochemical storage devices.

5.2 Motivation for Hybrid-Ion Capacitors

The main approach to meeting the requirements of renewable energy that can replace traditional fossil fuels, for tackling the ever-growing global challenges of energy scarcity and severe environmental deterioration is a high-efficiency energy storage system [2, 3, 15–17]. Electrical energy-storage devices are essential in consumer electronics, automotive, aerospace, hybrid electric vehicles, stationary markets, and power grid [18–21]. That is, energy storage provides the capacity to regulate the unpredictability of electrical energy since this is a fast-developing and currently underutilized market for energy storage. Portable electronic devices and electric vehicles are now being developed, and research into the development of energy storage devices in terms of energy density, power density, low-cost manufacturing, and safety has been approved [22, 23]. There are essentially two types of devices for reversible electrochemical energy storage among different energy storage technologies: secondary batteries and electrochemical capacitors such as ultracapacitors and supercapacitors. The former has high energy density, whereas the latter has high power and cyclability. The lithium-ion battery is the leading contemporary technology, with a lithium-containing metal oxide cathode and a graphite anode. Also, in lithium-ion batteries, charges are stored by a cation intercalation and deintercalation process which gives a high energy density. However, lithium-ion battery has disadvantages such as short life and poor power density, which make them unsuitable for use in power electronic advancements [18, 24]. For example, commercial lithium-ion batteries have a specific energy of up to 200 Wh kg^{-1} but a maximum specific power of less than 350 W kg^{-1}. In contrast, most commercial electrochemical capacitors, such as the electric double-layer capacitor (EDLC), have a high-power delivery capability of up to 10 kW kg^{-1} because charges are stored in their system via electrostatic ion adsorption in the Helmholtz double layer at the electrode–electrolyte interfaces [22, 25, 26]. When compared to lithium-ion batteries, the EDLC has a lower energy density (5–10 W kg^{-1}). However, batteries or capacitors cannot meet the energy and power density requirements of the aforementioned systems on their own because rechargeable batteries have a high energy density but a low power density, and capacitors have a high power density but a low capacity. Moreover, supercapacitors lack the intrinsic energy-storage density needed for cost-effective and space-efficient deployment. In effect, they are overly bulky in terms of weight, volume, and cost, i.e., kWh/$. Lithium-ion batteries, on the other hand, lack adequate power to capture even a significant proportion of the stopping energy, which necessitates a charging rate of up to 200 C, 1 C is 1 h charge, and 200 C is 18 s charge. The ability to supply both high energy and high power in a single system to reap the benefits of both is a new motivation for improved electrical energy-storage systems [27–31].

Hybrid-ion capacitors represent an emerging class of devices that may be able to bridge the performance gap between commercial EDLC ultracapacitors and traditional ion batteries. Hybrid-ion capacitor is a relatively new technology that is energy-wise midway between batteries and supercapacitors. Hybrid ion capacitors are one of the most significant sectors of hybrid supercapacitors and the entire energy storage research and development. This is due to their tunable and customizable performance properties, particularly in energy and power densities. Regenerative braking is a significant potential use for hybrid-ion capacitor devices. Because of the limitations of existing secondary batteries and supercapacitors i.e., electrochemical capacitors and ultracapacitor technologies, regenerative braking energy harvesting from trains, heavy vehicles, and eventually light vehicles represents a huge potential market that has yet to be fully exploited. The charge storage devices were developed by merging a capacitor-type cathode with a battery-type anode to create a hybrid supercapacitor that has both capacitor and battery advantages [32–34]. As a result, the motivation of hybrid ion capacitors in light of lithium-ion, sodium-ion, potassium-ion, magnesium-ion, calcium-ion, zinc-ion, and aluminum-ion capacitors has been addressed [35]. These are identified by their cations, for example, sodium-ion hybrid capacitor [36], lithium-ion hybrid capacitor [37, 38], and potassium-ion hybrid capacitor [39, 40]. The alkali metals lithium, sodium, and potassium are extremely reactive and require the use of an organic electrolyte. Lithium-ion capacitors have been studied to accomplish supercapacitor-type rapid charge–discharge operations as well as battery-type charge storage capacity [41–43]. In addition, sodium- or lithium-based ion capacitors may provide energy values that are 4 or 5 times greater than EDLC ultracapacitors. This better energy is obtained by using a carbonate-based battery electrolyte that provides a higher device voltage than conventional acetonitrile ($\sim$4.2 vs. $\sim$2.7 V) and having at least one of the electrodes store charges via bulk processes. As a result, the hybrid-ion capacitors appear to be well adapted to compete with electrochemical ultracapacitor technology. They also enable rapid surface or subsurface redox reactions and provide significantly higher energy density while maintaining the same power density as typical electrical double-layer capacitors, attracting special attention in recent research.

5.3 Brief History of Hybrid-Ion Capacitors

Electrochemical energy with high energy density, high power density and cycling stability is an indispensable system in consumer electronics, automotive, aerospace, and stationary markets [44, 45]. However, available electrochemical energy storage systems such as batteries and supercapacitors cannot independently satisfy this energy requirement; commercial lithium-ion batteries have high energy density about 150–200 W h kg^{-1} and more than 300 Wh Kg^{-1} in liquid and solid electrolytes, respectively, and low power density (< 350 W kg^{-1}), whereas, supercapacitors possess high power density ($\sim$10 kW kg^{-1}) and low energy density of $\leq$ 10 W h kg^{-1} in organic electrolyte and much lower in the aqueous electrolyte. In addition, the lifespan of lithium-ion batteries is unsatisfactory compared to supercapacitors that have a lifespan of more than 100,000 [44, 46–48]. Therefore, to satisfy this energy demand of our modern technology, the fabrication of electrochemical energy devices with both high energy and power densities, as well as high cycling stability was considered. Although batteries and supercapacitors exhibit different electrochemical behaviour, the similarity in their configuration such as anode, cathode, electrolyte, separator, and current collector makes it possible for them to be hybridized [45]. Energy storage that combines battery and supercapacitor electrodes incorporates the merit of secondary batteries into a supercapacitor [46]. The electrodes of hybrid capacitors are asymmetric and can be designed using electrodes with different storage mechanisms, as shown in Fig. 5.2 [49], typically, a capacitive double-layer electrode and a battery-type faradaic or pseudocapacitive electrode. The core of a classic hybrid-ion capacitor is the concept that an electrode that stores charge based on reversible surface adsorption of ions may be combined in a series with an electrode that stores charge by a bulk process [44].

Lithium-ion capacitors, which combine the lithium-ion intercalated electrode of lithium-ion batteries with the ion-adsorption electrode of EDLCs were proposed and developed when the demand for energy storage with high energy and power density, as well as high cyclability, increased [50, 51]. Lithium-ion batteries exhibit the highest energy and power densities compared to other commercially viable batteries, thus, lithium-ion chemistries for hybrid supercapacitors were the first option for appreciable energy and power characteristics [52].

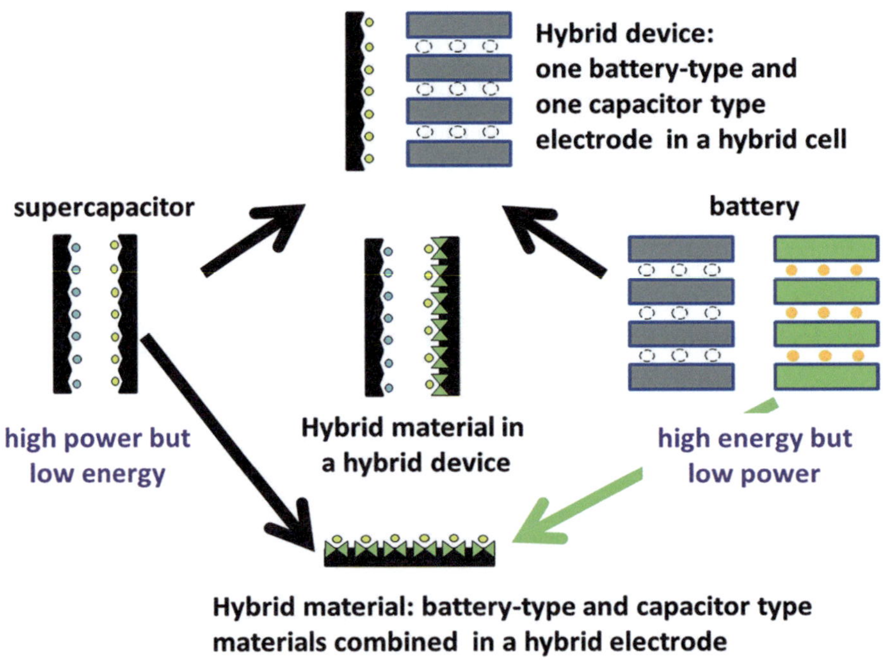

Fig. 5.2 Different possible hybridization approaches between supercapacitor and battery electrodes and materials. Reprinted with permission of the Royal Society of Chemistry from Dubal et al. [49]

The research on lithium-ion capacitors gained much attention when Yata and co-workers reported the insertion-disinsertion characteristics of lithium-ion in carbonaceous material, polyacenic semiconductive, synthesized through phenolic resin pyrolysis [35]. The first lithium-ion capacitors by hybridization of EDLCs and Li-ion battery electrodes were reported by Amatucci and co-workers. In 2001, they developed activated (AC+)/lithium titanate ($Li_4Ti_5O_{12}$ −) lithium-ion capacitors cell with 1-M lithium hexafluorophosphate ($LiPF_6$) in ethylene carbonate (EC)/dimethyl carbonate (DMC) (2:1, v/v) electrolyte. The energy density is three times that of a conventional carbon-based supercapacitor (20 W h kg^{-1}) [53]. Lithium titanate has a fascinating characteristic: zero-expansion electrode, high charge storage capacity of 100 mAh/g of charge, sloping voltage plateau between 2.75 and 1.5 V versus Li/Li^+, preventing solid-electrolyte interphase formation. In addition, the materials can be cycled between $Li_4Ti_5O_{12}$ and $Li_7Ti_5O_{12}$ [44]. These properties make $Li_4Ti_5O_{12}$ suitable for hybridization.

In 2005, Fuji Heavy Industry commercialized lithium-ion capacitors based on porous carbon positive electrodes and pre-lithiated negative electrodes with laminated structures. Commercialization of lithium-ion capacitors was also attempted by different companies based on various technologies (Fig. 5.3). The parameters of lithium-ion batteries, supercapacitors, and lithium-ion capacitors show that lithium-ion capacitors have the potential to bridge the gap between lithium-ion batteries and EDLCs [54].

The higher abundance rank that is lower availability and lower cost (USD kg^{-1}) of lithium compared to other charge carriers, such as sodium, potassium, magnesium, zinc, calcium, and aluminium (Fig. 5.4) shows that other charger carriers can be considered as low-cost alternatives to lithium-ion batteries for grid energy storage applications [35].

The next system of ion capacitors is sodium-ion capacitors. The shift from lithium- to sodium-ion capacitors is due to geo-economic factors, such as the uneven distribution of lithium reserves and the high cost of lithium caused by production scarcities [55]. Sodium is cheap and the performance of sodium-ion batteries is comparable to that of lithium-ion batteries, thus, sodium-ion capacitors were chosen as the next best alternative to lithium-ion capacitors [56, 57]. Moreover, the replacement of the lithium-ion capacitors copper-foil current collector with Al-foil current collector is possible in sodium-ion capacitors because the binary alloy of Al is formed with Li but not with Na at reduced voltages [35]. This current collector replacement contributes to the low cost of sodium-ion capacitors compared to lithium-ion capacitors. The first sodium-ion capacitor was designed in 2012, where V_2O_5-carbon nanotubes composite and activated carbon served as the anode and cathode, respectively, in sodium perchlorate/polycarbonate electrolyte. The assembled sodium-ion capacitors with a high cell voltage (2.8 V) displayed maximum energy and power densities of ~40 Wh kg^{-1} and 5 $kWkg^{-1}$. In addition, the capacity retention and cycling efficiency of the cell are 80% and 98%, respectively [58].

Sodium-ion capacitors could satisfy the high energy and power requirement of large-scale electrochemical storage because the radius of sodium-ion (1.02 Å) is larger than that of Li-ion (0.76 Å), which results in sluggish sodium-ion diffusion and

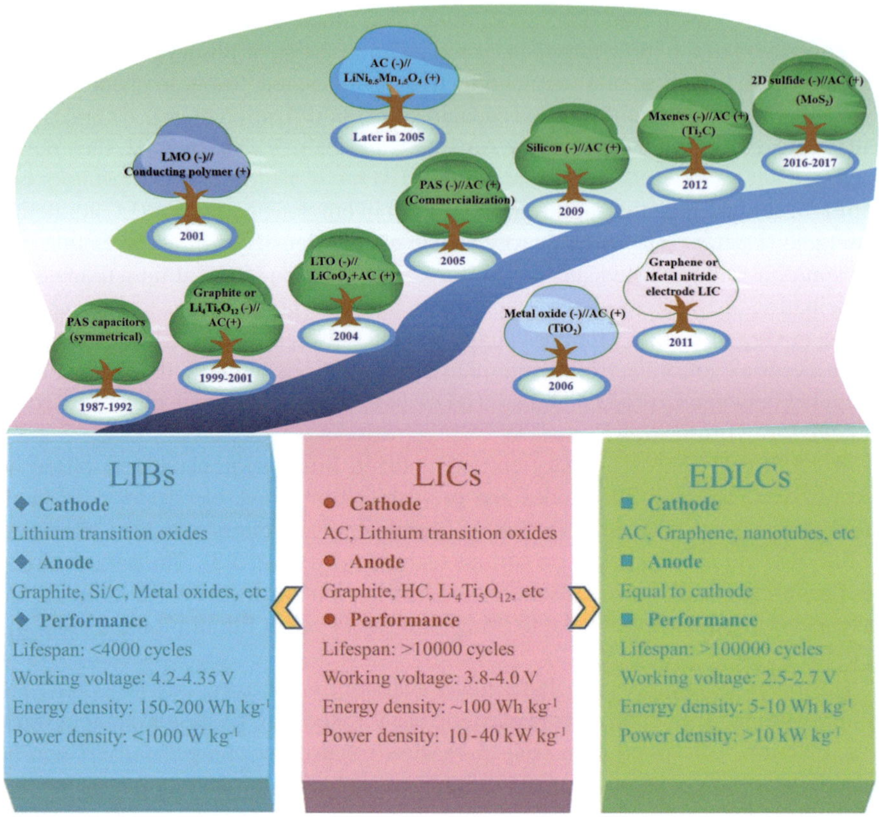

Fig. 5.3 History and evolution of various lithium-ion capacitor systems and several relevant parameters of lithium-ion batteries, supercapacitors, and lithium-ion capacitors. Reprinted with permission of the Royal Society of Chemistry from Li et al.[54]

Fig. 5.4 Relative abundance rank and cost of different charge carriers. Suggestions of other charger carriers that can be considered as low-cost alternatives to lithium-ion batteries. Reprinted with permission of Wiley from Naskar et al. [35]

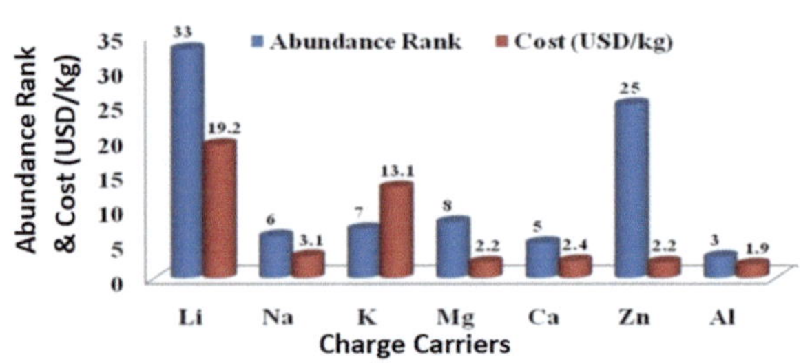

structural deformation in electrode materials during the charge and discharge process, thus, reduces the capacitors rate capacities and cycling properties [59]. The search for a better hybrid-ion capacitor led to the discovery of a potassium-ion capacitor. The small stoke radius of potassium in potassium carbonate-based electrolytes compared

Table 5.1 Stokes radium of different alkali metal [60]

Alkali	Stoke radius in potassium carbonate (nm) solvation
Lithium	0.408
Sodium	0.460
Potassium	0.360

to lithium and sodium (Table 5.1 contributed to the high specific capacitance of activated charcoal in potassium carbonate-based electrolytes. This knowledge was instrumental to the fabrication of a potassium-ion capacitor; graphite//0.8-M KPF6/PC//AC which exhibits approximately 35 mAh g^{-1} a thousand cycle [60]. Stable solid-electrolyte interphase layer formation contributed to the stability of this hybrid-ion capacitor over such a cycle.

Pasta et al. [61] constructed a cell with copper hexacyanoferrate (CuHCF), activated-polypyrrole, and 1 M potassium phosphate (pH = 1), as the positive electrode, negative electrode, and electrolyte, respectively. The faradaic charge storage mechanism of the positive electrode was associated with the Fe^{3+}/Fe^{2+} redox pair, and it is controlled by intercalation/deintercalation of potassium ions in and out of the host CuHCF structure. The first potassium-ion capacitor prototype was developed with good cycling stability (up to 55,000 cycles at a high charge–discharge rate) to replace the lithium-ion capacitor [55]. To achieve optimum energy density, power density, cycling stability and safety, and to replace the high-cost lithium-ion capacitor, hybrid-ion capacitors of other charger carriers, such as magnesium, zinc and aluminium were developed (magnesium ion capacitor—Patents WO-2019156219-A1, US10079116; [62, 63]).

5.4 Lithium- and Sodium-Ion Capacitors Fundamentals

Currently, secondary batteries and supercapacitors are the two primary electrochemical energy storage technologies [64]. Recently, there has been interest in developing outstanding energy storage devices with high energy and power density such as rechargeable batteries and supercapacitors [65]. They offer high power density and cycling ability but have a low energy density, which limits their application [66]. Whereas, secondary batteries, such as lithium-ion batteries, have the benefits of high energy density, high operating voltage range, and are environmentally friendly. However, their energy density and cycling ability are low [65]. Lithium-ion batteries have the potential to provide a high energy density of 150–200 Wh kg^{-1}. Thus, the low energy density typically 10 Wh kg^{-1} and accompanying considerable volumetric change during the charging-discharging process result in an unacceptable power density of 5 kW kg^{-1} and lengthy cycling stability over 100,000 cycles [67]. Thus, electrochemical energy storage cell designs that can concurrently provide high energy and power density are globally crucial [65] to address the limitations of batteries and supercapacitors. When it comes to meeting the demands for both high

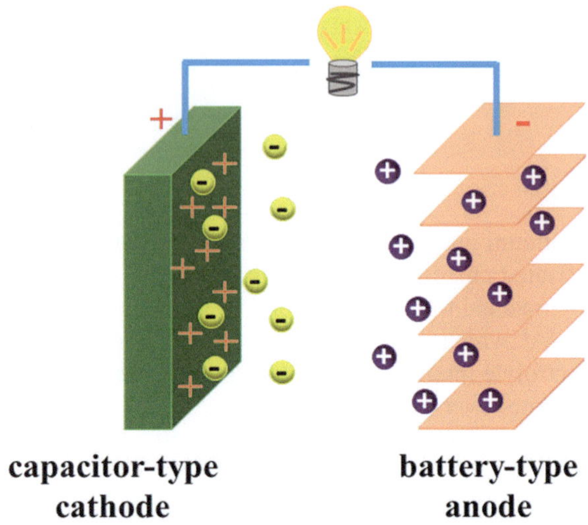

Fig. 5.5 The configuration of hybrid capacitors, metal-ion capacitors are a hybrid electrochemical energy storage device that combines the battery-type anode intercalation mechanism with the cathode double-layer mechanism

energy and power densities, a new and unique form of battery-like energy storage known as metal-ion capacitors or hybrid-ion capacitors has emerged as a promising candidates [44, 67–69]. Hybrid-ion capacitors, such as Li, Na, and K-ion capacitors, are a hybrid technology that incorporates the merits of metal-ion batteries and electrical double-layer capacitors (EDLCs) (Fig. 5.5) [70].

The metal-ion capacitors are a hybrid electrochemical energy storage device that combines the battery-type anode's intercalation mechanism with the cathode's double-layer mechanism (Fig. 5.5). Thus, hybrid-ion capacitors combine batteries' high energy density with supercapacitors high-power density [71, 72]. Unfortunately, modern hybrid capacitors cannot compete with lithium-ion batteries in terms of energy density. The kinetic gap between capacitor-type and battery-type electrodes must be bridged for high-performance hybrid capacitors. The kinetics of battery-type electrodes with Faradaic reaction are substantially worse than capacitor-type electrodes with physical desorption-adsorption. This kinetic gap can be bridged by producing battery-like materials with remarkable power [73]. In typical hybrid-ion capacitors, an electrode that stores charge by reversible surface adsorption is connected in series with an electrode that stores charge via a bulk process. One or more pure capacitive EDLCs, reversible surface redox processes, reversible ion adsorption at defects and heteroatoms, or reversible surface plating of metallic nanoparticles are some of the methods for storing charge on surfaces [44]. Hybrid capacitors can be categorized into three categories based on charge carrier migration during the electrochemical process, as shown in Fig. 5.6 [73].

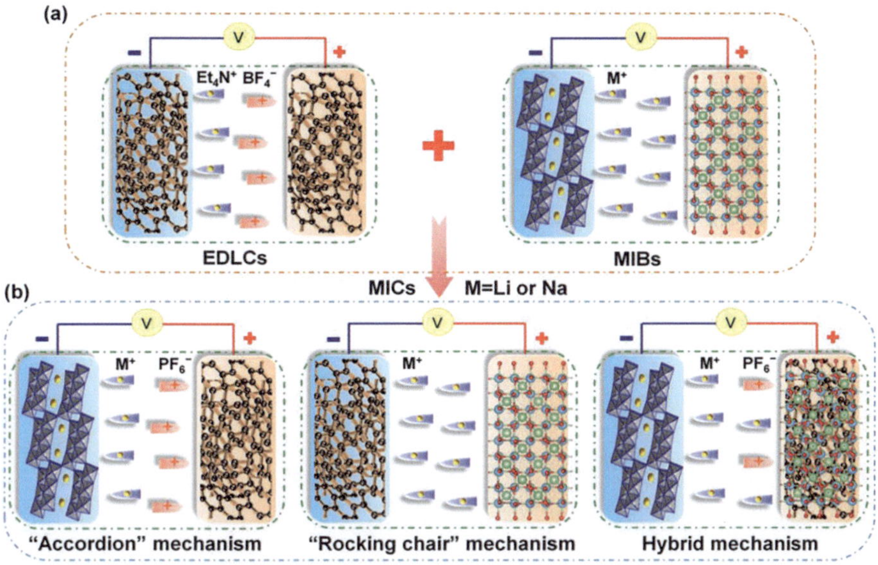

Fig. 5.6 Energy storage mechanisms for **a** electrical double-layer capacitors, lithium and sodium-ion batteries, and **b** lithium and sodium-ion hybrid capacitors with different operating mechanisms including the accordion, rocking chair, and hybrid mechanisms in sequence. Reprinted with permission of Wiley from Dong et al. [73]

5.4.1 The Accordion Mechanism

Electrochemical intercalation negative electrode and physical adsorption positive electrode are the most common configurations. During the charging process of this type of hybrid capacitor, cations and anions from the electrolyte are respectively segregated to the negative and positive electrodes [73].

5.4.2 The Rocking Chair Mechanism

Physical adsorption negative electrodes and electrochemical intercalation positive electrodes are the most common configurations. Deintercalation from the positive electrode and adsorption on the negative electrode surface occurs during charging of this hybrid capacitor type, and vice versa, which is similar to lithium-ion battery materials [73].

5.4.3 The Hybrid Mechanism

One or both electrodes in this type of hybrid capacitor contain both battery and capacitor-type components. The typical configuration is an electrochemical intercalation negative electrode with physical adsorption, with an electrochemical intercalation positive electrode. Cations deintercalation from the positive electrode during charging and the positive electrode's capacitor-type materials such as activated carbon absorb free anions from the electrolyte, whereas all cations deintercalated from the positive electrode and supplied by the electrolyte intercalate into the negative electrode [73]. Figure 5.6 presents the schematic diagrams of the various mechanisms as stated above.

5.5 Lithium-Ion Capacitors

Amatucci et al. [53] pioneered the use of a nanostructured $Li_4Ti_5O_{12}$ anode and an activated carbon cathode in a hybrid lithium-ion capacitor. They created the first lithium hybrid capacitor in 2001, using activated carbon as the cathode and nanostructured $Li_4Ti_5O_{12}$ as the anode [74]. This device's energy density is approximately three times that of conventional carbon-based supercapacitors implying excellent prospects [65]. A lithium-ion capacitor composed of a high-power EDLCs type positive electrode and a high-energy lithium insertion-desertion type negative electrode with a lithium-based organic electrolyte. A conceptual representation of lithium-ion capacitor production is illustrated in Fig. 5.7. Lithium-ion capacitors can be made by combining the anode from the lithium-ion battery and the cathode from the EDLCs in an organic electrolyte with a lithium-ion salt solution. The charge–discharge process of the lithium-ion capacitor involved a faradaic and non-faradaic electrochemical process. Lithium-ions are intercalated in the anode materials and anions $PF6^-$ are adsorbed on the surface of the activated cathode during charging, whereas the reverse occurs during discharging [68].

A high-capacity battery-type electrode and a high-rate capacitor-type electrode are used in most lithium-ion capacitor systems [38, 75]. Charging and discharging occur simultaneously and asymmetrically on the capacitor and battery electrodes, respectively, via surface ion adsorption–desorption and Li^+ intercalation-de-intercalation [69, 76, 77]. Among all battery-type materials, insertion-type materials typically exhibit relatively fast lithium-ion and sodium-ion transport within bulk electroactive materials as well as low crystal variation during the lithium-ion and sodium-ion insertion-disinsertion process, resulting in relatively high power output [78, 79]. Thus, insertion-type materials are frequently selected as the battery-type materials for hybrid capacitor applications [73]. Notably, the lithium-ion capacitors system's battery- and capacitor-type electrodes operate in various potential windows, extending the lithium-ion capacitors operational voltage range and contributing to high energy density [69]. The six important criteria for selecting high-performance

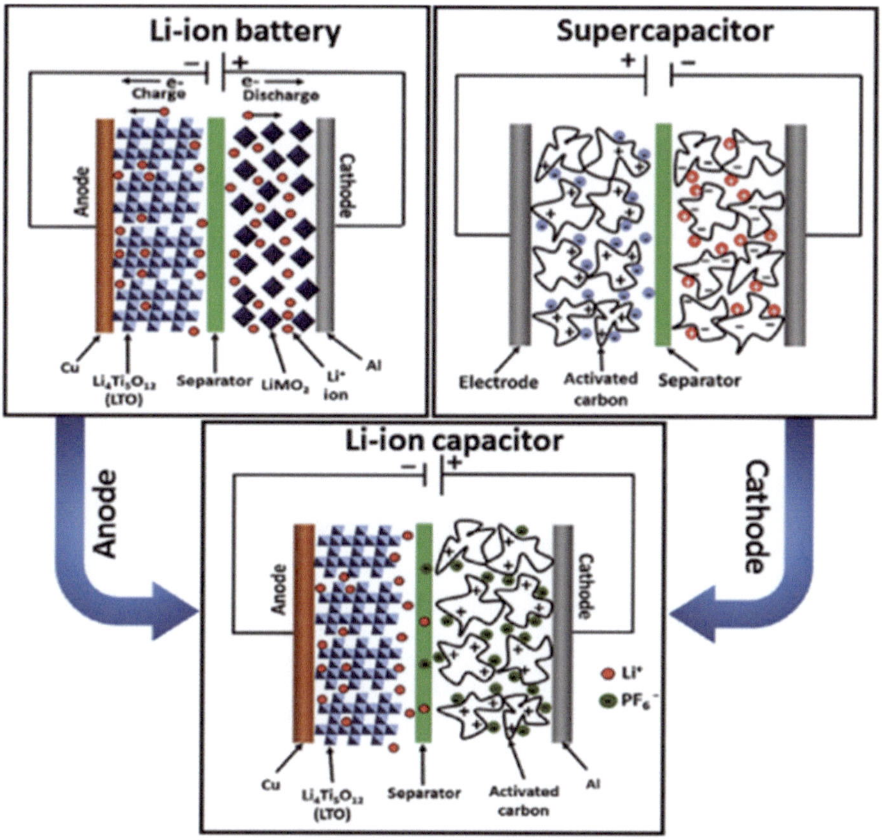

Fig. 5.7 Conceptual representation of lithium-ion capacitor production by combining the anode from the lithium-ion battery and the cathode from the EDLCs in an organic electrolyte with a lithium-ion salt solution. Reprinted with permission of Elsevier from Jagadale et al. [68]

electrochemical storage devices are observed in EDLCs, lithium-ion batteries, and sodium-ion capacitors as shown in Fig. 5.8.

5.5.1 Working Principle of Lithium-Ion Capacitors

Lithium-ion capacitors employ battery-type anodes that execute a reversible Li^+ insertion-desertion reaction in organic electrolyte and capacitor-type cathodes with high specific surface area and quick ions adsorption–desorption process. Lithium-ion capacitors can combine the physical charge storage of supercapacitors with the reversible chemical Li^+ insertion-desertion process of lithium-ion batteries to achieve high-performance levels. However, because the reaction rates in the anode

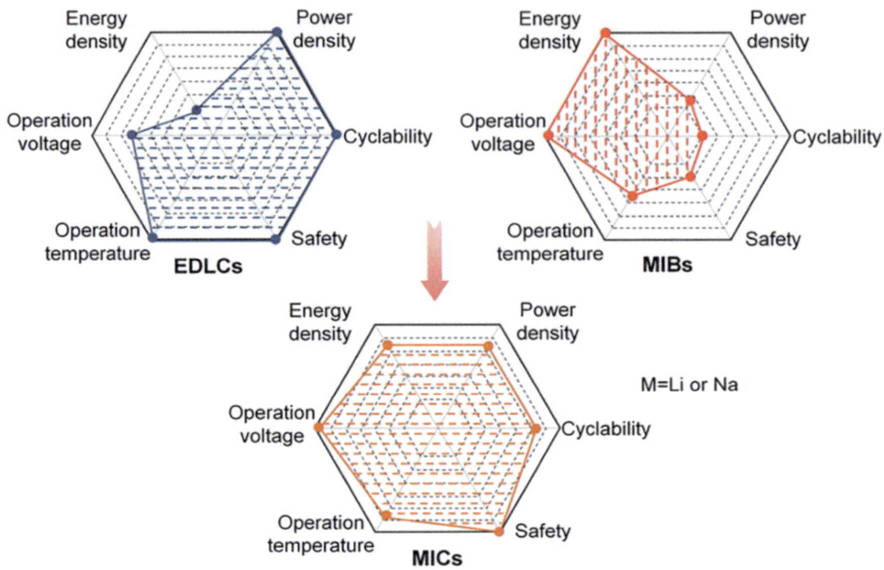

Fig. 5.8 Comparison of six important criteria of electrochemical energy storage devices such as EDLCs, lithium-ion batteries, sodium-ion batteries, and lithium-ion capacitors and sodium-ion capacitors (MIBs- M = Li or Na, MIBs – Metal-ion batteries, MICs- Metal-ion capacitors). Reprinted with permission of Wiley from Dong et al. [73]

and cathode are significantly uneven, the power density of a lithium-ion capacitor is determined by the rate of lithium-ion insertion-desertion and the energy density is determined by the specific capacity attained by the ions adsorption–desorption process. The anions in the electrolyte migrate rapidly to the surface of the capacitor-type cathode during the charging period, whereas lithium-ion in the electrolyte is embedded in the bulk phase of the battery-type anode, and electrons are transferred from cathode to anode via the external circuit to participate in the formation of Li-insertion compounds. Meanwhile, the process of discharge is the reverse of the aforementioned process. Thus, the charges within the lithium-ion capacitors maintain a dynamic balance during the reaction, ensuring that the electrolyte remains constantly neutral [80]. The anode and cathode materials that are used in lithium-ion capacitors can be categorized as shown in Fig. 5.9.

5.5.2 Sodium-Ion Capacitors

Sodium and lithium are both members of the first group of elements, and their chemical characteristics are similar. Although sodium ions have a bigger radius than lithium ions (0.102 vs. 0.076 nm), sodium's resources and availability on Earth much outweigh those of lithium [81, 82]. Additionally, sodium ions in solution are very

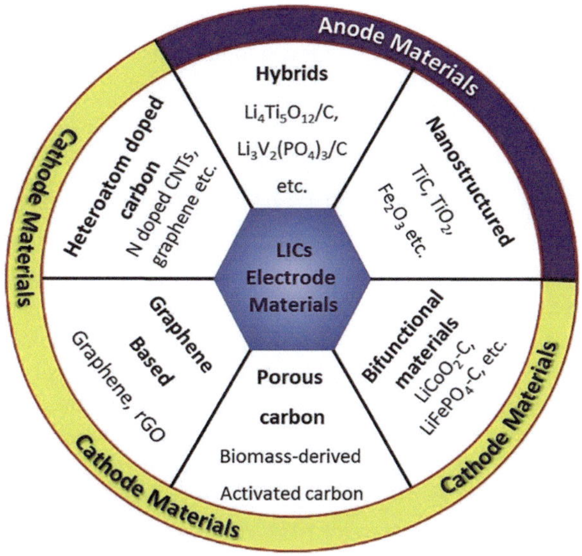

Fig. 5.9 Various anodes like nanostructured TiO_2 and cathodes like graphene materials are used as advanced electrode materials for lithium-ion capacitors. Reprinted with permission of Elsevier from Jagadale et al. [68]

tiny in size, in comparison to potassium ions and may diffuse more rapidly than other ions, sodium ions in solution have a higher ionic conductivity [65, 83]. A possible alternative to lithium-ion storage systems is sodium-ion storage, which includes both sodium-ion batteries and capacitors. This is because sodium is abundant and inexpensive and the cost of Na_2CO_3 is only 3% that of Li_2CO_3, and it has similar electrochemical properties to lithium only 0.3 V more positive than lithium [7]. As a developing technology, sodium-ion capacitors are viewed as a means of supplementing or potentially expanding existing lithium-ion capacitors [84]. Interestingly, sodium-ion capacitors have a higher energy density than conventional capacitors and a higher power density than conventional sodium-ion batteries [85]. As a result, sodium-ion capacitors can bridge the performance gap between standard supercapacitors and sodium-ion batteries, combining the advantages of both [65, 86]. Therefore, it is critical to characterize sodium-ion capacitors and define various energy storage strategies in contemporary circumstances [87, 88]. A thorough understanding of the mechanism is critical for the development of high-performance sodium-ion capacitor devices. Pseudocapacitive materials have been investigated in sodium ion capacitor devices to achieve excellent performance in cathodes and anodes [89, 90]. Increasing research has been directed toward exploiting pseudocapacitive materials to replace standard EDLC cathodes, as well as a further assembly with battery-type materials to create hybrid devices. Pseudocapacitive cathodes in sodium-ion capacitor devices fall between EDLCs and battery cathodes. Psudocapacitance can be considered a supplementary mechanism to EDLCs because this mechanism does not involve electrostatic adsorption but has comparable cyclic voltammogram curve forms and galvanostatic charge and discharge curve tendencies to EDLC [89]. Alternatively, sodium-ion capacitor devices are primarily interested in pseudocapacitive electrode materials such as anodes. Nonetheless, with the recent rapid advancement in nanotechnology and nanoscience in anodes, nanomaterial anodes have played a significant part in electrochemical energy systems [65].

5.5.3 Characteristics of Sodium-Ion Capacitors

Sodium-ion capacitors, incorporate cathode and anode materials, current collectors, electrolytes, separators, and a metal shell [91]. The electrode materials in sodium-ion capacitors function as the energy storage core and comprise battery- and capacitor-type electrodes [92]. Current collectors are responsible for transporting and collecting currents [93]. The operating voltage range of the electrolytes has a direct effect on the device's energy density [94]. Sodium-ion capacitors may use the same separator as sodium-ion batteries, such as glass fibers. Electrolytes operate as conductors of ions, and their operating temperature range affects the operating temperature limits of the devices in practice [95]. Aqueous and organic electrolytes are frequently used. Aqueous-based sodium-ion capacitors offer the advantages of increased stability, cost savings, and safety [96]. However, the operating potential region of the devices is limited, resulting in a low energy density. Fortunately, organic electrolytes have a wider operating voltage range [65]. A significant problem facing sodium-ion storage systems is the unavailability of acceptable anode materials [97]. Graphite, the most often used anode material in lithium-ion batteries, is thermodynamically unsuitable for sodium-ion batteries [20, 98].

5.6 Next-Generation Hybrid-Ion Capacitors Through Nanostructuring

The most prevalent nanostructured materials are:

1. Three-dimensional spatial ensembles of one-dimensional structured materials such as nanowires, nanotubes, nanorods or two-dimensional nanostructured materials such as nanosheets, nanoflakes nanostructures.
2. Three-dimensional nanoporous structures; and
3. Three-dimensional hierarchical nanostructured materials.

This type of material improves the performance of electrochemical energy conversion and storage devices [99]. Nanostructuring can enhance the rate capability and power density of a device by decreasing diffusion length for ions transport, which can enhance the rate capability and power density. The rate capability increases by increasing the electrode–electrolyte contact area to improve cycle life by accommodating strain during ion insertion-extraction and can ensure efficient electrochemical utilization of materials [98–100]. Three-dimensional nanostructures can display unique electronic, magnetic, and optical properties due to interactions and coupling between the excitons, magnetic moments, or surface plasmons of nanoscale units within the Three-dimensional ensemble [101]. Nanoparticles were used to solve the imbalance in the electrochemical kinetics and specific capacitance of lithium-ion capacitors since they reduced the distance for both electron and lithium-ion transport [46]. Nanostructuring facilitates fast kinetics of ion and electron transport, and hence, capacitor-like properties are achieved [35]. First, a $Li_4Ti_5O_{12}$-activated carbon system was introduced, however, the conventional $Li_4Ti_5O_{12}$ was affected by low power characteristics due to poor Li^+ diffusion coefficient ($< 10^{-6}$ cm^2 s^{-1}) and poor electronic conductivity ($<10^{-13}$ Scm^{-1}) [102]. Nanostructuring is needed to improve the performance of the cathode materials of Sodium-ion batteries [57]. The charge–discharge time attained for the nc-$Li_4Ti_5O_{12}$-carbon nanofiber composite and various $Li_4Ti_5O_{12}$ materials that have been reported so far are presented in Fig. 5.10.

The thin-film design, porosity enhancement, nanostructuring, morphology control, core–shell design, conducting composite design, etc. are the conventional routes of materials modifications for fast ions and electrons transport [35]. Pasquieret al. [103] developed lithium-ion batteries using nanostructured $Li_4Ti_5O_{12}$ negative and activated-$LiCoO_2$ composite positive electrodes with 2-M $LiBF_4$ in acetonitrile electrolyte. High power (4000 Wkg^{-1}) and energy (40 $Whkg^{-1}$) densities were achieved due to the activated carbon double-layer capacitors and nanostructured $Li_4Ti_5O_{12}$ negative electrode. In addition, high durability is attained with 80% capacity retention after 9000 cycles at full depth of discharge, which is quite superior compared to conventional lithium-ion batteries. Zhu et al. [104] synthesize a multi-component flower-like nanostructure by sulfurizing ternary transition metal (Ni–Co–Mo) oxides-hydroxides precursor (Fig. 5.11).

Nanostructuring of the ternary transition metal improved its electrical productivity, thereby promoting the Faradic redox reaction. The as-prepared porous flower-like Ni–Co–Mo–S/NF exhibited a capacity of 7.22 C cm^{-1} at 5 mA cm^{-1} and 96% retention after 5000 cycles. In addition, full cells assembled based on the material achieved energy and power densities of 54.54 Wh kg^{-1} and 540 W kg^{-1}, respectively and capacitance retention of 78% after 3500 cycles. Wang et al. [105] enhanced the specific capacitance, ions transport and excellent cycling stability by introducing nanopores into ruthenium oxide-multi-walled carbon nanocomposite (RuO_2/MWCNT). The (RuO_2/MWCNT) anode and ammonium pyrrolidine dithiocarbamate cathode achieved energy and power densities of 146 Wh kg^{-1} and 33 kW kg^{-1}, respectively. Goat hair-derived activated carbon (cathode) and MoO_2@reduced graphene oxide composite (anode) exhibited improved electrochemical performance for sodium-ion hybrid supercapacitors due to the nanosized MoO_2 particle distributed on the surface of reduced graphene oxide. The hybrid capacitor device exhibited energy and power densities of 79 W h kg^{-1} and 95 W kg^{-1}, respectively. In addition, the porous high surface area activated carbon creates a channel for ion adsorption and desorption, thus, improving the electrochemical performance of the cathode [106, 107] fabricated a hybrid nanoneedle-shaped manganese oxide and 55 spherical graphene oxide using an in-situ spray pyrolysis method. A high surface area (139 m^2g^{-1}) urchin-like morphology obtained from this process formed a network on the surface of the electrode. The hybrid material's electrochemical performance (425

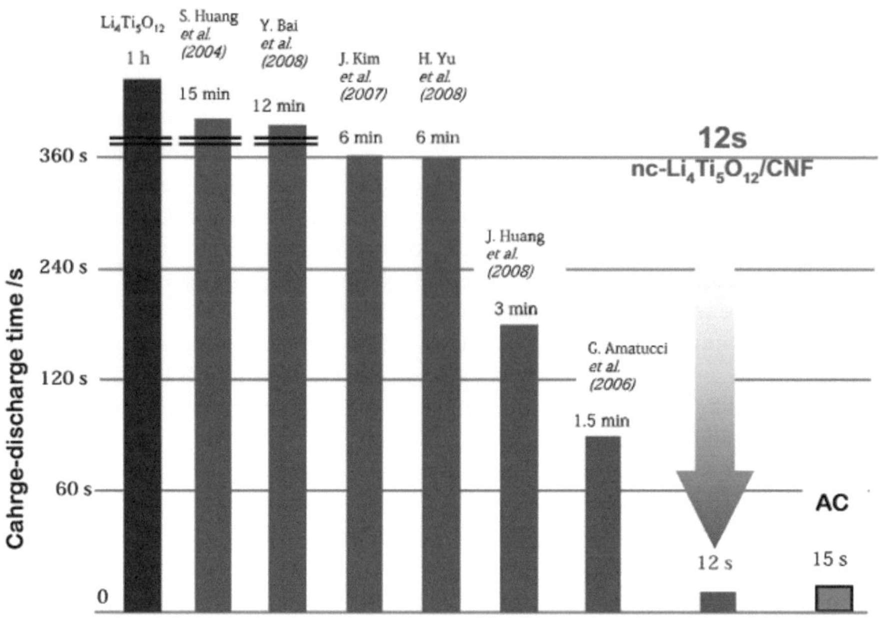

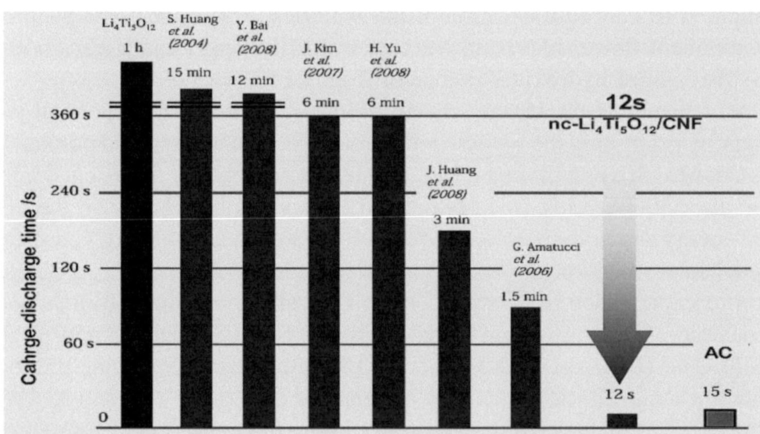

Fig. 5.10 Charge–discharge time attained for the Nc-$Li_4ti_5o_{12}$-caron nanofibers composite and various $li_4ti_5o_{12}$ materials reported in the literature so far. reprinted with permission of wiley from Naoi et al. [102]

F g^{-1} and 9.07 Wh kg^{-1}) is enhanced by the combination of reduced graphene oxide and manganese oxide. Nanostructured $Na_2Ti_9O_{19}$ exhibited excellent Na-ion kinetics with a capacitive retention of 86% at 1.0 mV s^{-1}, indicating that the material is suitable for sodium-ion hybrid capacitor anode. A full-cell hybrid sodium-ion capacitor is constructed using $Na_2Ti_9O_{19}$ and activated porous carbon as anode and cathode,

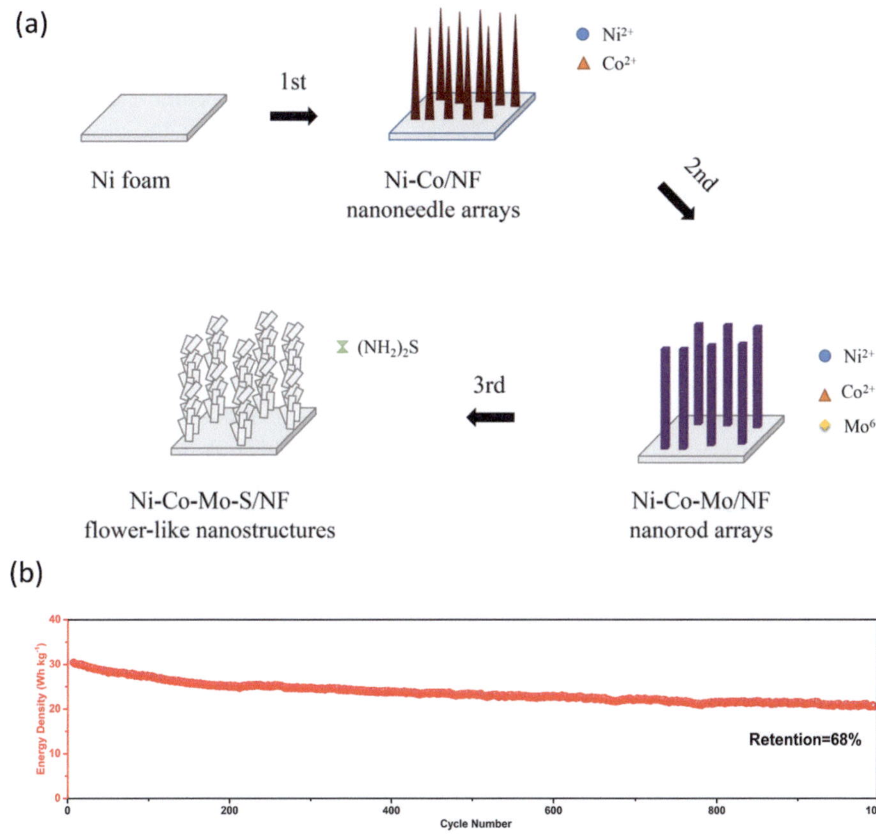

Fig. 5.11 **a** Synthesis process of flower-like Ni–Co–Mo–S-Nanofibers core–shell nanostructures **b** Long-cycle performance of device preparation using U-Co_2P@rGO-14//Activated carbon configuration. Reprinted with permission of the Royal Society of Chemistry from Wang et al. [105]

respectively, which display excellent electrochemical properties, with a maximum energy density of 54 Wh kg^{-1} and 5 kW kg^{-1}, respectively [108]. Wang et al. [105] prepared a novel U-Co_2P@rGO-14 nanocomposite potassium-ion battery anode, the electrode exhibits a high potassium ion storage capacity of 374 mAh g^{-1}, a remarkable rate capability of up to 141 mAh g^{-1} at 2 A g^{-1}, a high cycle efficiency. A sub-4 nm cobalt phosphate (Co_2P) nanorods with ultrafine orthorhombic nanostructure improved the transportation rate of potassium ions by offering a shorter solid-state diffusion length, along with buffering of the active material volume expansion. In addition, the asymmetric potassium-ion capacitor delivers a maximum energy density of approximately 87 $Whkg^{-1}$ and a power density of approximately 4260 Wkg^{-1}, as well as 68% retention (Fig. 11b) after 1000 cycles when paired with the activated sample.

5.6.1 Design and Performance of Electrode Materials

The electrode material is a critical aspect in influencing the specific capacitance, which has been the subject of much research [109]. Controlling the design and interface properties of electrode materials is critical for boosting the performance of high-performance supercapacitors, such as high specific capacitance, high-rate capability, and long cycling stability [110]. Fortunately, nanotechnology has pushed the boundaries of materials science and engineering and will be important in improving the development of charge storage to address the challenges of conventional electrode materials [111]. Over the last few decades, technological advancements in nanoscience have paved the path for the development of novel materials, and electrochemical performance has been enhanced by several orders of magnitude. Supercapacitors are predicted to advance with the development of an electrode material with a high surface area and good electrical conductivity [47]. A large surface area can provide additional active sites for EDLCs or pseudocapacitive reactions since the increased surface-to-volume ratio provides the maximum contact surface between the active materials and the electrolyte ion. The surface area of an electroactive material can be increased by manufacturing nanoscale or porous electroactive materials [112]. The morphology and chemical composition of the resulting materials are characterized using scanning electron microscopy, transmission electron microscopy, scanning transmission electron microscopy (STEM), powder X-ray diffraction, energy-dispersive X-ray spectroscopy, and X-ray photoelectron spectroscopy measurements, among others, are used to characterize electrode materials [113]. In addition, a porous electrolyte diffusion structure is advantageous because the porous structure ensures easy access for ion penetration while eliminating surface area losses owing to stacking or blocking of active sites. Furthermore, high electrical conductivity is required for rapid charge transport to achieve a high-rate capability and high-power density. The increased conductivity can be achieved by shrinking the electroactive material and/or combining it with additional materials having high conductivity [114]. An electrode is conventionally synthesized by grinding the electrode material, conductive agent, and binders into a homogenous slurry and then pasting it to current collectors [97]. Conductive agents are conductive carbon compounds, such as acetylene black, used to increase the electrode's electrical conductivity. Binders facilitate the linking of active materials and conductors and ensure their mechanical attachment to the current collector [113]. The two most frequently used binders for electrode preparation are carboxymethylcellulose (CMC) and polyvinylidene fluoride (PVDF) [64]. Binders are often electro-inert, which results in possibly higher internal resistance. By building an electrode material into three- or two-electrode systems, the performance of the electrode material can be examined. Thus far, flexible electrodes have developed as a novel type of electrode configuration due to their mechanical robustness and ability to achieve high specific gravimetric and volumetric capacitance. Flexible electrodes have been constructed using self-standing or integrated electrode materials with extreme flexibility on soft-matter substrates, eliminating the need for a conducting agent and binder [113, 115].

5.6.2 *Anode Materials for Hybrid-Ion Batteries*

To enhance the lithium-ion capacitors cycle performance and to increase their specific energy and power, many types of anode materials have been investigated [64]. The electrode potential of the anode material is the primary criterion for selection. The lower potential of the anodes in lithium-ion capacitors permits a large higher potential window for the entire cell. As a result, it is preferable to prepare electrodes with a low yet safe voltage and a high capacity, and is critical to consider the charge storage contributions of capacitive and battery mechanisms while designing high-performance anodes [68].

5.7 Lithium- and Sodium-Ion Capacitor Electrode Materials

Recently, various morphologies of nanostructured carbon materials have been reported, such as 0D nanoparticles (nanospheres), hollow or core–shell structures; one-dimensional nanorods, nanobelts, nanowires, and nanotubes; two-dimensional nanosheets and three-dimensional hierarchical architectures self-assembled from the above low- dimensional nanostructured building blocks via interactions such as van der Waals forces, as well as hydrogen, ionic, and covalent bonding.

5.7.1 *Activated Carbon*

Activated carbon, carbon nanospheres, and carbon nanoparticles are zero-dimensional carbon materials that are used as electrode materials in EDLCs [116–118]. Activated carbon is the most prominent carbon-based material utilized as cathode materials for lithium-ion capacitors and sodium-ion capacitors due to their low cost, nontoxicity, high electrical conductivity, and high specific surface area. Activated carbons are made from a variety of precursors, including polymers, biomass precursors such as wood, coal, nutshell, coal, metal–organic frameworks (MOFs), and organic salts [38, 119].

Activated carbons typically have a potential window of 1.5–4.2 V versus Na/Na^+. Furthermore, the specific capacity of commercially activated carbons ranges between 30 and 50 $mAhg^{-1}$ [94, 120]. The specific capacity of activated carbon is insufficient, which is why the mass and thickness of activated carbons compared to the battery-type electrode are enlarged to achieve charge balancing. It is worth noting that rational pore property design such as pore volume, specific surface area, and pore size distribution and surface modification of functional groups have proven viable techniques for improving the specific capacity of activated carbon electrodes. Activated carbon pores are broadly classified as micropores (<2 nm), mesopores

(2–50 nm), and macropores (more than 50 nm) [121]. To harvest the capacity, the micropores allow ions to pass through. Furthermore, mesopores and macropores can reduce the ion transport path to allow for mass transfer, resulting in high-rate capability. Aside from the surface rational design of pores, surface-functionalized activated carbons can display improved chemical binding with sodium ions, which helps to improve specific capacity.

Furthermore, three-dimensional structures such as hierarchical carbon have high chemical stability, a large specific surface area, and rapid electron transport kinetics. The advantages of activated carbons in energy storage may be fully used by producing a hierarchical porous structure. Xu et al. [80] have shown that a hierarchical porous structure with numerous mesopores, a high surface area, and a large pore volume is crucial in meeting the requirement for high-rate and high-capacity activated carbons. Li and co-worker synthesis a zeolitic imidazolate frameworks-8-derived porous carbon (ZDPC) with high specific capacities of 151.7 and 96.1 mAh g^{-1} at 0.1 and 10 A g^{-1}, respectively [75]. Furthermore, the ZDPC cathode electrode retained 87% of its capacity after 2000 cycles at 1 A g^{-1}, showing remarkable cycling stability within a voltage range of 1.5–4.0 V. It should be noted that the reversible capacity of ZDPC is significantly more than that of the commercial activated carbons listed above. The high surface area of 3738 m^2 g^{-1}, big pore volume (2.58 cm^3 g^{-1}), and mesopores (2–5 nm) all contribute to the superior electrochemical performance. Another method for increasing the capacity of activated carbons is surface modification of functional groups due to the pseudocapacitive reaction. Ding et al. [122] proved that strongly oxygen-doped functional groups (C=O quinone groups) had a higher capacity for the cathode electrode.

In addition, carbon generated from metal–organic frameworks with different configurations has also been extensively studied in lithium-ion capacitors. For example, high surface area (2714 m^2 g^{-1}) carbon cuboids were created by pyrolyzing the zinc-based MOF-5, which has a unique crumpled-sheet porous shape and the necessary micro and mesoporosity values [123]. Depending on the advanced structure, the metal–organic framework-dependent lithium-ion capacitor has a maximum effective energy density of 65 W h kg^{-1} and a good power capacity. Polyhedral hollow carbon generated from a metal–organic framework was also synthesized and utilized in lithium-ion capacitors [124]. Because of the presence of metal ions in the metal–organic framework precursor, it is necessary to remove metal ions during the production of metal–organic framework-derived carbon.

5.7.2 *Graphene*

Since 2004, graphene, a well-known two-dimensional material composed of carbon atoms in a hexagonal lattice, has attracted increasing attention due to its exceptionally high strength, the large theoretical specific surface area of ~2630 m^2 g^{-1}, stable chemical properties, and high intrinsic electrical conductivity of ~15,000 cm^2 V^{-1} S^{-1} [34]. Electrical conductivity and specific surface area are the two most important characteristics of energy-storage devices. Therefore, mechanical exfoliation, bottom-up synthesis, epitaxial growth, liquid-phase exfoliation, and graphene oxide reduction are some of the most common graphene production processes [125]. Because of the high yield, the most popular experimental technique for preparing graphene in energy-storage devices is the reduction of graphene oxide. Since graphene oxide is made by oxidizing graphite, it has an abundance of surface functional groups such as hydroxy, carboxyl, and carbonyl leading to poor electrical conductivity [126]. Alternatively, due to the poor chemical bonding between sodium and graphene [127], graphene typically exhibits a low specific capacity (40–50 mAh g^{-1}) toward becoming the cathode for sodium-ion capacitors, and surface modification of graphene such as heteroatom doping, thus plays an important role for graphene-based cathodes [67, 128]. To get improved electrochemical performance with high capacity and outstanding rate performance for the graphene-based cathode, graphene oxide as a precursor must be reduced to provide reduced graphene oxide. The decrease of graphene oxide is an important step in developing a viable graphene-based electrode that can not only achieve better electrical conductivity but also directly perform surface modification to provide psudocapacitance. The elaborate design of the graphene structure, in addition to surface modification, provides excellent support for increasing electron transmission and mass transfer. Shao and co-workers investigated surface modification [126], also, [67, 126, 129], found that surface oxygen functional groups inside the reduced graphene oxide and crumpled structures are advantageous for the high-capacity cathode. They found that the oxygen-functionalized crumpled graphene electrode provided a higher capacity of 105 mAh g^{-1} and after 3000 cycles, 91% capacity retention was obtained. However, the voltage window of 2.5–4.2 V limits future capacity growth. In particular, the positive electrode of oxygen-functionalized crumpled graphene may not fully use the oxygen functional groups such as C=O to provide pseudocapacitance. Nonetheless, the oxygen-functionalized crumpled graphene has been important in paving the way for the high-capacity cathode.

5.7.3 Titanium Oxide Compound-Based Lithium Anodes

Titanium dioxide (TiO_2) and lithium titanate are the most promising titanium-based anode materials for lithium storage [130]. These two materials have similar cost, stability, and electrochemical properties, as well as similar disadvantages, such as limited electrical conductivity. Both have high average working potentials (~ 1.5 V), which are well below the dangerous voltage zone associated with lithium plating. Their structures are highly stable and exhibit low strain when lithium is intercalated-extracted reversibly. Due to recent advancements in rate performance, particularly for lithium titanate, both TiO_2 and lithium titanate are commonly used as lithium-ion capacitor anodes. The breakthrough improvements in the rate capability of TiO_2 and lithium titanate anodes are based on increasingly fine nanoscopic mixes of oxides and high-conductivity carbons. It should be noted that nanostructuring each phase will not affect the bulk ionic conductivity of the materials but will increase the overall contribution of the surface and interface [131, 132]. Most of the lithium titanate-based lithium-ion capacitors have a charge–discharge time range of tens of seconds or even several seconds [133]. Additionally, because of the low volume change of the intercalation anodes and the absence of solid-electrode interphase layer development, the energy of the TiO_2- and lithium titanate-based lithium-ion capacitors may be well sustained during an extraordinarily long cycling period of 5000–20,000 cycles [44, 88].

5.7.4 Titanium Oxide-Based Anodes

Titanium-based materials comprise a sizable portion of sodium anodes. After carbons, the most thoroughly investigated systems for Na capacitors are those based on TiO_2 such as anatase, rutile, brookite, and amorphous. Researchers are discovering that TiO_2 behaves differently with Na in terms of redox chemistry and charge storage than its well-studied lithium counterparts. Charge storage in anatase TiO_2 may entail a combination of sodium ion intercalation into the lattice and a reversible conversion reaction to generate Ti and Na_2O in their terminally sodiated state [134, 135]. An advantage of sodium-ion capacitors over lithium-ion capacitors is that TiO_2 has a lower operating voltage against Na/Na^+ than it does against Li/Li^+ (ca. 0.75 V vs. ca. 1.5 V) [136]. This increases the energy of devices but introduces difficulties with solid-electrolyte interphase (SEI) production, which occurs below 1 V versus Na/Na^+. However, the issues limiting TiO_2's rate capacity persist and become considerably more difficult to overcome in the sodium system. Na ion transport, in particular, appears to be more sluggish in titania systems [137, 138]. The subsequent investigations devised ways for optimizing performance, particularly rate capability. While nanostructuring the active material is a well-established method for improving kinetics, a carbon support phase is also required for TiO_2 [44]. Numerous previous investigations have been conducted on extremely small TiO_2 particles, nanodots [139], nanorods, nanowires, and nanotubes [136], and thin TiO_2 sheets and sphere shells (2D) [134]. The reduction in the size of the TiO_2 can significantly reduce the sodium ion's diffusion distance within the material bulk. Introducing electronically conductive materials such as graphene or graphitic carbon into TiO_2 is beneficial to improve the conductivity [139, 140]. The intrinsic electrical conductivity of TiO_2 can also be altered through aliovalent doping with elements such as nitrogen [97, 141–143]. Based on these approaches, contemporary TiO_2 material was able to supply > 300 mAh.g^{-1} at a low rate, while maintaining > 50 mAh.g^{-1} capacity at current densities up to 30 A g^{-1}. Due to the significant increase in rate capability, TiO_2 is a potential anode material for sodium-ion capacitors [144].

5.7.5 Carbons as Anodes

Typically, activated carbon has been employed as the cathode in a lithium-ion capacitor that achieves high power density with the use of a capacitive EDLC-type absorbing-desorbing electrode. The basic nature of activated carbon is that it is porous, with a surface area of more than 1000 m^2/g. The cathode's storage capacity enables the final performance of the lithium-ion capacitor [68]. Carbon compounds may also serve as the active molecules in lithium-ion capacitors, as they include active lithium-ion intercalation-de-intercalation sites [38]. Lithium-ion capacitor counter-electrodes, unlike battery-type electrodes, are capacitor-type. As a result, various porous carbon materials with a large specific surface area, such as activated carbon and graphene, are promising candidates for lithium-ion capacitor-type electrodes [46]. Their capacitances are principally determined by the adsorption–desorption of ions on the surface of carbon-based electrodes [69, 145]. The specific capacitive performance of activated carbon is dependent on the surface area, the volume of the pores, and the distribution of the pore sizes. Additionally, the capacitance is dependent on inorganic contaminants and surface functional groups [69]. The surface area, pore volume, and pore size distribution are all dependent on the precursor type, the activation method, and the post-activation treatments. To optimize the capacitive performance of activated carbon-based cathode materials is required to adjust these characteristics during their synthesis [68]. As a result, porous structures with an adequate pore size distribution are critical for the electrochemical synthesis of carbon compounds in lithium-ion capacitors [146]. Typically, special compounds are included with carbonaceous materials to impart pseudo-capacity and therefore increase the capacitance of the capacitor-type electrode [69].

Carbon materials such as graphitized and non-graphitized carbon are employed in the anodes of lithium-ion capacitors. Additionally, non-graphitized carbon is classified as soft or hard carbon [75]. Soft carbons have the potential to be used as anodic materials in the fabrication of high-performance lithium-ion capacitors [147]. These materials have excellent performance at high current densities, low self-discharge, and long-term cyclability. Schroeder et al. reported a lithium-ion capacitor cell with modest specific energy and long cycle life [148], while [149] studied a hard carbon anode that retained 70% of its capacity at a rate of 40 C. Graphitized carbon, which allows for easy modification of the interlayer's spacing and microstructure, offers the most intriguing prospects for application in lithium-ion capacitor anodes. Due to the coexistence of graphical and amorphous structures, graphitic carbon can combine a high plateau potential with a higher rate of efficiency. Catalytic graphitization with the aid of metal ions is a critical technique for producing graphitic carbon at relatively low temperatures (1200 °C). Adjusting the temperature, precursor, and catalyst structures can control the degree of graphitization [69] (Tables 5.2 and 5.3).

Table 5.2 Specific capacity of lithium-ion capacitor electrode materials

Electrode materials	Electrode type	Specific capacity (m Ah g^{-1})	Current density (A g^{-1})	References
N-doped carbon nanopipes	Anode	166	20	[150]
Partially reduced graphene oxide	Cathode	92.3	8.7	[150]
TiO_2 nanoparticles embedded in mesoporous carbon nanofibers	Anode	125.7	10	[151]
Activated carbon	Cathode	55.6	10	[151]
Porous 3D graphene-based bulk material	Anode	372	3.72	[152]
Digital camera flash reduced graphene oxide	Cathode	220	Not provided	[152]
Electrochemical-exfoliated graphene nanosheets	Anode	85–162	0.35–1.5	[153]
Graphite	Cathode	372	Not provided	[154]
Nanocrystaline-$Li_4Ti_5O_{12}$ -CNF composite	Anode	158	Not provided	[102]
$Li_4Ti_5O_{12}$	Anode	58	0.1	[155]
Coconut shell-derived mesoporous carbon	Cathode	83	0.1S	[156]
Sodiated natural plant-based hard carbon	Anode	350	0.1	[156]
Boron and nitrogen dual-doped 3D carbon nanofibers	Cathode	113	0.1	[145]
Boron and nitrogen dual-doped 3D carbon nanofibers	Anode	1130	0.1	[145]
Nb_2O_5-rGO	Cathode	180	Not provided	[157]
Carbide-derived-carbon	Cathode	50	Not provided	[157]
Porous LTO (pLTO) nanoparticles	Anode	148	0.175	[158]

rGO—reduced graphene oxide, CNF—Carbon nanofiber, LTO- $Li_4Ti_5O_{12}$

5.8 Other Anode Materials

5.8.1 Silicon-Based Materials

Silicon-based materials are primarily composed of certain elements, such as silicon, germanium, and tin, that store charges via an alloy reaction with Li^+ and possess a high theoretical specific capacity. Silicon has a theoretical specific capacity of 4200 mAh g^{-1}, which is significantly greater than that of other materials. However, the electrode structure is prone to be destroyed and they detach more easily from the current collector when lithium-ion reversibly inserts-deserts into the lattice of these

Table 5.3 Specific capacity of sodium-ion capacitor electrode materials

Electrode materials	Electrode type	Specific capacity (m Ah g^{-1})	Current density (A g^{-1})	References
Activated carbon	Anode	30	Not provided	[159]
Sodium pre-doped hard carbon	Cathode	220	Not provided	[159]
Sodiated natural plant-based hard carbon	Anode	216	0.1	[156]
Sodium pre-doped hard carbon	Cathode	190	25	[159]
MoS_2-carbon monolayer inter overlapped superstructure	Anode	62	10	[38]
Polyaniline-derived porous carbon	Cathode	70	0.2	[38]
TiO_2-coated MoS_2@N-doped carbon composite	Anode	420.2	20	[160]
Activated carbon	Cathode	80.5	0.1	[160]
Peanut shell ordered carbon	Anode	315	0.1	[161]
Peanut shell nanosheet carbon	Cathode	73	25.6	[161]
Nb_2O_5@rGO	Anode	285	0.025	[162]
Commercial activated carbon (MSP-20)	Cathode	60	0.025	[162]
Alluaudite $Na_2Fe_2(SO_4)_3$	Anode	90	1.0	[163]
MXene Ti_2C	Cathode	40	5.0	[163]
Ti_3C_2Tx/ CNT-SA	Anode	179	20	[142]
$Na_{0.44}MnO_2$	Cathode	80	14	[142]
Titanium Carbide	Anode	170	10	[153]
pyridine-derived hierarchical porous nitrogen-doped carbon	Cathode	81.5	0.1	[153]
λ-MnO_2	anode	390	13.6	[164]
Activated carbon	Cathode	89.7	68	[164]

substances, resulting in extremely poor cycling stability for silicon-based electrodes during the charging-discharging process. Additionally, the rate of the insertion-desertion reaction is slower, resulting in a reduced rate capacity for silicon-based electrodes. As a result of the charge conservation rule between the cathode and anode of lithium-ion capacitors, it is extremely difficult to obtain cathodes compatible with silicon-based anodes [80].

5.8.2 Niobium Pentoxide (Nb_2O_5)

Niobium pentoxide is a promising anode material for high-performance lithium-ion capacitors with pseudocapacitive lithium-ion intercalation behaviour that has garnered considerable attention from researchers due to its defined crystal phases, low voltage plateau, and strong cycle stability. However, the restricted electrical conductivity and unknown mechanism of complicated phases preclude their implementation in the future [80].

5.8.3 Carbons for Sodium Electrodes

Microstructurally tailored carbons appear to be the most cost-effective and technically feasible options for sodium-ion capacitor applications, as Na kinetics in titania structures are often significantly slower than equivalent lithium kinetics (Sun et al., 2013), [138]. Notably, graphite does not intercalate Na significantly due to its higher ionic radius of 0.102 nm versus 0.076 nm and poorer chemical contact with graphene planes [44, 165]. Rather than graphitic carbons, various non-graphitic carbons, including N-doped carbons, have been used to store Na [156, 166]. In standard battery electrolytes, graphite, or graphite nanoflakes with an equilibrium graphite structure and layer spacing will not intercalate sodium reversibly. Thus, they have received little attention in the sodium literature, except for rare instances of sodium co-intercalation with a secondary ion [167].

Certain forms of reduced graphene oxide are used in the fabrication of sodium-ion capacitors and high-power sodium-ion battery electrodes [88, 136]. Reduced graphene oxide has a severely faulty structure, unlike graphite or well-ordered single-layer graphene. However, in terms of Raman I_G/I_D ratio, heteroatom content, surface area, and porosity, reduced graphene oxide is comparable to carbon nanosheets [153, 161]. Mitlin and coworkers were the first to do X-ray measurements throughout various stages of sodiation to examine the sodium charge-storage mechanisms in non-graphitic carbons, detecting and quantifying reversible Na intercalation at low voltages [6, 161, 168, 169]. At voltages less than 0.2 V versus Na/Na^+, it was established that successive sodiation results in a reversible shift of the (002) graphene peak to decreased 2 values. Mitlin and coworkers further suggested that the sloping section of the Na charge–discharge curve is not due to Na intercalation as proposed by Dahn and coworkers [170], but is rather due to the reversible adsorption of sodium on graphene defects such as divacancies and Stone Wales [6, 161, 168]. This was based on previous modelling studies by Shenoy and colleagues, and have shown that Na adsorption at faulty graphene sites results in sloping voltage versus capacity curves [171]. Subsequent studies verified both the low-voltage Na intercalation scenario, which necessitates sufficient ordering of the graphene planes and a dilated interlayer gap, and the higher-voltage reversible Na adsorption scenario [6, 44, 99, 160, 168, 172].

Recently, the emergence of new organic and inorganic nanomaterials, especially nanostructured carbons, conducting polymers, and metal oxides, has allowed the fabrication of different energy devices for high performance. Specific capacitance, energy density, power density, rate capability, and cycle stability are among the most important criteria for high-efficiency supercapacitors [126, 173, 174, 158], and these are largely determined by the configuration of electrode materials in the devices [45, 77]. Over the last decade, we have seen significant progress in the development of anode materials such as carbonaceous materials such as activated carbon, carbon black, single and multi-layer graphene, graphene oxide, and reduced graphene oxide for electrical double-layer capacitors (EDLC) as active materials. Biomass-derived electrode for next-generation lithium-ion capacitors [38, 175–177]. Conducting polymers such as polyaniline, polypyrrole, polythiophene and polythiophene-derivatives and different metallic compounds such as oxides, hydroxides, carbides, nitrides, phosphides, phosphates, phosphites, chalcogenides which are mostly used as pseudocapacitive materials [35, 75, 178–180]. However, the electrode materials for hybrid supercapacitors should potentially have the following properties [181].

Large specific surface area: As charge absorption occurs at the interface between the electrode and the electrolyte, the electrode with a higher surface area effectively provides more active sites for ionic adsorption, contributing to increased capacitance. Materials having nanostructures, in general, have a greater specific surface area than those with blocky structures.

Controllable porosity: A reasonable pore size allows electrolyte penetration and therefore increases the usage rate of electrode materials.

High electronic conductivity: High conductivity can improve the responsiveness of electrode materials to high scan rates or current densities, as well as assure an outstanding rate capability.

Desirable electroactive sites: The more electroactive sites, the more ion adsorption, which might lead to a high specific capacitance, such as when nitrogen-doping, sulphur-doping, or co-doping is used.

High thermal and chemical stability: In the cycle process, the insolubility and structural stability of electrode materials are essential for a long cycle life.

When compared to pseudocapacitive materials, EDLC materials, that is carbon-rich materials, have numerous advantages. For instance, carbon compounds are less costly than noble metal oxide utilized as pseudocapacitive materials due to a simple synthetic process and available raw materials. Also, the high conductivity of carbon materials leads to a quick reaction to high current density and increased power density. Carbon materials offer excellent physical and chemical stability in both strong acid and strong base electrolytes, as well as no phase change during the electrochemical process, resulting in a very long cycle life. The most significant aspect is that the microstructure of carbon materials, such as nanotubes and nanosheets, can be easily manipulated by physical and chemical techniques to provide rapid and reversible electrolyte ion adsorption–desorption [158].

5.9 Next-Generation of Lithium-Ion Batteries

5.9.1 Nanostructured Electrode Materials

Zero-dimensional Nanostructures: These nanostructures are spherical particles with three dimensions restricted to the nanoscale typically 1–100 nm. Examples include fullerenes, quantum dots, nano onions, and nanoparticles. Due to their unique features, such as low density, a high surface-to-volume ratio, and reduced paths for mass and charge transport, hollow Zero-dimensional nanostructures have been recognized as interesting options for electrode design [182].

One-dimensional Nanostructures: These nanostructures are one-dimensional slivers of material limited to the nanoscale [183]. Due to their unique and superior chemical and physical properties, they have been widely explored for use in the creation of electrode materials for supercapacitors [173, 184]. For example, the longitudinal axis of one-dimensional nanostructures can provide an effective transport channel for electrons and ions while retaining the benefits of zero-dimensional nanostructures in the two nanoscale dimensions [185]. Nanotubes, nanofibers, nanowires, nanopillars, nanoribbons, and nanobelts are a few examples [158].

Two-dimensional Nanostructures: The development of these materials, and the fact that they only have a surface area and therefore no bulk volume, has naturally attracted considerable interest, and increased the number of possible configurations of potential hybrid 2D configurations in electrode material design, given the importance of properties such as diffusion ion length or electrolyte contact area for the utilization of active material in electrochemical supercapacitors performance. There are unique qualities in Two-dimensional materials such as the capacity to bend and be transparent as well as the ability to withstand high temperatures and pressures. However, little is known about the unique behaviour and features of most Two-dimensional materials, and derived hybrids in aqueous or organic solvents will display in supercapacitors incorporating these structures. Two-dimensional materials have the potential to bring a paradigm shift to the field of electrochemical capacitors since capacitance is so closely tied to electrode surface area [158, 186].

Three-dimensional (3D) Nanostructures: the porous designs of active materials give a large surface area, well-defined pathways to the electrolyte, and mechanical stability for effective supercapacitor electrodes. Three-dimensional electrodes are frequently made using metal foam templates or by grouping active materials into three-dimensional nanostructures [158].

5.9.2 Pseudocapacitive Oxides

Due to their rapid charging and discharging kinetics, pseudocapacitive materials represent a substantial subclass of materials for lithium-ion capacitors, as they exhibit greater gravimetric and volumetric efficiency than real EDLC electrodes. For instance, vanadium oxide, V_2O_5 [69, 151, 187].

5.9.3 MXenes

The MXene family of metal carbides and carbonitrides is a two-dimensional family of metal carbides and carbonitrides [152]. Na^+, K^+, Mg^{2+}, and Al^{3+} can intercalate reversibly through the bulk of exfoliated multilayer $T_3C_2T_x$ MXene in an aqueous electrolyte. Li^+ ions reversibly intercalate between layers of MXene in organic electrolytes [188]. Due to its significant intercalation reaction, MXene has pseudocapacitor cyclic voltammetric and galvanostatic profiles [152, 188]. Li storage was also a reaction, rather than a diffusion-controlled Faradic cycle, unlike other pseudocapacitive materials. The cyclability of these materials is worth investigating [69].

5.9.4 Conversion Compounds

Reversible materials that rely on reaction conversion are another type of pseudocapacitive material that has gained favour as a substrate for lithium-ion capacitor electrodes [69]. A conversion electrode is defined in the lithium-ion battery literature as a crystalline or amorphous A_xB_y material that degrades reversibly into one or two lithium compounds upon charging [189]. Lithium-ion capacitor structures are found in a variety of materials, including Molybdenum disulfide (MoS_2), niobium nitride (NbN), vanadium nitride (VN), Manganese oxide (MnO), Iron(III)/(II) oxide or ferric oxide (Fe_2O_3/Fe_3O_4), nickel cobaltite ($NiCo_2O_4$), and other alloys [69]. While MoS_2 can intercalate lithium down to 1.1 V, reversible Mo and Li_2S conversion reaction down to 0 V provides the highest reversibility [6, 168]. MoS_2 is reported to have a reversible capacity of up to 1000 $mAh.g^{-1}$, which is significantly greater than the theoretical value [69].

5.9.5 Battery-Related Intercalation Ceramics

Researchers have developed a high-power lithium-ion battery with a more sloped tension profile by coupling ceramic lithium-ion battery cathode materials to carbon counter electrodes. Aravindan et al. [74] described intercalation-type materials for lithium-ion capacitor setups in detail. Many such systems, like batteries, demonstrate cyclability with many thousands of cycles possible. Layered oxides, spinel oxide, olivine, NASICON, and silicates are all examples [69].

5.10 Conclusion

In investigating energetic, powerful, and stable hybrid-ion capacitors, carbon has been the material of choice for the capacitor-type positive electrode, while the selection of the battery-type negative electrode has been under in-depth research. This chapter reviews some of the advances in anode and cathode materials for hybrid-ion capacitors. Great efforts have been made to use carbon-based materials as negative electrodes for hybrid-ion capacitors due to their simplicity, availability, low cost, and good electrical conductivity, which mainly cover graphite, hard carbon, soft carbon, nitrogen-doped carbon, and other carbon materials. Non-carbon materials have the advantage of stable structure and excellent cycling stability. There has been a significant focus on the cell design, working mechanisms, electrode materials and electrolyte systems. Nonetheless, areas that need to be improved upon such as the energy density, stable working voltage window and long cycle lifespan remain a challenge. Some perspectives and opportunities for future work are summarized as follows to achieve improved electrochemical performance. High-capacity capacitor-type materials are urgently needed and preparation of cathode-based materials at low cost is one of the critical factors for large-scale applications. In developing battery-type materials with a high rate and long-term stability, electrolytes that are stable with a wide voltage are also needed. Furthermore, the decomposition mechanism of the electrolytes presents an interesting area of research, lastly, nanomaterials could play a vital role in achieving high performance due to their outstanding electrochemical properties and hybrid-ion capacitors are witnessing tremendous research progress, especially in electrode materials. Their development steps are speeding up and large-scale applications will be realized soon.

Acknowledgements The authors graciously thank the African University of Science and Technology Abuja. for providing the platform and opportunity for this research study. Particularly, Charles Obi would like to acknowledge the support of the Partnership for Applied Science and Engineering Technology (PASET) and Research Innovation Fund (RSIF).

References

1. B.E. Conway, *Electrochemical Supercapacitors: Scientific Fundamentals and Technological Applications* (Kluwer Academic/Plenum, New York, 1999). http://books.google.co.za/books/about/Electrochemical_Supercapacitors.html?id=8yvzlr9TqI0C&pgis=1
2. B. Dunn, H. Kamath, J.-M Tarascon For the grid : a battery of choices **334**(November), 928–936 (2011a). https://doi.org/10.1126/science.1212741
3. B. Dunn, H. Kamath, J.-M. Tarascon, Electrical energy storage for the grid: a battery of choices. Science (New York, N.Y.) **334**(6058), 928–935 (2011b). https://doi.org/10.1126/science.1212741
4. Z. Lin, E. Goikolea, A. Balducci, K. Naoi, P.L. Taberna, M. Salanne, G. Yushin, P. Simon, Materials for Supercapacitors: When Li-Ion Battery Power Is Not Enough. Mater. Today **21**(4), 419–436 (2018). https://doi.org/10.1016/J.MATTOD.2018.01.035

5. Y. Shao, M.F. El-Kady, J. Sun, Y. Li, Q. Zhang, M. Zhu, H. Wang, B. Dunn, R.B. Kaner, Design and mechanisms of asymmetric supercapacitors. Chem. Rev. **118**(18), 9233–9280 (2018). https://doi.org/10.1021/ACS.CHEMREV.8B00252
6. J. Zhao, A.F. Burke, Electrochemical capacitors: materials, technologies and performance. Energy Storage Mater. **36**(April), 31–55 (2021). https://doi.org/10.1016/J.ENSM.2020.12.013
7. M. Armand, J.-M. Tarascon, Building better batteries. Nature **451**(7179), 652–657 (2008). https://doi.org/10.1038/451652a
8. J.B. Goodenough, K.-S. Sung Park, The li-ion rechargeable battery: a perspective. journal of the american chemical society, vol 135. American Chemical Society (2013). https://doi.org/10.1021/ja3091438
9. A. Manthiram, An outlook on lithium ion battery technology. ACS Cent. Sci. **3**(10), 1063–1069 (2017). https://doi.org/10.1021/acscentsci.7b00288
10. G. Armstrong, The li-ions share. Nat. Chem. **11**(12), 1076 (2019). https://doi.org/10.1038/s41557-019-0386-7
11. J.L. Brédas, J.M. Buriak, F. Caruso, K.S. Choi, B.A. Korgel, M. Rosa Palacín, K. Persson et al., An electrifying choice for the 2019 chemistry nobel prize: goodenough, whittingham, and yoshino. Chem. Mater. **31**(21), 8577–8581 (2019). https://doi.org/10.1021/acs.chemmater.9b04345
12. D. Castelvecchi, E. Stoye, World-changing batteries win nobel. Nature **574**, 308 (2019)
13. A.K. Stephan, The age of li-ion batteries. Joule **3**(11), 2583–2584 (2019). https://doi.org/10.1016/j.joule.2019.11.004
14. V.M. Tyutyunnik, Another breakthrough in power supply technology – lithium-ion batteries: 2019 nobel prize winners in chemistry john goodenough, Stanley Whittingham and Akira Yoshino. Image J. Adv. Mater. Technol. **6**(3), 163–66. https://doi.org/10.17277/jamt.2021.03.pp.163-166
15. J. Liu, J. Zhang, Zh. Yang, J.P. Lemmon, C. Imhoff, G.L. Graff, L. Li, et al. Materials science and materials chemistry for large scale electrochemical energy storage : from transportation to electrical grid, 929–946 (2013). https://doi.org/10.1002/adfm.201200690
16. M. North, Energy and environmental science 5692 (2010). https://doi.org/10.1039/b916098a
17. Z. Yang, J. Zhang, M.C.W. Kintner-Meyer, L. Xiaochuan, D. Choi, J.P. Lemmon, J. Liu, Electrochemical energy storage for green grid. Chem. Rev. **111**(5), 3577–3613 (2011). https://doi.org/10.1021/cr100290v
18. J.W. Choi, D. Aurbach, Promise and reality of post-lithium-ion batteries with high energy densities. Nat. Rev. Mater. **1**(4), 16013 (2016). https://doi.org/10.1038/natrevmats.2016.13
19. M. Salanne, B. Rotenberg, K. Naoi, K. Kaneko, P.-L. Taberna, C.P. Grey, B. Dunn, P. Simon, Efficient storage mechanisms for building better supercapacitors. Nat. Energy **61**(6), 1–10 (2016). https://doi.org/10.1038/nenergy.2016.70
20. M.D. Slater, D. Kim, E. Lee, C.S. Johnson, Sodium-ion batteries. Adv. Func. Mater. **23**(8), 947–958 (2013). https://doi.org/10.1002/adfm.201200691
21. N. Yabuuchi, K. Kubota, M. Dahbi, S. Komaba, Research development on sodium-ion batteries (2014)
22. P. Sharma, T.S. Bhatti, A review on electrochemical double-layer capacitors. Energy Convers. Manage. **51**(12), 2901–2912 (2010). https://doi.org/10.1016/j.enconman.2010.06.031
23. Y. Wang, Y. Song, Y. Xia, Characterization and applications chemical functional materials. Chem. Soc. Rev. Syst. 5925–5950 (2016). https://doi.org/10.1039/c5cs00580a
24. D. Qiu, C. Kang, M. Li, J. Wei, Z. Hou, F. Wang, Biomass-derived mesopore-dominant hierarchical porous carbon enabling ultra-Ef Fi cient lithium ion storage. Carbon **162**, 595–603 (2020). https://doi.org/10.1016/j.carbon.2020.02.083
25. C. Largeot, C. Portet, J. Chmiola, P.-L. Taberna, Y. Gogotsi, P. Simon, Relation between the ion size and pore size for an electric double-layer capacitor. J. Am. Chem. Soc. **130**(9), 2730–2731 (2008). https://doi.org/10.1016/j.carbon.2007.10.023.(15)
26. C. Noh, Y. Jung (2019) Understanding the charging dynamics of an ionic liquid electric double layer capacitor via molecular dynamics simulations **21**(13). https://doi.org/10.1039/c8cp07200k.

27. J. Cao, A. Emadi, S. Member, A New battery/ultracapacitor hybrid energy storage system for electric, hybrid, and plug-in hybrid electric vehicles **27**(1), 122–32 (2012)
28. D.P. Dubal, O. Ayyad, V. Ruiz, Chem soc rev and supercapacitor chemistries, 1777–90 (2015). https://doi.org/10.1039/c4cs00266k
29. D. Kaufman, K.L. Hudson, and R Mcclamrock, And Supercapacitors Begin? 343 (March) 1210–1212 (2014). https://doi.org/10.1126/science.1249625
30. L.T. Lam, R. Louey. Development of Ultra-battery for hybrid-electric vehicle applications **158**(March), 1140–48 (2006). https://doi.org/10.1016/j.jpowsour.2006.03.022
31. W.G. Pell, B.E. Conway, Peculiarities and requirements of asymmetric capacitor devices based on combination of capacitor and battery-type electrodes. J. Power. Sources **136**(2), 334–345 (2004)
32. N.R. Chodankar, H.D. Pham, A.K. Nanjundan, J.F.S. Fernando, K.Jayaramulu, D. Golberg, Y. Han, D.P. Dubal, True meaning of pseudocapacitors and their performance metrics : asymmetric versus hybrid supercapacitors 2002806, 1–35 (2020). https://doi.org/10.1002/smll.202002806.
33. T. Kshetri, D. Thanh, D. Chuong, N. Hoon, K. Lau, A. Materials, B.I.N. Convergence, Technology Bk, Plus Global, B.I.N. Convergence, Ternary Graphene-Carbon Nanofibers-Carbon Nanotubes Structure for Hybrid Supercapacitor. Chem. Eng. J. **380**(July 2019), 122543 (2020). https://doi.org/10.1016/j.cej.2019.122543
34. D.-G. Wang, Z. Liang, S. Gao, Q. Chong, R. Zou, Metal-organic framework-based materials for hybrid supercapacitor application. Coord. Chem. Rev. **404**, 213093 (2020). https://doi.org/10.1016/j.ccr.2019.213093
35. P. Naskar, D. Kundu, A. Maiti, P. Chakraborty, B. Biswas, A. Banerjee, Frontiers in hybrid ion capacitors: a review on advanced materials and emerging devices. ChemElectroChem **8**(8), 1393–1429 (2021). https://doi.org/10.1002/celc.202100029
36. E. Lim, C. Jo, M.S. Kim, M. Kim, J. Chun, H. Kim, J. Park, et al. High-performance sodium-ion hybrid supercapacitor based on Nb_2O_5 @ Carbon core—shell nanoparticles and reduced graphene oxide nanocomposites 3711–3719 (2016). https://doi.org/10.1002/adfm.201505548
37. X. Sun, X. Zhang, H. Zhang, X. Nansheng, K. Wang, Y. Ma, High performance lithium-ion hybrid capacitors with pre-lithiated hard carbon anodes and bifunctional cathode electrodes. J. Power. Sources **270**, 318–325 (2014). https://doi.org/10.1016/j.jpowsour.2014.07.146
38. R. Wang, S. Wang, X. Peng, Y. Zhang, D. Jin, P.K. Chu, L. Zhang, Elucidating the intercalation pseudocapacitance mechanism of moS2-carbon monolayer interoverlapped superstructure: toward high-performance sodium-ion-based hybrid supercapacitor. ACS Appl. Mater. Interfaces **9**(38), 32745–32755 (2017). https://doi.org/10.1021/acsami.7b09813
39. J. Chen, B. Yang, B. Liu, J. Lang, X. Yan, Science direct electrochemistry recent advances in anode materials for sodium - and potassium-ion hybrid capacitors. Curr. Opin. Electrochem. **18**, 1–8 (2019). https://doi.org/10.1016/j.coelec.2019.07.003
40. D. Qiu, J. Guan, M. Li, C. Kang, J. Wei, Y. Li, Z. Xie, Kinetics enhanced nitrogen-doped hierarchical porous hollow carbon spheres boosting advanced potassium-ion hybrid capacitors, 1903496 1–8 (2019). https://doi.org/10.1002/adfm.201903496
41. E. Frackowiak, F. Béguin Electrochimica acta electrochemical performance of a hybrid lithium-ion capacitor with a graphite anode preloaded from lithium bis (Trifluoromethane) sulfonimide-based electrolyte **86**, 282–286 (2012). https://doi.org/10.1016/j.electacta.2012.05.111
42. H. Gu, Y. Zhu, J. Yang, J. Wei, Z. Zhou Nanomaterials and technologies for lithium-ion hybrid supercapacitors, 578–87 (2016). https://doi.org/10.1002/cnma.201600068
43. Y. Ma, H. Chang, M. Zhang, Y. Chen, Graphene-based materials for lithium-ion hybrid supercapacitors, 5296–5308 (2015). https://doi.org/10.1002/adma.201501622
44. J. Ding, H. Wenbin, E. Paek, D. Mitlin, Review of hybrid ion capacitors: from aqueous to lithium to sodium. Review-article. Chem. Rev. **118**(14), 6457–6498 (2018). https://doi.org/10.1021/acs.chemrev.8b00116
45. W. Zuo, R. Li, C. Zhou, Y. Li, J. Xia, J. Liu, Battery-supercapacitor hybrid devices: recent progress and future prospects. Advanced Science **4**(7), 1600539 (2017). https://doi.org/10.1002/ADVS.201600539

46. P. Han, X. Gaojie, X. Han, J. Zhao, X. Zhou, G. Cui, Lithium Ion capacitors in organic electrolyte system: scientific problems, material development, and key technologies. Adv. Energy Mater. **8**(26), 1–30 (2018). https://doi.org/10.1002/aenm.201801243
47. S.W. Lee, B.M. Gallant, H.R. Byon, P.T. Hammond, Y. Shao-Horn, Nanostructured carbon-based electrodes: bridging the gap between thin-film lithium-ion batteries and electrochemical capacitors. Energy Environ. Sci. **4**(6), 1972 (2011). https://doi.org/10.1039/c0ee00642d
48. Y.S. Yun, S.Y. Cho, J. Shim, B.H. Kim, S.J. Chang, S.J. Baek, Y.S. Huh et al., Microporous carbon nanoplates from regenerated silk proteins for supercapacitors. Adv. Mater. **25**(14), 1993–1998 (2013). https://doi.org/10.1002/adma.201204692
49. P. Dubal, D., O. Ayyad, V. Ruiz, and P. Gómez-Romero, Hybrid energy storage: the merging of battery and supercapacitor chemistries. Chem. Soc. Rev. **44**(7), 1777–17790 (2015). https://doi.org/10.1039/C4CS00266K
50. G. Li, Y. Huang, Z. Yin, H. Guo, Y. Liu, H. Cheng, W. Yuping, X. Ji, J. Wang, Defective synergy of 2D graphitic carbon nanosheets promotes lithium-ion capacitors performance. Energy Storage Mater. **24**(July), 304–311 (2020). https://doi.org/10.1016/j.ensm.2019.07.046
51. Y. Liu, S. Wang, X. Sun, J. Zhang, F. uz Zaman, L. Hou, C. Yuan, Sub-nanoscale engineering of MoO_2 clusters for enhanced sodium storage. Energy and Environ. Mater. 0–2 (2021). https://doi.org/10.1002/eem2.12263
52. Y. Nishi, The development of lithium ion secondary batteries. Chem. Rec. **1**(5), 406–413 (2001). https://doi.org/10.1002/tcr.1024
53. G.G. Amatucci, F. Badway, A.D. Pasquier, T. Zheng, An asymmetric hybrid nonaqueous energy storage cell. J. Electrochem. Soc. **148**(8), A930 (2001). https://doi.org/10.1149/1.1383553
54. G. Li, Z. Yang, Z. Yin, H. Guo, Z. Wang, G. Yan, Y. Liu, L. Li, J. Wang, Non-aqueous dual-carbon lithium-ion capacitors: a review. J. Mater. Chem. A **7**(26), 15541–15563 (2019). https://doi.org/10.1039/c9ta01246j
55. A. Le Comte, Y. Reynier, C. Vincens, C. Leys, P. Azaïs, First prototypes of hybrid potassium-ion capacitor (KIC): an innovative, cost-effective energy storage technology for transportation applications. J. Power. Sources **363**, 34–43 (2017). https://doi.org/10.1016/j.jpowsour.2017.07.005
56. Z. Jian, Z. Xing, C. Bommier, Z. Li, X. Ji, Hard carbon microspheres : potassium-ion anode versus sodium-ion anode, 1–5 (2016). https://doi.org/10.1002/aenm.201501874
57. P. Veronica, S. Paula, V. Irune, H. Karina, C.-G. Javier, T. Rojo (2011) Na-Ion batteries , recent advances and present challenges to become low cost energy storage systems, pp 5884–5901. https://doi.org/10.1039/c2ee02781j
58. C. Zheng, A. Veronica, J. Xilai, X. Qiangfeng, D. Bruce, L. Yunfeng, High-performance sodium-ion pseudocapacitors based on hierarchically porous nanowire composites **5**, 4319–4327
59. J. Luo, C. Fang, C. Jin, H. Yuan, O. Sheng, R. Fang, W. Zhang et al., Tunable pseudocapacitance storage of MXene by cation pillaring for high performance sodium-ion capacitors. J. Mater. Chem. A **6**(17), 7794–7806 (2018). https://doi.org/10.1039/c8ta02068j
60. H. Wang, M. Yoshio, Short communication KPF 6 dissolved in propylene carbonate as an electrolyte for activated carbon/graphite capacitors **195**, 1263–1265 (2010). https://doi.org/10.1016/j.jpowsour.2009.08.073
61. M. Pasta, Wessells, C.D., Huggins, R.A., Cui, Y., Battery for grid-scale energy storage. **May** (2012). https://doi.org/10.1038/ncomms2139
62. J. Yin, W. Zhang, N.A. Alhebshi, N. Salah, H.N. Alshareef, Electrochemical zinc ion capacitors: fundamentals, materials, and systems. Adv. Energy Mater. **11**(21), 1–30 (2021). https://doi.org/10.1002/aenm.202100201
63. H.D. Yoo, S.D. Han, R.D. Bayliss, A.A. Gewirth, B. Genorio, N.N. Rajput, K.A. Persson, A.K. Burrell, J. Cabana, 'Rocking-chair'-type metal hybrid supercapacitors. ACS Appl. Mater. Interfaces **8**(45), 30853–30862 (2016). https://doi.org/10.1021/acsami.6b08367
64. M. Soltani, S. Hamidreza Beheshti, A comprehensive review of lithium ion capacitor: development, modelling, thermal management and applications. J. Energy Storage **34**(October), 102019 (2021). https://doi.org/10.1016/j.est.2020.102019

65. P. Cai, K. Zou, X. Deng, B. Wang, M. Zheng, L. Li, H. Hou, G. Zou, X. Ji, Comprehensive understanding of sodium-ion capacitors: definition, mechanisms, configurations, materials, key technologies, and future developments. Adv. Energy Mater. **11**(16), 1–56 (2021). https://doi.org/10.1002/aenm.202003804
66. A.-J. Jiao, J.-F. Gao, Z.-H. He, F.-F. Li, L.-B. Kong, Constructing high-performance li-ion capacitors via cobalt fluoride with excellent cyclic stability as anode and coconut shell biomass-derived carbon as cathode materials. Chem. Select **6**(32), 8349–8360 (2021). https://doi.org/10.1002/slct.202102420
67. K. Zou, P. Cai, B. Wang, C. Liu, J. Li, T. Qiu, G. Zou, H. Hou, X. Ji, Insights into enhanced capacitive behavior of carbon cathode for lithium ion capacitors: the coupling of pore size and graphitization engineering. Nano-Micro Letters **12**(1), 1–19 (2020). https://doi.org/10.1007/s40820-020-00458-6
68. A. Jagadale, X. Zhou, R. Xiong, D.P. Dubal, J. Xu, S. Yang Lithium ion capacitors (LICs): development of the materials. Energy Storage Mater. **19**(December 2018), 314–29. https://doi.org/10.1016/j.ensm.2019.02.031.
69. J.J. Lamb, O.S. Burheim, Lithium-ion capacitors: a review of design and active materials. Energies **14**(4), 1–27 (2021). https://doi.org/10.3390/en14040979
70. X. Sun, W. Chen, X. Li, J. Wang, H. Hu, G. Liang, Y. Huang, C. Wei, Performance and applications of lithium ion capacitors. Sci. Technol. Adv. Appl. Supercapacitors, 1–15 (2019). https://doi.org/10.5772/intechopen.80353.
71. C. Han, W. Xinyi, J. Peng, Q. Xia, S. Chou, G. Cheng, Z. Huang, W. Li, Recent progress on two-dimensional carbon materials for emerging post-lithium (Na+, K+, Zn2+) hybrid supercapacitors. Polymers **13**, 1–23 (2021)
72. P.K. Panda, A. Grigoriev, Y.K. Mishra, R. Ahuja, Progress in supercapacitors: roles of two dimensional nanotubular materials. Nanosc. Adv. **2**(1), 70–108 (2020). https://doi.org/10.1039/c9na00307j
73. S. Dong, N. Lv, W. Yulin, G. Zhu, X. Dong, Lithium-ion and sodium-ion hybrid capacitors: from insertion-type materials design to devices construction. Adv. Func. Mater. **31**(21), 1–29 (2021). https://doi.org/10.1002/adfm.202100455
74. V. Aravindan, J. Gnanaraj, Y.S. Lee, S. Madhavi, Insertion-type electrodes for nonaqueous li-ion capacitors. Chem. Rev. **114**(23), 11619–11635 (2014). https://doi.org/10.1021/cr5000915
75. H. Li, J. Lang, S. Lei, J. Chen, K. Wang, L. Liu, T. Zhang, W. Liu, X. Yan, A high-performance sodium-ion hybrid capacitor constructed by metal-organic framework–derived anode and cathode materials. Adv. Func. Mater. **28**(30), 1–12 (2018). https://doi.org/10.1002/adfm.201800757
76. G. Yu, X. Xie, L. Pan, Z. Bao, Y. Cui, Hybrid nanostructured materials for high-performance electrochemical capacitors. Nano Energy **2**(2), 213–234 (2013)
77. X.Y. Yu, X. Wen (David) Lou, Mixed metal sulfides for electrochemical energy storage and conversion. Adv. Energy Mater. **8**(3), 1–37 (2018). https://doi.org/10.1002/aenm.201701592
78. J. Maier, Thermodynamics of electrochemical lithium storage. Angewandte Chemie – Int. Edition **52**(19), 4998–5026 (2013). https://doi.org/10.1002/anie.201205569
79. M.R. Palacín, Recent advances in rechargeable battery materials: a chemist's perspective. Chem. Soc. Rev. **38**(9), 2565–2575 (2009). https://doi.org/10.1039/b820555h
80. J. Xu, Z. Liu, F. Zhang, J. Tao, L. Shen, X. Zhang, Bacterial cellulose-derived carbon nanofibers as both anode and cathode for hybrid sodium ion capacitor. RSC Adv. **10**(13), 7780–7790 (2020). https://doi.org/10.1039/c9ra10225f
81. F. Meng, T. Long, X. Bin, Y. Zhao, H. Zexuan, L. Zhang, J. Liu, Electrolyte technologies for high performance sodium-ion capacitors. Front. Chem. **8**(August), 1–6 (2020). https://doi.org/10.3389/fchem.2020.00652
82. M. Wang, W. Yaling, M. Qiu, X. Li, C. Li, R. Li, J. He et al., Research progress in electrospinning engineering for all-solid-state electrolytes of lithium metal batteries. J. Energy Chem. **61**, 253–268 (2021). https://doi.org/10.1016/j.jechem.2021.02.023
83. M. Ge, C. Cao, J. Huang, S. Li, Z. Chen, K.-Q. Zhang, S.S. Al-Deyab, Y. Lai, A review of one-dimensional TiO_2 nanostructured materials for environmental and energy applications. J. Mater. Chem. A Chem. 1–26 (2013). https://doi.org/10.1039/x0xx00000x

84. L. Zhou, J. Mao, S. Yi Ren, T. Han, V.A.L. Roy, Y. Zhou, Recent advances of flexible data storage devices based on organic nanoscaled materials. Small **14**(10), 1–27 (2018). https://doi.org/10.1002/smll.201703126
85. Q. Ren, N. Qin, B. Liu, X. Yuan Yao, Z.D. Zhao, Y. Li et al., An oxygen-deficient vanadium oxide@n-doped carbon heterostructure for sodium-ion batteries: insights into the charge storage mechanism and enhanced reaction kinetics. J. Mater. Chem. A **8**(6), 3450–3458 (2020). https://doi.org/10.1039/c9ta11965e
86. P. Simon, Y. Gogotsi, B. Dunn, Where do batteries end and supercapacitors begin? Sci. Magaz. **343**. 1210 (2014).
87. J. Ding, H. Zhou, H. Zhang, T. Stephenson, Z. Li, D. Karpuzov, D. Mitlin, Exceptional energy and new insight with a sodium-selenium battery based on a carbon nanosheet cathode and a pseudographite anode. Energy Environ. Sci. **10**(1), 153–165 (2017). https://doi.org/10.1039/c6ee02274j
88. J. Xu, M. Wang, N.P. Wickramaratne, M. Jaroniec, S. Dou, L. Dai, High-performance sodium ion batteries based on a 3d anode from nitrogen-doped graphene foams. Adv. Mater. **27**(12), 2042–2048 (2015). https://doi.org/10.1002/adma.201405370
89. M. Han, S.B. Srivastava, E. Yildiz, R. Melikov, S. Surme, I.B. Dogru-Yuksel, I.H. Kavakli, A. Sahin, S. Nizamoglu, Organic photovoltaic pseudocapacitors for neurostimulation. ACS Appl. Mater. Interfaces **12**(38), 42997–42998 (2020). https://doi.org/10.1021/acsami.0c11581
90. C.-z Zhao, B.-C. Zhao, C. Yan, X.-Q. Zhang, J.-q Huang, Y. Mo, X. Xiaoxiong, H. Li, Q. Zhang, Liquid phase therapy to solid electrolyte-electrode interface in solid-state li metal batteries: a review. Energy Storage Mater. (2019). https://doi.org/10.1016/j.ensm.2019.07.026
91. J. Liu, X. Ying Ge, L.B. Kong, Synthesis of polyvalent ion reaction of MoS_2/CoS_2-RGO anode materials for high-performance sodium-ion batteries and sodium-ion capacitors. J. Colloid Interface Sci. **575**, 42–53 (2020). https://doi.org/10.1016/j.jcis.2020.04.074
92. W. Shao, H. Fangyuan, T. Zhang, S. Liu, C. Song, N. Li, Z. Weng, J. Wang, X. Jian, Engineering ultramicroporous carbon with abundant C=O as extended 'slope-dominated' sodium ion battery anodes. ACS Sustainable Chem. Eng. **9**(29), 9727–9739 (2021). https://doi.org/10.1021/acssuschemeng.1c01885
93. Y. Wu, L. Xie, H. Ming, Y. Guo, J.Y. Hwang, W. Wang, X. He et al., An empirical model for the design of batteries with high energy density. ACS Energy Lett. **5**(3), 807–816 (2020). https://doi.org/10.1021/acsenergylett.0c00211
94. H. Zhang, Hu. Mingxiang, Q. Lv, Z.H. Huang, F. Kang, R. Lv, Advanced materials for sodium-ion capacitors with superior energy-power properties: progress and perspectives. Small **16**(15), 1–22 (2020). https://doi.org/10.1002/smll.201902843
95. L. Liu, Z. Zhao, H. Zhengqiang, L. Xiangjun, S. Zhang, L. Huang, Y. Zheng, H. Li, Designing uniformly layered $FeTiO_3$ assemblies consisting of fine nanoparticles enabling high-performance quasi-solid-state sodium-ion capacitors. Front. Chem. **8**(May), 1–9 (2020). https://doi.org/10.3389/fchem.2020.00371
96. M. Minakshi, D.R.G. Mitchell, A.R. Munnangi, A.J. Barlow, M. Fichtner, New insights into the electrochemistry of magnesium molybdate hierarchical architectures for high performance sodium devices. Nanoscale **10**(27), 13277–13288 (2018). https://doi.org/10.1039/c8nr03824d
97. F. Paquin, J. Rivnay, A. Salleo, N. Stingelin, C. Silva, Multi-phase semicrystalline microstructures drive exciton dissociation in neat plastic semiconductors. J. Mater. Chem. C **3**(207890), 10715–10722 (2015). https://doi.org/10.1039/b000000x
98. P. Zhao, B.J. Kim, H.S. Jung, Passivation in perovskite solar cells: a review passivation in perovskite solar cells : a review. Materials Today Energy **7**(March), 267–286 (2018). https://doi.org/10.1016/j.mtener.2018.01.004
99. H. Zhao, Y. Lei, 3D nanostructures for the next generation of high-performance nanodevices for electrochemical energy conversion and storage. Adv. Energy Mater. **10**(28), 1–8 (2020). https://doi.org/10.1002/aenm.202001460
100. A. Manthiram, A. Vadivel Murugan, A. Sarkar, T. Muraliganth, Nanostructured electrode materials for electrochemical energy storage and conversion. Energy Environ. Sci. **1**(6), 621–638 (2008). https://doi.org/10.1039/b811802g

101. Z. Nie, A. Petukhova, E. Kumacheva, Properties and emerging applications of self-assembled structures made from inorganic nanoparticles. Nat. Nanotechnol. **5**(1), 15–25 (2010). https://doi.org/10.1038/nnano.2009.453
102. K. Naoi, S. Ishimoto, Y. Isobe, S. Aoyagi, High-rate nano-crystalline $Li_4Ti_5O_{12}$ attached on carbon nano-fibers for hybrid supercapacitors. J. Power. Sources **195**(18), 6250–6254 (2010). https://doi.org/10.1016/j.jpowsour.2009.12.104
103. A. Pasquier, I. Du, J. Plitz, F.B. Gural, G.G. Amatucci, Power-ion battery: bridging the gap between li-ion and supercapacitor chemistries. J. Power. Sources **136**(1), 160–170 (2004). https://doi.org/10.1016/j.jpowsour.2004.05.023
104. J. Zhu, Y. Wang, X. Yan, S. Li, J. Ren, W. Cai, Synthesis of porous flower-like Ni–Co–Mo–S nanostructures on ni foam for battery-supercapacitor hybrid devices. Int. J. Energy Res. **44**(4), 2864–2874 (2020). https://doi.org/10.1002/er.5102
105. P. Wang, G. Zhang, H. Jiao, W. Chen, L. Liu, X. Wang, Q. Chen, X. Deng, X. Zheng, Advanced li-ion hybrid capacitors based on the nanostructured ruthenium oxide on MWCNTs. Appl. Surf. Sci. **466**, 724–729 (2019). https://doi.org/10.1016/j.apsusc.2018.10.105
106. A.T. Chidembo, S.H. Aboutalebi, K. Konstantinov, C.J. Jafta, H.K. Liu, K.I. Ozoemena, In situ engineering of urchin-like reduced graphene oxide-mn $2O_3$-Mn_3O_4 nanostructures for supercapacitors. RSC Adv. **4**(2), 886–892 (2014). https://doi.org/10.1039/c3ra44973d
107. K. Ramakrishnan, C. Nithya, R. Karvembu, High-performance sodium ion capacitor based on MoO_2@rGO nanocomposite and goat hair derived carbon electrodes. ACS Appl. Energy Mater. **1**(2), 841–850 (2018). https://doi.org/10.1021/acsaem.7b00284
108. S.S.M. Bhat, B. Babu, M. Feygenson, J.C. Neuefeind, M.M. Shaijumon, Nanostructured Na2Ti9O19 for hybrid sodium-ion capacitors with excellent rate capability. ACS Appl. Mater. Interfaces **10**(1), 437–447 (2018). https://doi.org/10.1021/acsami.7b13300
109. C. Liu, F. Li, L.-P. Ma, H.-M. Cheng, Advanced materials for energy storage. Adv. Mater. **22**(8), E28–E62 (2010)
110. C. Ghosh, Rajib, S. Paria, Core/shell nanoparticles: classes, properties, synthesis mechanisms, characterization, and applications. Chem. Rev. **112**(4), 2373–2433 (2012). https://doi.org/10.1021/cr100449n
111. P.J. Hall, S. Mojtaba Mirzaeian, I. Fletcher, F.B. Sillars, A.J.R. Rennie, O. Gbolahan, G.W. Shitta-Bey, A. Cruden, R. Carter, Energy storage in electrochemical capacitors: designing functional materials to improve performance. Energy Environ. Sci. **3**(9), 1238–1251 (2010). https://doi.org/10.1039/c0ee00004c
112. X. Bai, M. Hou, Y. Zhaoyu, C. Liu, H. Cao, D. Wang, F. Junjie, An Optimized 3D carbon matrix for high rate silicon anodes. RSC Adv. **7**(53), 33521–33525 (2017). https://doi.org/10.1039/c7ra05647h
113. Z. Wu, L. Li, J. Min Yan, X.B. Zhang, Materials design and system construction for conventional and new-concept supercapacitors. Adv. Sci. **4**(6) (2017). https://doi.org/10.1002/advs.201600382
114. T.M. Arruda, M. Heon, V. Presser, P.C. Hillesheim, S. Dai, Y. Gogotsi, S.V. Kalinin, N. Balke, In situ tracking of the nanoscale expansion of porous carbon electrodes. Energy Environ. Sci. **6**(1), 225–231 (2013). https://doi.org/10.1039/c2ee23707e
115. W.-W. Liu, X.-B. Yan, J.-W. Lang, C. Peng, Q.-j Xue, Flexible and conductive nanocomposite electrode based on graphene sheets and cotton cloth for supercapacitor. Mater. Chem. 207890, 1–39 (2012). https://doi.org/10.1007/978-1-4020-6120-2
116. D.S. Jung, T.H. Hwang, J.H. Lee, H.Y. Koo, R.A. Shakoor, R. Kahraman, Y.N. Jo, M.S. Park, J.W. Choi, Hierarchical porous carbon by ultrasonic spray pyrolysis yields stable cycling in lithium-sulfur battery. Nano Lett. **14**(8), 4418–4425 (2014). https://doi.org/10.1021/nl501383g
117. V.G. Pol, P. Thiyagarajan, Remediating plastic waste into carbon nanotubes. J. Environ. Monit. **12**(2), 455–459 (2010). https://doi.org/10.1039/b914648b
118. P. Simon, Y. Gogotsi, Carbon à electrolyte systems 46(5) (2013)
119. D. Saurel, B. Orayech, B. Xiao, D. Carriazo, X. Li, T. Rojo, From charge storage mechanism to performance: a roadmap toward high specific energy sodium-ion batteries through carbon

anode optimization. Adv. Energy Mater. **8**(17), 1–33 (2018). https://doi.org/10.1002/aenm.201703268

120. J. Ajuria, E. Redondo, M. Arnaiz, R. Mysyk, T. Rojo, E. Goikolea, Lithium and sodium ion capacitors with high energy and power densities based on carbons from recycled olive pits. J. Power. Sources **359**, 17–26 (2017). https://doi.org/10.1016/j.jpowsour.2017.04.107
121. P. Cai, K. Zou, G. Zou, H. Hou, X. Ji, Quinone/ester-based oxygen functional group-incorporated full carbon li-ion capacitor for enhanced performance. Nanoscale **12**(6), 3677–3685 (2020). https://doi.org/10.1039/c9nr10339b
122. J. Ding, H. Wang, Z. Li, K. Cui, D. Karpuzov, X. Tan, A. Kohandehghan, D. Mitlin, Peanut shell hybrid sodium ion capacitor with extreme energy-power rivals lithium ion capacitors. Energy Environ. Sci. **8**(3), 941–955 (2015). https://doi.org/10.1039/C4EE02986K
123. A. Banerjee, K.K. Upadhyay, D. Puthusseri, V. Aravindan, S. Madhavi, S. Ogale, MOF-derived crumpled-sheet-assembled perforated carbon cuboids as highly effective cathode active materials for ultra-high energy density li-ion hybrid electrochemical capacitors (Li-HECs). Nanoscale **6**(8), 4387–4394 (2014). https://doi.org/10.1039/c4nr00025k
124. M. Cai, X. Sun, W. Chen, Z. Qiu, X. Long Chen, J.W. Li, Z. Liu, Y. Nie, Performance of lithium-ion capacitors using pre-lithiated multiwalled carbon nanotubes/graphite composite as negative electrode. J. Mater. Sci. **53**(1), 749–758 (2018). https://doi.org/10.1007/s10853-017-1524-5
125. H. Huang, H. Shi, P. Das, J. Qin, Y. Li, X. Wang, S. Feng et al., Porous graphene materials: the chemistry and promising applications of graphene and porous graphene materials (Adv. Funct. Mater. 41/2020). Adv. Func. Mater. **30**(41), 2070275 (2020). https://doi.org/10.1002/adfm.202070275
126. J. Zhang, W. Lv, Y. Tao, Y.B. He, D.W. Wang, C.H. You, B. Li, F. Kang, Q.H. Yang, Ultrafast high-volumetric sodium storage of folded-graphene electrodes through surface-induced redox reactions. Energy Storage Mater. **1**, 112–118 (2015). https://doi.org/10.1016/j.ensm.2015.08.006
127. Y. Liu, B.V. Merinov, W.A. Goddard, Origin of low sodium capacity in graphite and generally weak substrate binding of na and mg among alkali and alkaline earth metals. Proc. Natl. Acad. Sci. U.S.A. **113**(14), 3735–3739 (2016). https://doi.org/10.1073/pnas.1602473113
128. J. Zhang, W. Lv, D. Zheng, Q. Liang, D.-W. Wang, F. Kang, Q.-H. Yang, Sodium ion capacitors: the interplay of oxygen functional groups and folded texture in densified graphene electrodes for compact sodium-ion capacitors (Adv. Energy Mater. 11/2018). Adv. Energy Mater. **8**(11), 1870050 (2018). https://doi.org/10.1002/aenm.201870050
129. S. Dong, Y. Xu, L. Wu, H. Dou, X. Zhang, Surface-functionalized graphene-based quasi-solid-state na-ion hybrid capacitors with excellent performance. Energy Storage Mater. **11**(September 2017), 8–15 (2018). https://doi.org/10.1016/j.ensm.2017.09.006
130. T. Yuan, Z. Tan, C. Ma, J. Yang, Z.F. Ma, S. Zheng, Challenges of spinel $Li_4Ti_5O_{12}$ for lithium-ion battery industrial applications.. Adv. Energy Mater. **7** (12) (2017). https://doi.org/10.1002/aenm.201601625
131. A.G. Dylla, G. Henkelman, K.J. Stevenson, Lithium insertion in nanostructured TiO_2(B) architectures. Acc. Chem. Res. **46**(5), 1104–1112 (2013). https://doi.org/10.1021/ar300176y
132. E. Pohjalainen, T. Rauhala, M. Valkeapää, J. Kallioinen, T. Kallio, Effect of $Li_4Ti_5O_{12}$ particle size on the performance of lithium ion battery electrodes at high c-rates and low temperatures. J. Phys. Chem. C **119**(5), 2277–2283 (2015). https://doi.org/10.1021/jp509428c
133. L. Gao, D. Huang, Y. Shen, M. Wang, Rutile-TiO_2 Decorated Li_4Ti_5O12 nanosheet arrays with 3D interconnected architecture as anodes for high performance hybrid supercapacitors. J. Mater. Chem. A **3**(46), 23570–23576 (2015). https://doi.org/10.1039/c5ta07666h
134. D. Su, S. Dou, G. Wang, Anatase TiO_2: better anode material than amorphous and rutile phases of TiO_2 for na-ion batteries. Chem. Mater. **27**(17), 6022–6029 (2015). https://doi.org/10.1021/acs.chemmater.5b02348
135. Y. Xu, E.M. Lotfabad, H. Wang, B. Farbod, X. Zhanwei, A. Kohandehghan, D. Mitlin, Nanocrystalline anatase TiO_2: a new anode material for rechargeable sodium ion batteries. Chem. Commun. **49**(79), 8973–8975 (2013). https://doi.org/10.1039/c3cc45254a

136. K.T. Kim, G. Ali, K.Y. Chung, C.S. Yoon, H. Yashiro, Y.K. Sun, L. Jun, K. Amine, S.T. Myung, Anatase titania nanorods as an intercalation anode material for rechargeable sodium batteries. Nano Lett. **14**(2), 416–422 (2014). https://doi.org/10.1021/nl402747x
137. H. Kang, Y. Liu, K. Cao, Y. Zhao, L. Jiao, Y. Wang, H. Yuan, Update on anode materials for na-ion batteries. J. Mater. Chem. A **3**(35), 17899–17913 (2015). https://doi.org/10.1039/c5ta03181h
138. X. Wang, L. Fan, D. Gong, J. Zhu, Q. Zhang, L. Bingan, Core-shell Ge@Graphene@TiO_2 nanofibers as a high-capacity and cycle-stable anode for lithium and sodium ion battery. Adv. Func. Mater. **26**(7), 1104–1111 (2016). https://doi.org/10.1002/adfm.201504589
139. H. Liu, K. Cao, X. Xiaohong, L. Jiao, Y. Wang, H. Yuan, Ultrasmall TiO_2 Nanoparticles in situ growth on graphene hybrid as superior anode material for sodium/lithium ion batteries. ACS Appl. Mater. Interfaces **7**(21), 11239–11245 (2015). https://doi.org/10.1021/acsami.5b02724
140. M.N. Tahir, B. Oschmann, D. Buchholz, X. Dou, I. Lieberwirth, M. Panthöfer, W. Tremel, R. Zentel, S. Passerini, Extraordinary performance of carbon-coated anatase TiO_2 as sodium-ion anode. Adv. Energy Mater. **6**(4), 1–9 (2016). https://doi.org/10.1002/aenm.201501489
141. H. Usui, S. Yoshioka, K. Wasada, M. Shimizu, H. Sakaguchi, Nb-doped rutile TiO_2: a potential anode material for na-ion battery. ACS Appl. Mater. Interfaces **7**(12), 6567–6573 (2015). https://doi.org/10.1021/am508670z
142. X. Xie, M.Q. Zhao, B. Anasori, K. Maleski, C.E. Ren, J. Li, B.W. Byles, E. Pomerantseva, G. Wang, Y. Gogotsi, Porous heterostructured mxene/carbon nanotube composite paper with high volumetric capacity for sodium-based energy storage devices. Nano Energy **26**, 513–523 (2016). https://doi.org/10.1016/j.nanoen.2016.06.005
143. Y. Yang, X. Ji, M. Jing, H. Hou, Y. Zhu, L. Fang, X. Yang, Q. Chen, C.E. Banks, Carbon dots supported upon N-Doped TiO_2 nanorods applied into sodium and lithium ion batteries. J. Mater. Chem. A **3**(10), 5648–5655 (2015). https://doi.org/10.1039/c4ta05611f
144. Z. Le, F. Liu, P. Nie, X. Li, X. Liu, Z. Bian, G. Chen, W. Hao Bin, L. Yunfeng, Pseudocapacitive sodium storage in mesoporous single-crystal-like TiO_2-graphene nanocomposite enables high-performance sodium-ion capacitors. ACS Nano **11**(3), 2952–2960 (2017). https://doi.org/10.1021/acsnano.6b08332
145. Q. Xia, H. Yang, M. Wang, M. Yang, Q. Guo, L. Wan, H. Xia, Y. Yan, High energy and high power lithium-ion capacitors based on boron and nitrogen dual-doped 3d carbon nanofibers as both cathode and anode. Adv. Energy Mater. **7**(22), 1–9 (2017). https://doi.org/10.1002/aenm.201701336
146. F. Sun, X. Liu, W. Hao Bin, L. Wang, J. Gao, H. Li, L. Yunfeng, In situ high-level nitrogen doping into carbon nanospheres and boosting of capacitive charge storage in both anode and cathode for a high-energy 4.5 v full-carbon lithium-ion capacitor. Nano Lett. **18**(6), 3368–3376 (2018). https://doi.org/10.1021/acs.nanolett.8b00134
147. S.R. Sivakkumar, A.G. Pandolfo, Evaluation of lithium-ion capacitors assembled with pre-lithiated graphite anode and activated carbon cathode. Electrochim. Acta **65**, 280–287 (2012). https://doi.org/10.1016/j.electacta.2012.01.076
148. M. Schroeder, M. Winter, S. Passerini, A. Balducci, On the cycling stability of lithium-ion capacitors containing soft carbon as anodic material. J. Power. Sources **238**, 388–394 (2013). https://doi.org/10.1016/j.jpowsour.2013.04.045
149. J.H. Kim, J.S. Kim, Y.G. Lim, J.G. Lee, Y.J. Kim, Effect of carbon types on the electrochemical properties of negative electrodes for li-ion capacitors. J. Power. Sources **196**(23), 10490–10495 (2011). https://doi.org/10.1016/j.jpowsour.2011.08.081
150. D.P. Dubal, P. Gomez-Romero, All nanocarbon li-ion capacitor with high energy and high power density. Mater. Today Energy **8**(June), 109–117 (2018). https://doi.org/10.1016/j.mtener.2018.03.005
151. L. Li, S. Peng, H.B. Wu, L. Yu, S. Madhavi, X. Wen, D. Lou, A flexible quasi-solid-state asymmetric electrochemical capacitor based on hierarchical porous V 2 O 5 Nanosheets on Carbon Nanofi Bers (2015). https://doi.org/10.1002/aenm.201500753
152. M.R. Lukatskaya, O. Mashtalir, C.E. Ren, Y. Dall'Agnese, P. Rozier, P. Louis Taberna, M. Naguib, P. Simon, M.W. Barsoum, Y. Gogotsi, Cation intercalation and high volumetric

capacitance of two-dimensional titanium carbide. Science **341**, 1501–1505 (2013). https://doi.org/10.1126/science.1241488
153. H. Wang, Y. Zhang, H. Ang, Y. Zhang, H.T. Tan, Y. Zhang, Y. Guo, J.B. Franklin, W. Xing Long, M. Srinivasan, A high-energy lithium-ion capacitor by integration of a 3D interconnected titanium carbide nanoparticle chain anode with a pyridine-derived porous nitrogen-doped carbon cathode. Adv. Func. Mater. **26**, 3082–3093 (2016). https://doi.org/10.1002/adfm.201505240
154. J.J. Ren, L.W. Su, X. Qin, M. Yang, J.P. Wei, Z. Zhou, P.W. Shen, Pre-lithiated graphene nanosheets as negative electrode materials for li-ion capacitors with high power and energy density. J. Power. Sources **264**, 108–113 (2014). https://doi.org/10.1016/j.jpowsour.2014.04.076
155. V. Aravindan, D. Mhamane, W.C. Ling, S. Ogale, S. Madhavi, Nonaqueous lithium-ion capacitors with high energy densities using trigol-reduced graphene oxide nanosheets as cathode-active material. Chemsuschem **6**(12), 2240–2244 (2013). https://doi.org/10.1002/cssc.201300465
156. S. Madhavi, Li-Ion vs. Na-Ion capacitors: a performance evaluation with coconut shell derived mesoporous carbon and natural plant based hard carbon. Chem. Eng. J. (2017). https://doi.org/10.1016/j.cej.2017.01.108
157. C. Lai, D. Ashby, M.K. Moz, Y. Gogotsi, L. Pilon, B.S. Dunn, Designing pseudocapacitance for Nb_2O_5/carbide-derived carbon electrodes and hybrid devices. Am. Chem. Soc. no. May: 1–36 (2017)
158. Z. Yu, L. Tetard, L. Zhai, J. Thomas, Supercapacitor electrode materials: nanostructures from 0 to 3 dimensions. Energy Environ. Sci. **8**(3), 702–730 (2015). https://doi.org/10.1039/C4EE03229B
159. K. Kuratani, M. Yao, H. Senoh, N. Takeichi, T. Sakai, T. Kiyobayashi, Na-Ion capacitor using sodium pre-doped hard carbon and activated carbon. Electrochim. Acta **76**, 320–325 (2012). https://doi.org/10.1016/j.electacta.2012.05.040
160. Y. Li, Y. Hu, M. Titirici, L. Chen, X. Huang, Hard carbon microtubes made from renewable cotton as high-performance anode material for sodium-ion batteries (2016). https://doi.org/10.1002/aenm.201600659.
161. J. Ding, H. Wang, Z. Li, K. Cui, D. Karpuzov, X. Tan, A. Kohandehghan, D. Mitlin, Peanut shell hybrid sodium ion capacitor with extreme energy—power rivals lithium ion capacitors. Energy Environ. Sci. 1–46. https://doi.org/10.1039/C4EE02986K
162. E. Lim, J. Changshin, M.S. Kim, M.H. Kim, J. Chun, H. Kim, J. Park et al., High-performance sodium-ion hybrid supercapacitor based on Nb_2O_5@carbon core-shell nanoparticles and reduced graphene oxide nanocomposites. Adv. Func. Mater. **26**(21), 3711–3719 (2016). https://doi.org/10.1002/adfm.201505548
163. X. Wang, S. Kajiyama, H. Iinuma, E. Hosono, S. Oro, I. Moriguchi, M. Okubo, A. Yamada, Pseudocapacitance of MXene nanosheets for high-power sodium-ion hybrid capacitors. Nat. Commun. **6**(January), 6544 (2015). https://doi.org/10.1038/ncomms7544
164. Y. Zhang, C. Yuan, K. Ye, X. Jiang, J. Yin, G. Wang, D. Cao, An aqueous capacitor battery hybrid device based on na-ion insertion-deinsertion in λ-MnO_2 positive electrode. Electrochim. Acta **148**, 237–243 (2014). https://doi.org/10.1016/j.electacta.2014.10.052
165. P. Ge, M. Fouletier, Electrochemical intercalation of sodium in graphite pascal ge and mireille fouletier **30**, 1172–1175 (1988)
166. S. Komaba, W. Murata, T. Ishikawa, N. Yabuuchi, T. Ozeki, Electrochemical na insertion and solid electrolyte interphase for hard-carbon electrodes and application to na-ion batteries. no. Dmc: 3859–3867 (2011). https://doi.org/10.1002/adfm.201100854
167. A.P. Cohn, K. Share, R. Carter, L. Oakes, C.L. Pint, Ultrafast solvent-assisted sodium ion intercalation into highly crystalline few-layered graphene (2015) https://doi.org/10.1021/acs.nanolett.5b04187.
168. R. Ding, L. Qi, H. Wang, Electrochimica acta an investigation of spinel $NiCo_2O_4$ as anode for na-ion capacitors **114**, 726–35 (2013). https://doi.org/10.1016/j.electacta.2013.10.113

169. E.M. Lotfabada, P. Kalisvaarta, A. Kohandehghana, D. Mitlina, Origin of non-SEI related coulombic efficiency loss in carbons tested against Na and Li. J. Mater. Chem. A 1–32 (2014). https://doi.org/10.1039/C4TA04995K
170. D.A. Stevens, J.R. Dahn, High capacity anode materials for rechargeable sodium-ion batteries. J. Electrochem. Soc. **147**(4), 1271–1273 (2000)
171. D. Datta, J. Li, V.B. Shenoy, Defective graphene as a high-capacity anode material for na- and ca-ion batteries (2014)
172. Y. Cao, L. Xiao, M.L. Sushko, W. Wang, B. Schwenzer, J. Xiao, Z. Nie, L.V. Saraf, Z. Yang, J. Liu, Sodium ion insertion in hollow carbon nanowires for battery applications (2012)
173. X. Lu, G. Wang, T. Zhai, M. Yu, J. Gan, Y. Tong, Y. Li, Hydrogenated TiO_2 nanotube arrays for supercapacitors (2012)
174. J. Yan, Q. Wang, T. Wei, Z.J. Fan, Recent advances in design and fabrication of electrochemical supercapacitors with high energy densities. Adv. Energy Mater. **4**(4) (2014). https://doi.org/10.1002/aenm.201300816
175. Group, Francis. 2006. Carbon Nanomaterials © 2006.
176. W. Long, B. Fang, A. Ignaszak, W. Zhuangzhi, Y.J. Wang, D. Wilkinson, Biomass-derived nanostructured carbons and their composites as anode materials for lithium ion batteries. Chem. Soc. Rev. **46**(23), 7176–7190 (2017). https://doi.org/10.1039/c6cs00639f
177. P. Sennu, V. Aravindan, M. Ganesan, Y.G. Lee, Y.S. Lee, Biomass-derived electrode for next generation lithium-ion capacitors. Chemsuschem **9**(8), 849–854 (2016). https://doi.org/10.1002/cssc.201501621
178. J. Yang, Y. Liu, S. Liu, L. Li, C. Zhang, T. Liu, Conducting polymer composites: material synthesis and applications in electrochemical capacitive energy storage. Mater. Chem. Front. **1**(2), 251–268 (2017). https://doi.org/10.1039/c6qm00150e
179. C. An, Y. Zhang, H. Guo, Y. Wang, Metal oxide-based supercapacitors: progress and prospectives. Nanoscale Adv. **1**(12), 4644–4658 (2019). https://doi.org/10.1039/c9na00543a
180. , S. Yuan, S.Y. Pang, J. Hao, 2D transition metal dichalcogenides, carbides, nitrides, and their applications in supercapacitors and electrocatalytic hydrogen evolution reaction. Appl. Phys. Rev. **7**(2) (2020). https://doi.org/10.1063/5.0005141
181. J. Jiang, Y. Zhang, P. Nie, X. Guiyin, M. Shi, J. Wang, W. Yuting, F. Ruirui, H. Dou, X. Zhang, Progress of nanostructured electrode materials for supercapacitors. Adv. Sustainable Syst. **2**(1), 1–24 (2018). https://doi.org/10.1002/adsu.201700110
182. E. Environ, X. Lai, E. Halpert, D. Wang Environmental science recent advances in micro-/ nano-structured hollow spheres for energy applications : from simple to complex systems 5604–5618 (2012). https://doi.org/10.1039/c1ee02426d
183. B.Y. Xia, P. Yang, Y. Sun, Y. Wu, B. Mayers, B. Gates, Y. Yin, F. Kim, H. Yan, One-dimensional nanostructures : synthesis. Characterization Appl. **5**, 353–389
184. Z. Yua, J. Moorea, B. Duongb, C. Lia, J. Thomasa, PAN@AuPd@MnO_2 core-shell nanopillars for high-performance electrochemical energy storage. ECS Trans. **61**(18), 49–53 (2014)
185. H. Wang, Q. Gao, L. Jiang, Facile approach to prepare nickel cobaltite nanowire materials for supercapacitors 2454–2459 (2011). https://doi.org/10.1002/smll.201100534
186. R. Mas-Balleste, C. Gomez-Navarro, J. Gomez-Herrero, F. Zamora, 2D materials: to graphene and beyond. Nanoscale **3**(20), 20–30 (2011). https://doi.org/10.1039/c0nr00323a
187. Q. Qu, Y. Zhu, X. Gao, Y. Wu, Core—shell structure of polypyrrole grown on V 2 O 5 nanoribbon as high performance anode material for supercapacitors 1–6 (2012). https://doi.org/10.1002/aenm.201200088
188. J. Come, M. Naguib, P. Rozier, M.W. Barsoum, Y. Gogotsi, P. Taberna, A non-aqueous asymmetric cell with a Ti_2 C-based two-dimensional negative electrode. J. Electrochem. Soc. **159**(8), 1368–1373 (2012). https://doi.org/10.1149/2.003208jes
189. B.J. Cabana, L. Monconduit, D. Larcher, M.R. Palacín, Beyond intercalation-based li-ion batteries : the state of the art and challenges of electrode materials reacting through conversion reactions 170–92 (2010). https://doi.org/10.1002/adma.201000717

Chapter 6
Nanohybrid Materials in Flexible and Wearable Supercapacitors

Rajendran Suresh Babu, Rajangam Vinodh, Rajendran Raja, Adhigan Murali, Mani Devendiran, Leandro Marques Samyn, and Ana Lucia Ferreira de Barros

Abstract The rapid progress of flexible and wearable electronic devices has gained considerable attention for developing electrochemical sensors, flexible displays, and health monitoring devices. In recent years, the development of flexible and wearable supercapacitors has received extensive consideration owing to their simplicity of construction, inexpensive, flexible integration into textiles, lightweight, long-term stability, rapid charge-discharge, higher efficiency, and ability to bridge the power-energy gap between traditional capacitors and fuel cells and batteries. The appropriate design and synthesis of new nanostructured hybrid materials for electrode fabrication on a range of flexible-based current collectors, for example, carbon-cloth, carbon fiber, graphene-coated fabric, silver-nickel-coated fabric, copper-nickel-coated polyester fabric, and so on are reviewed in this context. An analysis of the most current representative methods and active materials for newly developed high-performance wearable and flexible supercapacitors is presented. Finally, the current challenges for optimizing and enhancing the efficiency of energy storage and function

R. S. Babu (✉) · L. M. Samyn · A. L. F. de Barros
Laboratory of Experimental and Applied Physics, Centro Federal de Educação Tecnológica Celso Suckow da Fonseca, Rio de Janeiro, Brazil
e-mail: suresh.rajendran@aluno.cefet-rj.br

R. Vinodh
Department of Electronics Engineering, Pusan National University, Geumjeong-gu, Busan, Republic of Korea
e-mail: vinoth6482@gmail.com

R. Raja
Institute of Nanoscience and Department of Physics, National Chung Hsing University, Taichung, Taiwan

A. Murali
School for Advanced Research in Polymers (SARP)-ARSTPS, Central Institute of Plastics Engineering & Technology (CIPET), Chennai, India

M. Devendiran
Central Instrumentation Laboratory, Vels Institute of Science, Technology & Advanced Studies (VISTAS), Pallavaram, Chennai, Tamil Nadu, India

© The Author(s), under exclusive license to Springer Nature Switzerland AG 2026
K. Raju et al. (eds.), *Hybrid Nanostructures for Advanced Supercapacitors*, Modern Aspects of Electrochemistry 63, https://doi.org/10.1007/978-3-032-04593-5_6

of flexible and wearable supercapacitors for real-time applications are addressed, as well as future research directions and perspectives.

Keywords Energy storage · Nanohybrid materials · Flexible electrodes · Wearable electronics · Supercapacitors

6.1 Introduction

In recent years, the fast diminishing of fossil fuels, environmental pollution aggravation, global warming, and the urgent need for eco-friendly devices in modern electronics. There are several attempts have been devoted to designing and developing clean and sustainable renewable energy storage systems with excellent electrochemical performances [5, 69, 70]. Recently, a bunch of effort has gone into investigating and developing flexible and wearable energy storage systems, with the main objective of using flexible electronic devices like laptops, power backup, flexible screens, wearable devices, portable electronics, health monitors, electronic sensors, cell phones, and so on [10, 42, 58, 65]. The majority of existing rechargeable energy storage systems markets are dominated by the design and manufacturing of electrochemical energy storage systems with durability, high energy, and power densities. Due to the issue of being heavy, bulky, and having poor capacity, traditional lithium-ion batteries are not well suited for portable and wearable electronics. Furthermore, the heat produced by commercially available lithium-ion batteries can cause skin irritation, limiting their use in wearable systems [49, 50]. As a result, supercapacitors are seen as a viable alternative to traditional lithium-ion batteries [12, 63]. Supercapacitors are rapidly growing in the field of energy storage devices because of their fast charging-storage capacity, i.e., low discharging time: supercapacitor: 1–10 s versus Li-ion battery: 10–60 min and improved cyclic stability of > 30,000 h (battery > 500 h), high-power density, environmentally safe, and easy to operate at different temperatures [36–38, 40, 81]. Unfortunately, supercapacitors have lower energy densities than traditional batteries, which have limited their use in electronic appliances. Lower energy density is primarily due to a lower cut-off window and specific capacitance ($E = ½ CV^2$). Supercapacitors would be commercialized and commonly used if the above narrow slabs were resolved. While some recent publications claim to have increased the energy densities of supercapacitors to 100–150 Wh Kg^{-1} and is nearly equivalent to lithium-ion batteries (120–170 Wh Kg^{-1}) [13, 75, 86, 87]. As a result, designing and developing new anode and cathode materials to construct versatile supercapacitors with higher energy densities is highly desirable.

The electrochemical characteristics of the electrode materials, the selection of electrolytes, and the cut-off window of the electrodes all influence the energy storage efficiency of supercapacitors. The design and development of advanced electrode materials for flexible supercapacitors with appropriate structural properties to allow efficient electron transport and ionic diffusion have received a lot of attention [17, 21, 59, 71]. By categorizing different electrode materials of flexible anode and cathode, we offer a detailed vision of current progress and developments in flexible and wearable supercapacitors. Finally, various future research patterns and directions are explored to aid future research and development. This analysis aims to offer readers a thorough understanding of the fundamentals of flexible and wearable supercapacitors, as well as new electrode materials and product designs.

6.2 Flexible and Wearable Electrodes

The rapid growth of wearable electronics has increased the development of wearable energy storage devices. Due to their high-power density, wearable supercapacitors have attracted a lot of attention. However, the traditional supercapacitors are large and hard, which cannot meet the requirements of application in the field of wearable electronic devices. Hence ultrahigh-performance handy wearable supercapacitors have been developed. Such wearable supercapacitors have the characteristics of good volumetric energy density, lightweight and flexibility, which are extremely significant for wearable electronic devices [24, 57]. The electrode materials are cost-effective, ecological, and flexible with remarkable stability; excellent electrochemical behavior and exceptional mechanical stability are key characteristics of flexible and wearable supercapacitors. The key components of supercapacitors are current collectors, two active electrode materials, an electrolyte, and a separator. Based on the charge-storage mechanism, supercapacitors are divided into three types: electric double-layer capacitors (EDLCs), pseudocapacitors, and hybrid supercapacitors [4]. Pseudocapacitors store charge through electrochemical Faradaic redox processes, whereas EDLCs store charge electrostatically at the electrode-electrolyte interface. Carbonaceous materials for example including porous activated carbon, various dimensions of carbon nanotubes, two-dimensional graphene and graphitic carbon are used primarily in EDLCs [6, 60, 72], whereas transition metal oxide, for example, RuO_2, MnO_2, Fe_2O_3 and conducting polymers, for instance, poly(3,4-ethylene dioxythiophene), polyaniline, polypyrrole, etc., are used in pseudocapacitors [36–38, 40]. According to the working mechanism, EDLCs have higher power densities and excellent stability in the long-term however a lower specific capacitance, whereas pseudocapacitors do the opposite. The device of the hybrid supercapacitor, which is made up of the integration of electrode materials like EDLC and pseudocapacitor, has a wider operating cut-off range, high-energy density, and higher specific capacitances [74, 78, 89]. EDLCs show an exceptional benefit for use in smart garment applications compared to batteries as they can be made entirely of non-toxic materials, do not ignite if punctured, and can be paired with textile-piezoelectrics, thermoelectrics, or solar panels to efficiently harvest energy from body movements, body heat or sunlight, respectively. EDLCs can be used to power wearable biosensors and antennas and have the potential to be used in flexible robotic prosthetics for rapid movements due to their high power and rapid discharge [44].

Figure 6.1a shows the typical constituents of traditional supercapacitors are the outer packages, outer current collectors (metal foil), positive and negative electrodes,

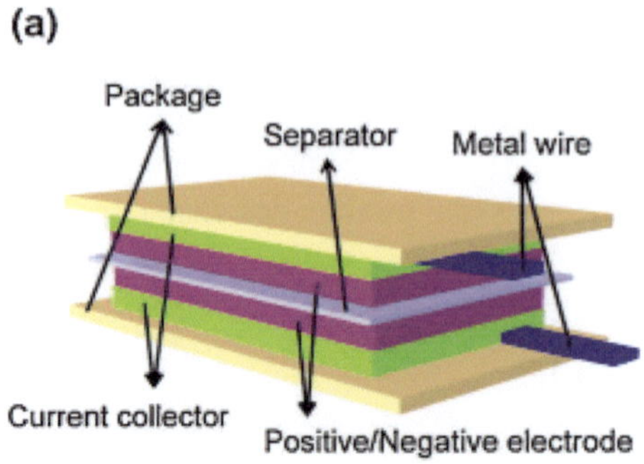

Fig. 6.1 Fabrication process and the arrangement of electrode constituents in a conventional supercapacitor (**a**) and a flexible supercapacitor (**b**). Reprinted with the permission of Elsevier from Shi et al. [62]

an electrolyte, and a separator. The carbon network with its high conductivity and flexibility works as both the active materials and the current collector, the structure of flexible supercapacitors is simplified, with no need for separate current collectors and binders (Fig. 6.1b). Polybag, ethylene, and vinyl acetate copolymer (EVA) film, polyethene terephthalate (PET), PDMS, and Teflon substrate have all been widely used as packaging materials [8, 16, 20, 51]. In comparison with conventional supercapacitors, the streamlined architecture is more versatile and lightweight.

Carbon materials are used to make flexible electrodes. Depending on the design, single carbon electrodes such as carbon fabric, fiber, cloth, film, coating, paper, textiles, or composite electrodes act as positive or negative electrodes. The supercapacitor can be classified as a symmetric supercapacitor, in which both electrodes are identical, or an asymmetric supercapacitor, in which the two electrodes are different, depending on the electrode materials utilized. The asymmetric design's higher cut-off voltage leads to a high-energy density. For example, a flexible asymmetric supercapacitor used MnO_2-SWNT as the cathode and In_2O_3-SWNT electrodes as the anode and exhibited a specific capacitance of 184 F g^{-1} with an energy density of 25.5 Wh/kg in an aqueous electrolyte [7, 9]. High-performance as-assembled asymmetric supercapacitor device fabricated using graphene-MnO_2-textile as the positive electrode and SWNT-textile as the negative electrode [88]. The specific capacitance achieved up to 315 F g^{-1}. In recent times, an asymmetric supercapacitor with excellent flexibility and electrochemical performance attained a specific capacitance of 120 F g^{-1} and a high-energy density of 66.7 Wh/kg [88]. Besides the positive and negative electrodes, the other essential constituent is the electrolyte. The electrolytes used in flexible supercapacitors can be classified into two types: liquid electrolytes such as aqueous and organic electrolytes and solid electrolytes, for example, polymer gel electrolytes. More details about electrolytes were discussed in Sect. 6.2.

6.2.1 *Anode Materials for Flexible and Wearable Supercapacitors*

The design of flexible and wearable supercapacitors is mainly lightweight and flexible current collectors. Carbon-based materials, for example, carbon nanotubes, graphene, carbon fiber, activated carbon and graphitic carbon have a large surface area, high stability, excellent conductivity, and strong mechanical behaviors that attract a lot of interest as flexible electrodes. Carbon foam, carbon nanofibers, carbon cloth, titanium foil, graphite foam flax fiber textile, Ni-cotton fabrics, and graphene sheets are used as flexible electrodes due to their lightweight, high durability, and wide accessible surface area.

Flexible electrode materials were prepared by the chemical vapor deposition method proposed [91–93]. Increasing carbon nanotubes directly on carbonized natural flax fabric enhanced the mechanical stability, high surface area and stable high-rate performance showing promising applications in flexible supercapacitors. The carbon nanotube growth on the carbonized fabric varies at 10, 15, and 20 min in the chemical vapor deposition process. The electrochemical performance was measured by cyclic voltammetry and galvanostatic charge-discharge of the carbon fibers-carbon nanotubes---1, 2 and 3 in the 6 M KOH electrolyte. The cyclic voltammetry plot shows a rectangular shape, indicating good capacitive behavior as shown in Fig. 6.2. Furthermore, the charge-discharge profiles of electrodes show symmetrical linear charge-discharge activity. As compared to the capacitance of a pure carbon fiber electrode, the carbon nanotube loaded on carbonized fabric showed higher specific capacitance by increasing the electrodes' specific surface area. Furthermore, the cycling stability of carbon fiber-carbon nanotubes-2 demonstrated a very stable capacitance with a high retention of 96% after 5000 cycles.

As binder-free electrodes, carbon cloth & carbon nanotubes yarns have been used. Qiu et al. [54] claimed that the synthesis of vertical-grown carbon nanotubes (VCNTs) over carbon nanofibers was efficient and easy. PAN (polyacrylonitrile)/PVP (polyvinylpyrrolidone)/Ni in various weight ratios (85:10:5, 75:20:5, 70:20:10)/$Ni(acetate)_2$ is electro-spun to produce carbon nanofibers. $Ni(Ac)_2$ was reduced to Ni nanoparticles with carbon species and catalyzed the creation of carbon nanotubes. According to the electron microscopic images, morphology shows multi-walled nanotubes with a diameter of 30–60 and 10–20 nm thickness. The cyclic voltammetric curves of the symmetric two-electrode system exhibited a nearly rectangular shape in a 1 M NaOH electrolyte, revealing ideal capacitive behavior. The VCNTs-carbon nanofibers composite at a scan rate of 10 mV/s demonstrated the specific capacitance values of 213.6 and 128.0 F/g at a scan rate of 1000 mV/s. When employed with ionic liquid electrolyte, $EMIMBF_4$, the specific capacitances of 146.8, 132.1, 94.8, 76.9, and 63.8 F/g are achieved even at higher rates of 0.5, 1, 2, 5, and 10 A/g. The electrochemical double-layer capacitor provided a high specific energy of 70.7 Wh/kg at a current density of 0.56 A/g due to the higher voltage window of the ionic liquid electrolyte. The fabricated flexible capacitor illuminated a blue LED indicator (3.2 V) as shown in the inset of Fig. 6.3f.

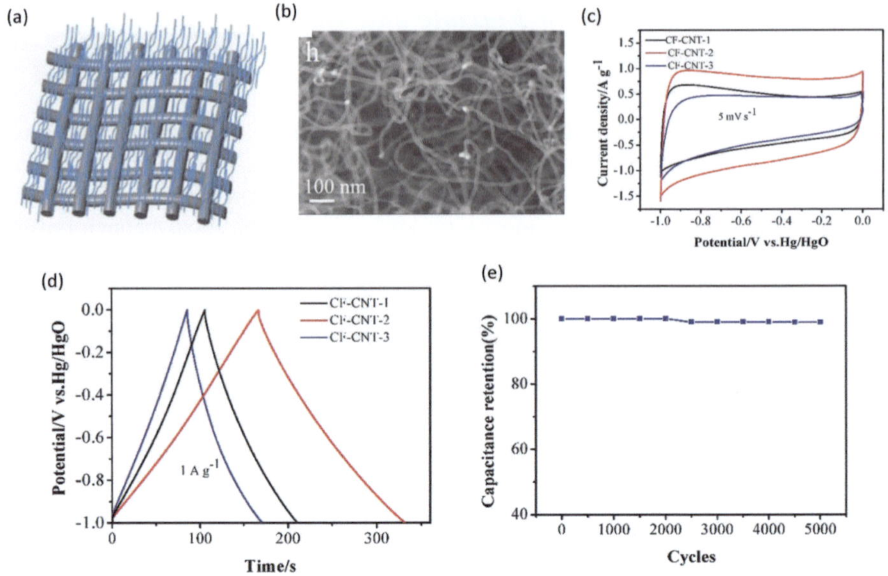

Fig. 6.2 **a** Constructed carbon fiber-carbon nanotube electrode. **b** Scanning electron microscopic image, **c** cyclic voltammetry curves at a sweep rate of 5 mV s^{-1}, **d** charge-discharge profiles of carbon fiber-carbon nanotube electrode at a current density of 1 A g^{-1} **e** long-term stability of carbon fiber-carbon nanotube electrode over 5000 cycles. Reproduced with the permission of Wiley from Zhang et al. [91–93].

Yang et al. [85] proposed a composite fabric electrode prepared by a filtration process using Ni-coated cotton fabric to grow multi-walled carbon nanotubes (MWCNT) and reduced graphene oxide. A conductive Ni layer was typically first coated on pre-cleaned cotton fabric using a polymer-assisted metal deposition (PAMD) method. Following that, MWCNT and reduced graphene oxide were coated on the surface of Ni-cotton using vacuum filtration, within this the Ni-cotton acted as a filtering membrane. One MWCNT bilayer was formed on cotton using one alternating filtration cycle, and ten bilayers were produced by repeating the alternating filtration cycle. Reduced graphene oxide A wearable and waterproof all-solid-state symmetric supercapacitor was assembled using flexible MWCNT-reduced graphene oxide fabric with the aid of a PVA/LiCl polymer gel electrolyte. All the devices produced nearly rectangular cyclic voltammetric curves and normal triangular shapes of charge-discharge profiles. The capacitance of the fabricated flexible supercapacitor was 2.7 F cm^2 at 20 $mAcm^2$ while using a ten-bilayered MWCNT-reduced graphene oxide hybrid fabric electrode. Furthermore, the capacitance of the versatile supercapacitor increased by 3.2 F cm^2 over 10,000 cycles. The recent advancements in flexible anode materials with their corresponding electrochemical performance are listed in Table 6.1.

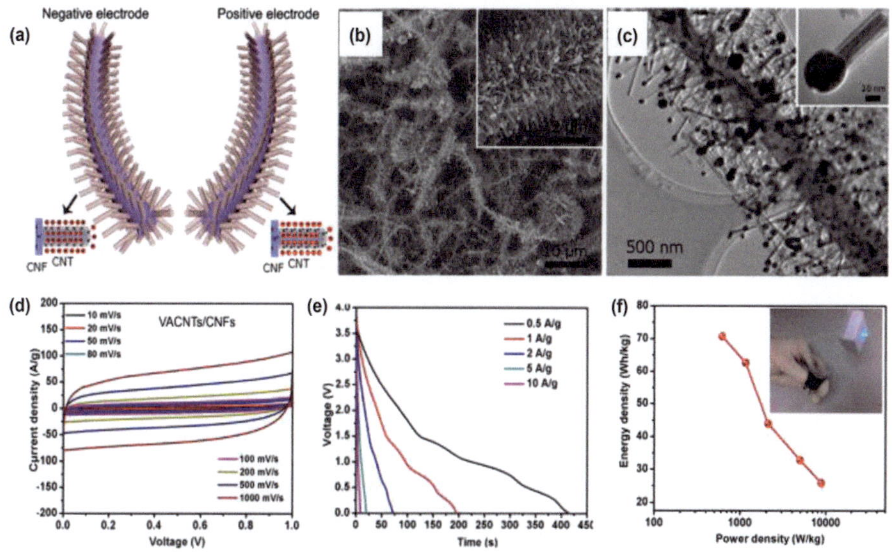

Fig. 6.3 **a** The VACNTs—carbon nanofibers electrode structure. **b**, **c** Scanning electron microscopic and transmission electron microscopic image of VACNTs-carbon nanofibers electrode. **d** Cyclic voltammetric curves at different sweep rates. **e**, **f** Charge-discharge and Ragone plots of VACNTs-carbon nanofibers based double layer capacitors in $EMIMBF_4$ ionic liquid electrolyte. Inset image of VACNTs-carbon nanofibers based flexible electrochemical double layer capacitor and a lighted LED indicator. Reproduced with the permission of the American Chemical Society from Qiu et al. [54]

6.2.2 Cathode Materials for Flexible and Wearable Supercapacitors

Massive efforts have been made in the development of metal-based materials such as metal oxides, nitrides, sulfides, carbides and hexacyanoferrates. Metal-based materials, such as pseudocapacitive materials (MnO_2 and RuO_2) and battery-type materials (NiO, CoO, Co_3O_4, $Cu_2Co_2O_4$, $NiCo_2O_4$, and $Ni(OH)_2$), outperformed carbon-based materials in terms of energy storage due to surface and diffusion-controlled electrochemical redox reactions with electrolyte ions [1, 22, 53]. Generally, for fabricating pseudocapacitive materials using porous metals such as nickel foam, carbon foam, nanoporous, gold and silver, and stainless-steel mesh as substrates, however, their rigidity makes devices less flexible and reduces energy density. To overcome this problem, carbon fiber, carbon cloth, carbon nanotubes and nanofibers, copper and nickel-coated polyester fabric, silver-sputtered textile cloth, Cu-PET fiber, soft platinum foil, and stainless-steel mesh are attractive materials for constructing flexible electrodes owing to their higher flexibility and strength. So far, carbon cloth or carbon fiber is an extremely appropriate current collector of flexible electrode material owing to its excellent electrical conductivity, good mechanical strength,

Table 6.1 Summary of different anode materials for flexible and wearable supercapacitors and their electrochemical performances

Electrode materials	Current density	Specific Capacitance	Potential window	Energy and power density	Capacitance retention	References
Carbonzed fabric-CNT	0.1 A/g	191 F g^{-1}	1.0 V	3.3 kW kg^{-1}	96% (5000 cycles)	Zhang et al. [91–93]
CF-MSP	1.0 A g^{-1}	374.7 F g^{-1}	1.0 V	250 kW kg^{-1}		Zhang et al. [91–93]
VCNT/CNF	0.5 A g^{-1}	146.8 F g^{-1}	1.0 V	70.7 Wh kg^{-1}	97% (20,000 cycles)	Qiu et al. [54]
MWCNT/RGO	20 mA cm^2	2.7 F cm^2	0.8 V	3.2 F cm^2	(10,000 cycles)	Yang et al. [85]
NSCP-CNF	0.5 A g^{-1}	103 F g^{-1}	1.0 V	9 Wh kg^{-1}	107% (3000 cycles)	Li et al. [28–30]
3D-GCAs	0.4 A g^{-1}	4.76 F g^{-1}	0.8 V	4079.9 W/kg	95.5% (10,000 cycles)	Zhu et al. [96]
M-NGM	0.5 A g^{-1}	364 F g^{-1}	1.0 V	12.2 Wh kg^{-1}	100% (10,000 cycles)	Wang et al. [73, 76]
Fe_2O_3@AC	1 mF cm^2	1565 mF cm^2	1.8 V	9.2 mWh cm^3	95% (4000 cycles)	Li et al. [28–30]
Fe_2O_3-P@carbon cloth	1 mA cm^2	1 mA cm^2	1.6 V	0.42 mWh cm^{-3}	83% (9000 cycles)	Liang et al. [31]
VNNDs/CNSs	1.1 A cm^2	1203.6 Fcm^{-3}	1.1 V	30.9 Wh L^{-1}	91% (10,000 cycles)	Li et al. [27]
TiN@VN NWAs	1 mA cm^2	1255 mF cm^2	1.6 V	36.0 mWh cm^{-3}	91.5% (5000 cycles)	Li et al. [28–30]
GF-CNT@Fe_2O_3	20 mA cm^2	470.5mF cm^2	1.6 V	37.3 Wh kg^{-1}	95.4% (50,000 cycles)	Guan et al.[18]

CNS Carbon nanosheet, *GF* Graphene fibers, *CNT* Carbon nanotubes, *TiN* Titanium nitride, *VN* Vanadium nitride, *NWs* Nanowires, *NDs* Nanodots, *GC* Carbon cloth, *NGM* N-doped graphene monolith, *VCNT* Vertically grown carbon nanotubes, *NSCP* Nitrogen and sufur doped carbon polyhedra, *CF-MSP* Carbon fiber multiscale pores, *AC* Activated carbon chemical stability, and flexibility. Furthermore, clean carbon cloth has no pseudocapacitive properties. Therefore, it is very prudent and viable to prepare flexible electrode materials by loading transition metal-based compounds with pseudocapacitive and battery properties on carbon cloth or fiber. Babu et al. [2] and Maier et al. [43] prepared the unique binder-free polyaniline-metal hexacyanoferrates (M = Mn, Ni, and Co) nanoparticles (PANI/MHCFs) on carbon fiber substrate by facile electrodeposition method. The unique core-shell ZnO-C-$NiCo_2O_4$ nanorod sheet arrays are grown on carbon cloth current collector by hydrothermal and electrodeposition methods [90]. The composite electrode achieved a higher energy storage performance (2650 F g^{-1} at 5 A g^{-1}), thanks to the benefits from the introduction of carbon obtained from metal-organic frameworks and the core-shell structure.

6.2.2.1 Metal-Organic Frameworks as Positive Electrodes

Metal-organic framework electrodes are appealing materials for supercapacitor electrodes due to their large surface area and porosity and structural tailorability as well as their supramolecular construction. Furthermore, the unique structure of metal-organic frameworks enables an increase in pseudocapacitive redox-active centers, which is a prerequisite for good supercapacitor materials. Pristine metal-organic frameworks were used to create flexible supercapacitors. Yan et al. [83] developed a flexible supercapacitor using an accordion-like nickel-based metal-organic framework material, to create this Ni-MOF ($[Ni(OH)_2(C_8H_4O_4)_2(H_2O)_4 2H_2O]$, P-benzene dicarboxylic acid was used as an organic linker and nickel nitrate hexahydrate was used as a metal precursor. The specific surface area of this amorphous structural metal-organic framework was 117.42 m^2/g. These Ni-metal-organic frameworks are made up of several layered microsheets with a wide active region for electrochemical ion intercalation and deintercalation.

Ni-metal-organic frameworks as a supercapacitor material yielded a specific capacitance of 988 F/g at a current density of 1.4 A/g, with a capacitance retention of 96.5% after 5000 cycles. In the construction of a flexible supercapacitor, this accordion-like Ni-metal-organic framework was used as a positive electrode and activated carbon as a negative electrode. This flexible supercapacitor achieved an areal capacitance value of 230 mF/cm^2 at 1 mA/cm^2 with excellent capacitance retention of 92.85% even after 5000 cycles for a current density of 5 mA/cm^2. This asymmetric supercapacitor had a maximum energy density of 4.18 mWh/cm^3 and a maximum power density of 231.2 mW/cm^3. According to electrochemical impedance spectroscopy, this accordion-like metal-organic framework has a charge transfer resistance of 6.3 Ω. Despite their high performance, one of these devices' most noticeable shortcomings has been their poor conductivity. The conductivity of the metal-organic framework structure can be enhanced by using highly conductive materials and effectively controlling the morphology.

Metal-organic frameworks-derived composites or hybrids can suffer from massive agglomeration in their structure during high-temperature annealing due to their powder form and the high surface energy of the nanoparticles. For optimizing surface active sites and equal volume transportation between electrodes and electrolytes, the direct construction of metal-organic frameworks-derived composites over conductive templates is an effective strategy. To make phosphorous-doped cobalt oxide (P–Co_3O_4), [34] used a Co-based metal-organic framework hybrid as a prototype, which was then hybridized with phosphorous nitrogen co-doped carbon (PNC) to make a P-Co_3O_4-PNC flexible electrode. At 1A/cm^2, this electrode produced a high areal capacitance of 1535 mF/cm^2. The complete cell P-Co_3O_4-PNC/PNC, which used phosphorous nitrogen co-doped carbon as a negative electrode, exhibited the highest energy density of 69.6 Wh/kg at a powder density of 750 W/kg after 10,000 cycles, with 96.8% cyclic stability.

6.2.2.2 Metal-Organic Framework-Derived Hybrids as Positive Electrodes

Metal oxide or metal-organic frameworks can maximize both the merits of metal-organic frameworks and the merits of the other materials which are composited by hybridizing metal-organic frameworks with suitable carbon-based materials. Xu et al. [82] used a three-dimensional metal-organic system with conductive polypyrrole tubes to incorporate versatility into the supercapacitor structure to investigate applications in flexible supercapacitors. ZIF-67 was used as a source metal-organic framework for the supercapacitor electrode in the creation of this 3D structure. This metal-organic framework was made using cobalt ion as a metal precursor and 2-methylimidazole as an organic linker. The existence of conductive polypyrrole tubes, as well as the 3D metal-organic framework network, was a key feature of this supercapacitor system structure, as it provided stability while also improving the structure's electrical conductivity and enhancing electron transfer. The specific capacitance of ZIF-polypyrrole networks with 597.6 F/g was found to be significantly higher than that of pristine ZIF-67 with 99.2 F/g. at a current density of 0.4 mA/cm^2, the mounted symmetric supercapacitor unit ZIF-PPy/ZIF-PPy shows a real capacitance of 225.8 mF/cm^2. This supercapacitor system achieved a potential window of 0 to 0.6 V, with a maximum energy density of 0.0113 mWh/cm^2 at a power density of 0.12 mW/cm^2 and a maximum power density of 1.44 mW/cm^2 at an energy density of 0.0076 mWh.cm^2. Pyrolysis, a high-temperature annealing process in a specific atmosphere is an effective method for the preparation of metal-organic frameworks-derived materials. This fact also makes the development of next-generation supercapacitor devices easier.

Zhou et al. [95] used a hydrothermal method to establish a reasonable design and synthesis of electrode materials with uniformly stacked lamella multi-layer micro and nanostructure (Co_9S_8) on carbon cloth. In the first step, $CoCO_3$-carbon cloth was made by dissolving cobalt nitrate hexahydrate, sodium molybdate dehydrate, and urea in water. A piece of carbon cloth was immersed in the solution and then moved to a stainless autoclave lined with Teflon and heated to 180 °C. The second stage was to make Co_9S_8-carbon cloth using the hydrothermal process as shown in Fig. 6.4. The $Co_3{}^{2-}$ ions in the $CoCO_3$ are replaced by S^{2-} ions when sodium sulfide monohydrate is used. The mass loading of Co_9S_8 on carbon cloth is ~ 0.37 mg cm^{-2}. The Co_9S_8-carbon cloth electrode is a versatile material that can be tailored, bent, and twisted in any direction without losing its electrochemical properties. At various current densities, the charge-discharge profiles of Co_9S_8-carbon cloth display a linear and nearly symmetric existence. The electrodes' specific capacitance is 1475.4 F/g at 1 A/g, indicating a reasonable rate capacity of 80.2% capacitance retention at 20 A/g and excellent cyclic stability with 92.9% retention over 5000 cycles. The fabricated all-solid-state asymmetric supercapacitor demonstrated a high-energy density of 20.3 Whkg^{-1} at a power density of 221,796 W k/g.

To fabricate a versatile binder-free electrode for supercapacitors, Huang et al. [23] used a two-step hydrothermal route to develop hierarchical core-shell $NiCo_2O_4$-$NiMoO_4$ nanowires directly on carbon cloth as demonstrated in Fig. 6.5a. The composite electrodes demonstrate a synergistic effect of $NiCo_2O_4$ and $NiMoO_4$ that improves electrochemical efficiency and provides rapid electron transfer pathways. A high areal capacitance of 2.917 F cm^{-2} at a current density of 2 mA cm^{-2} (Fig. 6.5b–d). In the KOH electrolyte, the cyclic voltammetric curves display a redox peak, indicating reversible faradaic redox reactions of Ni–Co–O, and Ni–Co–O–OH with OH^- ions. Figure 6.5e–g shows the core-shell nanostructure is more stable; exhibits excellent cycling stability at high current density and shows 90.6% retention over 2000 cycles at 20 mA cm^{-2}.

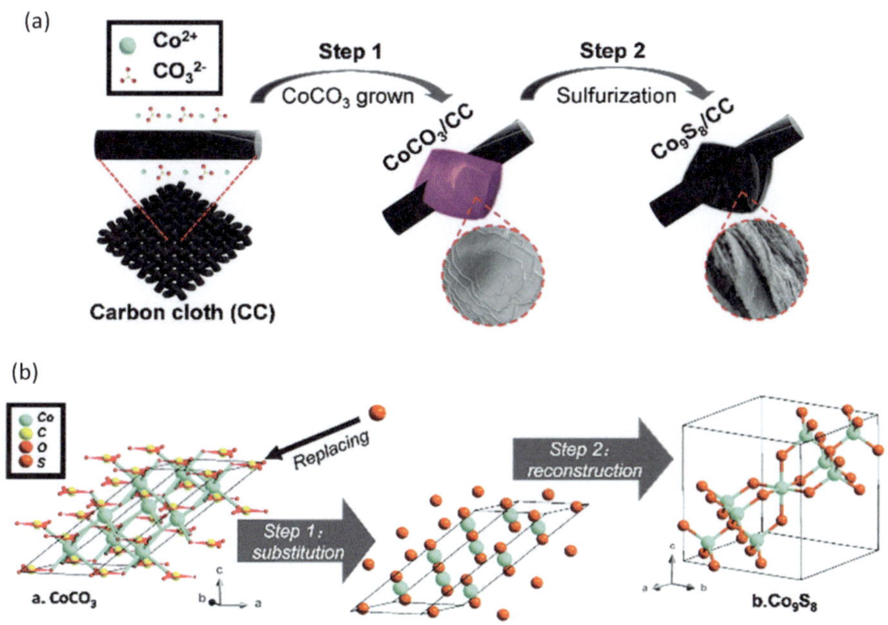

Fig. 6.4 **a** A step-by-step fabrication process of Co_9S_8 on carbon cloth using a hydrothermal route **b** the structure of rhombohedral $CoCO_3$ and cube Co_9S_8. Reproduced with the permission of the Royal Society of Chemistry from Zhou et al. [95]

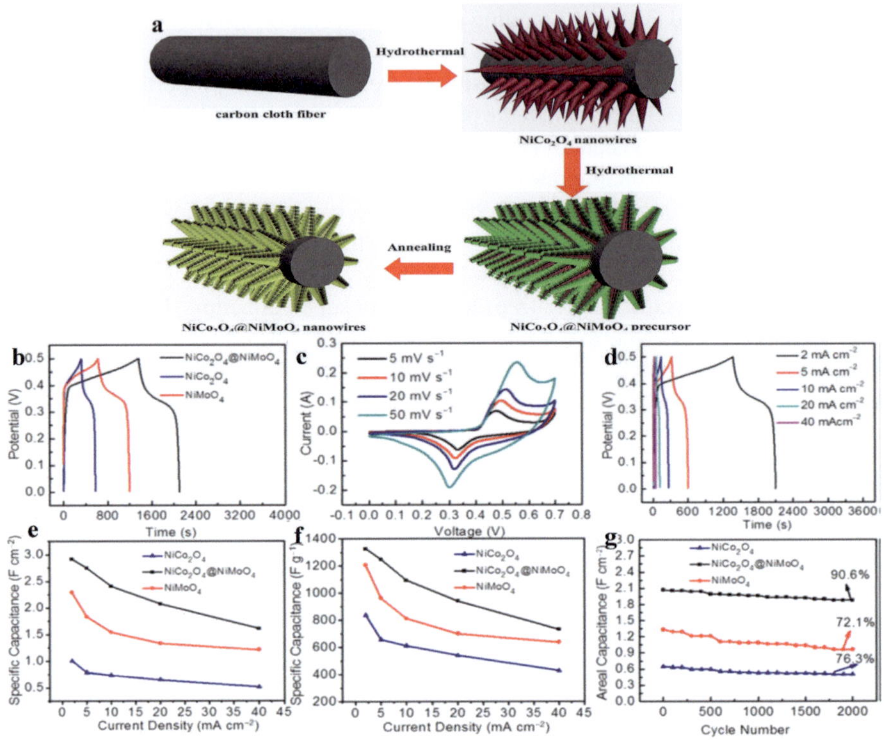

Fig. 6.5 **a** The formation process of carbon cloth-$NiCo_2O_4$-$NiMoO_4$ integrated electrode. **b** Charge-discharge curves of the carbon cloth-$NiCo_2O_4$-$NiMoO_4$, CC-$NiCo_2O_4$, CC-$NiMoO_4$ at current density of 2 mA cm^{-2} **c** cyclic voltammetric curves of CC-$NiCo_2O_4$-$NiMoO_4$ electrode at various scan rates in the voltage window of 0–0.7 V **d** charge-discharge curves of CC-$NiCo_2O_4$-$NiMoO_4$ electrode at different current densities. **e, f** Areal and mass-specific capacitance of CC–$NiCo_2O_4$–$NiMoO_4$, CC–$NiCo_2O_4$, CC–$NiMoO_4$ electrode at different scan rates. **g** Cycling stability of CC–$NiCo_2O_4$–$NiMoO_4$, CC–$NiCo_2O_4$, and CC–$NiMoO_4$ electrode at 20 $mAcm^{-2}$. Reproduced with the permission of Springer from Huang et al. [23]

Liang et al. [32] developed a PH_3 plasma-assisted route for fabric ternary NiCoP nanoplates on a flexible carbon paper substrate as shown in Fig. 6.6. To increase the surface hydrophilicity of the carbon paper layer, it was first heated to 700 °C. In the second stage, a solution of nickel nitrate hexahydrate, cobalt nitrate, and hexamethylenetetramine was dissolved in water and transferred to a stainless-steel autoclave lined with Teflon. The carbon paper was placed in the autoclave, sealed, and heated at 100 °C for 10 h. The layer was stripped and cleaned with ethanol and water after cooling. The metal hydroxide-coated carbon paper substrate was placed in a plasma-enhanced chemical vapor deposition (PECVD) method, followed by thermal phosphorization at 250 °C with PH_3 plasma. NiCoP has a hierarchical structure of nanosheets on a carbon paper substrate, according to the morphology. In 1 M KOH electrolyte, the electrochemical performance of NiCoP electrodes was investigated using a three-electrode system, which showed high electrochemical activity and good stability. The cyclic voltammetric curve shows redox peaks relating to the Faradaic redox reactions. The NiCoP flexible electrode shows high specific capacity of 194 mAh/g at 1 A/g. the asymmetric supercapacitor NiCoP//graphene films device exhibited a high-energy density of 32.9 Wh k/g at a power density of 1301 W k/g with high cyclic stability of 83% capacity retention after 5000 cycles at 20 A/g as shown in Fig. 6.6.

Xia et al. [79] developed a facile high-temperature carbonization process and an electrochemical deposition system to fabricate $Ni(OH)_2$ nanostructures on carbonized cotton fabric (CCF) with better pores and petal arrangement for wearable supercapacitors. The electrochemical performance of $Ni(OH)_2$–CCF//$Ni(OH)_2$-Conductive cotton fabric devices with symmetrical two-electrode systems was investigated. Even at a high scan rate, the cyclic voltammetry performs both EDLC and pseudocapacitance behaviors with no distortion. Unlike EDLC capacitors, the galvanostatic charge-discharge curves showed distinct non-linearities, revealing the pseudocapacitance behaviors of $Ni(OH)_2$. The symmetric supercapacitor shows a high specific capacitance of 131.43 F/g at 0.25 A/g with a high-energy density of 35.78 Wh k/g at a power density of 0.35 kW k/g as shown in Fig. 6.7. The recent

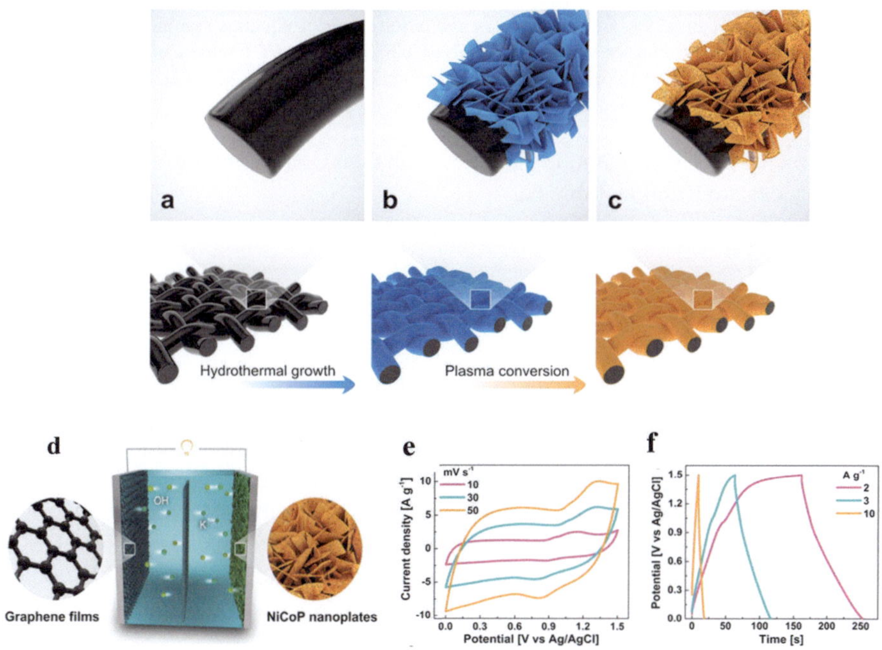

Fig. 6.6 Synthetic process of metal phosphide nanostructures. **a** Carbon paper substrate. **b** Hydrothermal growth of metal hydroxide precursor on carbon paper **c** plasma conversion of hydroxide nanoplates into phosphide porous nanoplates. **d** Schematic illustration of the asymmetric supercapacitors **e**, **f** cyclic voltammetric and charge-discharge curves at different scan rates and current densities. Reproduced with the permission of Elsevier from Liang et al. [32]

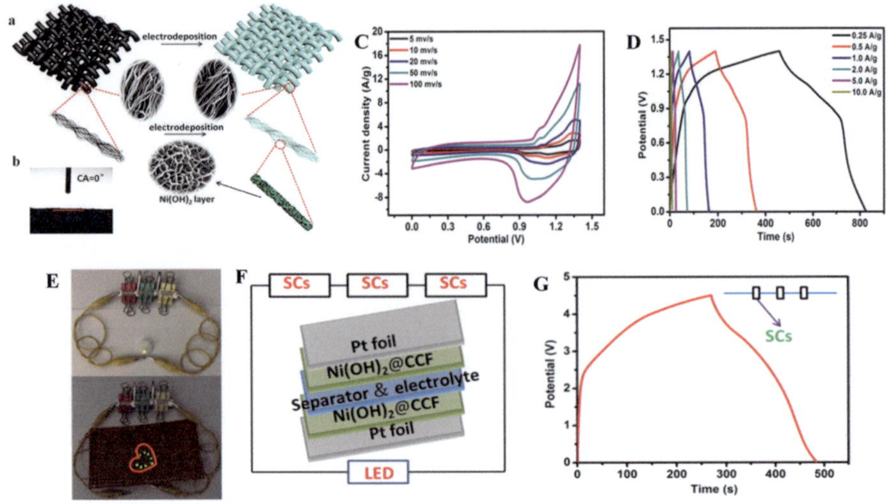

Fig. 6.7 **a**, **b** The preparation process and photo for the water contact angle measurement of $Ni(OH)_2$-CCF. **c** Cyclic voltammetric curves of $Ni(OH)_2$-CCF-200// $Ni(OH)_2$-CCF-200 at different scan rates in 6 M KOH. **d** Charge-discharge curves of $Ni(OH)_2$-CCF-200//$Ni(OH)_2$-CCF electrode at different current densities in 6 M KOH. **e** Green LEDs powered by the $Ni(OH)_2$-CCF//$Ni(OH)_2$-CCF supercapacitor in series. **f** Schematic for the assembling of aqueous symmetric supercapacitor. **g** Charge-discharge curves of three $Ni(OH)_2$-CCF//$Ni(OH)_2$-CCF supercapacitors in series. Reprinted with the permission of the American Chemical Society from Xia et al. [79]

advancements in flexible cathode materials with their corresponding electrochemical performances are listed in Table 6.2.

6.3 Electrolyte Materials for Flexible and Wearable

Generally, a supercapacitor contains a positive electrode, negative electrode and separator or electrolyte using various conductive additives materials. In addition to the positive and negative electrodes, the electrolyte is very important to the flexible supercapacitors. Recently, two kinds of electrolytes have been used namely, liquid electrolytes such as aqueous and organic and solid electrolytes such as polymer gel electrolytes. In general, aqueous electrolytes can be in the form of acids like H_2SO_4 and H_3PO_4 or neutral like KCl, LiCl, KNO_3, Na_2SO_4, and Li_2SO_4 or alkaline like LiOH, KOH, and NaOH whereas organic electrolytes such as ionic liquids or organic solvents with $NaClO_4$ or $NaPF_6$.

A solid-state electrolyte can be in the form of gel or polymer mixtures. Usually, gel agents contain different polymer mixtures such as polyvinyl alcohol (PVA), poly (vinylidine fluoride) (PVDF), and poly (vinylidine fluoride-co-hexafluoropropylene) (PVDF-HF). The flexible solid polymer electrolyte can be easily synthesized from various conductive polymer mixtures for example, polyvinyl alcohol is dissolved in

Table 6.2 Different cathode materials for flexible and wearable supercapacitors and their electrochemical performance

Electrode materials	Current density	Potential window	Specific capacitance	Energy density and power density	Capacitance retention	References
Co_9S_8/CC	1.0 A g^{-1}	1.5 V	1475.4 F/g	20.3 Wh kg^{-1}	92.9% (5000 cycles)	Zhou et al. [95]
$NiCo_2O_4$@$NiMoO_4$/	2 mA cm^{-2}	0.7 V	2.917 F cm^{-2}	–	90.6% (2000 cycles)	Huang et al. [23]
NiCoP	1.0 A g^{-1}	1.5 V	194 mAh g^{-1}	32.9 Wh kg^{-1} 1301 W kg	87% (5000 cycles)	Liang et al. [32]
$Ni(OH)_2$@CCF	0.2 A g^{-1}	1.4 V	131.43 F g^{-1}	18.28 Wh kg^{-1} 0.35 kW kg^{-1}		Xia et al. [79]
$CuCo_2S_4$/CF	1.0 A g^{-1}	1.6 V	875 C g^{-1}	65.1 Wh kg^{-1} 800 Wkg^{-1}	90.6% (5000 cycles)	Liu et al. [36–38, 40]
$CoMoO_4$@ $NiMoO_4.xH_2O$	2.5 mA cm^{-2}	0.8 V	1039 F g^{-1}	–	75.1% (1000 cycles)	Liu et al. [35]
MNCS-NTAs	1.0 mA cm^{-2}	1.6 V	2554.5 F cm^3	52.4 mWh cm^{-3}	92.9% (5000 cycles)	Xiaona et al. [80]
$FeCo_2S_4$/$NiCo_2S_4$/	5.0 mA cm^{-2}	1.0 V	519 F g^{-1}	46 Wh kg^{-1} 4723 W kg^{-1}	192% (3000 cycles)	Zhu et al. [97]

CC Carbon Cloth, *CF* Carbon fiber, *CCF* Carbonised cotton fabric, *MNCS* Manganese–nickel–cobalt sulfide, *NTAs* Nanotube arrays a water medium then further H_3PO_4 is added into the polyvinyl alcohol solution to afford the flexible solid polymer electrolyte [7, 9, 45]. The polymer electrolytes-based energy storage devices do not require any high safety standard materials and different geometric shapes, which can bring a new design for energy storage supercapacitor devices to shortly wearable electronics field including batteries [41]. More importantly, the electrolyte is very critical for smart electrochemical capacitors to introduce self-healability into hydrogel-based polymer electrolytes, due to the restored capacitive behavior while the polymer undergoes physical damage or leakage or breaking [66]. To develop a thin and flexible high-performance self-healable hydrogel electrolyte, for example, polyvinyl alcohol is a better choice as a flexible solid electrolyte for smart supercapacitors due to its high chemical stability, and environmentally benign [36–38, 40].

6.3.1 Development of Self-healable Hydrogel-Based Polymer Electrolyte

The self-healing hydrogel electrolyte can be made in a few simple steps. First, dissolve polyvinyl alcohol in double distilled water by stirring at 90 °C until it becomes a clear solution. Next, add the graphite oxide solution to the polyvinyl alcohol solution. Then, slowly add the potassium solution and borax solution while stirring. Finally, adjust the pH to 9 by adding a small amount of 25 wt% ammonia solution drop by drop. This creates the self-healing B-PVA/KCL/GO flexible electrolyte as given in Fig. 6.8.

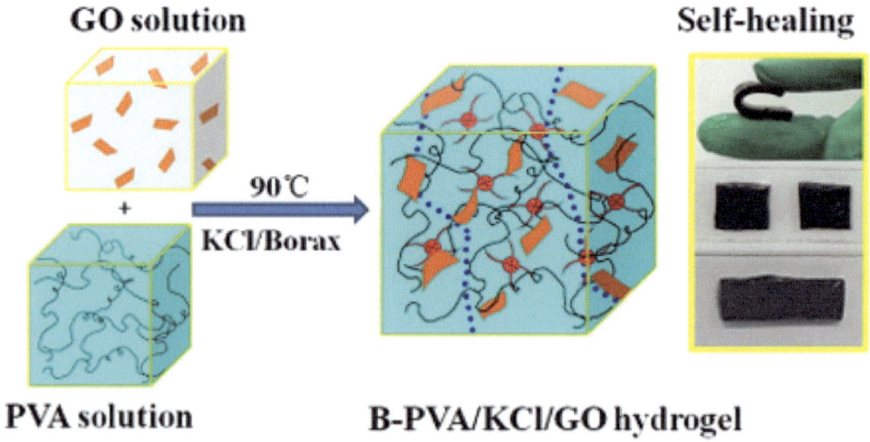

Fig. 6.8 Preparation of self-healable hydrogel electrolyte of B-PVA/KCL/GO with step-by-step processes, PVA-polyvinyl alcohol, KCl-potassium chloride, and GO-graphene oxide. Reprinted with permission of Elsevier from Peng et al. [52]

6.3.1.1 Polyvinyl Alcohol-Based Solid Polymer Electrolyte

Chen et al. [7, 9] reported the SWCNTs-based thin film electrode has been printed on a flexible substrate and cloth fabrics by an inkjet printer. The electrodes were sandwiched between printable polymer electrolytes to form a flexible and wearable supercapacitor, which shows excellent capacitive even after 1000 charge-discharge cycles. Moreover, they have impregnated the RuO_2 nanowire with SWCNTs which enhance their specific capacitance (138 F/g), power density (96 kW/kg) and energy density (18.8 Wh/kg).

The SWCNTs were formed as tangled, dense, and well-dispersed networks with bundle lengths of 0.2–1.8 μm and a diameter of 9–20 nm by an inkjet printer. The printed SWCNT film on cloth fabric was used as electrodes without any treatment and it was sandwiched between polymer electrolytes to form a flexible supercapacitor as shown in Fig. 6.9a. RuO_2 nanowire/SWCNTs film on polyethene terephthalate substrate was sandwiched between PVA/H_3PO_4 polymer electrolyte and studied their electrochemical potential by cyclic voltammetric with different scan rates from 50 to 500 mV/s. It shows quasi rectangular shape with a higher current density as shown in Fig. 6.9b. The shape of the cyclic voltammetric curves is different from printed SWCNTs-based supercapacitors due to the pseudocapacitance behavior, which is contributed by RuO_2 nanowires through electrochemical protonation [56]. The SWCNTs-based flexible devices on plastic, it has acted as both electrodes and charge collectors for flexible supercapacitors. The SWCNTs-based flexible devices show very high energy and power densities (6 Wh/kg and 70 kW/kg) and these printable materials lead to a new class of flexible and printable storage devices [25]. Poly (3,4, -ethylenedioxythiophene) based solid polymer electrolyte for flexible supercapacitor [11]. They have fabricated the flexible supercapacitor using poly (3,4,-ethylenedioxythiophene)/carbon paper (PEDOT/CP) as an electrode with 1-butyl-3-methylimidazole tetrafluoroborate/PVA/H_2SO_4 based gel polymer electrolyte. It exhibited better electrochemical performance of specific capacitance (86.81 F/g at 1 mA/cm^2) and energy density (176.90 Wh/kg) [11]. The electrolyte and electrode materials are the very important components to fabricate the flexible supercapacitor.

Generally, various types of electrolytes have been utilized for supercapacitors such as liquid, solid electrolyte, and gel electrolytes. Gel electrolytes and polymer electrolytes have some advantages including avoiding corrosion and liquid leakage, and better electrochemical stability. For instance, the PEDOT-carbon paper electrode and gel electrolyte were assembled face to face in a solid-state supercapacitor as shown in Fig. 6.10, and exhibited a high specific capacitance of126.24 F/g with better cyclic stability over 5000 cycles [11].

Recently, green electrolytes or ionic liquid-based electrolytes have been used for supercapacitors due to their non-volatile, high electrochemical performance, high conductivity, low viscosity, and simple synthesis process. Generally, such kinds of green electrolytes can be prepared from copolymer gelators such as blocks, stars, and random copolymers. Wang et al. [77] reported the poly(methoxy/hexadecyl-poly(ethylene glycol) methacrylate-based copolymer gel electrolyte by the solution

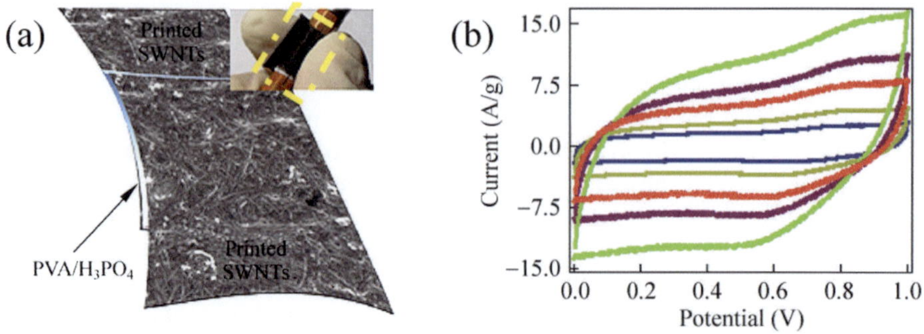

Fig. 6.9 **a** Inkjet printed SWCNTs capacitor using PVA-H_3PO_4 as a solid electrolyte, **b** cyclic voltammetry of RuO_2 nanowire-SWCNTs based supercapacitor using PVA-H_3PO_4 as a solid polymer electrolyte with scan rates of 50, 100, 200, 300 and 500 mV/s. Reprinted with the permission of Springer from Chen et al. [7, 9].

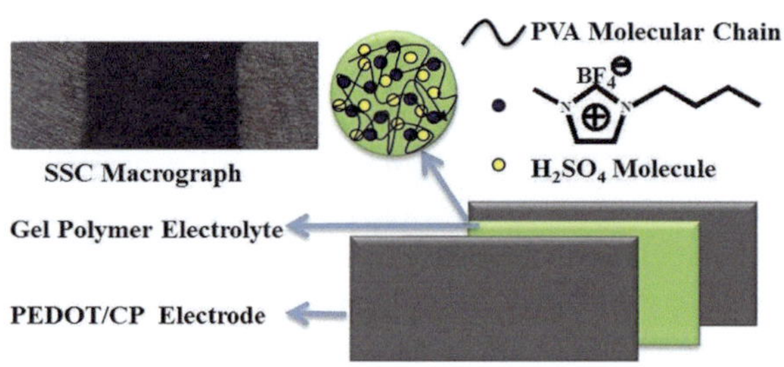

Fig. 6.10 Assembly of PEDOT-Carbon paper electrode and polyvinyl alcohol, PVA-H_2SO_4 based ionic liquid gel polymer electrolyte sandwiched face to face in a symmetric solid-state supercapacitor

casting technique, which shows excellent thermal stability, and mechanical and ionic conductivity.

Carboxymethyl cellulose-based water-soluble binder was used as electrode materials, two major problems are being used in the energy storage devices: (i) relatively low mass loading in electrode and (ii) brittleness. Varzi et al. [67] dissolved nitrocellulose in 1-ethyl-3-methylimidazolium acetate (EmimAc) and they demonstrated that nitrocellulose has excellent stability at higher voltages and high specific capacitance of 13 F/g at 10 mA cm^{-2}. Moreover, several eco-friendly binders such as potato starch, sucrose, polysaccharides, and cellulose-based electrolytes have been used in supercapacitors [68]. Spina et al. [64] reported green supercapacitors from green polymers, they chose water-soluble natural polymers for electrical double-layered capacitors. Pullulan-based ionic liquid EDLCs have been fabricated by electrospinning technique and enhanced their specific energy (7.2 Wh kg^{-1}) and power density (3.7 kW kg^{-1}) with good cyclic stability (5000 cycles) [64].

Na et al. [46] reported that egg white and rice husk waste-based polymer gel electrolytes have been used for green solid supercapacitors. These egg-white gel electrolytes have many advantages such as high retention, high ionic conductivity, and excellent mechanical, and thermal stability. The electrolyte demonstrated an excellent specific capacitance of 214.3 Fg^{-1} at 0.2 A g^{-1}. The overall fabrication

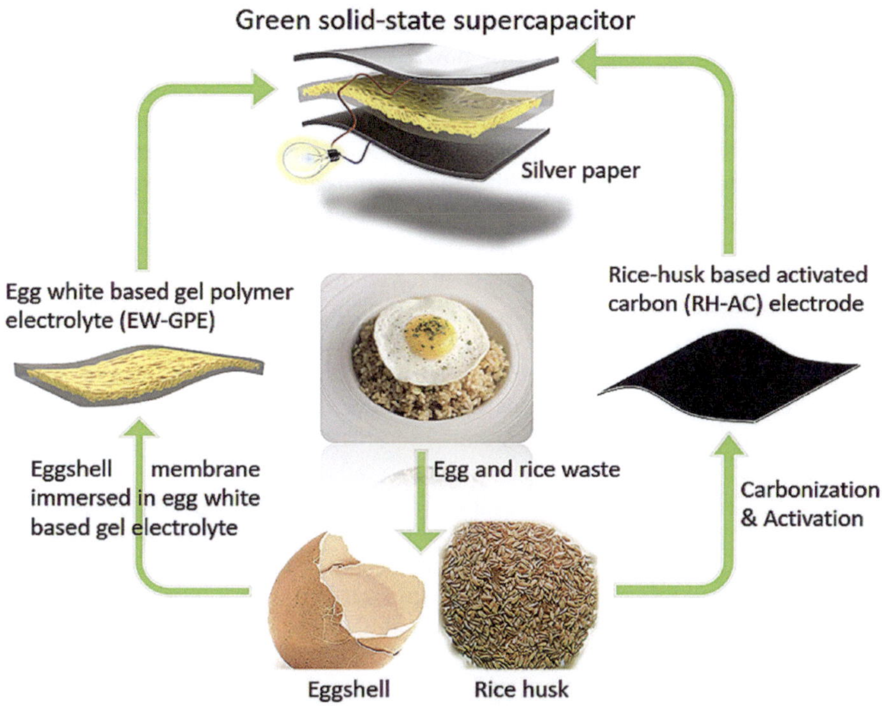

Fig. 6.11 Green solid-state supercapacitor, the fabrication steps of green solid-state supercapacitors using eggshell and rice husk-based gel polymer electrolyte. Reprinted with permission of Elsevier from Na et al. [46]

step of green solid-state supercapacitors using eggshell and rice husk was represented in Fig. 6.11. The electrolyte materials from rice husk and egg were synthesized by carbonization followed by the activation method. A common solid polymer electrolyte has several practical difficulties such as low ionic conductivity and poor conductance between the polymer electrolyte and electrode. To resolve those problems, researchers have developed a porous polymer electrolyte to enhance their ionic transport during charge-discharge cycles. Such kind of porous membrane-based gel electrolyte has good retention.

6.4 Next-Generation Flexible and Wearable Materials for Hybrid Supercapacitor

Owing to their high-power density, rapid charge-discharge rates and excellent cycling performance, supercapacitors have gained special consideration as an excellent energy storage device for multifarious applications. Supercapacitors may be more appealing options for next-generation high-performance energy storage devices due to their higher energy density, higher power density, and ultra-cycling capability. On the other hand, the lower energy density of supercapacitors than the rechargeable batteries hints at the commercial viability of most supercapacitor materials. According to the equation of energy density ($E = 0.5\ CV^2$), the energy density could be prolonged by increasing either the specific capacitance or operating potential window. To address this key issue and make the supercapacitors for commercial applications, the fabrication of a battery-supercapacitor hybrid or hybrid supercapacitor device constructed with battery-type and pseudocapacitive positive and carbon-based negative materials is key to gaining both high-energy density and power densities. This efficient strategy enhances the potential window and capacitance, hence improving the energy density by the assembly of intrinsic charge-storage capabilities of the cathode electrode, i.e., pseudocapacitive-battery-type materials and porous anode electrode, i.e., carbon-based material for battery-supercapacitor hybrid or hybrid supercapacitor device.

Recently, Wang and co-workers rationally designed and synthesized moss-like nickel-cobalt bimetallic phosphide (Ni–Co–P) nanosheet arrays on carbon fabrics, which are directly used as positive electrodes and commercial activated carbon as negative electrode. The electrochemical behavior of the hybrid supercapacitors was studied with the cut-off voltage range between 0 and 1.7 V and exhibit higher specific capacitance of 136 F g^{-1} at 1 A g^{-1}, the excellent energy density of 48.4 Wh kg^{-1}at 811.2 W kg^{-1}, and remarkable cyclic stability of ~ 88.8% after 80,000 cycles which depicted in Fig. 6.12. In addition, the assembled hybrid supercapacitors show outstanding mechanical flexibility and stability without decay during different bending angles.

The hybrid electrodes of free-standing Ni–Mo–S and Ni–Fe–S nanosheets on carbon cloth by a simple hydrothermal method followed by sulfurization exhibit high-performance flexible asymmetric supercapacitors, and achieved a high volumetric capacity of ~ 1.9 mAh cm^3, the exceptional energy density of ~ 82.13 Wh kg^{-1} at 0.561 kW kg^{-1}, outstanding power density (~ 13.103 kW kg^{-1} at 61.51 Wh kg^{-1}) and remarkable cycling stability with retention of $\approx$ 95.86% of initial capacity over 10,000 cycles [3, 15] successfully constructed the flexible solid-state asymmetric supercapacitor using Ni–Co double hydroxides -Ni-pen ink-carbon fiber as the positive electrode and pen ink coated on highly conductive nickel/carbon fiber as negative electrodes. Liu et al. [33, 39] fabricated the flexible quasi-solid-state asymmetric supercapacitors using F-Co_2MnO_{4-x}/carbon fiber and Fe_2O_3-CF as the positive and negative electrodes, respectively, and PVA-KOH as the gel electrolyte. The as-assembled asymmetric supercapacitor device attained a high-energy density of 64.4 Wh kg^{-1} at 800 Wkg^{-1}. [84] also constructed a high-performance flexible solid-state asymmetric supercapacitor with α-MnO_2 nanowires-carbon fabric and amorphous Fe_2O_3 nanotubes-carbon fabric as the positive and negative electrodes, respectively. The solid-state asymmetric supercapacitor achieved a volumetric capacitance of 1.5 F/cm^3 at 2 mA/cm^{-2} and a high-energy density of 0.55 mWh cm^{-3}.

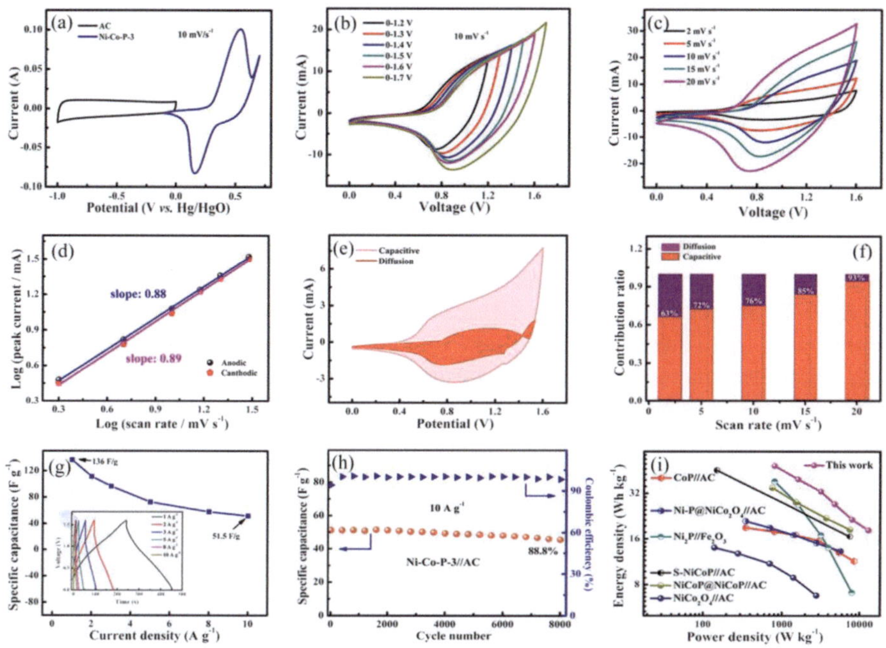

Fig. 6.12 **a** Three electrode configurations cyclic voltammetric profiles of the activated carbon and Ni–Co–P-3 at 10 mV s^{-1}; **b** cyclic voltammetric profiles of the as-assembled supercapacitor 10 mV s^{-1} **c** various sweep rates; **d** the relationship between the peak current and sweep rate; **e** the contributions of capacitance and diffusion for cyclic voltammetric curve at 2 mV s^{-1}; **f** the relative ratio of capacitance contribution to diffusion contribution at various sweep rates; **g** specific capacitances, and the inset is the charge-discharge curves; **h** long-term cycling performance at 10 A g^{-1}; **i** Ragone plot of the as-assembled supercapacitor. Reprinted with permission of Elsevier from Qu et al. [55]

Owing to their low cost, flexibility and highly porous structures, paper and textiles have been employed as a medium of substrate which can absorb electroactive materials. Conductive materials, for instance, graphene ink, carbon nanotubes or silver nanowires, have been generally incorporated into paper or textile substrates to fabricate flexible and wearable devices. Ko et al. [26] demonstrated the hybrid metallic paper-based asymmetric supercapacitors composed of alternating MnO-Au nanoparticles as a positive electrode and Fe_3O_4-Au as the negative electrode, respectively, and attained a maximum specific capacitance of 222 F g^{-1} at 1 mA cm^{-2} with an excellent energy density of 121.5 Wh kg^{-1}. Moreover, metal wires, such as Ni and Cu wire, and stainless steel have also been investigated as substrates for flexible and wearable supercapacitors owing to their reasonable conductivity. Recently, Wang and co-workers fabricated flexible asymmetric supercapacitors consisting of $NiCo_2O_4$-ultrafine nickel wire and Fe_3O_4-ultrafine nickel wire as the positive and negative electrodes, respectively, and the flexible asymmetric device demonstrated an energy density of 32.6 Wh kg^{-1} at a power density of 846 Wkg^{-1}.

The most important concern for the environment and one of the fastest-growing waste streams in the world in recent decades is electronic waste (e-waste) for instance expired domestic electrical devices such as televisions, fans, heaters, electrical cable wires and information technology electronic devices such as mobile phones, circuit boards, computers. Meanwhile, these e-wastes are valuable, including considerable amounts of precious metals. Among the various types of e-waste, electrical cable wires are generally made up of metallic fibers such as copper and aluminum, which can be used in a variety of electronic and energy-related applications. As a result, recycling these e-waste cable wires for the fabrication of fiber-type supercapacitor applications reduces not only the fabrication cost of supercapacitors but also the depletion of fossil fuels, providing new options for real-world applications while also being environmentally friendly. Inspired by the e-waste recycling approach. Nagaraju et al. [47, 48] fabricated the high-performance fiber-based hybrid supercapacitor using the fibers of Cu from waste cable wires as a low-cost substrate and grew the core-shell-like nanostructures of NiO NSs–CNTs–CuO, which then used as a positive electrode to fabricate the solid-state supercapacitor against activated carbon coated carbon fibers in a gel electrolyte. The schematic representation of the fabrication process of flexible devices is illustrated in Fig. 6.13a. Cyclic voltammetry curves at various sweep rates with a cut-off range of 0–1.55 V as shown in Fig. 6.13b. The cyclic voltammetry curves displayed excellent capacitive performance relative to the redox peaks, which is due to the presence of EDLC material, Furthermore, the charge and discharge profile showed symmetric durations, indicating the device's outstanding rate capability (Fig. 6.13c). More significantly, the flexible device attained a maximum energy density of 26.32 Wh kg^{-1} at a power density of 219.03 W kg^{-1} (Fig. 6.13d). Two flexible devices are connected serially and attached to human shirts to operate the LED with a multifunction electronic display (Fig. 6.13e, f).

Gao et al. [14] demonstrated another flexible device using nickel fiber from waste, cellular copper foam (CF), and graphene sheets (GSs), as well as layered double hydroxides as shown in Fig. 6.13g, which consists of Ni–Co LDH–GSs–CF–NF and activated carbon as the positive and negative electrodes, respectively, and PVA-KOH as the gel electrolyte. Cyclic voltammetry curves at different sweep rates showed both battery-type behavior and double-layer capacitor features, indicating efficient transportation of ions (Fig. 6.13h). Moreover, the charge-discharge profiles attained almost linear and symmetric behavior, demonstrating remarkable electrochemical reversibility (Fig. 6.13i) and the areal capacitance was found to be 350.9 mF cm^{-2} at a current density of 1 mA cm^{-2}, the energy density of 109.6 μWh cm^{-2} at a power density of 749.5 μW cm^{-2} and retained 64.2 μWh cm^{-2} even at a power density of 14997.6 μW cm^{-2} (Fig. 6.13j). Authors further evidenced the flexibility by measuring the capacitance at different bending angles from 0 to 180°, and the device exhibited only 5% retention of decay (Fig. 6.13k). Moreover, the flexible device connected to a variety of devices weaved into the cotton shirt and even flexed to fit our body shape. Besides the flexible device curved and twined around a finger and successfully powered a digital watch and integrated with a headphone as shown in Fig. 6.13l and m. Furthermore, the FFSC was connected to lab cloth to power electronics while

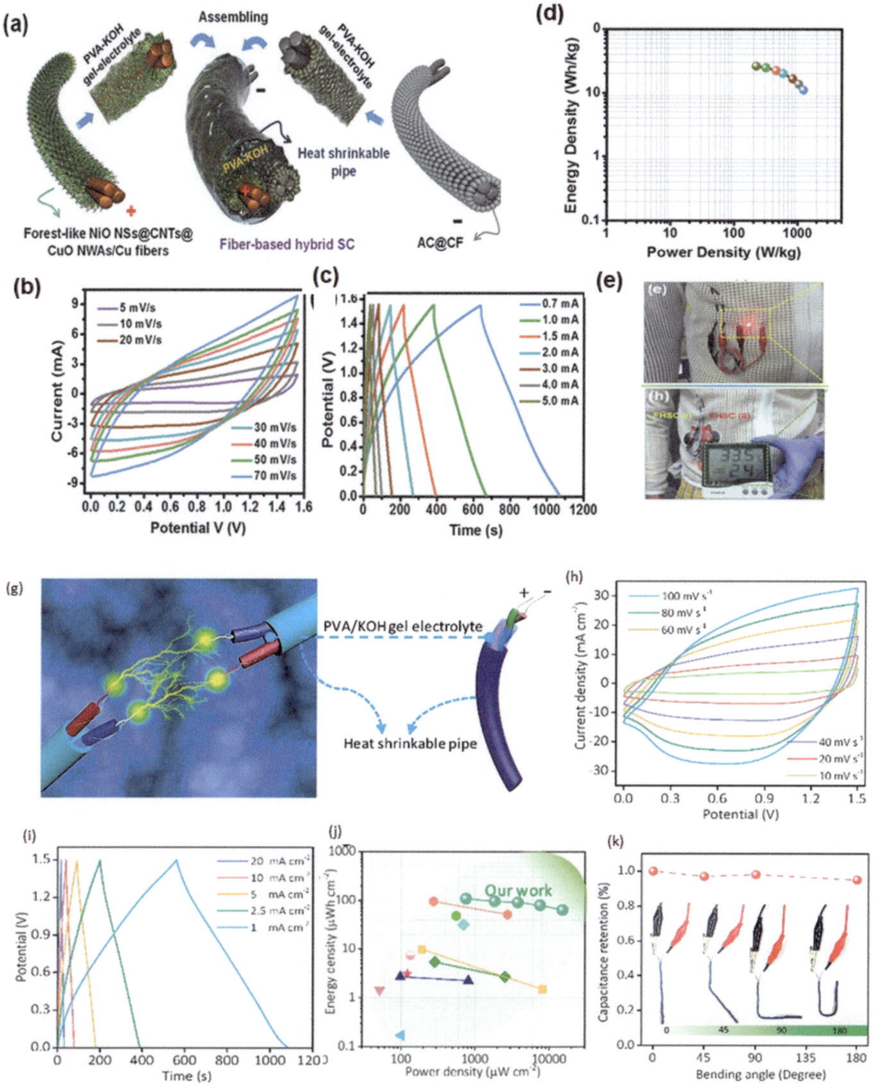

Fig. 6.13 **a** The construction of PVA-KOH electrolyte-coated fiber-based hybrid supercapacitor. **b** Cyclic voltammetric profiles, **c** charge-discharge profiles and **d** Ragone plot of flexible solid-state hybrid supercapacitor. **e**, **f** Two flexible solid-state hybrid supercapacitors are serially connected and power LED and electronic display [46]. **g** As-assembled flexible fiber-based solid asymmetrical supercapacitor device (NiCoLDH-GSs-CF-NF//AC). **h** Cycle voltammetric profiles, **i** charge-discharge profiles and **j** Ragone plot of the fibre-based solid asymmetrical supercapacitor **k** capacitance retention of the device at different bending angles. Practical application of the supercapacitor packaged with different electronic devices. The fibre-based solid asymmetrical supercapacitor **l** powered a digital watch, **m** integrated with a headphone, **n** attached to lab cloth to power electronics, and **o** connected to a motor fan Reprinted with the permission of Elsevier from Qu et al. [55], Gopi et al. [17]

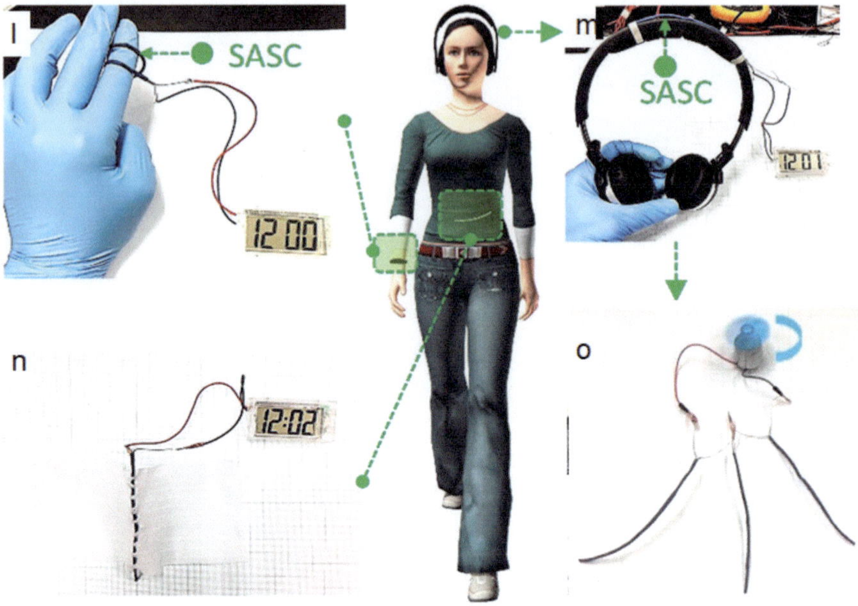

Fig. 6.13 (continued)

bending (Fig. 6.13n). The motor fan was also powered with the serially connected flexible device as shown in Fig. 6.13o. This design idea opens new possibilities for a fiber-shaped energy storage device to achieve high-energy output for a wide range of applications.

Table 6.3 shows the summary of the electrochemical performances of flexible and wearable hybrid supercapacitors, asymmetric supercapacitors, or battery-supercapacitor hybrid designs based on various positive and negative electrodes with their corresponding electrolytes.

With exceptional capabilities of energy storage such as fast charge-discharge rates, higher power density, reasonable energy density and excellent cycling performance, the flexible supercapacitor can be used for some practical applications, for example, piezoelectric supercapacitors, shape memory supercapacitors, photo-supercapacitors, electrochromic supercapacitors, microbial supercapacitors, self-healing supercapacitors, the integrated device of supercapacitor-sensor, and so on [19, 33, 39, 47, 48, 61, 73, 76, 86, 87, 91–94].

Table 6.3 Various flexible and wearable hybrid supercapacitors asymmetric supercapacitors or battery-supercapacitor hybrids and their electrochemical performances

Electrode materials	Current collector	Current density	Specific capacitance	Potential window	Energy density and power density	Capacitance retention	References
NC LDH NSs@ag//AC	Carbon cloth	1 mA cm^{-2}	230.2 mF cm^{-2}	1.6 V	78.8 μ Wh cm^{-2} 785 μ W cm^{-2}	88.1% 2000 cycles	Sekhar et al. [61]
Co_3O_4nanosheets// MnO nanosheets	Carbon cloth	2 A g^{-1}	166 F g^{-1}	1.7 V	59.6 Wh kg^{-1} 1529.8 W kg^{-1}	84.6% 5000 cycles	Yu et al. [86, 87]
SiC NWs@$NiCo_2O_4$@ $Ni(OH)_2$ HNAs//SiC NWs@Fe_2O_3 NNAs	Carbon cloth	4 A g^{-1}	242 F g^{-1}	1.7 V	103 Wh kg^{-1} 3.5 kW kg^{-1}	86.6% 5000 cycles	Zhao et al. [94]
Co_9S_8@$NiCo_2O_4$//AC	Carbon cloth	1 A g^{-1}	250 F g^{-1}	1.75 V	86 Wh kg^{-1} 11,360 W kg^{-1}	84.2% 10,000 cycles	Liu et al. [33, 39]
Ni-Mo-S NS//Ni-Fe-S NS	Carbon cloth	2 mA cm^{-2}	103 mA h g^{-1}	1.6 V	82.13 Wh kg^{-1} 0.561 W kg^{-1}	95.86% 10,000 cycles	Balamurugan et al. [3]
Ni-co DHs/pen ink/nickel//pen ink/nickel	Carbon fiber	0.5 A g^{-1}	22.94 F g^{-1}	1.5 V	7.66 Wh kg^{-1} 393.94 W kg^{-1}	86% 5000 cycles	Gao et al. [15]
F-Co_2MnO_{4-x}/CF//Fe_2O_3/CF	Carbon fiber	1 A g^{-1}	180 F g^{-1}	1.6 V	64.4 Wh kg^{-1} 800 W kg^{-1}	89.9% 2000 cycles	Liu et al. [33, 39]
MnO_2 NWs//Fe_2O_3 NTs	Carbon fiber	2 mA cm^{-2}	1.5 F cm^{-3}	1.6 V	0.55 mWh cm^{-3}	84% 5000 cycles	Yang et al. [84]
Zinc-nickel-cobalt ternary oxides//vanadium nitride	Carbon nanotubes fibers	0.1 mA cm^{-2}	50 F cm^{-3}	1.6 V	17.78 mWh cm^{-3} 80 mW cm^{-3}	91% 3000 cycles	Guo et al. [19]
MnO_2@PEDOT:PSS@OCNTF// MoS_2@CNTF	Carbon nanotubes fibers	0.6 mA cm^{-2}	278.6 mF cm^{-2}	1.8 V	125.37 μWh cm^{-2} 540 μ W cm^{-2}	90.7% 5000 cycles	Zhang et al. [91–93]

(continued)

Table 6.3 (continued)

Electrode materials	Current collector	Current density	Specific capacitance	Potential window	Energy density and power density	Capacitance retention	References
$NiCo_2O_4$//Fe_3O_4	Nickel wire	1 A g^{-1}	81.6 F g^{-1}	1.7 V	32.6 Wh kg^{-1} 846 W kg^{-1}	94.8% 20,000 cycles	Wang et al. [73, 76]
Core-shell like NC LDH NFAs@NSs/Ni fabric //AC@CF	Polyester fabric/Ni	3 mA cm^{-2}	1147.23 mF cm^{-2}	1.6 V	0.392 mWh cm^{-2} 2.353 mW cm^{-2}	86.49% 2000 cycles	Nagaraju et al. [47, 48]

NC Nanocomposite, *NS* Nanosheet, *AC* Activated carbon, *LDH* Layered double hydroxide, *CF* Carbon fibers,

6.5 Conclusion

Nanohybrid materials offer great opportunities for the construction of flexible and wearable supercapacitors. Portable and wearable electronics, electronic skins, implantable medical devices, and smart clothes are just a few of the technological uses for flexible supercapacitors. This review discussed recent advancements in flexible substrates for the growth of diverse materials of anode and cathode, as well as unique flexible supercapacitor cell designs. Currently, several researchers are working on new electroactive materials development for flexible supercapacitors, as well as cell designs such as symmetric, asymmetric, and hybrid. Furthermore, recent study findings of flexibly supported novel electroactive materials with regulated structures hold tremendous potential for improved electrochemical performance in terms of higher energy, power density, and long-term cycling stability. The electrochemical performance of flexible supercapacitors is usually measured at gravimetric and volumetric scales, which encourages the development of innovative electrode designs such as flexible substrate-supported and free-standing electrodes. To achieve high-energy storage performance, it is critical to produce flexible, lightweight electrode materials that utilize the entire electrode materials, avoiding the unnecessary weight of substrates, binders, or additives, which are delicate and difficult to handle.

To achieve high-performance flexible supercapacitors, a variety of flexible substrates such as carbon cloth, metal foils and wires, textiles, carbon nanotube nanofibers, polyethylene terephthalate, and others have been investigated. Carbon nanotube fibers and carbon-based substrates have gotten a lot of attention for flexible supercapacitor applications because of their excellent electrical conductivity, large surface area, lightweight, and flexibility. The electrochemical stability of nanostructures on their ligaments, on the other hand, is not satisfactory, and carbon-based current collectors are quite expensive. As a result, metal deposition over textiles is used to produce conductive textile-based current collectors. Furthermore, significant efforts have been made to investigate new anode and cathode electroactive materials that can improve overall flexible supercapacitor performance. Numerous novel cathode and anode materials because of their good conductivity and high specific capacitance for example N, S, dual doped porous carbon polyhedral embedded carbon nanofibers (NSCPCNFs), porous Fe_2O_3 nanospheres@ACC, carbon foam-multiscale pores (CF-MSP), VNNDs/CNSs, TiN NWAs@VN NSs, $NiCo_2O_4$@$NiMoO_4$, MNCS, multi tripod NTAs, $FeCo_2S_4$–$NiCo_2S_4$, microsphere-like $CoFe_2Se_4$-$CoNiSe_2$ have been found to be promising electroactive materials for flexible supercapacitor application. More crucially, high-performance anode and cathode materials are combined to create a flexible battery-supercapacitor hybrid, hybrid supercapacitor, or asymmetric supercapacitor with increased energy density that may be used in a variety of applications. This is an important step in the direction of smart supercapacitors. Furthermore, we believe that future research on high-energy-density flexible supercapacitor applications should focus on the following concerns.

- The mass and volume of the entire device, including the package and non-active material substrates are used to calculate the energy density of flexible supercapacitors in practical applications. In comparison to electroactive material weight, the non-active material substrate is heavy in mass and large in volume. To obtain high energy and power densities, it is critical to reduce the weight and volume portion of electrochemically inactive components in flexible supercapacitors.
- The electroactive materials determine the performance of a flexible supercapacitor. To achieve excellent electrochemical performance, careful selection of electrochemically active materials is required.
- Better energy storage performance was accomplished through the integration of several anode or cathode materials, which included a higher energy density and a wider voltage window. As a result, more work should be put into elucidating new synthesis processes for composite electrode materials that meet industrial specifications.
- In a flexible supercapacitor, the electrolytes can be optimized. It is necessary to place a greater emphasis on improving device performance in terms of capacitance, energy density, stability, and operating cell voltage by using new redox-active electrolytes. Electrolytes with higher ion conductivity and larger operational voltage are worth developing. The development of electrolytes with better ion conductivity and a higher operational voltage is worthwhile.
- To obtain stable performance, the flexible supercapacitor must be sealed. In this regard, research should also concentrate on supercapacitor sealing.
- Flexible supercapacitors have good flexibility and mechanical strength must be thoroughly investigated. Many researchers examine the electrochemical properties of produced flexible supercapacitors at various bending angles, but it's unclear how many cycles flexible supercapacitors can withstand.
- For real-time applications, cycle stability is also a critical parameter. Most of the research only assessed cyclic stability over 3000–10,000 cycles. As a result, researchers should create a very stable flexible supercapacitor that can withstand more cycles.
- Another concern is the high expense of manufacturing the flexible supercapacitor for practical use. As a result, future research should concentrate on developing highly flexible and energetic supercapacitors using inexpensive raw materials.
- For real-time applications, the integration of flexible supercapacitors with different gadgets is receiving a lot of interest. New device designs that are flexible, bendable, foldable, and stretchable must be investigated as a result.
- Integration of flexible supercapacitors with other functional electronic devices such as lithium-ion batteries, solar cells, and sensors in a single component is a hot topic. Smart supercapacitors would play a critical role in the development of lightweight, flexible, and wearable capacitive electronics soon if recommended improvement efforts are implemented.

Acknowledgements Dr. R. Suresh Babu would like to acknowledge the financial support from the Coordenação de Aperfeiçoamento de Pessoal de Nível Superior (CAPES, Grant No. 88887.798322/

2022-00) and Fundação de Amparo à Pesquisa do Estado do Rio de Janeiro (FAPERJ) (E-26/202.465/2023 (293937)), Brazil. Dr. A.L.F. de Barros gratefully acknowledges the Brazil government funding agencies including CNPq (CNPq (Bolsa de Produtividade 30.4511/2022-6), FINEP (1455/24), and FAPERJ (E-26/201.801/2021, E-26-210.965/2021, E-26-210.801/2021, E-26-200.320/2023, E-26/202.601/2023, E-26/201.622/2023 and E-26/203.465/2023).

References

1. R.S. Babu, R. Vinodh, A.L.F. de Barros, L.M. Samyn, K. Prasanna, M.A. Maier, C.H.F. Alves, H.-J. Kim, Asymmetric supercapacitor based on carbon nanofibers as the anode and two-dimensional copper cobalt oxide nanosheets as the cathode. Chem. Eng. J. **366**, 390–403 (2019). https://doi.org/10.1016/j.cej.2019.02.108
2. R.S. Babu, A.L.F. de Barros, M.A. Maier, D.M. Sampaio, J. Balamurugan, J.H. Lee, Novel polyaniline/manganese hexacyanoferrate nanoparticles on carbon fiber as binder-free electrode for flexible supercapacitors. Compos. Part B Eng. **143**, 141–147 (2018). https://doi.org/10.1016/j.compositesb.2018.02.007
3. J. Balamurugan, C. Li, V. Aravindan, N.H. Kim, J.H. Lee, Hierarchical Ni-Mo-S and Ni-Fe-S nanosheets with ultrahigh energy density for flexible all solid-state supercapacitors. Adv. Funct. Mater. **28**, 1803287 (2018). https://doi.org/10.1002/adfm.201803287
4. F. Bonaccorso, L. Colombo, G. Yu, M. Stoller, V. Tozzini, A.C. Ferrari, R.S. Ruoff, V. Pellegrini, Graphene, related two-dimensional crystals, and hybrid systems for energy conversion and storage. Science **347**, 1246501 (2015). https://doi.org/10.1126/science.12465
5. D. Cai, D. Wang, B. Liu, Y. Wang, Y. Liu, L. Wang, H. Li, H. Huang, Q. Li, T. Wang, Comparison of the electrochemical performance of $NiMoO_4$ nanorods and hierarchical nanospheres for supercapacitor applications. ACS Appl. Mater. Interfaces **5**, 12905–12910 (2013). https://doi.org/10.1021/am403444v
6. Y. Cao, X. Wang, K. Wang, Z. Gu, Q. Fan, W. Gibbons, J.D. Hoefelmeyer, P.R. Kharel, M. Shrestha, Hierarchical porous activated carbon for supercapacitor derived from corn stalk core by potassium hydroxide activation. Electrochim. Acta **212**, 839–847 (2016). https://doi.org/10.1016/j.electacta.2016.07.069
7. P. Chen, H. Chen, J. Qiu, C. Zhou, Inkjet printing of single-walled carbon nanotube/RuO_2 nanowire supercapacitors on cloth fabrics and flexible substrates. Nano Res. **3**, 594–603 (2010a). https://doi.org/10.1007/s12274-010-0020-x
8. Y.C. Chen, Y.K. Hsu, Y.G. Lin, Y.K. Lin, Y.Y. Horng, L.C. Chen, K.H. Chen, Highly flexible supercapacitors with manganese oxide nanosheet/carbon cloth electrode. Electrochim. Acta **56**, 7124–7130 (2011). https://doi.org/10.1016/j.electacta.2011.05.090
9. P.C. Chen, G. Shen, Y. Shi, H. Chen, C. Zhou, Preparation and characterization of flexible asymmetric supercapacitors based on transition-metal-oxide nanowire/single-walled carbon nanotube hybrid thin-film electrodes. ACS Nano **4**, 4403–4411 (2010b). https://doi.org/10.1021/nn100856y
10. C. Choi, H.J. Sim, G.M. Spinks, X. Lepro, R.H. Baughman, S.J. Kim, Elastomeric and dynamic MnO_2/CNT core-shell structure coiled yarn supercapacitor. Adv. Energy Mater. **6**, 1502119 (2016). https://doi.org/10.1002/aenm.201502119
11. H. Du, Z. Wu, Y. Xu, S. Liu, H. Yang, Poly(3,4-ethylenedioxythiophene) based solid-state polymer supercapacitor with ionic liquid gel polymer electrolyte. Polymers **12**, 297 (2020). https://doi.org/10.3390/polym12020297
12. D.P. Dubal, N.R. Chodankar, D.H. Kim, P. Gomez-Romero, Towards flexible solid-state supercapacitors for smart and wearable electronics. Chem. Soc. Rev. **47**, 2065–2129 (2018). https://doi.org/10.1039/C7CS00505A

13. J.X. Feng, S.H. Ye, A.L. Wang, X.F. Lu, Y.X. Tong, G.R. Li, Flexible cellulose paper-based asymmetrical thin film supercapacitors with high-performance for electrochemical energy storage. Adv. Funct. Mater. **29**, 1905171 (2014). https://doi.org/10.1002/adfm.201905171
14. L. Gao, J. Song, J.U. Surjadi, K. Cao, Y. Han, D. Sun, X. Tao, Y. Lu, Graphene-bridged multifunctional flexible fiber supercapacitor with high energy density. ACS Appl. Mater. Interfaces **10**, 28597–28607 (2018). https://doi.org/10.1021/acsami.8b08680
15. L. Gao, J.U. Surjadi, K. Cao, H. Zhang, P. Li, S. Xu, C. Jiang, J. Song, D. Sun, Y. Lu, Flexible fiber-shaped supercapacitor based on nickel-cobalt double hydroxide and pen ink electrodes on metallized carbon fiber. ACS Appl. Mater. Interfaces **9**, 5409–5418 (2017). https://doi.org/10.1021/acsami.6b16101
16. J. Ge, G. Cheng, L. Chen, Transparent and flexible electrodes and supercapacitors using polyaniline/single-walled carbon nanotube composite thin films. Nanoscale **3**, 3084–3088 (2011). https://doi.org/10.1039/C1NR10424A
17. C.V.V.M. Gopi, R. Vinodh, S. Sambasivam, I.M. Obaidat, H.-J. Kim, Recent progress of advanced energy storage materials for flexible and wearable supercapacitor: From design and development to applications. J. Energy Storage **27**, 101035 (2020). https://doi.org/10.1016/j.est.2019.101035
18. C. Guan, J. Liu, Y. Wang, L. Mao, Z. Fan, Z. Shen, H. Zhang, J. Wang, Iron oxide-decorated carbon for supercapacitor anodes with ultrahigh energy density and outstanding cycling stability. ACS Nano **9**, 5198–5207 (2015). https://doi.org/10.1021/acsnano.5b00582
19. J. Guo, Q. Zhang, J. Sun, C. Li, J. Zhao, Z. Zhou, B. He, X. Wang, P. Man, Q. Li, J. Zhang, L. Xie, M. Li, Y. Yao, Direct growth of vanadium nitride nanosheets on carbon nanotube fibers as novel negative electrodes for high-energy-density wearable fiber-shaped asymmetric supercapacitors. J. Power Sources **382**, 122–127 (2018). https://doi.org/10.1016/j.jpowsour.2018.02.034
20. L. Hu, W. Chen, X. Xie, N. Liu, Y. Yang, H. Wu, Y. Yao, M. Pasta, H.N. Alsharee, Y. Cui, Symmetrical MnO_2-carbon nanotube-textile nanostructures for wearable pseudocapacitors with high mass loading. ACS Nano **5**, 8904–8913 (2011). https://doi.org/10.1021/nn203085j
21. Y. Huang, F. Cui, J. Bao, Y. Zhao, J. Lian, T. Liu, H. Li, $MnCo_2S_4$/$FeCo_2S_4$ "lollipop" arrays on a hollow N-doped carbon skeleton as flexible electrodes for hybrid supercapacitors. J. Mater. Chem. A **7**, 20778–20789 (2019). https://doi.org/10.1039/C9TA04457D
22. X. Huang, J. Wu, R. Guo, Y. Lin, P. Zhang, Aligned nickel-cobalt oxide nanosheet arrays for lithium ion battery applications. Int. J. Hydrog. Energy **39**, 21399–21404 (2014). https://doi.org/10.1016/j.ijhydene.2014.01.098
23. L. Huang, W. Zhang, J. Xiang, H. Xu, G. Li, Y. Huang, Hierarchical core-shell $NiCo_2O_4$@$NiMoO_4$ nanowires grown on carbon cloth as integrated electrode for high-performance supercapacitors. Sci. Rep. **6**, 31465 (2016). https://doi.org/10.1038/srep31465
24. K. Jost, D. Stenger, C.R. Perez, J.K. McDonough, K. Lian, Y. Gogotsi, G. Dion, Knitted and screen printed carbon-fiber supercapacitors for applications in wearable electronics. Energy Environ. Sci. **6**, 2698–2705 (2013). https://doi.org/10.1039/C3EE40515J
25. M. Kaempgen, C.K. Chan, J. Ma, Y. Cui, G. Gruner, Printable thin film supercapacitors using single-walled carbon nanotubes. Nano Lett. **9**, 1872–1876 (2009). https://doi.org/10.1021/nl8038579
26. Y. Ko, M. Kwon, W.K. Bae, B. Lee, S.W. Lee, J. Cho, Flexible supercapacitor electrodes based on real metal-like cellulose papers. Nat. Commun. **8**, 536 (2017). https://doi.org/10.1038/s41467-017-00550-3
27. Q. Li, Y. Chen, J. Zhang, W. Tian, L. Wang, Z. Ren, X. Ren, X. Li, B. Gao, Spatially confined synthesis of vanadium nitride nanodots intercalated carbon nanosheets with ultrahigh volumetric capacitance and long life for flexible supercapacitors. Nano Energy **51**, 128–136 (2018). https://doi.org/10.1016/j.nanoen.2018.06.053
28. J. Li, Y. Wang, W. Xu, Y. Wang, B. Zhang, S. Luo, X. Zhou, C. Zhang, X. Gu, C. Hu, Porous Fe_2O_3 nanospheres anchored on activated carbon cloth for high-performance symmetric supercapacitors. Nano Energy **57**, 379–387 (2019a). https://doi.org/10.1016/j.nanoen.2018.12.061

29. Q. Li, Q. Zhang, J. Sun, C. Liu, J. Guo, B. He, Z. Zhou, P. Man, C. Li, L. Xie, All hierarchical core-shell heterostructures as novel binder-free electrode materials for ultrahigh-energy-density wearable asymmetric supercapacitors. Adv. Sci. Lett. **6**, 1801379 (2019b). https://doi.org/10.1002/advs.201801379
30. Y. Li, G. Zhu, H. Huang, M. Xu, T. Lu, L. Pan, A N, S dual doping strategy via electrospinning to prepare hierarchically porous carbon polyhedra embedded carbon nanofibers for flexible supercapacitors. J. Mater. Chem. **7**, 9040–9050 (2019c). https://doi.org/10.1039/C8TA12246F
31. H. Liang, C. Xia, A.H. Emwas, D.H. Anjum, X. Miao, H.N. Alshareef, Phosphine plasma activation of α-Fe_2O_3 for high energy asymmetric supercapacitors. Nano Energy **49**, 155–162 (2018). https://doi.org/10.1016/j.nanoen.2018.04.032
32. H. Liang, C. Xia, Q. Jiang, A.N. Gandi, U. Schwingenschlögl, H.N. Alshareef, Low temperature synthesis of ternary metal phosphides using plasma for asymmetric supercapacitors. Nano Energy **35**, 331–340 (2017). https://doi.org/10.1016/j.nanoen.2017.04.007
33. Q. Liu, X. Hong, X. Zhang, W. Wang, W. Guo, X. Liu, M. Ye, Hierarchically structured Co9S8@$NiCo_2O_4$ nanobrushes for high-performance flexible asymmetric supercapacitors. Chem. Eng. J. **356**, 985–993 (2019a). https://doi.org/10.1016/j.cej.2018.09.095
34. S. Liu, L. Kang, J. Zhang, E. Jung, S. Lee, S.C. Jun, Structural engineering and surface modification of MOF-derived cobalt-based hybrid nanosheets for flexible solid-state supercapacitors. Energy Storage Mater. **32**, 167–177 (2020). https://doi.org/10.1016/j.ensm.2020.07.017
35. M.-C. Liu, L.-B. Kong, C. Lu, X.-J. Ma, X.-M. Li, Y.-C. Luo, L. Kang, Design and synthesis of $CoMoO_4$-$NiMoO_4 \cdot xH_2O$ bundles with improved electrochemical properties for supercapacitors. J. Mater. Chem. A **1**, 1380–1387 (2013). https://doi.org/10.1039/C2TA00163B
36. K. Liu, X. Pan, L. Chen, L. Huang, Y. Ni, J. Liu, S. Cao, H. Wang, Ultrasoft self-healing nanoparticle-hydrogel composites with conductive and magnetic properties. ACS Sustain. Chem. Eng. **6**, 6395–6403 (2018a). https://doi.org/10.1021/acssuschemeng.8b00193
37. C. Liu, X. Yan, F. Hu, G. Gao, G. Wu, X. Yang, Toward superior capacitive energy storage: Recent advances in pore engineering for dense electrodes. Adv. Mater. **30**, 1705713 (2018b). https://doi.org/10.1002/adma.201705713
38. S. Liu, Y. Yin, K.S. Hui, K.N. Hui, S.C. Lee, S.C. Jun, High-performance flexible quasi-solid-state supercapacitors realized by molybdenum dioxide@nitrogen-doped carbon and copper cobalt sulfide tubular nanostructures. Adv. Sci. **5**, 1800733 (2018c). https://doi.org/10.1002/advs.201800733
39. S. Liu, Y. Yin, D. Ni, K.S. Hui, M. Ma, S. Park, K.N. Hui, C.Y. Ouyang, S.C. Jun, New insight into the effect of fluorine doping and oxygen vacancies on electrochemical performance of Co_2MnO_4 for flexible quasi-solid-state asymmetric supercapacitors. Energy Storage Mater. **22**, 384–396 (2019b). https://doi.org/10.1016/j.ensm.2019.02.014
40. L. Liu, H. Zhao, Y. Lei, Review on nanoarchitectured current collectors for pseudocapacitors. Small Methods **3**, 1800341 (2018d). https://doi.org/10.1002/smtd.201800341
41. G. Ma, M. Dong, K. Sun, E. Feng, H. Peng, Z. Lei, A redox mediator doped gel polymer as an electrolyte and separator for a high performance solid state supercapacitor. J. Mater. Chem. A **3**, 4035–4041 (2015). https://doi.org/10.1039/C4TA06322H
42. L. Ma, Y. Zhao, X. Ji, J. Zeng, Q. Yang, Y. Guo, Z. Huang, X. Li, J. Yu, C. Zhi, A usage scenario independent "air chargeable" flexible zinc ion energy storage device. Adv. Energy Mater. **9**, 1900509 (2019). https://doi.org/10.1002/aenm.201900509
43. M.A. Maier, R.S. Babu, D.M. Sampaio, A.L.F. de Barros, Binder-free polyaniline interconnected metal hexacyanoferrates nanocomposites (metal = Ni, co) on carbon fibers for flexible supercapacitors. J. Mater. Sci. Mater. Electron. **28**, 17405–17413 (2017). https://doi.org/10.1007/s10854-017-7674-z
44. L. Manjakkal, A. Pullanchiyodan, N. Yogeswaran, E.S. Hosseini, R. Dahiya, A wearable supercapacitor based on conductive PEDOT:PSS-coated cloth and a sweat electrolyte Adv. Dent. Mater. **32**, 1907254 (2020). https://doi.org/10.1002/adma.201907254
45. C. Masarapu, L.P. Wang, X. Li, B. Wei, Tailoring electrode/electrolyte interfacial properties in flexible supercapacitors by applying pressure. Adv. Energy Mater. **2**, 546–552 (2012). https://doi.org/10.1002/aenm.201100529

46. R. Na, X. Wang, N. Lu, G. Huo, H. Lin, G. Wang, Novel egg white gel polymer electrolyte and a green solid-state supercapacitor derived from the egg and rice waste. Electrochim. Acta **274**, 316–325 (2018). https://doi.org/10.1016/j.electacta.2018.04.127
47. G. Nagaraju, S.C. Sekhar, L.K. Bharat, J.S. Yu, Wearable fabrics with self-branched bimetallic layered double hydroxide coaxial nanostructures for hybrid supercapacitors. ACS Nano **11**, 10860–10874 (2017a). https://doi.org/10.1021/acsnano.7b04368
48. G. Nagaraju, S.C. Sekhar, J.S. Yu, Utilizing waste cable wires for high-performance fiber-based hybrid supercapacitors: An effective approach to electronic-waste management. Adv. Energy Mater. **8**, 1702201 (2017b). https://doi.org/10.1002/aenm.201702201
49. W.T. Navaraj, C.G. Núñez, D. Shakthivel, V. Vinciguerra, F. Labeau, D.H. Gregory, R. Dahiya, Nanowire FET based neural element for robotic tactile sensing skin. Front. Neurosci. **20**, 1–20 (2017). https://doi.org/10.3389/fnins.2017.00501
50. A. Nazari, S. Farhad, Heat generation in lithium-ion batteries with different nominal capacities and chemistries. Appl. Therm. Eng. **125**, 1501–1517 (2017). https://doi.org/10.1016/j.applthermaleng.2017.07.126
51. M. Pasta, F. La Mantia, L. Hu, H.D. Deshazer, Y. Cui, Aqueous supercapacitors on conductive cotton. Nano Res. **3**, 452–458 (2010). https://doi.org/10.1007/s12274-010-0006-8
52. H. Peng, Y. Lv, G. Wei, J. Zhou, X. Gao, K. Sun, G. Ma, Z. Lei, A flexible and self-healing hydrogel electrolyte for smart supercapacitor. J. Power Sources **431**, 210–219 (2019). https://doi.org/10.1016/j.jpowsour.2019.05.058
53. S.J.R. Prabakar, R.S. Babu, M. Oh, M.S. Lah, S.C. Han, J. Jeong, M. Pyo, Dense CoO/graphene stacks via self-assembly for improved reversibility as high performance anode in lithium ion batteries. J. Power Sources **272**, 1037–1045 (2014). https://doi.org/10.1016/j.jpowsour.2014.09.044
54. Y. Qiu, G. Li, Y. Hou, Z. Pan, H. Li, W. Li, M. Liu, F. Ye, X. Yang, Y. Zhang, Vertically aligned carbon nanotubes on carbon nanofibers: A hierarchical three-dimensional carbon nanostructure for high-energy flexible supercapacitors. Chem. Mater. **27**, 1194–1200 (2015). https://doi.org/10.1021/cm503784x
55. G. Qu, P. Sun, G. Xiang, J. Yin, Q. Wei, C. Wang, X. Xu, Moss-like nickel-cobalt phosphide nanostructures for highly flexible all-solid-state hybrid supercapacitors with excellent electrochemical performances. Appl. Mater. Today **20**, 100713 (2020). https://doi.org/10.1016/j.apmt.2020.100713
56. M. Ramani, B.S. Haran, R.E. White, B.N. Popov, Synthesis and characterization of hydrous ruthenium oxide-carbon supercapacitors. J. Electrochem. Soc. **148**, A374 (2001). https://doi.org/10.1149/1.1357172
57. M. Ren, J. Di, W. Chen, Recent progress and application challenges of wearable supercapacitors. Batter Supercaps **4**, 1279–1290 (2021). https://doi.org/10.1002/batt.202000333
58. L.M. Samyn, R.S. Babu, M. Devendiran, A.L.F. de Barros, Electropolymerization of p-phenylenediamine films on carbon fiber fabrics electrode for flexible supercapacitors: Surface and electrochemical characterizations. Ionics **26**, 3041–3050 (2020). https://doi.org/10.1007/s11581-020-03562-0
59. L.M. Samyn, R.S. Babu, M. Devendiran, A.L.F. de Barros, One-step electropolymerization of methylene blue films on highly flexible carbon fiber electrode as supercapacitors. Micro Nano Syst. Lett. **9**, 3 (2021). https://doi.org/10.1186/s40486-021-00130-7
60. S. Sankar, H. Lee, H. Jung, A. Kim, A.T.A. Ahmed, A.I. Inamdar, H. Kim, S. Lee, H. Im, D.Y. Kim, Ultrathin graphene nanosheets derived from rice husks for sustainable supercapacitor electrodes. New J. Chem. **41**, 13792–13797 (2017). https://doi.org/10.1039/C7NJ03136J
61. S.C. Sekhar, G. Nagaraju, J.S. Yu, Conductive silver nanowires-fenced carbon cloth fibers-supported layered double hydroxide nanosheets as a flexible and binder-free electrode for high-performance asymmetric supercapacitors. Nano Energy **36**, 58–67 (2017). https://doi.org/10.1016/j.nanoen.2017.04.019
62. S. Shi, C. Xu, C. Yang, J. Li, H. Du, B. Li, F. Kang, Flexible supercapacitors. Particuology **11**, 371–377 (2013). https://doi.org/10.1016/j.partic.2012.12.004

63. B. Song, L. Li, Z. Lin, Z.K. Wu, K.S. Moon, C.P. Wong, Water-dispersible graphene/polyaniline composites for flexible micro-supercapacitors with high energy densities. Nano Energy **16**, 470–478 (2015). https://doi.org/10.1016/j.nanoen.2015.06.020
64. G.E. Spina, F. Poli, A. Brilloni, D. Marchese, F. Soavi, Natural polymers for green supercapacitors. Energies **13**, 3115 (2020). https://doi.org/10.3390/en13123115
65. J.Y. Tang, P. Yuan, C.L. Cai, Y.B. Fu, X.H. Ma, Combining nature-inspired, graphene-wrapped flexible electrodes with nanocomposite polymer electrolyte for asymmetric capacitive energy storage. Adv. Energy Mater. **6**, 1600813 (2016). https://doi.org/10.1002/aenm.201600813
66. F. Tao, L. Qin, Z. Wang, Q. Pan, Self-healable and cold-resistant supercapacitor based on a multifunctional hydrogel electrolyte. ACS Appl. Mater. Interfaces **9**, 15541–15548 (2017). https://doi.org/10.1021/acsami.7b03223
67. A. Varzi, A. Balducci, S. Passerini, Natural cellulose: A green alternative binder for high voltage electrochemical double layer capacitors containing ionic liquid-based electrolytes. J. Electrochem. Soc. **161**, A368 (2014). https://doi.org/10.1149/2.063403jes
68. A. Varzi, S. Passerini, Enabling high areal capacitance in electrochemical double layer capacitors by means of the environmentally friendly starch binder. J. Power Sources **300**, 216–222 (2015). https://doi.org/10.1016/j.jpowsour.2015.09.065
69. G.K. Veerasubramani, A. Chandrasekhar, M.S.P. Sudhakaran, Y.S. Mok, S.J. Kim, Liquid electrolyte mediated flexible pouch-type hybrid supercapacitor based on binderless core-shell nanostructures assembled with honeycomb-like porous carbon. J. Mater. Chem. A **5**, 11100–11113 (2017). https://doi.org/10.1039/C7TA01308F
70. M. Vinayagam, R.S. Babu, A. Sivasamy, A.L.F. de Barros, Biomass-derived porous activated carbon from Syzygium cumini fruit shells and Chrysopogon zizanioides roots for high-energy density symmetric supercapacitors. Biomass Bioenergy **143**, 105838 (2020). https://doi.org/10.1016/j.biombioe.2020.105838
71. M. Vinayagam, R.S. Babu, A. Sivasamy, A.L.F. de Barros, Biomass-derived porous activated carbon nanofibers from Sapindus trifoliatus nut shells for high-performance symmetric supercapacitor applications. Carbon Lett. **31**, 1133-1143 (2021). https://doi.org/10.1007/s42823-021-00235-4
72. R. Vinodh, R.S. Babu, C. Deviprasath, C.V.V.M. Gopi, R. Atchudan, L.M. Samyn, A.L.F. de Barros, M.S. Yi, H.J. Kim, Influence of annealing temperature in nitrogen doped porous carbon balls derived from hypercross-linked polymer of anthracene for supercapacitor applications. J. Energy Storage **28**, 101196 (2020). https://doi.org/10.1016/j.est.2020.101196
73. X. Wang, Y. Ding, F. Chen, H. Lu, N. Zhang, M. Ma, Hierarchical porous N-doped graphene monoliths for flexible solid-state supercapacitors with excellent cycle stability. ACS Appl. Energy Mater. **1**, 5024–5032 (2018a). https://doi.org/10.1021/acsaem.8b01011
74. L.B. Wang, H.L. Yang, X.X. Liu, R. Zeng, M. Li, Y.H. Huang, X.L. Hu, Constructing hierarchical tectorum-like α-Fe_2O_3/PPy nanoarrays on carbon cloth for solid-state asymmetric supercapacitors. Angew. Chem. Int. Ed. **56**, 1105–1110 (2017). https://doi.org/10.1002/anie.201609527
75. Y. Wang, X. Yang, A.G. Pandolfo, J. Ding, D. Li, High-rate and high-volumetric capacitance of compact graphene-polyaniline hydrogel electrodes. Adv. Energy Mater. **6**, 1600185 (2016). https://doi.org/10.1002/aenm.201600185
76. N. Wang, P. Zhao, B. Sun, M. Yao, W. Hu, S. Komarneni, Electrodeposition preparation of $NiCo_2O_4$ mesoporous film on ultrafine nickel wire for flexible asymmetric supercapacitors. Chem. Eng. J. **345**, 31–38 (2018b). https://doi.org/10.1016/j.cej.2018.03.147
77. L. Wang, H.J. Zhu, W. Zhai, F. Cai, X.M. Liu, H. Yang, Study of a novel gel electrolyte based on poly-(methoxy/hexadecyl-poly(ethylene glycol) methacrylate) co-polymer plasticized with 1-butyl-3-methylimidazolium tetrafluoroborate. RSC Adv. **4**, 36357–36365 (2014). https://doi.org/10.1039/C4RA03661A
78. L.Y. Wen, M. Zhou, C.L. Wang, Y. Mi, Y. Lei, Nanoengineering energy conversion and storage devices via atomic layer deposition. Adv. Energy Mater. **6**, 1600468 (2016). https://doi.org/10.1002/aenm.201600468

79. T. Xia, X. Zhang, J. Zhao, Q. Li, C. Ao, R. Hu, Z. Zheng, W. Zhang, C. Lu, Y. Deng, Flexible and conductive carbonized cotton fabrics coupled with a nanostructured $Ni(OH)_2$ coating for high performance aqueous symmetric supercapacitors. ACS Sustain. Chem. Eng. **7**, 5231–5239 (2019). https://doi.org/10.1021/acssuschemeng.8b06150
80. W. Xiaona, Z. Qichong, S. Juan, Z. Zhenyu, L. Qiulong, H. Bing, Z. Jingxin, L. Weibang, W. Ching-ping, Y. Yao, Facile synthesis of hierarchical porous manganese nickel cobalt sulfide nanotube arrays with enhanced electrochemical performance for ultrahigh energy density fiber-shaped asymmetric supercapacitors. J. Mater. Chem. A **6**, 8030–8038 (2018). https://doi.org/10.1039/C8TA01440J
81. W. Xu, Z. Jiang, Q. Yang, W. Huo, M.S. Javed, Y. Li, L. Huang, X. Gu, C. Hu, Approaching the lithium-manganese oxides' energy storage limit with Li_2MnO_3 nanorods for high-performance supercapacitor. Nano Energy **43**, 168 (2018). https://doi.org/10.1016/j.nanoen.2017.10.046
82. X. Xu, J. Tang, H. Qian, S. Hou, Y. Bando, M.S.A. Hossain, L. Pan, Y. Yamauchi, Three-dimensional networked metal-organic frameworks with conductive polypyrrole tubes for flexible supercapacitors. ACS Appl. Mater. Interfaces **9**, 38737–38744 (2017)
83. Y. Yan, P. Gu, S. Zheng, M. Zheng, H. Pang, H. Xue, Facile synthesis of an accordion-like Ni-MOF superstructure for high-performance flexible supercapacitors. J. Mater. Chem. A **4**, 19078–19085 (2016). https://doi.org/10.1039/C6TA08331E
84. P. Yang, Y. Ding, Z. Lin, Z. Chen, Y. Li, P. Qiang, M. Ebrahimi, W. Mai, C.P. Wong, Z.L. Wang, Low-cost high-performance solid-state asymmetric supercapacitors based on MnO_2 nanowires and Fe_2O_3 nanotubes. Nano Lett. **14**, 731–736 (2014). https://doi.org/10.1021/nl404008e
85. Y. Yang, Q. Huang, L. Niu, D. Wang, C. Yan, Y. She, Z. Zheng, Waterproof, ultrahigh areal-capacitance, wearable supercapacitor fabrics. Adv. Mater. **29**, 1606679 (2017). https://doi.org/10.1002/adma.201606679
86. M. Yu, D. Lin, H. Feng, Y. Zeng, Y. Tong, X. Lu, Boosting the energy density of carbon-based aqueous supercapacitors by optimizing the surface charge. Angew Chem. Int. Ed **56**, 5454–5459 (2017a). https://doi.org/10.1002/anie.201701737
87. N. Yu, K. Guo, W. Zhang, X. Wang, M.-Q. Zhu, Flexible high-energy asymmetric supercapacitors based on MnO@C composite nanosheet electrode. J. Mater. Chem. A **5**, 804–813 (2017b). https://doi.org/10.1039/C6TA08330G
88. G. Yu, L. Hu, M. Vosgueritchian, H. Wang, X. Xie, J.R. McDonough, Solution-processed graphene/MnO_2 nanostructured textiles for high-performance electrochemical capacitors. Nano Lett. **11**, 2905–2911 (2011). https://doi.org/10.1021/nl2013828
89. N. Yu, H. Yin, W. Zhang, Y. Liu, Z.Y. Tang, M.Q. Zhu, High-performance fiber-shaped all-solid-state asymmetric supercapacitors based on ultrathin MnO_2 nanosheet/carbon fiber cathodes for wearable electronics. Adv. Energy Mater. **6**, 1501458 (2016). https://doi.org/10.1002/aenm.201501458
90. W. Zeng, L. Wang, H. Shi, G. Zhang, K. Zhang, H. Zhang, F. Gong, T. Wang, H. Duan, Metal-organic-framework-derived $ZnO@C@NiCo_2O_4$ core–shell structures as an advanced electrode for high-performance supercapacitors. J. Mater. Chem. A **4**, 8233–8241 (2016). https://doi.org/10.1039/C6TA01510G
91. Q. Zhang, J. Sun, Z. Pan, J. Zhang, J. Zhao, X. Wang, C. Zhang, Y. Yao, W. Lu, Q. Li, Y. Zhang, Z. Zhang, Stretchable fiber-shaped asymmetric supercapacitors with ultrahigh energy density. Nano Energy **39**, 219–228 (2017a). https://doi.org/10.1016/j.nanoen.2017.06.052
92. F. Zhang, T. Liu, M. Li, M. Yu, Y. Luo, Y. Tong, Y. Li, Multiscale pore network boosts capacitance of carbon electrodes for ultrafast charging. Nano Lett. **17**, 3097–3104 (2017b). https://doi.org/10.1021/acs.nanolett.7b00533
93. Y. Zhang, T. Mao, H. Wu, L. Cheng, L. Zheng, Carbon nanotubes grown on flax fabric as hierarchical all-carbon flexible electrodes for supercapacitors. Adv. Mater. Interfaces **4**, 1601123 (2017c). https://doi.org/10.1002/admi.201601123
94. J. Zhao, Z. Li, X. Yuan, Z. Yang, M. Zhang, A. Meng, Q. Li, A high-energy density asymmetric supercapacitor based on Fe_2O_3 nanoneedle arrays and $NiCo_2O_4/Ni(OH)_2$ hybrid nanosheet arrays grown on SiC nanowire networks as free-standing advanced electrodes. Adv. Energy Mater. **8**, 1702787 (2018). https://doi.org/10.1002/aenm.201702787

95. Y. Zhou, N. Li, L. Sun, X. Yu, C. Liu, L. Yang, S. Zhang, Z. Wang, Multi-layer-stacked Co9S8 micro/nanostructure directly anchoring on carbon cloth as a flexible electrode in supercapacitors. Nanoscale **11**, 7457–7464 (2019). https://doi.org/10.1039/C9NR00828D
96. C. Zhu, T. Liu, F. Qian, T. Han, E. Duoss, J. Kuntz, C. Spadaccini, M. Worsley, Supercapacitors based on three-dimensional hierarchical graphene aerogels with periodic macropores. Nano Lett. **16**, 3448–3456 (2016). https://doi.org/10.1021/acs.nanolett.5b04965
97. J. Zhu, S. Tang, J. Wu, X. Shi, B. Zhu, X. Meng, Wearable high-performance supercapacitors based on silver-sputtered textiles with $FeCo_2S_4$-$NiCo_2S_4$ composite nanotube-built multitripod architectures as advanced flexible electrodes. Adv. Energy Mater. **7**, 1601234 (2017). https://doi.org/10.1002/aenm.201601234

Chapter 7
Polymer-Based Hybrid Nanostructures for Supercapacitors

Rajangam Vinodh, K. V. G. Raghavendra, Rajendran Suresh Babu, Moonsuk Yi, and Hee-Je Kim

Abstract Electrochemical energy storage devices play a crucial role because of the intermittent nature of renewable sources to the end user from the point of generation for the construction of sustainable energy storage systems. Also, the main crucial factor for the long-life cycle is to optimize the energy and power densities of electrochemical energy storage devices, in addition, to meeting the demand for next-generation electronic applications. To augment the overall performance of electrochemical energy storage devices, greater efforts were made devoted toward the search for new materials. Although more research is going on in this field, their performance has not met up with the commercialization level. Further, the development of new electrode materials and the understanding of the mechanism of charge storage required are essential, the focus of the current book chapter reveals the recent progress in polymer-based hybrid nanostructures for supercapacitor applications. The various types of conducting polymers, polyaniline, polypyrrole, polythiophene and poly(3,4-ethylenedioxythiophene), as well as their composites with carbon-based materials and metal oxides and sulfides involved in the performance of the supercapacitor are well documented in detail. We have also demonstrated the recent advancements in hyper cross-linked polymer research and their electrochemical performance studies. Finally, we have summarized the possible challenges and future developments in this research field.

Keywords Supercapacitors · Specific capacitance · Conducting polymers · Metal–organic framework · Metal oxides · Dichalcogenides · Hypercross-linked polymers

---

R. Vinodh (✉) · M. Yi
Department of Electronics Engineering, Pusan National University, Geumjeong-Gu, Busan, Republic of Korea
e-mail: vinoth6482@gmail.com

K. V. G. Raghavendra · H.-J. Kim
School of Electrical and Computer Engineering, Pusan National University, Geumjeong-Gu, Busan, Republic of Korea

R. S. Babu
Laboratory of Experimental and Applied Physics, Centro de Educacao Technologica Celso da Fonseca, Rio de Jeneiro, Brazil

© The Author(s), under exclusive license to Springer Nature Switzerland AG 2026
K. Raju et al. (eds.), *Hybrid Nanostructures for Advanced Supercapacitors*, Modern Aspects of Electrochemistry 63, https://doi.org/10.1007/978-3-032-04593-5_7

7.1 Introduction

Environmental issues and the serious energy crisis have increasingly attracted strong demand for energy storage devices which can achieve sustainable, clean, efficient energy storage, and release [28, 83]. Supercapacitors are promising power source devices because of their long durability, high-power density, and low-cost maintenance with the safe operation [115, 145]. Recently, significant efforts have been made to explore new materials or micro and nanostructures for supercapacitor applications that hope for further improvement in their performances [20, 107, 150, 152]. The process consists of various principles which can be followed. At first, the electrode material should be made more active electrochemically which contains a pseudocapacitive component, which can lead to remarkably high specific capacitance because of its faradaic charge transfer reactions of or through the reversible redox [133, 159, 160]. Secondly, a porous structure of electrode material should develop with high conductivity that can promote rapid charge transport and ion diffusion [102, 161, 163]. Hence, these factors are crucial to achieving a high capacitive performance.

7.1.1 Ragone Plot

Figure 1a represents the clear picture of energy and power densities of various energy storage devices known as the Ragone plot [158]. From the plot in Fig. 1a, supercapacitor technology can bridge the gap between batteries and capacitors in terms of both power and energy densities. Furthermore, supercapacitors have longer cycle life than batteries because the chemical phase changes in the electrodes of a supercapacitor are much less than that in a battery during continuous charging-discharging [170]. These key attributes make supercapacitors more attractive and versatile as high-powered energy storage. The US Department of Energy has spotlighted batteries and supercapacitors as major future energy storage technologies [51].

7.1.2 Types of Supercapacitors

Supercapacitors are classified into three main types based on their difference in the storage mechanism of electric energy, electrochemical double-layer capacitor (EDLC), pseudocapacitor, and battery hybrid supercapacitor [96] depicted in Fig. 1b. The EDLC is like a traditional capacitor, the mechanism of EDLC arises from charge accumulation in the electric double-layer formed at the electrode–electrolyte interface, but the specific capacitance of an EDLC is several orders of magnitude higher than a traditional capacitor.

Carbon-based materials are typical electrode materials for EDLCs, including activated carbon, carbon nanotube, mesoporous carbon and graphene owing to

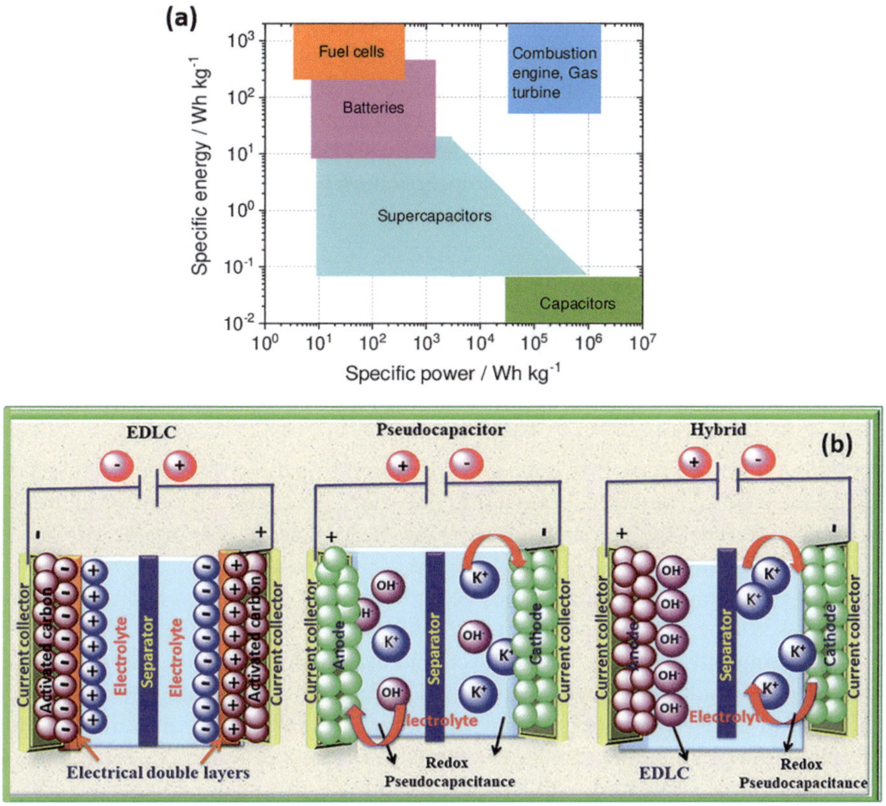

Fig. 7.1 **a** Ragone plot of various electrochemical energy conversion and storage devices. Reprinted with permission of the American Chemical Society from Winter and Brodd [158]. **b** Type of supercapacitors: EDLC, pseudocapacitor and hybrid. Reprinted with permission of Elsevier from Pal et al. [96]

their high surface area and low matrix resistivity [21, 106]. On the other hand, pseudocapacitors are based on fast and reversible redox reactions, also known as Faradaic charge transfer reactions, on the surface of electroactive species in materials such as *transition metal oxides* such as $CoWO_4$, $CoMn_2O_4$, $NiMoO_4$, and V_2O_5 [9, 52], transition metal sulfides such as CuS and $MnCo_2S_4$ [108, 109], carbon materials possessing oxygen- and nitrogen-containing surface functional groups, and *conducting polymers* e.g., *polyaniline*, *polythiophene*, *polypyrrole*, and derivatives [123].

The electrochemical behaviors of some metal oxides, metal hydroxides and metal sulfides are a semi-infinite diffusion-limited reaction process, which works according to a linear relationship of the current peak (i_p) with the square root of the scan rate ($v^{1/2}$) [29]. The third type is battery hybrid supercapacitor, and the working principle is governed by a combination of the EDLC (non-faradaic) and pseudocapacitor (faradaic) charge storage principles. In other words, battery hybrid supercapacitors can try to utilize the relative advantages and mitigate the relative disadvantages of EDLCs and pseudocapacitors to realize better electrochemical performance or charge storage performance. Exploiting both non-faradaic and faradaic process charge storage principles, hybrid supercapacitors have achieved higher energy and power densities than EDLCs and pseudocapacitors without compromising cycling stability and rate capability.

7.2 Conducting Polymers

Conducting polymers stand out as promising materials for metal oxide-based devices due to their unique physical and chemical properties in a wide range of conductivity, flexibility, processability, low cost, and lightweight qualities for manufacturers at a large scale [90]. Therefore, conducting polymers are chemically and structurally customized to meet the demands of any kind of application. The major difference between conducting polymers-based supercapacitors and EDLCs is presented in Fig. 7.2 [1]. Owing to its aided structure and fast ionic adsorption and desorption, carbon-based supercapacitors (i.e., EDLCs) will exhibit high-power capabilities [98].

Conducting polymers are mainly used and expected to improve energy storage devices. As a result, the redox reaction will undergo to store charge from the bulk material and increase the energy stored. However, the slow diffusion of ions within the bulk of the conducting polymer electrode leads to lower power i.e., lower rate of charge–discharge compared with EDLCs. Despite these conducting polymers have more advantages which can counteract its drawbacks and are expected to bridge the gap between batteries and EDLCs and exhibit faster electron transfer kinetics than other materials like metal oxides [98]. As previously discussed with properties, in addition, conducting polymers have the potential to be engineered with specific nanostructures that optimize the electrodes to uptake at maximum capacity. Hence, such optimization can be achieved by manipulating the surface morphology and surface area per weight of the conducting polymer electrodes, as the capacitance for any given capacitor is proportional to the surface area of the electrode (Eq. 7.1).

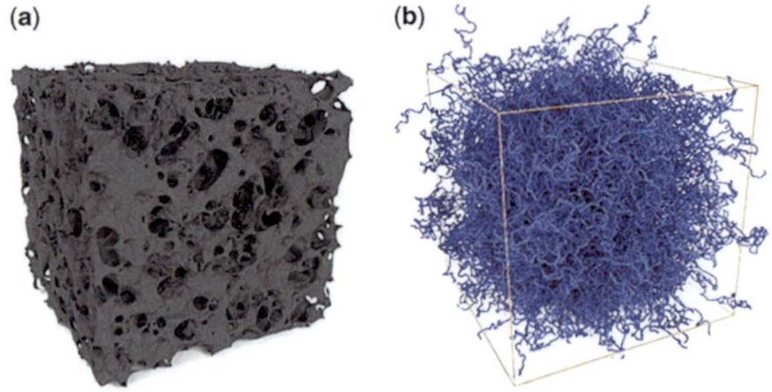

Fig. 7.2 Charging mechanism of **a** porous carbon-based electrochemical double-layer capacitors (EDLCs) and **b** conducting polymers-based pseudocapacitors. Reprinted with permission of Wiley from Abdelhamid and Snook [1]

$$C = \varepsilon_0 \varepsilon_\gamma \frac{A}{d} \tag{7.1}$$

where ε_0 is the permittivity of a vacuum (8.85 pF m^{-1}), ε_γ is the relative permittivity of the dielectric material, *A* is the surface area of the electrode, and *d* is the thickness of the dielectric material, Hence, it is very essential that the electrode material with a high surface area [42] can also be achieved using conducting polymers. The properties of various conducting polymers are represented in Table 7.1.

7.2.1 Polyaniline

Polyaniline (PANI) can be polymerized with aniline monomer through different methods, which has more advantages, easy synthesis, simple acid/base doping—de-doping chemistry and environmental stability [80, 81] to become one of the most active promising materials that suited for a pseudocapacitor electrode. The surface morphologies of PANI nanostructures play a crucial role in influencing the electrochemical properties to finding a convenient high-efficiency synthesis method for the preparation of PANI nanostructures is essential. The fact is that PANI can be relatively easy to synthesize with chemical or electrochemical polymerization. During chemical oxidative polymerization, PANI preferentially forms as nanofibers in an aqueous solution [58].

More varieties with different kinds of polymerization methods were obtained to form PANI nanofibers [27, 59, 60]. Interfacial polymerization is one of the common methods which are relatively very easy and less expensive. Sivakumar et al., have synthesized the PANI nanofibers via the interfacial polymerization method and evaluated their electrochemical properties in an aqueous redox supercapacitor constituted with a two-electrode cell setup [121]. This supercapacitor has exhibited a high initial specific capacitance of 554 F g^{-1} at a current density of 1.0 A g^{-1}, meanwhile, their cyclic stability was very poor due to a rapid decrease in initial capacitance value.

Li et al. [80, 81] have examined the experimental and theoretical specific capacitance of PANI in a sulfuric acid medium. The theoretical maximum specific capacitance of PANI is 2000 F g^{-1}, on the other hand, the experimental values were evaluated in different ways which are much lower than that of the theoretical values as the PANI has made only a small amount of contribution to the percentage of effective PANI and capacitance ability which depend upon both diffusion of counter-anions and the conductivity of PANI. Hu and Chu have prepared the PANI electrodes over a graphite substrate through a cyclic voltammetry technique containing 0.5 M HCl with 0.2 M aniline by sweeping the potential range between − 200 and 800 mV [56]. The PANI electrode prepared shows a maximum specific capacitance of 168 mF cm^{-2} at 200 μA cm^{-2}. Dhawale et al. [38] have reported a stable nanostructured polyaniline electrode through the chemical bath deposition method for the application of supercapacitors. The fabricated electrode attains a maximum specific capacitance of 503 F g^{-1} with a specific energy and a specific power of 96.23 Wh

Table 7.1 Properties of various conducting polymers

Conducting polymer	Conductivity (mS cm^{-1})	Structure (one repeating unit)	Advantages	Disadvantages
Polyaniline (PANI)	10^2 to 10^8	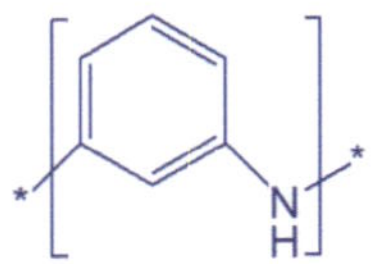	High theoretical specific capacitance, ease of synthesis; high conductivity and stability, flexibility	Poor cycling stability and specific capacitance much depends on the preparation protocol, which only applies to proton-based electrolytes
Polypyrrole (PPy)	10^3 to 5×10^4	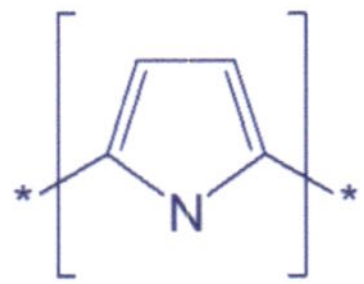	Simple fabrication process, good specific capacitance, long cyclic life, appropriate to neutral electrolytes, flexibility	The tedious process of doping, de-doping, poor specific capacitance per unit mass, only applicable to cathode materials
Polythiophene (PTh)	10^{-1} to 10^{-4}	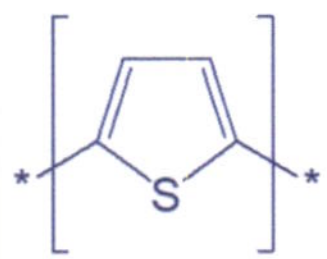	Flexibility, facile synthesis methodology, promising durability, and eco-friendliness	Relatively low specific capacitance and deprived conductivity
Poly(3,4-ethylenedioxy thiophene) (PEDOT)	3×10^5 to 5×10^5	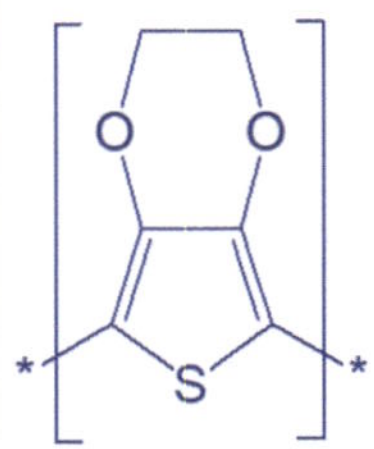	High electrical conductivity and flexibility, wider operating voltage window, potential window, moderate theoretical capacitance of 210 F g^{-1}, and excellent stability compared to other conducting polymers	The synthesis process is too long, most supercapacitor electrodes are based on vapor phase or electrochemical deposition methods. The synthesis process is too long and relatively high cost, higher temperatures are required in the synthesis process

kg^{-1} and 8.88 kW kg^{-1}, respectively. The authors have also stated furthermore, that the PANI electrodes show an outstanding electrochemical performance with a very good stable life cycle.

Guan et al., have fabricated PANI nanofibers via interfacial polymerization method in the presence of p-phenylenediamine (PPD) [53]. PANI nanofibers were prepared in the presence of p-phenylenediamine at a constant current discharge current density of 0.18 A g^{-1} and the specific capacitance value (C_s) of 548 F g^{-1}, a specific power value of 127 W kg^{-1} and a specific energy value of 36 Wh kg^{-1} have been obtained. Furthermore, the reduction of PANI-PPD after 1000 cycles has a specific capacitance of only 25%.

Rajendra Prasad and Munichandraiah have prepared PANI deposited on stainless-steel substrates potentiodynamically and characterized to yield a gel polymer electrolyte. Hence through the charge–discharge studies of solid-state PANI capacitors, a specific capacitance of 250 F g^{-1} has been obtained with a specific power of 7.5 kW kg^{-1} of PANI [110].

In summary, a lot of pure PANI-applied supercapacitors have been studied, especially their electrochemical performance and cyclic stability were still not meet the practical application requirements. A short cycle life existed due to the poor cycling stability of the supercapacitor leading to a rapid decrease in the specific capacitance. Therefore, the researchers have tried their best to prepare various PANI-based composites via mixing PANI with various carbon materials and/or metal oxides mainly to improve the supercapacitor performance, especially for the electrochemical properties [89, 162].

7.2.2 Polypyrrole

Polypyrrole is a well-known conducting polymer which consists of five-membered heterocyclic rings with good chemical stability, high electrical conductivity, and low oxidative potential with no toxicity. It has a theoretical specific capacitance of about 620 F g^{-1} with good cyclic stability [61]. In general, polypyrrole can be synthesized via electrochemical polymerization, interfacial polymerization, chemical oxidative polymerization, radiolytic polymerization, enzyme-catalyzed polymerization, and layer-by-layer self-assembly. Among these, chemical oxidative polymerization is the best method, owing to their suitability and economic feasibility to scale up synthesis commercially [4].

Although the development of a variety of polypyrrole with hierarchical nanostructures such as nanoparticles, nanorods, nanoribbons, nanowires, or nanotubes has already been reported. Especially with a hierarchical structure, the nanostructured polypyrrole with specific morphology possesses several advantages in fabricating nanodevices because of its advantages like fast charge–discharge kinetics, low cost, and fast doping-undoing processes [92]. In comparison with bulk polypyrrole, the polypyrrole with hierarchical nanostructures was expected to display improved performance in the view of technological applications owing to its unique properties that have been arising from the nanometer-scaled size: large surface area, high conductivity, and lightweight [125]. Therefore, the hierarchical nanostructured polypyrrole is recognized to be one of the most promising electrode materials for supercapacitors due to its distinctive characteristics of conducting pathways, high electrical conductivity in the doped state, surface interactions, and nanoscale dimensions with good thermal and chemical stability.

Dubal et al. [41] examined that the polypyrrole nanobricks were electrodeposited on a stainless-steel substrate with an aqueous bath that contains 0.05 M pyrrole + 0.1 M KNO_3. The electrochemical performance of the polypyrrole electrode was evaluated using cyclic voltammetry and galvanostatic charge–discharge techniques. The potential ranges from − 0.4 to + 0.6 V in 0.5 M H_2SO_4 solution possessing a high specific capacitance of 476 F g^{-1}. Furthermore, the polypyrrole electrode has also exhibited a high discharge–charge efficiency of 89%. Ingram et al. [63] studied that the polypyrrole was deposited onto a 2 mm diameter platinum microelectrode of thick films, typically 15–20 μm in depth potentiostatically. The electrochemical performance has also been evaluated using cyclic voltammetry, and the specific capacitance achieved was 255 F cm^{-2}. Fan and Maier have prepared highly active polypyrrole electrodes by electrodeposition for redox supercapacitors over a Ti foil by cyclic voltammetry at 200 mV s^{-1} in an oxalic acid solution medium [43]. The PPy-Ti electrode exhibited a very high specific capacitance of about 480 ± 50 F g^{-1} with excellent cycle stability.

Han et al. [55] investigated the conducting polypyrrole with a nanoscale hierarchical structure was synthesized chemically in an aqueous solution in the presence of cetyltrimethylammonium bromide (CTAB) and 1,5-naphthalene disulfonic acid. The fabricated polypyrrole electrode shows good electrochemical reversibility and exhibited a better specific capacitance of 158 F g^{-1}.

In summary, polypyrrole exhibits a better cyclic stability when compared to PANI, However, still polypyrrole needs to reach the value of theoretical specific capacitance. Various other modifications like surface functionality, the formation of composites, and core-shell nanoparticles were carried out to improve the electrochemical performance of polypyrrole.

7.2.3 Polythiophene

Among the conducting polymers, polythiophene having good inherent electrochemical properties like high electrical conductivity and stability, plays a critical role in hybrid capacitors [88]. Importantly, polythiophene thin films and bulk polythiophene as electrodes in hybrid capacitors showing different physical and chemical properties. For instance, Fu et al., reported the electrodeposited polythiophene thin film demonstrated the specific capacitance of 110 F g^{-1} whereas Patil et al. [100] prepared polythiophene electrode via successive ionic layer adsorption and reaction (SILAR) technique exhibited the maximum specific capacitance of 252 F g^{-1}. There are many synthesis routes such as chemical, electrochemical, photoelectrochemical, ultrasonic-assisted electrochemical, and template for the synthesis of polythiophene.

Laforgue et al. [76] synthesized polythiophene and polyparafluorophenylthiophene (PFPT) by chemical method for use as supercapacitor electrode material, The prepared electrode showed a specific capacity of 40 mAh g^{-1} for PTh and 7 mAh g^{-1} for PFPT, respectively. Furthermore, the specific capacitance of 260 F g^{-1} was obtained with polythiophene and 110 F g^{-1} with PFPT.

Patil et al. [101] prepared polythiophene thin films by chemical bath deposition method in an ambient atmosphere through oxidative polymerization of thiophene using ammonium peroxodisulfate as an oxidant. The fabricated polythiophene thin film exhibited the highest specific capacitance with 0.1 M $LiClO_4$/propylene carbonate electrolyte showing about 300 F g^{-1} at 5 mV s^{-1}. Furthermore, the asymmetric supercapacitor was assembled with polythiophene film and graphite showed supercapacitive behavior beneficial in energy storage applications. Straightforward derivatives of thiophene, such as poly-3-methyl thiophene and polyethylene dioxythiophene [2, 17, 73, 114, 122, 124, 166] were also could be used as prospective electrode material for power storage applications.

In summary, there is a considerable effort to produce thiophene monomers complex offshoots. Modifying thiophene monomers helps in enhancing the chemical and mechanical strength. The inadvertent consequence immensely increases the cost and the greater molecular weight per monomer unit reduces the gravimetric capacitance.

7.2.4 Poly(3,4-Ethylenedioxythiophene) (PEDOT)

The conducting polymer, PEDOT has gained more interest as electrode material due to its high electrical conductivity (> 500 S cm^{-1}), mechanical flexibility, large electroactive potential window, lightweight, high charge storage capabilities, good processability, and sustainability as well as reasonable theoretical capacitance (210 F g^{-1}) [10, 31, 45, 77, 123, 172]. However, the problem associated with PEDOT is, that most of the electrode materials have been prepared by electrochemical deposition or by vaporization [11, 32, 97, 168] which is costly, high working temperature and needs longer reaction time during synthesis. Furthermore, the yield produced from these synthesis processes has a thickness of around a few micrometers to several hundred nanometers [75, 78] producing low capacitances is not suitable for up-scaled industrial production. To obtain high energy storage capacity with power capability, the high surface is thin film PEDOT electrodes is important.

Recently, Wang et al. [156] reported highly flexible and conductive nanostructured PEDOT paper by solution-based reactions and exhibit high surface area (137 m^2 g^{-1}) as well as low sheet resistance (1.4 Ω $square^{-1}$). The fabricated symmetric PEDOT paper supercapacitors showed a high specific capacitance of 90 F g^{-1} and it exhibited outstanding cyclic stability of 93% capacitance retention over 15,000 cycles. Liu et al., prepared PEDOT microspheres with good processing capability, and intact morphology, without the use of surfactant and hazardous chemicals [84, 85]. The prepared PEDOT-MSs exhibited a superior specific capacitance of 91.2 F g^{-1}, which is 1.51 times greater than that of randomly structured PEDOT. Furthermore, they concluded that this novel synthesis method as water-based conducting polymers could be fabricated as printable electronic devices.

Wang et al., prepared PEDOT-Polypyrrole composite electrode by electropolymerization of 3,4-ethylene dioxythiophene (EDOT) on the surface of polypyrrole. The PEDOT-PPY composite film possesses a very good porosity with 230 F g^{-1} of specific capacitance in 1 M $LiClO_4$ and even produces a good capacitance of 290 F g^{-1} in 1 M KCl aqueous solutions [148]. Also, the composite reached a power density of 13 kW kg^{-1} with outstanding cyclic stability.

Very recently, Karri and Srinivasan reported PEDOT: PSS (poly (3,4-ethylenedioxythiophene: polystyrene sulfonate) based electrode using benzoyl peroxide as an oxidant for supercapacitor applications [70]. The fabricated composite electrode showed a rectangular shape in a cyclic voltammogram with a specific capacitance of 150 F g^{-1}. Julio et al., described evaporative vapor phase polymerization (EVPP) of nano fibrillar PEDOT for supercapacitors [31]. They even mentioned the deposition of nanofibers was effortlessly developable into a supercapacitor without the use of any organic additives or binders. Furthermore, the fabricated EVPP-PEDOT exhibited a gravimetric capacitance of 175 F g^{-1} and maintained its initial capacitance of 92% over 10,000 cycles.

Figure 3a represents the pictorial representation of the supercapacitor module fabricated with a PEDOT electrode. Figure 3b describes the cyclic voltammetry curve of the EVPP-PEDOT electrode at 2 mV s^{-1}. The cyclic voltammetric curve showed a good rectangular shape and possessed a high current density when compared with PEDOT: PSS electrode. Figure 3c shows the cyclic voltammetric curve of EVPP-PEDOT revealing the notable specific capacitance of 70 F g^{-1} when compared to PEDOT: PSS having only 24 F g^{-1}. Figure 3d shows charge–discharge curves having a current density of 0.2 mA cm^{-2} for EVPP-PEDOT supercapacitor and Fig. 3e indicates the specific capacitance values of the tested supercapacitor at different charge–discharge rates. The short ionic diffusion pathway enhances the rate capability of the EVPP-PEDOT during the charge–discharge cycle (Fig. 3f). Figure 3g shows the Ragone plot. When compared to the conventional PEDOT: PSS electrode, the energy and power densities are much higher in the nanostructured electrode. Furthermore, the EVPP-PEDOT electrode showed outstanding cyclic stability even after 10,000 cycles, retaining 92% of its initial capacitance (Fig. 3h). The image in Fig. 3i revealed that the practical application of EVPP-PEDOT supercapacitor in light emitting diodes (LED) for 5 min. Figure 3j and k represent the surface morphology of the prepared composite electrode on the carbon paper with different magnifications and it showed nanofibrillar structures.

However, pure conducting polymers cannot be utilized as electrode active materials even with special properties. The other active materials like metal oxides, metal–organic framework, metal sulfides and carbon-based materials have been studied

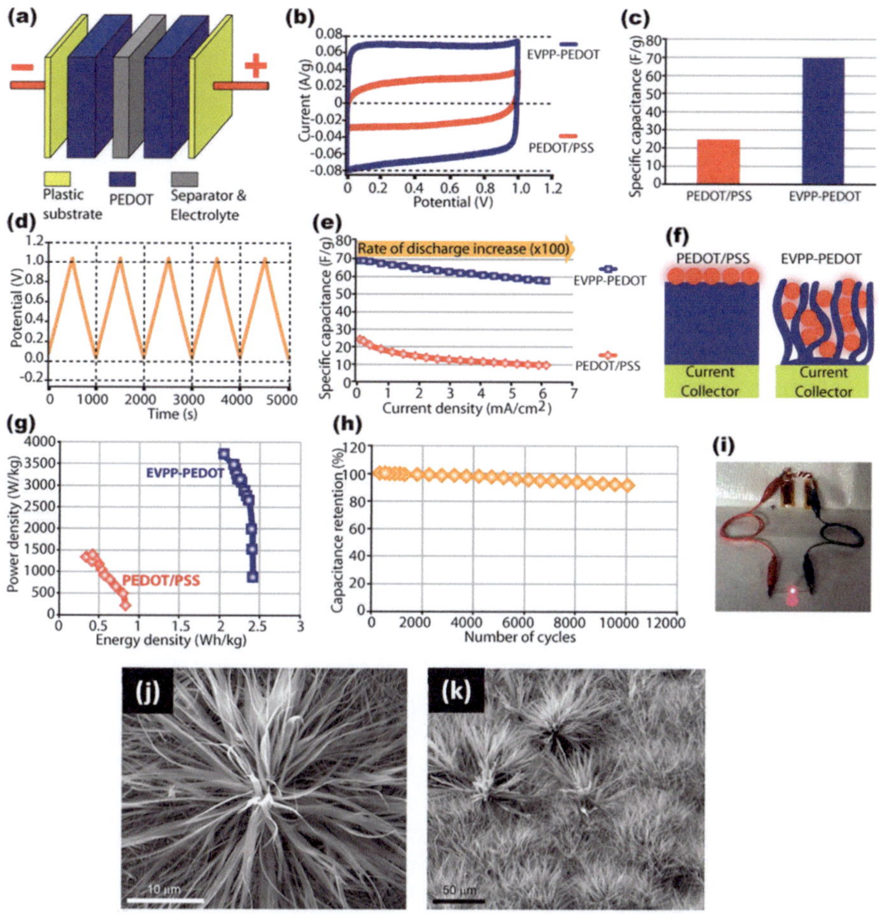

Fig. 7.3 **a** Supercapacitor setup; **b** cyclic voltammetry of EVPP-PEDOT and PEDOT-PSS at 2 mV s^{-1}; **c** specific capacitance values obtained; **d** charge–discharge profile of EVPP-PEDOT at 0.2 mA cm^{-2}; **e** specific capacitance values at different charge–discharge rates; **f** sketch of ion transport mechanism in EVPP-PEDOT; **g** Ragone plot; **h** cycling stability of EVPP-PEDOT supercapacitor at 1.8 mA cm^{-2}; **i** two supercapacitors coupled in series and charged up at a constant potential of 2 V to light up light emitting diodes (LED) for 5 min; **j** and **k** depicts scanning electron microscopic images of EVPP-PEDOT with low and high magnification. Reprinted with permission of American Chemical Society from D'Arcy et al. [31]

and experimented with by the researchers to develop ternary and binary composites to improve the electrochemical performance and stability of the conducting polymers-based supercapacitors.

7.3 Conducting Polymer Composites

Pure conducting polymer composites as electrode material possess poor cyclic stability even though they have numerous exclusive properties. Meanwhile, the conducting polymer composites have been studied to improve the cyclic stability and electrochemical performance conducting polymer composites generally branched out as two types of conducting polymer composites, which include pseudocapacitive and electrochemical double-layer capacitance materials. A suitable mixture of conducting polymer composites can efficiently enhance the efficiency of the specific capacitance by redox tuning, as well as improving the cyclic durability, by moderating their surface morphology and intrinsic elastic nature of the polymer. Furthermore, conducting polymer composites could also allow the possibility of enhancement of other physical properties, which greatly hinge on their equivalents and mutual interaction between them.

Carbon-based composites typically improve the surface area of an electrode, enhancing the contribution of EDLC [82, 94, 120], whereas composites with metal oxides introduce an added charge storage mechanism to the system [44]. Furthermore, as conferred in the following sections, composites can also enhance the material exploitation efficiency, conductivity or cycling stability of an electrode. Experimental specific capacitance values of various conducting polymers and conducting polymer composites are depicted in Fig. 7.4. Figure 7.4 clearly illustrates the conducting polymer composites enhanced specific capacitance when compared to pure conducting polymers. Furthermore, metal oxides or hydroxides composite with conducting polymer composites exhibit high specific capacitances when compared to carbon-based materials. The use of these conducting polymer composite electrode materials opens a new perspective on the field of energy storage. Furthermore, we discuss metal oxide-based, carbon-based, cross-linked polymers and metal–organic framework-based conducting polymer electrode materials.

7.3.1 Conducting Polymer/Carbon Composites

Carbon-based electrodes have been fascinating for many years before. However, their poor capacitive performance restricts their usage in the field of supercapacitors. Meanwhile, carbon materials have numerous excellent characteristics, such as large surface area, better electrical conductivity, exceptional mechanical properties, and good durability etc. Numerous studies have confirmed that conducting polymer/carbon composites showed excellent electrochemical behavior when compared with virgin conducting polymers.

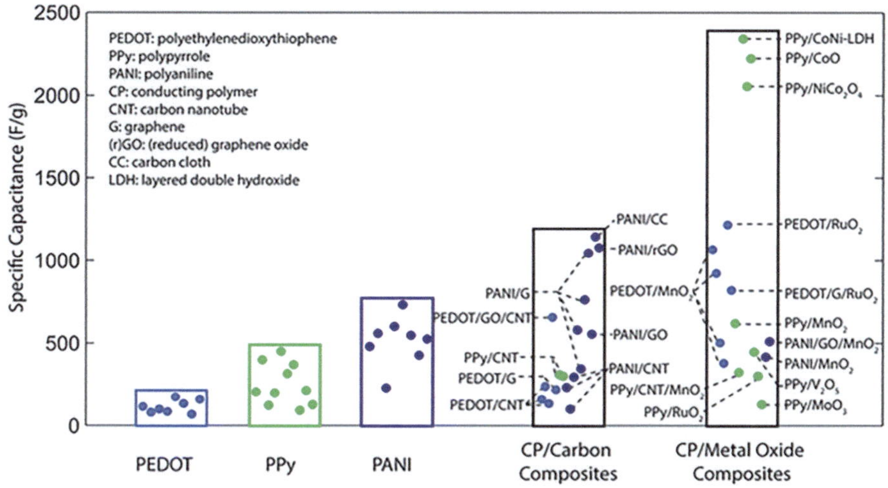

Fig. 7.4 Comparative chart of the specific capacitance of various conducting polymers and their composites used as electrode materials for supercapacitor applications. Reprinted with permission of the Royal Society of Chemistry from Fong et al. [44]

7.3.2 *PANI-Carbon-Based Binary and Ternary Composites*

Several PANI-carbon nanocomposite electrode materials have been explored for supercapacitor applications, including PANI-graphene [30, 150, 152], PANI-carbon nanotubes [62, 95], PANI-carbon nanofibers [24], PANI-graphene oxide or reduced graphene oxide [127, 129], PANI-carbon spheres [119] and PANI-carbon particles [71]. For example, Zhang et al., synthesized PANI-graphene composite via an in situ polymerization method with graphene oxide in the acid. The resulting composite electrode revealed a high capacitive value of 480 F g^{-1} at 0.1 A g^{-1} [176]. Furthermore, they declared the high specific capacitance attributed to the doping of bulky PANI with graphene-graphene oxide exhibited good cyclic stability. Gomez et al. [50] formulated PANI-graphene nanocomposite material by chemical precipitation method. They customized the properties of capacitance by adjusting the elements having 300–500 F g^{-1} at 0.1 A g^{-1}.

Wang et al. [146] fabricated fibrillar PANI doped graphene oxide sheet showed high conductivity (10 S cm^{-1}) and improved specific capacitance (531 F g^{-1} at 0.2 A g^{-1}), greater than bare PANI (216 F g^{-1}), revealed that the incorporation of graphene oxide sheets has a synergistic impact on PANI. Sun et al., synthesized in situ polymerization of a self-standing 3D PANI-reduced graphene oxide foam. The prepared PANI-reduced graphene oxide nanocomposite exhibited a maximum specific capacitance of 701 F g^{-1} at 1 A g^{-1} current density and further, it had a retention percentage of 92% of its initial specific capacitance over 1000 cycles [127, 129]. Mi et al. [91] developed PANI-MWCNTs composite via a facile and microwave-assisted polymerization. It showed a specific capacitance (322 F g^{-1})

and a high energy density (22 Wh kg^{-1}), which was about 12 times greater than that of MWCNTs. Zhang et al. [176] reported PANI-graphene oxide or reduced graphene oxide composite and it showed lower specific capacitance than that of fibrous PANI (420 F g^{-1}) because of the presence of insulating graphene oxide sheets in the composite that contributes tiny to the total specific capacitance of the composites.

Bharatiya et al. [18] synthesized graphene oxide wrapped PANI composite shown in (Fig. 5a) and they enhanced the d-spacing of the graphene oxide sheets from 12.76 Å for PANI-graphene oxide monolayer to 13.46 Å for PANI-graphene oxide multilayer by the wrapping of mono and multilayer of graphene oxide sheets on PANI nanorods respectively. From Fig. 5b, it can be noticed that PANI-graphene oxide multilayer composites exhibited by wrapping of graphene oxide on these nanorods by varying nanorods aspect ratio. The black space seen in the images illustrates the existence of multilayer graphene oxide sheets on the monolayer of graphene oxide sheet. Figure 5c showed the cyclic voltammetric performance of PANI-graphene oxide multilayer nanocomposites with varying scan rates from 1 to 20 mV s^{-1}. The cathodic peak with slightly higher potential and anodic peak with lower potential was observed which assures the excellent stability of the nanocomposite due to the intense interface of PANI with graphene oxide. PANI-graphene oxide multilayer nanocomposite showed a specific capacitance of 276 F g^{-1} by cyclic voltammetry while it was 246 F g^{-1} from charge–discharge profile at a current density of 0.1 A g^{-1} (Fig. 5d). Agyemang et al. [5] fabricated carbon nanofiber (CNF)-CNTs-PANI composite as an assuring electrode material for supercapacitor applications. The prepared tri-composite demonstrated higher specific capacitance than any of the composites, CNF-PANI (794 F g^{-1}), CNF-CNT (320 F g^{-1}) and carbon nanotubes (278 F g^{-1}). Kumar et al. [74] synthesized PANI-CNT-reduced graphene oxide ternary nanocomposite with a specific capacitance of 312 F g^{-1}. Chang et al. [23] studied the novel PANI-MWCNTs composites with biomimetic hierarchical structure from fresh plant leaves as a pattern template via a nano-casting method with higher specific capacitance than that of PANI (420 F g^{-1}).

7.3.3 Polypyrrole-Carbon-Based Binary and Ternary Composites

Xu et al. [161, 163] reported a bendable, twistable, and conductive polypyrrole-carbon nanotube composite fiber as a core–shell nanostructure for high-performance supercapacitor electrodes. The polypyrrole-carbon nanotube composite exhibited a high specific capacitance of 350 F g^{-1} and good cyclic stability. Furthermore, they revealed that the superior properties come from highly conductivity core–shell nanostructured polypyrrole-carbon nanotube composite delivering the charge transfer easily among carbon nanotube and polypyrrole.

Bhardwaj et al. [19] discussed the fabrication of supercapacitor electrode material consisting of surfactant doped polypyrrole carbon materials united with graphene,

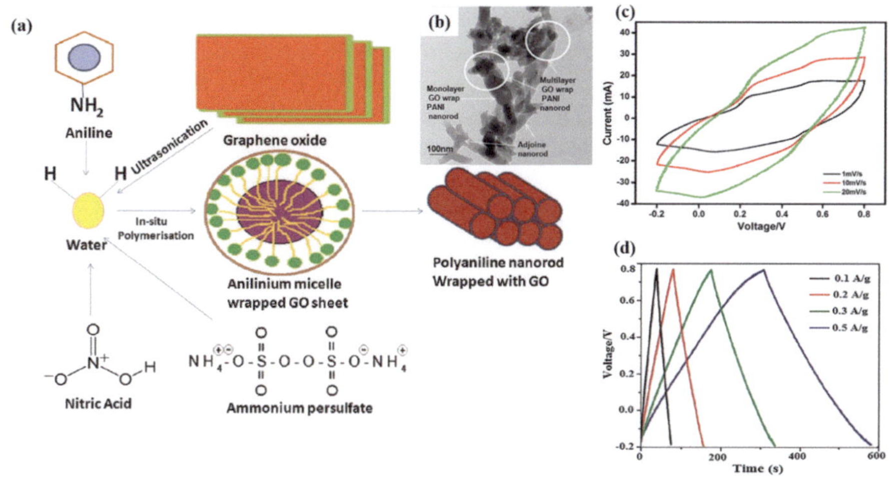

Fig. 7.5 **a** Formation of graphene oxide wrapped PANI nanorod, **b** transmission electron microscopic image of PANI-GO_{10} nanohybrid; **c** cyclic voltammetric curve of PANI-GO_{10} nanohybrids at varying scan rate (1, 10 and 20 mV s^{-1}); **d** charge–discharge plot of PANI-GO_{10} nanohybrids at different current density (0.1, 0.2, 0.3 and 0.5 A g^{-1}). Reprinted with permission of Wiley from Bharadiya et al. [18]

amine-functionalized MWCNTs and hybrid carbon assembly (SDBSDPPy-HC). The composite, SDBSDPPy-hybrid carbon holds a specific capacitance of 1086 F g^{-1} at 5 mV s^{-1} in a 6 M KOH electrolyte. Furthermore, it loses only 10% of its initial capacitance after 2500 cycles. Jyothibasu et al. [68] studied paper-like composite film from carbon nanotubes (f-CNTs) developed from the porous fibrous matrix of cellulose followed by in situ chemical polymerization of polypyrrole. The fabricated PPy-f-CNT-cellulose composite film displayed an excellent areal capacitance of 2147 mF cm^{-2} at a current density of 1 mA cm^{-2}. The authors revealed that the extraordinary specific capacitance of the composites attributed to the following important parameters exceptional microstructure with high electrical conductivity, super hydrophilic and a porous construction helps in charge–discharge of charge carriers and for the facile diffusion of electrolyte ions. Furthermore, the flexibility, eco-friendly nature, and biodegradability of the PPy/f-CNT-cellulose composite films could be an eco-friendly electrode material for flexible supercapacitor electrodes.

Sung and Lin proposed the synthesis of polypyrrole on carbon fiber using the electropolymerization method for flexible supercapacitor electrodes [130]. The polypyrrole-carbon fiber electrode showed a maximum specific capacitance of 308.2 F g^{-1}. Furthermore, they assembled a flexible supercapacitor with polypyrrole electrodes, and a gel electrolyte of polyvinyl alcohol-H_3PO_4 exhibited a capacitance of 30 F g^{-1} and an energy density of 5.87 Wh kg^{-1} at the power density of 60 W kg^{-1}. Furthermore, the flexible device delivered excellent cyclic stability with 70% retention and 98% coulombic efficiency over 3000 charge–discharge cycles. To enhance the specific capacitive properties of the electrode material, Zhang et al. [173] fabricated a polypyrrole-carbon nanotube electrode with a three-dimensional ordered macroporous (3-DOM) architecture having about 427 F g^{-1} at 5 mV s^{-1}. Furthermore, they concluded that 3-DOM polypyrrole-carbon nanotube films with mesopores structure had greater ion exchange when compared to nanoporous and planer films.

An et al. [8] synthesized a polypyrrole-carbon aerogel composite with varying polypyrrole contents by chemical oxidative polymerization via ultrasonic irradiation. The composite attained a maximum capacitance of 433 F g^{-1} when compared to the carbon aerogel which has 174 F g^{-1}. In the polypyrrole-carbon aerogel electrode, there is preliminary capacitance loss due to the unstable polypyrrole and the capacitance stabilizes after 500 cycles. Very recently, Jyothibasu et al. [67] reported a facile and eco-friendly method was employed to prepare a freestanding electrode material for supercapacitor applications. In brief, the first polypyrrole nanoparticles were synthesized by in situ chemical polymerization with curcumin as a bioavailable template. Second, PPy/f-CNTs are prepared by blending the mixture of polypyrrole and functionalized carbon nanotubes. The prepared freestanding PPy/f-CNT composite delivered a maximum areal capacitance of 4585 mF cm^{-2} and an equivalent volumetric capacitance of 176.35 F cm^{-3} at 2 mA cm^{-2}. Furthermore, they assembled symmetrically all-solid-state supercapacitors, and they exhibited high areal energy and power density of 129.24 μWh cm^{-2} and 12.5 mW cm^{-2}, respectively. Also, it possesses 79.03% capacitance retention after 10,000 charge–discharge cycles. Zhu et al. [181] controlled the cyclic stability and rate capability by synthesizing PPy-graphene-based composite electrodes and controlled the amount of graphene as the composite and turning the cauliflower structured morphology of polypyrrole into the homogeneous nanosheet and obtained extraordinary energy storage performance. This performance hike is due to the structural integrity of the nanosheets, and fast diffusion intercalation of the electrode was provided by the high conductivity electrode with unhindered pathways.

Wu et al. [159, 160] synthesized a 3D core–shell graphene oxide-polypyrrole nanomaterial in which polypyrrole nanospheres were coated on graphene oxide sheets via an in situ surface-initiated polymerization route. The composite exhibited an outstanding capacitance of 370 F g^{-1} at 0.5 A g^{-1} and superior cycle stability of 91.2% capacitive retention over 4000 cycles. Warren et al. [157] also probed the hybrid electrode materials from the PPy-CNT composite, and it showed that the electrochemical performance of the PPy-CNT retained around 92% of its initial capacitance over 3000 cycles.

7.3.4 *Polythiophene-Carbon-Based Binary and Ternary Composites*

Recently, many researchers have turned their attention to composite electrodes made by mixing polythiophene and carbon materials to expect the composite to have enhanced supercapacitor performance. Micro and nanostructures of carbon composites have a significant impact on the properties of polythiophene-carbon nanocomposites as carbon materials with different micro and nanostructures are often used to form polythiophene-based composites. Balakrishnan et al. [16] reported carbonaceous polythiophene nanofibers (C-PTh) for asymmetric supercapacitor applications. A brief synthesis methodology is first, highly porous polythiophene nanofibers were prepared by surfactant-assisted dilute polymerization method using ferric chloride as the oxidant. Second, the prepared polythiophene nanofibers were calcined at 1400 °C under an inert atmosphere for 2 h to attain C-PTh nanofibers. Furthermore, the asymmetric supercapacitor was assembled using C-PTh nanofibers as the positive electrode and polythiophene nanofibers as the negative electrode material in a 6 M KOH aqueous electrolyte. The fabricated asymmetric supercapacitor device unveiled a maximum operating voltage of 1.2 V and reached a higher capacitance of 252 F g^{-1} at 5 mV s^{-1}. The energy density and power density were measured at 54.6 Wh kg^{-1} and 1.7 kW kg^{-1}, respectively, and the fabricated asymmetric supercapacitor provided excellent cyclic stability by retaining 93% of its initial capacitance over 2000 cycles. Zhang et al. [174] synthesized polythiophene via the electrochemical polymerization method onto the MWCNT-modified carbon (PTh/MWCNT) by galvanostatic technique in oil-in-ionic liquid microemulsion. The fabricated PTh/MWCNT electrode displayed a maximum specific capacitance of 216 F g^{-1} at 1 A g^{-1} in 1 M sodium sulfate solution. Additionally, the composite exhibited excellent cyclic stability with a very low degrading rate of capacitance after 500 cycles.

Zhou et al. [179] studied polythiophene grafted on SWCNTs via oligo (ethylene oxide) linkages (PD2ET-g-SWCNTs) as an esterification reaction for supercapacitor applications (Fig. 6a). The PD2ET-g-SWCNTs electrode has a high specific capacitance of 399 F g^{-1} at 1 A g^{-1} by galvanostatic charge–discharge measurement. The authors set the symmetric supercapacitor module by using PD2ET-g-SWCNTs electrode with PVA-KOH as a gel electrolyte and separator. As illustrated in Fig. 6b, a device displayed a clear rectangular voltammogram curve between the potential of − 0.5 and + 0.3 V at a higher scan rate (100 mV s^{-1}). The cyclic voltammetry for the tandem device reached a maximum operating potential of 1.6 V, whereas the current output for the parallel device doubles as a single supercapacitor, which clearly explains why the individual device retain their supercapacitive performance in the series and parallel configuration. Figure 6c shows the galvanostatic charge–discharge curves at a current density of 0.625 A g^{-1}, the tandem device reaching an output voltage of 1.6 V with equivalent charge and discharge time, while the parallel devices have almost twice the charge and discharge time with an operating voltage (0.8 V). Furthermore, the assembled solid-state capacitor (PD2ET-g-SWCNTs/PD2ET-g-SWCNTs) has higher energy densities (22.5 Wh kg^{-1}) and power densities (500 W kg^{-1}) than other polythiophene-based composites (Fig. 6d). They concluded that the improved electrochemical performance of the fabricated solid-state capacitor device may be due to the influence of the oligo (ethylene oxide) linkage in polythiophene-grafted SWCNTs, which facilitates ion diffusion and allows a faster charge process. After 5 months, they found the stability of the fabricated device by the galvanostatic charge–discharge method; the device lost only 12.5% of its initial specific capacitance (Fig. 6e).

Thakur et al. [133, 134] discussed the synthesis of polythiophene-carbon nanotube composite by an easy in situ polymerization method. The binary composite of polythiophene-carbon nanotubes displayed a specific capacitance of 125 F g^{-1}. Moreover, the authors concluded that the carbon nanotubes not only enhance the conductivity of the hybrid but also maintain the mechanical stability of the material so that the specific capacitance is up to 125 F g^{-1} for the composite. Azimi et al. [12] synthesized polythiophene-graphene oxide and polythiophene-reduced graphene oxide by emulsion polymerization. The polythiophene, polythiophene-graphene oxide and polythiophene-reduced graphene oxide electrodes showed specific capacitance of 3.5, 16.39 and 28.68 F g^{-1}, respectively, at a scan rate of 5 mV s^{-1}. Furthermore, the energy density of the fabricated composite electrodes like polythiophene-graphene oxide and polythiophene-reduced graphene oxide showed an excellent performance than that of the bare polythiophene electrode at 5 mV s^{-1}.

Alvi et al. [7] synthesized nanocomposites of PEDOT-graphene and polythiophene-graphene electrode material for supercapacitor applications. The specific capacitance of the prepared composites was measured charge–discharge profile in a 2 M HCl medium, and it was determined between 160 and 400 F g^{-1}. Fu et al. [46] prepared a PTh-MWCNT composite by electropolymerization method in an ionic liquid medium. The fabricated hybrid electrode showed better capacitive performance of 110 F g^{-1} at a 50 mV s^{-1} than MWCNT and bare polythiophene. Moreover, the device holds 90% of its initial capacitance over 1000 cycles. Zhang et al. (2016) also fabricated PTh-MWCNT via an electrochemical polymerization technique. The composite displayed a capacitance of 216 F g^{-1} at 1 A g^{-1}. Wang et al. [155] reported the electrochemical deposition of polythiophene with a series of composite electrode materials on a monolithic coral-like porous carbon surface. The polythiophene-monolithic coral-like porous carbon surface hybrid exhibited a high capacitance of 720 F g^{-1} at 0.5 A g^{-1} and held 79.4% of its initial value at 1.0 A g^{-1} over 1000 cycles.

In summary, polythiophene-carbon-based composites have better electrochemical performance and cyclic stability compared to pristine polythiophene, and these properties are affected by several parameters, including the method of synthesis, the effective deposition of polythiophene on carbon materials, the type of carbon materials and micro and nanostructures.

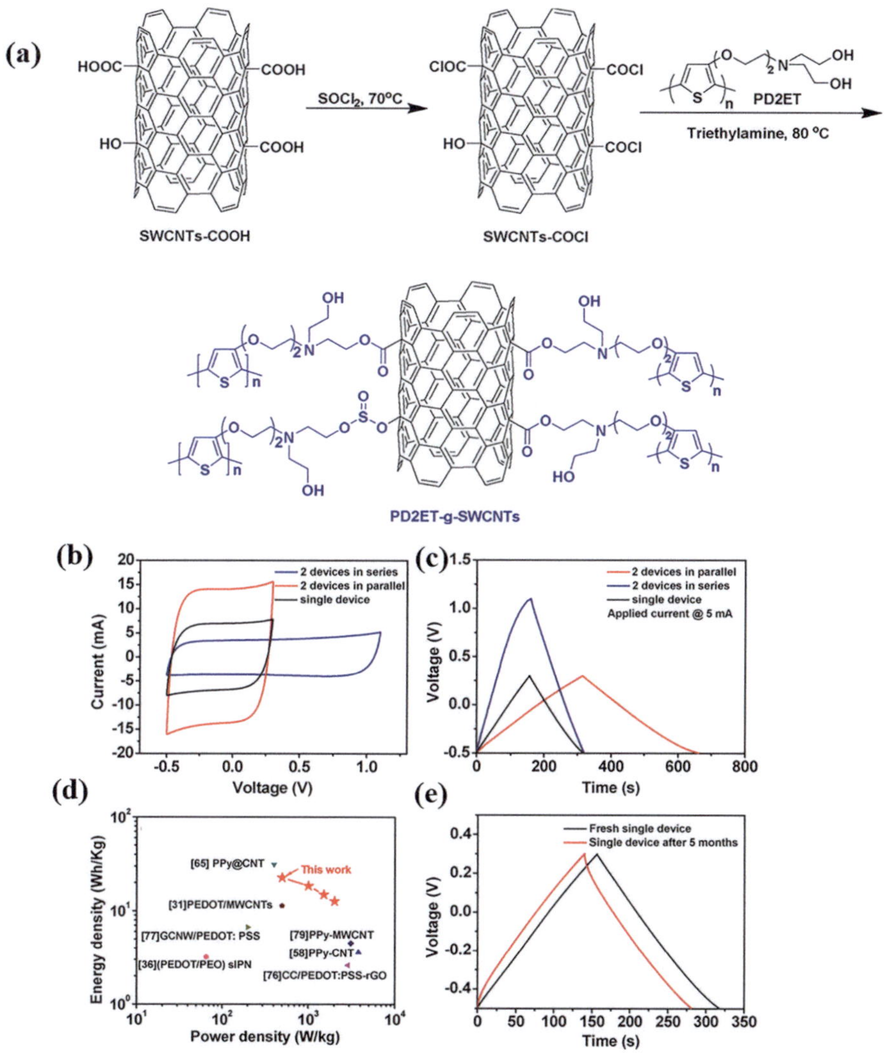

Fig. 7.6 **a** Reaction protocol for the synthesis of PD2ET-g-SWCNTs hybrids; **b** and **c** represents cyclic voltammetry and galvanostatic charge–discharge curves of single supercapacitors, two supercapacitors connected in series and parallel, scan rate: 100 mV s^{-1}; current density: 0.625 A g^{-1}; **d** Ragone plot of a single supercapacitor with the reported literature; **e** comparative galvanostatic charge–discharge curves of a fresh device and 5 months old device. Reprinted with permission of Elsevier from Zhou et al. [179]

7.3.5 PEDOT-Carbon-Based Binary and Ternary Composites

Abidin et al. [3] studied MWCNT, graphene oxide and nanocrystalline cellulose as a doped material in the synthesis of PEDOT hybrid nanocomposites for supercapacitor applications. They found that the morphological capacitance properties were mostly affected by the choice of carbon materials. For instance, PEDOT-graphene oxide displayed a compressed surface paper-like structure, whereas PEDOT-MWCNT and PEDOT-nanocrystalline cellulose possess network-like and rod-like structures, respectively. Among the composites, PEDOT-graphene oxide exhibited a high specific capacitance of 120.13 F g^{-1} due to its high surface area and retained 87.99% of its initial specific capacitance over 1000 cycles.

Jacob et al. [64] reported the electrochemical performances of PEDOT-graphene hybrid electrodes. A brief experimental procedure is as follows: First, graphene was deposited on an indium tin oxide (ITO) plate by electrophoretic deposition technique followed by electropolymerization of EDOT monomer. The highest specific capacitance of 1410 F g^{-1} was visualized at 40 mV s^{-1} in the PEDOT-graphene composite. Sun et al., also produced a PEDOT-graphene hybrid by microwave-assisted in situ synthesis method where in situ polymerization of EDOT and the reduction of graphene oxide were conducted concurrently (Fig. 7a) [126, 128]. The voltammograms displayed an almost rectangular shape with different scan rates as shown in Fig. 7b, which illustrates the optimal capacitive performance of the material, the galvanostatic charge–discharge profiles of the prepared composite are depicted in Fig. 7c. It showed almost similar charge and discharge profile suggesting the reversible charge–discharge capacitive property of the electrode material. Figure 7d represents a plot of specific capacitance versus current density. Due to the interaction between graphene and PEDOT, the hybrid has a high specific capacitance value (270 F g^{-1}) and capacitance retention (93% over 10,000 cycles) owing to the contact between graphene and PEDOT. The Ragone plot is illustrated in Fig. 7e and was found that the energy density is greatly reduced by increasing the power density. In contrast, Zhang and Zhao [175] prepared PEDOT-graphene nanocomposite by graphene oxide reduction followed by in situ polymerizations of EDOT. However, this composite only showed a specific capacitance of 108 F g^{-1}. The poor performance of the composites is mainly attributed to the aggregation of reduced graphene oxide that leads to meager distribution and insufficient contact with PEDOT.

Yu et al. [170] also reported PEDOT-graphene on carbon cloth (PEDOT-graphene-CC) composite materials as flexible electrodes for energy storage and conversion applications. The hybrid has a maximum specific capacitance of 714.93 F g^{-1}, while the case of PEDOT-CC exhibited a capacitance of 118.4 F g^{-1}. The superior performance of the composite electrode is due to the high surface area of the graphene material,

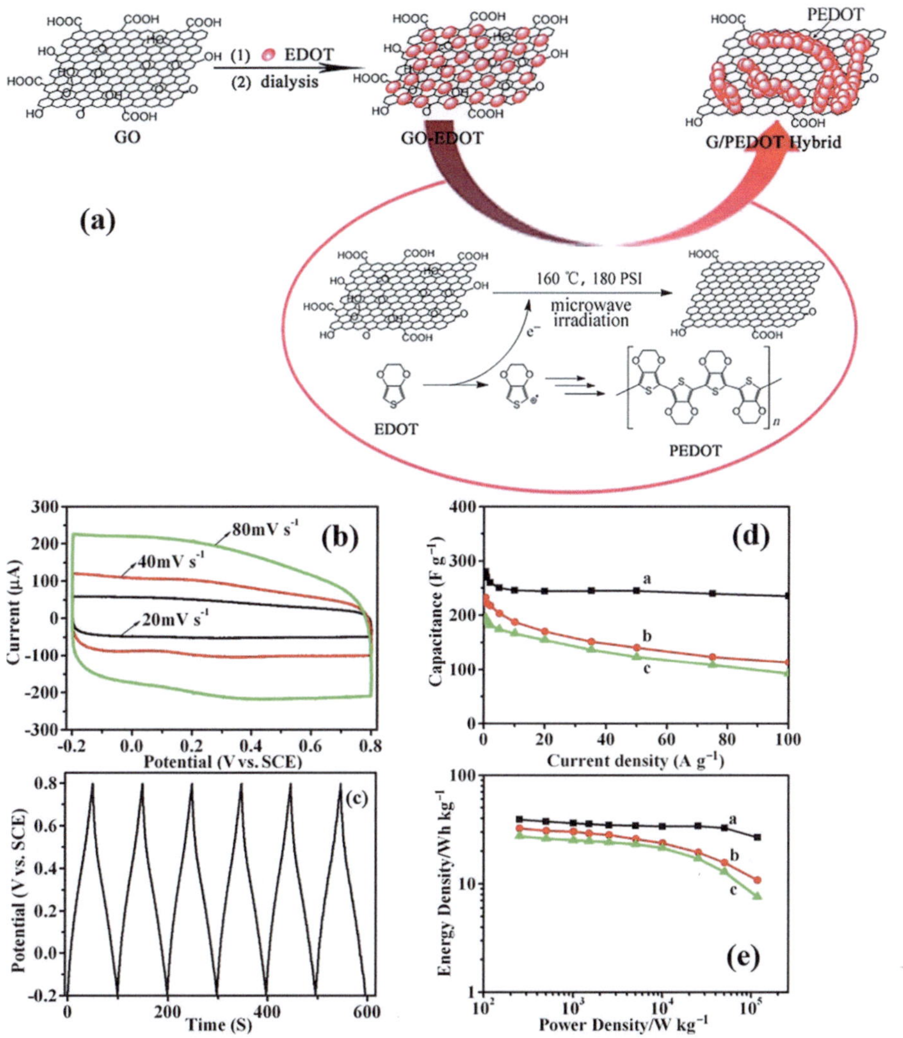

Fig. 7.7 **a** Synthesis process of graphene-PEDOT hybrid, **b** cyclic voltammetry at different scan rates, **c** galvanostatic charge–discharge profile at 5 A g^{-1}; **d** specific capacitance vs discharge current density of different graphene-EDOT hybrids in various electrolytes; **e** Ragone plot of different graphene-PEDOT hybrids in various electrolytes: **a** and **c** in 1 M H_2SO_4; **b** in 1 M $LiClO_4$ solution. Reprinted with permission of Wiley from Sun et al. [126, 128]

7.3.6 Conducting Polymer-Transition Metal Oxides (TMOs)

Metal oxides, especially transition metal oxides, have been exploited as electrode material in rechargeable batteries and pseudocapacitors owing to their extraordinary physicochemical characteristics of cheap price, larger surface area, excellent charge storing properties, and environmentally friendliness for many years [54, 127, 129, 171]. While their usage is greatly limited by their poor conductivity and acid electrolyte instability [147, 153]. To improve its cyclic stability and electrochemical performance, many researchers have united conducting polymers with various metal oxides, such as Co_3O_4, $NiCo_2O_4$, MnO_2, Fe_3O_4, and V_2O_5. For example, most recently Roy et al. [113] produced a PANI-MnO_2 hybrid of varying compositions of MnO_2 by in situ chemical oxidative polymerization of aniline monomer in an acidic medium. The specific capacitance of the PANI-MnO_2 hybrid was as high as 241 F g^{-1} and stable up to 1000 cycles at 0.25 A g^{-1}.

Recently Ma et al. [87] have reported 3D core–shell Fe_3O_4-PANI coaxial heterogeneous nanonets via magnetic field-induced self-assembly and in situ polymerization method for supercapacitor electrode material, The supercapacitor exhibited a specific capacitance of 620 F g^{-1} at 1.0 A g^{-1} in 1 M H_2SO_4 solution and 85% capacitance retention at 2.0 A g^{-1} after 2000 cycles. Very recently, Huang et al. [57] prepared ternary composite electrode material comprised of conducting polymer (PANI) which provides a large surface area, a fine dispersed NiO, and sulfonated graphene as a substrate material. The invented ternary composite, PANI-NiO-SGO displayed a high specific capacitance of 1350 F g^{-1} at 1 A g^{-1} and a 92.23% capacity retention rate over 5000 cycles. Additionally, assembled all-solid states asymmetric supercapacitors (PANI-NiO-SGO//AC) delivered a high specific capacity of 308.8 F g^{-1}, energy density of 109.8 Wh kg^{-1}, power density of 0.8 kW kg^{-1} and stable cycle life with a capacitance retention rate of 91.15% after 10,000 consecutive cycles. Roy et al. [112] synthesized V_2O_5-PANI-based nanocomposite through in situ intercalation and polymerization reaction of aniline with V_2O_5. The electrode with optimum content shows the highest capacitance of 529.56 F g^{-1} at 2 mV s^{-1}. Due to the organic–inorganic compatibility, V_2O_5-PANI exhibits good electrochemical charge storage in a wide operating voltage window of 1.6 V, ranging between – 0.4 V to 1.2 V with excellent stability of 84.2% capacitance retention after 5000 charge–discharge cycles. Furthermore, the asymmetric supercapacitor is composed of V_2O_5-PANI composite and $NiMn_2O_4$ electrodes, which provide significant energy and power density.

Zhao et al. [178] constructed a hierarchical polypyrrole-MnO_2 on carbon fabric by a simple chemical vapor deposition -polymerization was subsequently employed to deposit polypyrrole films on MnO_2 nanowires (PPy-MnO_2-CF). The fabricated exceptional 1D structure was advantageous owing to its rapid electrochemical kinetics, robust electrical conductivity, and structure stabilization. The PPY-MnO_2-CF showed a maximum areal capacitance of 1.24 F cm^{-2} at 1 mA cm^{-2}. A flexible solid-state asymmetric supercapacitor was assembled using PPY-MnO_2-CF and Fe_2O_3-carbon fiber as the anode and cathode materials, respectively. The flexible device provided a maximal energy density of 1.93 mWh cm^{-3} at a power density of 9.8 mW cm^{-3} and a long cycle life with 92.6% of initial capacitance after 10,000 cycles. Furthermore, the device exhibited exceptional flexibility and excellent mechanical stability after a 200-bending cycle test, suggesting that it is close to practical application in flexible electronics. Yang et al. [167] synthesized a polypyrrole-cobalt oxyhydroxide-cellulose fiber hybrid electrode through a liquid phase reduction strategy in an ambient atmosphere. The hybrid electrode exhibited excellent electrochemical properties, with a certain specific capacitance and capacitance retention. The very high specific capacitance of the hybrid electrode does not have a significant loss, which shows a high cyclic stability of 93.02% after 1000 cycles. Furthermore, the incorporation of cobalt oxyhydroxide can be said to have excellent electrochemical properties, maintaining the volumetric change of polypyrrole in the electrochemical redox process and promoting the rapid displacement of electrons.

Gao et al. [47] developed an urchin-like MgC_2O_4-PPy-NF core–shell structure composite by a hydrothermal method and in situ chemical oxidative polymerization for supercapacitor application. Figure 8a shows the schematic representation of the preparation of $MgCo_2O_4$-PPy-NF. From the schematic, the electrons complete the electron transfer through successive nano fiber, $MgCo_2O_4$ and polypyrrole layers. The field emission scanning electron microscope images are shown in Fig. 8b and c, the nanowires of polypyrrole were completely shielded with $MgCo_2O_4$. The fabricated composite, $MgCo_2O_4$-PPy-NF showed a high specific capacitance of 1079.6 F g^{-1} at 1 A g^{-1} and it has 97.4% of excellent cycling stability after 1000 cycles. Furthermore, an all-solid-state asymmetric supercapacitor was assembled using $MgCo_2O_4$-PPy-NF as the positive electrode and activated carbon as the negative electrode (Fig. 8d). The $MgCo_2O_4$-PPy-NF//AC asymmetric supercapacitor exhibits a high specific capacitance of 94 F g^{-1} at 0.4 A g^{-1}. The Ragone plot showed a high energy density of 33.4 Wh kg^{-1} at a power density of 320 W kg^{-1}, a high volumetric energy density of 17.18 mWh cm^{-3} at a volumetric power density of 0.16 W cm^{-3} (Fig. 8e). Additionally, the two built-in asymmetric supercapacitors with two light emitting diodes (LEDs) of one white and one blue can operate for up to 20 min (Fig. 8f–i).

Yin et al. [169] have synthesized a hybrid of PEDOT-Ni-Mn-Co–O and examined the influence of different fixing anions on its supercapacitor performance. The hybrid exhibited overwhelming electrochemical properties with a specific capacitance of 540 F g^{-1} at 1 A g^{-1}, and a good capacitance retention of 98.3% after 5000 cycles at a high current density of 5 A g^{-1}. Furthermore, the impact of the structure, molecular chain length, and the degree of conjugation of the doping anions on the doping level and electrochemical performance of the prepared materials were investigated. Gao et al. [48] have reported an advanced method to prepare PEDOT-MnO_2 composite electrodes without adding any binder, conductive agent, or catalyst. The PEDOT-MnO_2 hybrid electrode exhibits excellent performance with a high specific capacitance of 575 F g^{-1} at 8 A g^{-1}. The surface area of the composite offers an energy density of 14.6 Wh kg^{-1} with a high-power density of 1365 Wh kg^{-1}. Moreover, the PEDOT-MnO_2 supercapacitor electrodes can be prepared from flexible substrates, the specific capacitance of which still reaches 352.9 F g^{-1} at 5 A g^{-1}.

In brief, conducting polymers and metal oxide-based composite electrodes regularly displayed excellent electrochemical performance comprising good specific capacitance, high specific energy and specific power, and excellent durability and rate capability owing to the synergistic factor of the two materials.

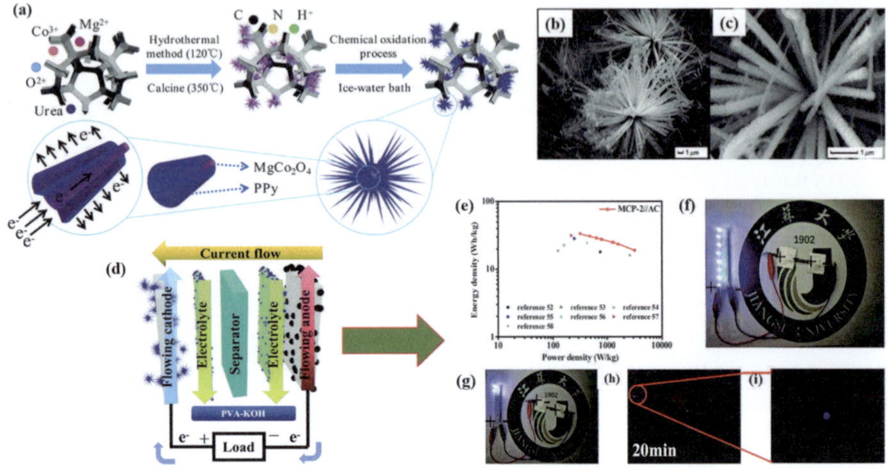

Fig. 7.8 **a** sketch of the reaction protocol and electron transfer of $MgCo_2O_4$-PPy-NF; **b** and **c** scanning electron microscope pictures of $MgCo_2O_4$-PPy-NF; **d** assembled asymmetric supercapacitor device; **e** Ragone plot of the asymmetric supercapacitor device; **f** light emitting diodes of two asymmetric supercapacitors after charging; **g** lighting of two light emitting diodes by two asymmetric supercapacitors after charging; **h** photo image of the discharge of **g** after 20 min and **i** partial expansion of **h**. Reprinted with permission of the Royal Society of Chemistry from Gao et al. [47]

7.3.7 Conducting Polymer-Metal–Organic Frameworks Composite

Metal–organic frameworks are porous material types formed by the interconnection of metal ions with organic links and have drawn the attention of various research fields [93, 126, 128]. In recent times, metal–organic frameworks have been applied more dramatically in energy storage systems and are found to be promising because of enhanced surface area, metal ion binding sites and porous. Even though one reason for their poorer conductive nature demerits their electrochemical features, the conducting polymer—composites were much investigated and adopted for the application of flexible supercapacitors. For example, Shao et al. [118] described the PANI molecular chains grown in the UiO-66 pores as one of the metal–organic frameworks (PANI-UiO-66) to form a rigid interpenetrating network structure by employing the highly stable porous MOFs using different synergistic effects to increase the conductive nature and electrochemical efficacy. They reported that PANI-UiO-66 showed excellent specific capacitance of 1015 F g^{-1} at 1 A g^{-1} by electrochemical analysis. Furthermore, they constructed the symmetric solid-state flexible supercapacitors and depicted a satisfactory capacitance of 647 F g^{-1} at 1 A g^{-1} and a high cycling stability of 91% capacitance retention after 5000 cycles. The bending test was also conducted to assess fabricated supercapacitor flexibility and showed its performance only decreased by 10% after 800 bending cycles with a bending angle of 180. Recently, Zhu et al. [180] developed hierarchical ZnO-MOF-PANI core–shell nanorod arrays on carbon cloth by joining hydrothermal and electrodeposition methods. The obtained results showed that the ZnO-MOF-PANI composite with 13 cyclic voltammetric cycles of PANI deposition illustrated the maximum capacitance of 340.7 F g^{-1} at 1.0 A g^{-1}, decent capacitance retention with 84.3% from 1 to 10 A g^{-1} and outstanding cyclic life of 82.5% capacitance retention later 5000 cycles at higher current density of 2.0 A g^{-1}. Besides, they inferred that the augmented core–shell nanoarchitectures awarded the composite electrode with rapid ion and electron transfer, short ion diffusion pathway and utilization of active materials that depicted outstanding electrochemical performance of the ternary composite.

Liu et al. [84, 85] synthesized a nanocomposite of metal–organic framework-polypyrrole as the positive electrode of a supercapacitor via electrodeposition and smear routes. The fabricated positive electrode has an excellent specific capacitance of 284.3 F g^{-1} at 1 mA cm^{-2}. This electrode shows an excellent capacitance retention of 100.7% over 40,000 cycles. Xu et al. [164, 165] studied the cumulative conductivity and the metal–organic framework performance; 3D metal–organic frameworks are generally made by the conductive tubes of the polypyrrole as the cluster for metal–organic framework growth. The capacitance of zeolitic imidazole framework (ZIF67-PPy) 597.6 F g^{-1} improved almost 5 times its pristine ZIF-67 (99.2 F g^{-1}), illustrating the reputation of the construction of the ZIF67-PPy microstructure. Furthermore, ZIF-PPy networks-based flexible supercapacitors show an excellent areal capacitance of 225.8 mF cm^{-2}, which is much higher than that of existing MOFs-based supercapacitors.

Jiao et al. [65] produced conductive polypyrrole-anchored bimetallic organic frameworks to achieve efficient supercapacitor electrodes (Zn-Ni-MOF-PPy). Fascinatingly, they realized that the Zn-Ni-MOF-PPy alters the polymerization of pyrrole into polypyrrole by using oxygen in air acting as a green oxidant. The conductive polymer chains not only enhance the Zn-Ni-MOF layer spacing but also offer favorable charge transport channels. The Zn-Ni-MOF-PPy showed excellent electrochemical nature, and a "trade-off effect" between ion diffusion kinetics and electrical conductivity was discovered with different loads of polypyrrole. During this period, a coin-type hybrid supercapacitor was manufactured using Zn-Ni-MOF-PPy and CNTs-COOH showing a high specific energy of 50.9 Wh kg^{-1} and a specific power of 1338 W kg^{-1} concurrently. Surprisingly, the hybrid supercapacitor shows dominant capacitance retention after 5000 cycles of charge–discharge. In brief, the conducting polymer-based metal–organic framework nanostructures have dramatically changed the portable energy storage features because of their enhanced performance.

7.3.8 *Conducting Polymer-Transition Metal Sulfides and Dichalcogenides*

Metal sulfides have dramatically changed the supercapacitor applications because of their electrochemical features and many researchers have investigated the conducting polymers with the various MSs like Nis, CoS and CuS. The transition metal dichalcogenides resemble the layer-like structures with the empirical formula of MX_2, here M is the transition metal obtained from the groups 4, 5 or 6 and X is the chalcogen generally referring as the sulfur (S), selenium (Se) or tellurium (Te). One of the best examples of the family is the molybdenum sulfide (MoS_2) which has driven much interest in supercapcitor applications because of their abundant availability and higher theoretical capacitances. The enhanced surface area sheets of MoS_2 generally afford greater energy storage and the pseudocapacitance material-like behavior because Mo offers many kinds of oxidation states.

For instance, Ren et al. [111] developed a MoS_2-PANI of 3D tubular hybrid electrodes with varying content of PANI for higher supercapacitor performance. They obtained a high specific capacitance of 552 F g^{-1} at 0.5 A g^{-1} for the loading amount of 60% PANI. The illustration for the alarming rate performance is assigned the design of the synergetic effects of the 3D tubular MoS_2 and the PANI arrays of the nanowires. Likely, the as-assembled nanostructures of MoS_2-PANI symmetric supercapacitor exhibit a capacitance retention of 79% over 6000 cycles. Wang et al. [154] reported the PANI-MoS_2 by electrodeposition route. In short, the first PANI was electrodeposited on a glassy carbon electrode (GCE) via the chronopotentiometry method supported by the electrodeposition of MoS_x. The as-prepared composite (PANI-MoS_x) displayed a capacitance of 48.64 mF cm^{-1} at 0.1 mA cm^{-2} and retained long-term stability with no deprivation for 150 cycles.

Diggikar et al. [39] quantified silver vanadium sulfide ($Ag_2V_2S_4$) affixed PANI nanocomposite electrode for supercapacitors and found that the capacitance of the 440 F g^{-1} obtained at the 500 mV s^{-1} for the PANI-$Ag_2V_2S_4$ composite structure. Very recently, Chen et al. [25] described ternary flexible composite electrodes (MoS_2-PANI-CNT) with MoS_2, PANI and carbon nanotubes as depicted in Fig. 9a. The as-constructed ternary flexible electrode showed a good capacitance of 204.2 F g^{-1} at 0.1 A g^{-1}. Together with two slightly disconnected MoS_2-PANI-CNT composite paper electrodes solidified in a PVA-H_2SO_4 gel electrolyte solution, an all-solid-state supercapacitor was fabricated and possesses the great ability to drive ultrathin electronics (Fig. 9b and c). The binder-free supercapacitor exhibited a classic dual-layer capacitive performance at scan rate varying from 1 to 200 mV s^{-1} (Fig. 9d). The evaluation of the suppleness can be identified with the symmetric cell of the all-solid-state supercapacitor with the different bending angles as shown in Fig. 9e. Figure 9f shows cyclic voltammetric curves that continue nearly unaffected as the supercapacitor was twisted from 0 to 120° which suggests that the supercapacitor displayed an exceptional amount of elasticity. The supercapacitor displays an admirable intensity of stability for the charge–discharge cycles due to the gel electrolyte solution with a retaining value of 80% after 1000 cycles (Fig. 9g). Figure 9h indicates that the solution resistance (R_s) and charge transfer resistance (R_{CT}) values of 19.09 and 13.13 Ω, respectively. The obtained results were passionately interesting due to their electrical conductivity and faster ion transportation at the greater level of execution. As shown in Fig. 9i, the voltage window was widened from 1.0 V for a single solid-state supercapacitor to 3.0 V for a sequence of three sold-state supercapacitors. Additionally, the voltage window was seemingly broader ranging from the 1.0 V and the 3.0 V for all-solid-state supercapacitors. The sequence of the three solid-state supercapacitors represents the identical charging and discharging cycles at constant currents, generally suggesting the supercapacitors are consistent. To reveal the realistic handling of these supercapacitors, linking two supercapacitors in series can effectively operate a light emitting diode (LED) as shown in Fig. 9j, indicating the huge potential of such a supercapacitor as a flexible power source. Recently Zhang et al. [177] prepared MoS_2 grown on the surface of MWCNTs-PANI by an in situ polymerization followed by a hydrothermal method. The electrochemical performance is assessed by the cyclic voltammetric and the charge–discharge calibrations. The assembled ternary hybrid MoS_2-MWCNTs-PANI composite depicted the specific capacitance of 542.56 F g^{-1} at 0.5 A g^{-1}. Simultaneously, the hybrid ternary composite exhibited excellent rate capability and magnificent cyclic stability.

Bai et al. [15] reported a ternary composite material of MoS_2-rGO-PANI for supercapacitor applications by implementing the two-step method via in situ polymerization and hydrothermal method. They described MoS_2-rGO-PANI composites with an 80% weight showing the highest capacitance of 570 F g^{-1} at 1 A g^{-1} and 78.6 preserved its original capacitance after 500 cycles at 1 A g^{-1}. Ultimately, they determined the excellent electrochemical behavior is ascribed to its 3D system formation and the synergistic impacts among the three elements that make up the composites achieve equal pseudocapacitance and double-layer capacitance.

Very recently, Wang et al. [149] developed ultra-steady tri-nickel disulfide (Ni_3S_2 -PANI) electrodes on carbon fibers (CF-Ni_3S_2-PANI) for an efficient flexible asymmetric supercapacitor. The built CF-Ni_3S_2-PANI electrode demonstrated a considerably improved specific capacity of 715.3 F g^{-1} at 1 A g^{-1} contrasted with the CF-Ni_3S_2 electrode with a capacitance of 318 F g^{-1} at 1 A g^{-1}. Further crucially, the constructed flexible asymmetric supercapacitor design produced a remarkable specific energy of 35.7 Wh kg^{-1} at a specific power of 850 W kg^{-1}. They determined that the excellent supercapacitor behavior was ascribed to the conductive matrix of carbon fiber, the defensive layer of PANI and the rigid bridges of N-Ni bonds. Ba et al. [13] set fiber-shaped supercapacitors derived from PANI-CuS-poly (ethylene terephthalate) [PANI-CuS-PET] yarn electrodes because of their excellent potential for small and wearable electronics due to their outstanding flexibility, compactness, and good decay. In short, poly (ethylene terephthalate) fibers were effectively transformed into an electrically conductive and electrochemically active yarn by establishing CuS and PANI through easy chemical bath deposition and electrochemical deposition. The fiber-shaped supercapacitor displays a high capacitance of 29 mF cm^{-2}, and 116 mF cm^{-2} for a particular electrode, and excellent cyclic stability with 93.1% retention after 1000 cycles. Moreover, the present scheme offers

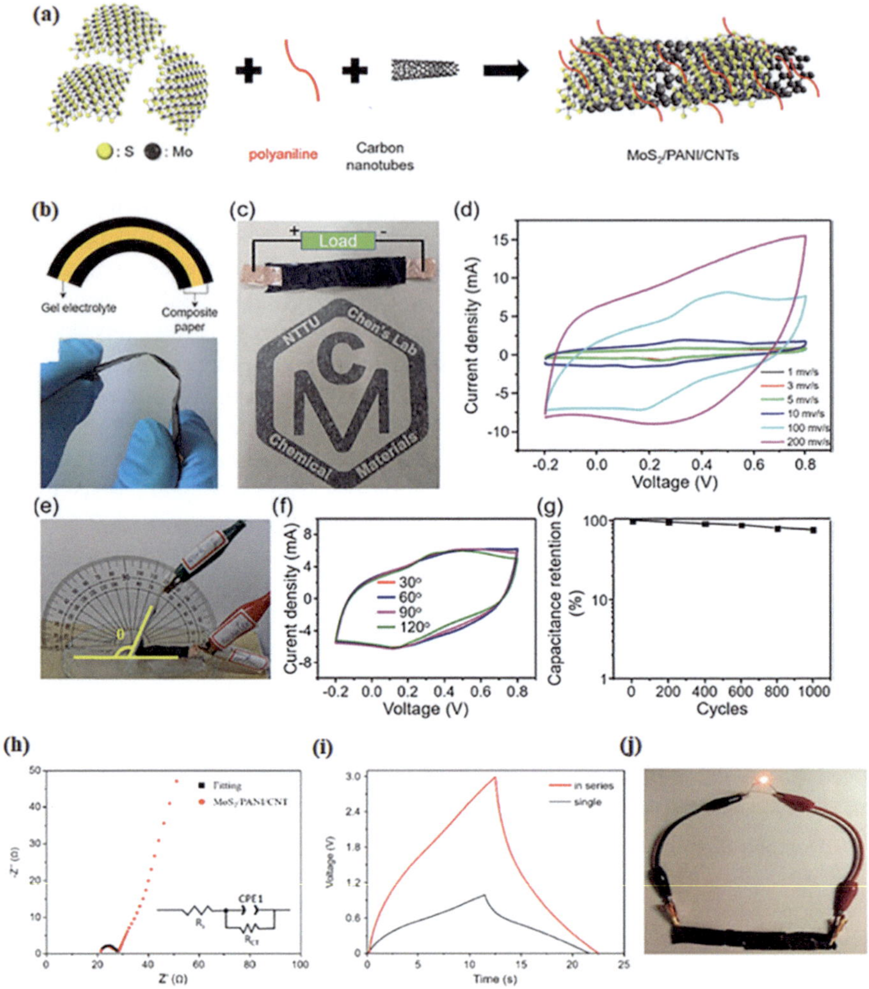

Fig. 7.9 **a** Composite fabrication of MoS_2-PANI-CNT; **b** and **c** side and top view of the supercapacitor; **d** cyclic voltammetry curves at varying scan rate; **e** photo image of the bendable supercapacitor; **f** cyclic voltammetric curves at different bending angles; **g** stability test of supercapacitor by charge–discharge; **h** electrochemical impedance spectroscopic profiles of all-solid-state supercapacitor; **i** charge–discharge profile of single all-solid-state supercapacior and a series of three all-solid-state supercapacitors at 0.45 A and **j** light emitting diodes (LED) powered by a series of two all-solid-state supercapacitors. Reprinted with permission of the American Chemical Society from Chen et al. [25]

a reference for the manufacture of compact and wearable energy storage devices as easy formation technique and low-price raw materials. Peng et al. [103] established high-performance supercapacitors centered on multi-structured CuS-PPy composites by in situ oxidative polymerization. The CuS-PPy displayed the highest specific capacitance of 427 F g^{-1} at 1 A g^{-1} and 88% of its original capacitance held after 1000 cycles. Chang et al. [22] produced MoS_2-PPy nanocomposites utilizing polypyrrole nanotubes and coated MoS_2 by a hydrothermal method. The capacitance of the MoS_2-PPy nanocomposites was evaluated up to 307.5 F g^{-1}, which was higher than that of virgin polypyrrole with capacitance of 106.3 F g^{-1} and MoS_2 with capacitance of 138.5 F g^{-1} at 1.0 A g^{-1} in 1 M KCl. Additionally, holds 96.47% after 1000 charge–discharge cycles offering the potential candidate for hybrid supercapacitor applications.

Similarly, Thakur et al. [133, 134] arranged MoS_2-PPy nanocomposites through an in situ polymerization technique with differing load percentages of metal sulfide. Among the nanocomposite materials, MoS_2-PPy-25% demonstrated a maximum capacitance of 400 F g^{-1} at 1 A g^{-1} which was greater than that of bare polypyrrole. MoS_2-PPy-25% have a significant capacitance retaining of 92.8% after 5000 cycles. Moreover, the MoS_2-PPy-25% nanocomposite electrode exhibited far greater specific energy (8.88 Wh kg^{-1}) with specific power (2286 W kg^{-1}). Chen et al. [26] made 3D tubular MoS_2-PPy by an in situ oxidative polymerization technique in the existence of the MoS_2. The fabricated MoS_2-PPy composite exhibited a capacitance of 462 F g^{-1} at 1 A g^{-1} and possesses good cyclic stability of 82% after 2000 cycles at 3 A g^{-1}. Additionally, the composite electrode material produces a specific energy as high as 25.5 Wh kg^{-1} at a specific power of 266.3 W kg^{-1} and retains 14.5 Wh kg^{-1} at a high specific power of 4002 W kg^{-1}.

Recently Wang et al. [151] described in situ electrodeposition of polypyrrole onto tantalum diselenide ($TaSe_2$) nanobelt quasi-arrays for high specific capacitance supercapacitor applications by one-step surface-assisted chemical vapor transport (SACVT) technique. $TaSe_2$ as a component of transition metal dichalcogenides, draws a fantastic deal of notice as a coated electric conductor with minimal dimension and metallic nature. The schematic plans for the fabrication of $TaSe_2$ nanobelts, $TaSe_2$-PPy composite, and $TaSe_2$-PPy supercapacitor are shown in Fig. 10a. The digital descriptions of the preserved ampoule before and after SACVT are shown in Fig. 10b. The digital images of the $TaSe_2$ nanobelts before and after the electrodeposition are shown in Fig. 10c. Figure 10h signifies a symmetric supercapacitor built from the composites validates high areal capacitance of 835 mF cm^{-2} at 2 mV s^{-1}, wide potential window of 1.2 V, and exceptional cyclic stability with 98.7% capacitance retaining after 10,000 cycles. Figure 10i and j confirm the areal capacitance versus scan rate and charge–discharge measurement, correspondingly of $TaSe_2$-PPy composites. Also, they took scanning electron microscope images before and after the stability test (Fig. 10d–g). However, the structure of $TaSe_2$-PPy composites is nearly unbroken even after 10,000 cycles. Consequently, the fabrication method of this supercapacitor is very easy since no current collector, binder, or conductive additive is required.

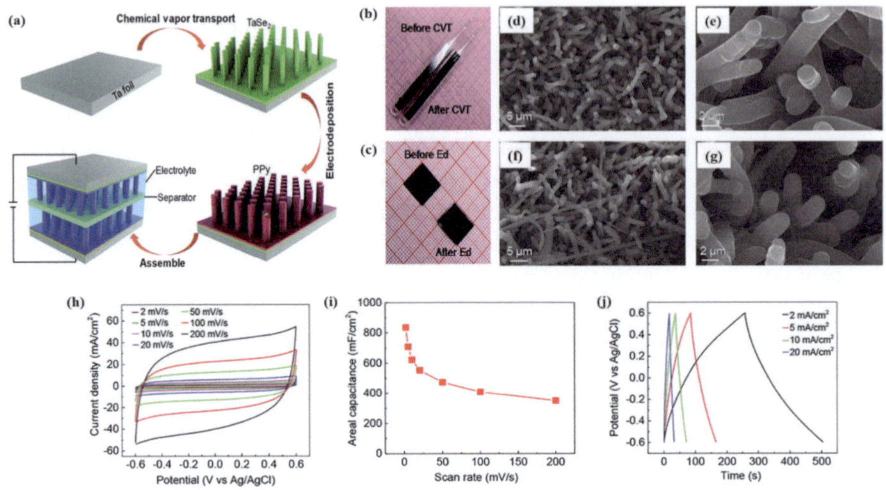

Fig. 7.10 **a** graphics of $TaSe_2$ nanobelts, $TaSe_2$-PPy hybrid and $TaSe_2$-PPy supercapacitor fabrication; **b** Photo image of Ta foil before and after chemical vapor transport; **c** photographs of $TaSe_2$ nanobelts on a Ta foil before and after electrodeposition; **d–g** represents scanning electron microscopic images of $TaSe_2$-PPy hybrid before and after cycling test; **h** cyclic voltammetric curves at different scan rate; **i** areal capacitances derived from cyclic voltammetry and **j** charge–discharge curves at different current density. Reprinted with permission of the Royal Society of Chemistry from Wang et al. [151]

Very recently, Karimi et al. [69] reported PPy-FeCoS-rGO over the Ni foam surface as a novel electrode material for the asymmetric supercapacitor and the device yields a capacitance of 3178 F g^{-1} at 2 A g^{-1} with a decent rate capability. The excellent electrochemical behavior is elucidated with the enhanced surface area, greater conductivity, and minimized contact resistance among the PPy-FeCoS-rGO because of the excellent faradaic charge storage phenomenon. Additionally, a hybrid asymmetric supercapacitor device was fabricated using PPy-FeCoS-rGO-NF as the positive electrode and reduced graphene oxide-coated nickel foam as the negative electrode. It showed a maximum capacitance of 94 F g^{-1} at 1 A g^{-1} and a specific energy of 28.3 Wh kg^{-1} at a specific power of 810 W kg^{-1}. Peng et al. [104] declared nanofibrous hybrid electrodes of PPy-CoS-Bacterial cellulose with superior flexibility for supercapacitor usage. Bacterial cellulose has attracted much attention due to its cheap price, high surface area and adequate porosity. The specific capacitance of PPy-CoS-bacterial cellulose hybrid can reach approximately 614 F g^{-1} at 0.8 mA cm^{-2}, maintaining 62.4% of the original value after 300 cycles.

Peng et al. [105] produced another composite membrane from polypyrrole, nickel sulfide and bacterial cellulose, PPy-NiS-bacterial cellulose with a high initial capacitance of 713 F g^{-1} at 0.813 mA cm^{-2}. However, the capacitance is dramatically reduced during the charge–discharge cycles, leaving only 24.2% of its initial value after 300 cycles. Dong et al. [40] synthesized MnO_2-MoS_2-PPy ternary composite by the hydrothermal and simple oxidation process. The fabricated ternary hybrid has a high specific capacitance of 490 F g^{-1} at 1 A g^{-1}. In addition, they appealed for a simpler way to produce effective materials for flexible pseudocapacitance electrode materials for high energy storing devices.

Recently, Sarmah and Kumar [116] produced an ion beam-modified MoS_2-rGO-PPyNTs ternary composite for supercapacitor electrodes. The specific capacitance of the ternary hybrid enhanced from 1561 to 1875 F g^{-1} after swift heavy ion surface transformation. The increase in capacitance response can be attributed to increased crystallinity, improved electroactive surface, and enhanced electrode kinetics for swift heavy ion irradiation. In addition, since the oxidation state of PPyNTs was stabilized by increased contact with the swift heavy ion radiation, the irradiated electrode appeared to be more stable than the pristine electrode and showed a cyclic stability of 91% after 10,000 charge–discharge cycles.

Alamro and Ram [6] prepared poly ethylenedioxythiophene and MoS_2 (PEDOT-MoS_2) nanocomposite for supercapacitor applications. For the prepared composite, a maximum specific capacitance of 361 F g^{-1} was achieved and no stability test was found. Wang et al. [147, 153] also reported PEDOT-MoS_2 nanocomposite via in situ polymerization in the presence of ammonium persulfate. The fabricated hybrid showed improved electrochemical performance (405 F g^{-1}) over bare PEDOT. Also, the hybrid capacitance retention rate is 90% after 1000 cycles. Ge et al. [49] formed a strong hybrid film, MoS_2 anchored with PEDOT: PSS. The prepared film has a good volumetric capacitance of 141.4 mF cm^{-3} at 0.2 A g^{-1} and a capacitance retention of 98.6% after 5000 charge–discharge cycles. Table 7.2 comprises different types of chalcogenides and dichalcogenide-based conducting polymers for supercapacitor applications.

7.4 Hyper Cross-Linked Polymers

Hypercross-linked polymers are a sub-class of conjugated microporous polymers and were first developed by Davankov [33, 37, 136]. The first hyper cross-linked polymers were prepared by Tsyurupa and Davankov in the late 1970s [137], in which linear polystyrene or gel-type swollen polystyrene-co-divinylbenzene were cross-linked by cross-linking agents in the presence of Lewis acid catalyst and an appropriate solvent [34–36]. Since the finding of Davankov resins, there has been a prompt bust of hyper cross-linked polystyrene in their strategy and composition, as well as their prominent usages. More recently, hyper-linked polymers have gained rapid progress due to the noteworthy merits of different synthesis tactics, simplicity of functionalization, greater surface area, lower cost of chemicals and, mild working atmosphere [132, 138–140, 142–144].

Generally, hyper cross-linked polymers are synthesized via Friedel–Crafts alkylation with simple aromatic monomers and inexpensive crosslinkers [99, 131, 135], resulting in the production of high-yielding materials which contest with conventional cheaper microporous materials such as activated carbon.

Table 7.2 Conducting polymers-chalcogenide-based electrode material for supercapacitor

S. No.	Material	Specific capacitance, current density/scan rate	Cycling stability (capacitance retention % and number of cycles)	Reference
1	3D tubular MoS_2/PANI	552 F g^{-1} at 0.5 A g^{-1}	79%; 6000 cycles	Ren et al. [111]
2	PANI/MoS_x	48.64 mF cm^{-2} at 0.1 mA cm^{-2}	100%; 150 cycles	Wang et al. [154]
3	PANI/$Ag_2V_2S_4$	440 F g^{-1} at 500 mV s^{-1}	–	Diggikar et al. [39]
4	MoS_2/PANI/CNT	204.2 F g^{-1} at 0.1 A g^{-1}	80%; 1000 cycles	Chen et al. [25]
5	MWCNT/PANI/MoS_2	542.56 F g^{-1} at 0.5 A g^{-1}	73.71%; 3000 cycles	Zhang et al. [177]
6	MoS_2/rGO/PANI	570 F g^{-1} at 1 A g^{-1}	78.6%; 500 cycles	Bai et al. [15]
7	CF/Ni_3S_2@PANI	715.3 F g^{-1} at 1 A g^{-1}	93.14%; 2500 cycles	Wang et al. [149]
8	PANI/CuS/PET	29 mF cm^{-2}	93.1%; 1000 cycles	Ba et al. [13]
9	CuS/PPy	427 F g^{-1} at 1 A g^{-1}	88%; 1000 cycles	Peng et al. [103]
10	PPy/MoS_2	307.5 F g^{-1} at 1 A g^{-1}	96.47%; 1000 cycles	Chang et al. [22]
11	PPy/MoS_2	400 F g^{-1} at 1 A g^{-1}	92.8%; 5000 cycles	Thakur et al. [133, 134]
12	MoS_2/PPy nanowire	462 F g^{-1} at 1 A g^{-1}	82%; 2000 cycles	Chen et al. [26]
13	$TaSe_2$/PPy	835 mF cm^{-2} at 2 mV s^{-1}	98.7%; 10,000 cycles	Wang et al. [151]
14	PPy/FeCoS-rGO	3178 F g^{-1} at 2 A g^{-1}	97.5%; 2500 cycles	Karimi et al. [69]
15	PPy/CoS/BC	614 F g^{-1} at 0.7 A g^{-1}	62.4%; 300 cycles	Peng et al. [104]
16	PPy/NiS/BC	713 F g^{-1} at 0.7 A g^{-1}	24.2%; 300 cycles	Peng et al. [105]
17	MnO_2@MoS_2/PPy	490 F g^{-1} at 1 A g^{-1}	90%; 1000 cycles	Dong et al. [40]
18	MoS_2-rGO/PPy	1875 F g^{-1} at 1 A g^{-1}	91%; 10,000 cycles	Sarmah and Kumar [116]
19	PEDOT/MoS_2	361 F g^{-1}	–	Alamo and Ram [6]

(continued)

Table 7.2 (continued)

S. No.	Material	Specific capacitance, current density/scan rate	Cycling stability (capacitance retention % and number of cycles)	Reference
20	PEDOT/MoS_2	405 F g^{-1}	90%; 1000 cycles	Wang et al. [147, 153]
21	MoS_2/PEDOT: PSS	141.4 F cm^{-3} at 0.2 A g^{-1}	98.6%; 5000 cycles	Ge et al. [49]

PANI polyaniline, *PPy* polypyrrole, *PEDOT* poly ethylenedioxythiophene, *CNT* carbon nanotubes, *BC* bacterial cellulose, *PSS* polystyrene sulfonate, *rGO* reduced graphene oxide, *CF* carbon fibers

Figure 11a represents the hyper cross-linked polymers synthesized from their corresponding monomers by a simple Friedel–Crafts alkylation method. In brief, known quantities of monomer, 1,3,5-trioxane (crosslinker) and 1,2-dichloroethane were taken in a round bottom flask coupled with a water condenser. After the complete dissolution of the mixture, ferric chloride (catalyst) was added to it and the reaction was continued for 24 h at 80 °C. The brown precipitate was collected, filtered, and washed with a mixture of water, methanol, and acetone [143]. Figure 11b shows the schematic illustration of the hyper cross-linked polymers. Figure 11c represents the schematic sketch of hyper cross-linked polymers from naphthalene, and α,α′-dichloro-p-xylene reaction protocol. Babu et al. [14] have studied a porous carbon nanofiber obtained from hyper cross-linked polymers (HCP-CNF) and 2D copper-cobalt oxide nanosheets (CCO-NS) as the anode and cathode materials, respectively, for asymmetric supercapacitor application. Furthermore, the fabricated asymmetric supercapacitor device displayed a good Cs of 244 F g^{-1} at 1 A g^{-1}, which is mainly attributed to the porous structure of CCO-NS and the well-organized pores-enriched carbon framework of HCP-CNF along with large surface area. Additionally, the asymmetric supercapacitor delivered a high energy density of 25.1 Wh kg^{-1} at a power density of 400 W kg^{-1} with a maximum operation voltage (1.6 V). The electrode displayed excellent capacitance maintenance of 91.1% over 5000 cycles.

Vinodh et al. [138, 139, 142, 144] reported a porous carbon electrode material for supercapacitor applications from hyper cross-linked polymers of phloroglucinol and α,α′-dichloro-p-xylene. Recently, Kim et al. [72] made a porous carbon supercapacitor electrode material from p-xylene via hyper cross-linking followed by pyrolysis at 800 °C. Furthermore, it has a maximum specific capacitance of 242.5 F g^{-1} at 1.25 A^{-1} in a three-electrode configuration. Recently, Lu et al. [86] synthesized heteroatom-enriched hierarchical porous carbon obtained by hyper cross-linked polymerization tactics from pyrrole and thiophene monomers. Significantly, the heteroatom-incorporated porous carbon in a solid-state capacitor has a capacitance of 464 F g^{-1} at 0.2 A g^{-1}, and a high energy density of 16.2 Wh kg^{-1} at 50 W kg^{-1}. In addition, the capacitance retention of the electrode material over 10,000 cycles was more than 92%. Lee et al. [79] explored hyper cross-linked polymers that were synthesized from aromatic precursors such as pyrrole, benzene, pyrrole, aniline, or thiophene consuming a one-step knitting technique. Among them, N-Py-700 unveiled

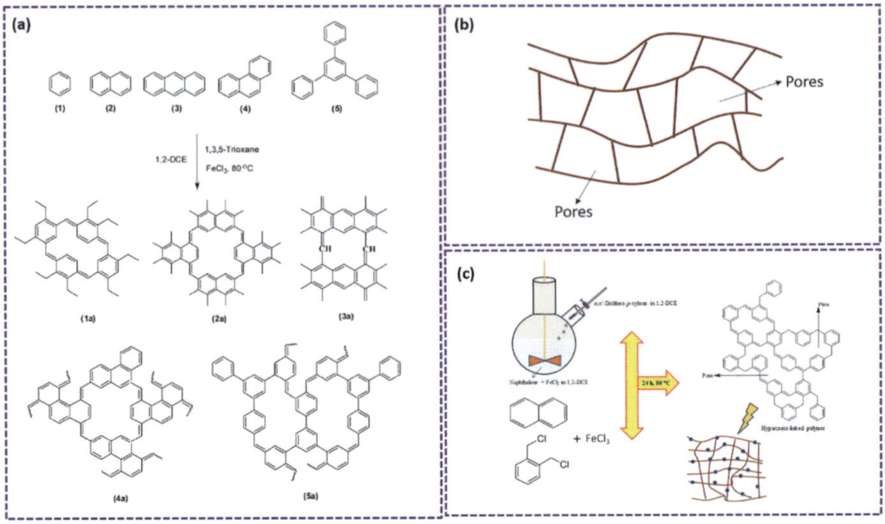

Fig. 7.11 **a** The preparation of different hyper cross-linked polymers from their respective monomer unit via the Friedel Crafts alkylation route. Reproduced with permission of the Royals Society of Chemistry from Vinodh et al. [143]. **b** Pores-enriched hyper cross-linked polymers, **c** the sketch of hyper cross-linked polymers from α, α′-dichloro-*p*-xylene and naphthalene. Reproduced with permission of Elsevier from Babu et al. [14]

an excellent supercapacitive behavior, with the highest capacitance being 374 F g^{-1} at 0.1 A g^{-1} and retaining 96% of its initial capacitance over 15,000 charge–discharge cycles. Vinodh et al. [141] fabricated nitrogen-anchored carbon balls from hyper cross-linked polymers of anthracene monomer via the Friedel–Crafts reaction and following pyrolysis by changing the carbonization temperature between 600 and 800 °C. Additionally, the fabricated N-HCP-700 displayed a maximum specific capacitance of 412.8 F g^{-1} at 0.2 A g^{-1} in a three-electrode setup. Furthermore, the assembled solid-state capacitor possesses a high specific capacitance of 206 F g^{-1} at 0.2 A g^{-1} along with excellent rate capability and exhibited a cyclic stability of 96.5% capacitance retention over 5000 cycles.

Xu et al. [164, 165] prepared nitrogen-anchored porous carbon electrode material from conjugated microporous polymer. The obtained NPCM-1 showed a moderate specific capacitance of 264 F g^{-1} at 0.1 A g^{-1} and extraordinary cycling stability. Jung et al. [66] have made hybrid electrode material from the electrochemical polymerization of aniline and hyper cross-linked polymers of pyrene. The hybrid displayed a maximum capacitance of 15.83 F g^{-1} at 0.1 A g^{-1} in an assembled solid-state capacitor device. Sarno et al. [117] synthesized two different concentrations of graphite nanoplatelets loaded in hyper cross-linked styrene with a slight modification of the Davankov procedure. The obtained hybrid composite unveiled high specific capacitance of 52.1 F g^{-1} and 60.4 F g^{-1}. Furthermore, the prepared nanohybrid was shown to have a supreme capacitance maintenance of up to 97% over 10,000 cycles.

Table 7.3 summarizes hyper cross-linked polymers obtained from porous carbon electrode materials for supercapacitor applications.

In short, the commonly used method to synthesize hyper cross-linked polymers is the Friedel–Crafts alkylation reaction. Dimethoxy methane is the most widely used crosslinker to obtain a hypercross-linked structure. Other crosslinkers, such as paraformaldehyde, dimethyl acetal, and dimethyl amine can also be used to obtain hyper cross-linked polymers. The main advantage of the Friedel–Crafts reaction is the need for inexpensive catalysts, such as $FeCl_3$ and $ZnCl_2$. In addition, only cheap monomers such as pyrrole, thiophene, naphthalene, benzene, and anthracene are needed to synthesize different types of hypercross-linked polymers. However, the rate capability and capacitance have been relatively low when compared to other electrode materials. Therefore, to achieve a high-performance electrode material from hyper cross-linked polymers more research is still required, which confirms extraordinary specific capacitance, wide cyclic life, and extreme rate capability.

7.5 Conclusion

Research and developments in polymer-based nanostructures for supercapacitors have been rapidly advancing over the past numerous years. Several attempts have been made toward supercapacitors based on conducting polymers as a new avenue for flexible energy storage devices. In this book chapter, we first discussed the synthesis of pure conducting polymers of polyaniline, polypyrrole, polythiophene and poly ethylenedioxythiophene and their exploration toward supercapacitor applications. However, bare capacitors have poor cyclic stability, low specific capacitance, and low rate capability which limits their use in practical applications. We demonstrated the preparation and supercapacitor fabrication of conducting polymers with different carbon-based derivatives, as well as metal oxides, metal sulfides like chalcogenides, and metal–organic frameworks of binary and ternary composites. These composite electrode materials enhance the mechanical strength, effective surface area, structural stability, and porosity and thus significantly boost the specific capacitance, specific energy, specific power, cyclic stability, and rate capability. The enhanced electrochemical properties are primarily assigned to the combinational characteristics of their equivalents, along with the synergistic interaction between them. Then, we discussed porous carbon derived from supercapacitor applications. Compared to other electrode materials, the specific capacitance and rate capability of the hyper cross-linked polymers obtained in carbon electrodes are relatively less. Therefore, further study is hyper cross-linked polymers necessary for cross-linked polymers -based porous carbon electrodes, which guarantees high specific energy, excellent stability, good specific capacitance, and high-rate capability.

Recent developments from the point of design and effective conducting polymer-based pseudocapacitor electrode materials for flexible supercapacitor devices have been explained in detail. However, various problems continue to be posed to reach the performance of the device, like power density, energy density and cyclic stability,

Table 7.3 Hypercross-linked polymers derived porous carbon electrode materials for supercapacitor applications

S. No.	Material, synthesis condition and pyrolysis temperature	Surface area (m^2g^{-1})	Electrolyte	Specific capacitance (F g^{-1})	Current density (A g^{-1})	Energy density (Wh kg^{-1})	Power density (W kg^{-1})	Reference
1	Porous carbon nanofiber obtained from HCPs of naphthalene and α,α′-dichloro-p-xylene (HCP-CNF) [$FeCl_3$ as a catalyst; pyrolysis temperature 700 °C]	745.24	3 M KOH	244*	1.0	25.1	400	Babu et al. [14]
2	Porous carbon electrode obtained from HCPs of phloroglucinol and α,α′-dichloro-p-xylene [$FeCl_3$ as a catalyst; pyrolysis temperature 700 °C]	596.22	1 M Na_2SO_4	86**	0.1	–	–	Vinodh et al. [138, 139, 142, 144]
3	Porous carbon derived from hyper cross-linking of p-xylene [$FeCl_3$ as a catalyst; dimethoxy methane as the crosslinker; pyrolysis temperature 800 °C]	489	3 M KOH	242.5**	1.25	–	–	Kim et al. [72]

(continued)

Table 7.3 (continued)

S. No.	Material, synthesis condition and pyrolysis temperature	Surface area (m^2g^{-1})	Electrolyte	Specific capacitance (F g^{-1})	Current density (A g^{-1})	Energy density (Wh kg^{-1})	Power density (W kg^{-1})	Reference
4	Hetero atom (nitrogen and sulfur) anchored porous carbon obtained from pyrrole and thiophene-based HCP [$FeCl_3$ as a catalyst; dimethoxy methane as the crosslinker; pyrolysis temperature varying from 300 to 800 °C]	1339	6 M KOH	464*	0.2	16.2	50	Lu et al. [86]
5	N-doped carbon derived from HCP of pyrrole (N-Py-700) [$FeCl_3$ as a catalyst; dimethoxy methane as the crosslinker; pyrolysis temperature 800 °C]	329	1 M H_2SO_4	317*	0.1	–	–	Lee et al. [79]

(continued)

Table 7.3 (continued)

S. No.	Material, synthesis condition and pyrolysis temperature	Surface area (m^2g^{-1})	Electrolyte	Specific capacitance (F g^{-1})	Current density (A g^{-1})	Energy density (Wh kg^{-1})	Power density (W kg^{-1})	Reference
6	N-anchored porous carbon electrode obtained from HCP of anthracene [$FeCl_3$ as a catalyst; dimethoxy methane as the crosslinker; urea was used for nitrogen source; pyrolysis temperature 600 to 800 °C]	928	6 M KOH	206.4*	1.0	28.6	200	Vinodh et al. [141]
7	N-doped porous carbon obtained from conjugated microporous polymer [pyrolysis temperature 700 °C]	718	6 M KOH	264*	0.1	3.6	5 k	Xu et al. [164, 165]

(continued)

Table 7.3 (continued)

S. No.	Material, synthesis condition and pyrolysis temperature	Surface area (m^2g^{-1})	Electrolyte	Specific capacitance (F g^{-1})	Current density (A g^{-1})	Energy density (Wh kg^{-1})	Power density (W kg^{-1})	Reference
8	Composite electrode material derived from HCP of pyrene and electrochemical polymerization of aniline ($FeCl_3$ as a catalyst; paraformaldehyde as the crosslinker; pyrolysis temperature 800 °C)	827	6 M KOH	16.83*	0.1	3.72	65.07	Jung et al. [66]
9	Hypercross-linked styrene resins occupied with graphite nanoplatelets	1900	0.5 M H_2SO_4	60.4**, €	–	–	–	Sarno et al. [117]

Note *—2-electrode system; **—3-electrode system; €—specific capacitance calculated from cyclic voltammetry; HCP—hyper cross-linked polymers; CNF—carbon nanofibers; KOH—potassium hydroxide appropriate for the commercial aspects. Further improvements are needed to attain a high specific power and specific energies using a wider operating voltage window in an environmentally friendly ionic liquid electrolyte system and their application in the large-scale industrial sector. Additional efforts are also needed to promote large-scale commercial electronics applications. Greater knowledge and healthier experience are the key essential needs to be addressed to overcome the challenges offered by conducting polymer materials and their composites by implementing the novel synthesis and fabrication method.

Acknowledgements This work was supported by BK21 Plus Creative Human Resource Education and Research Programs for ICT Convergence in the 4th Industrial Revolution, Pusan National University, Busan, South Korea.

References

1. M. Abdelhamid, G. Snook, in *Encyclopedia of Polymer Science and Technology* ed. by A. Seidel. Conducting Polymers and Their Application in Supercapacitor Devices (Wiley), p. 20. https://doi.org/10.1002/0471440264.pst666
2. M.E. Abdelhamid, G.A. Snook, A.P. O'Mullane, Electropolymerisation of catalytically active PEDOT from an ionic liquid on a flexible carbon cloth using a sandwich cell configuration. ChemPlusChem **80**, 74–82 (2015). https://doi.org/10.1002/cplu.201402235
3. S.N.J.S.Z. Abidin, N.H.N. Azman, S. Kulandaivalu, Y. Sulaiman, Poly(3,4-ethylenedioxythiophene) doped with carbon materials for high-performance supercapacitor: a comparison study. Nanomaterials (2017) (Article ID 5798614, 13 pp) (2017). https://doi.org/10.1155/2017/5798614
4. A. Afzal, F.A. Abuilaiwi, A. Habib, M. Awais, S.B. Waje, M.A. Atieh, Polypyrrole/carbon nanotube supercapacitors: technological advances and challenges. J. Power. Sources **352**, 174–186 (2017). https://doi.org/10.1016/j.jpowsour.2017.03.128
5. F.O. Agyemang, G.M. Tomboc, S. Kwofie, H. Kim, Electrospun carbon nanofiber-carbon nanotubes coated polyaniline composites with improved electrochemical properties for supercapacitors. Electrochim. Acta **259**, 1110–1119 (2018). https://doi.org/10.1016/j.electacta.2017.12.079
6. T. Alamro, M.K. Ram, Polyethylenedioxythiophene and molybdenum disulfide nanocomposite electrodes for supercapacitor applications. Electrochim. Acta **235**, 623–631 (2017). https://doi.org/10.1016/j.electacta.2017.03.102
7. F. Alvi, M.R. Ram, P. Basnayaka, E. Stefanakos, Y. Goswami, A. Hoff, A. Kumar, Electrochemical supercapacitors based on graphene-conducting polythiophenes nanocomposite. J. Electrochem. Soc. **35**, 167 (2011). https://doi.org/10.1149/1.3654215
8. H. An, Y. Wang, X. Wang, L. Zheng, X. Wang, L. Yi, L. Bai, X. Zhang, Polypyrrole/carbon aerogel composite materials for supercapacitor. J. Power. Sources **195**, 6964–6969 (2010). https://doi.org/10.1016/j.jpowsour.2010.04.074
9. T. Anitha, A.E. Reddy, R. Vinodh, H.J. Kim, Y.R. Cho, Preparation and characterization of $CoWO_4/CoMn_2O_4$ nanoflakes composites on Ni foam for electrochemical supercapacitor applications. J Energy Storage **30**, 101483 (2020). https://doi.org/10.1016/j.est.2020.101483
10. B. Anothumakkool, R. Soni, S.N. Bhange, S. Kurungot, Novel scalable synthesis of highly conducting and robust PEDOT paper for a high performance flexible solid supercapacitor. Energy Environ. Sci. **8**, 1339–1347 (2015). https://doi.org/10.1039/C5EE00142K

11. B. Anothumakkool, A.T. Torris, S.N. Bhange, M.V. Badiger, S. Kurungot, Electrodeposited polyethylenedioxythiophene with infiltrated gel electrolyte interface: a close contest of an all-solid-state supercapacitor with its liquid-state counterpart. Nanoscale **6**, 5944–5952 (2014). https://doi.org/10.1039/C4NR00659C
12. M. Azimi, M. Abbaspour, A. Fazli, H. Setoodeh, B. Pourabbas, Investigation on electrochemical properties of polythiophene nanocomposite with graphite derivatives as supercapacitor material on breath figure-decorated PMMA electrode. J. Electron. Mater. **47**, 2093–2102 (2018). https://doi.org/10.1007/s11664-017-6016-5
13. Y. Ba, S. Zhou, S. Jiao, W. Pan, Fabrication of polyaniline/copper sulfide/poly(ethylene terephthalate) thread electrode for flexible fiber-shaped supercapacitors. J. Appl. Polym. Sci. **135**, 46769 (2018). https://doi.org/10.1002/app.46769
14. R.S. Babu, R. Vinodh, A.L.F. de Barros, L.M. Samyn, K. Prasanna, M.A. Maier, C.H.F. Alves, H.J. Kim, Asymmetric supercapacitor based on carbon nanofibers as the anode and two-dimensional copper cobalt oxide nanosheets as the cathode. Chem. Eng. J. **366**, 390–403 (2019). https://doi.org/10.1016/j.cej.2019.02.108
15. L.Z. Bai, Y.H. Wang, S.S. Cheng, F. Li, Z.Y. Zhang, Y.Q. Liu, Synthesis and electrochemical performance of molybdenum disulfide-reduced graphene oxide-polyaniline ternary composites for supercapacitors. Front. Chem. **6**, 218 (2018). https://doi.org/10.3389/fchem.2018.00218
16. K. Balakrishnan, M. Kumar, A. Subramania, Synthesis of polythiophene and its carbonaceous nanofibers as electrode materials for asymmetric supercapacitors. Adv. Mater. Res. **938**, 151–157 (2014)
17. A. Balducci, W.A. Henderson, M. Mastragostino, S. Passerini, P. Simon, F. Soavi, Cycling stability of a hybrid activated carbon//poly(3-methylthiophene) supercapacitor with N-butyl-N-methylpyrrolidinium bis(trifluoromethanesulfonyl)imide ionic liquid as electrolyte. Electrochim. Acta **50**, 2233–2237 (2005). https://doi.org/10.1016/j.electacta.2004.10.006
18. P. Bharadiya, R. Jain, V. Chaudhari, S. Mishra, Graphene oxide-wrapped polyaniline nanorods for supercapacitor applications. Polym. Compos. **40**(S2), E1716–E1724 (2019). https://doi.org/10.1002/pc.25129
19. P. Bhardwaj, S. Singh, P.R. Kharangarh, A.M. Grace, Surfactant decorated polypyrrole-carbon materials composites electrodes for supercapacitor. Diam. Relat. Mater. **108**, 107989 (2020). https://doi.org/10.1016/j.diamond.2020.107989
20. M. Boota, B. Anasori, C. Voigt, M.Q. Zhao, M.W. Barsoum, Y. Gogotsi, Pseudocapacitive electrodes produced by oxidant-free polymerization of pyrrole between the layers of 2D titanium carbide (MXene). Adv. Mater. **28**, 1517–1522 (2016). https://doi.org/10.1002/adma.201504705
21. T. Cetinkaya, R.A.W. Dryfe, Electrical double layer supercapacitors based on graphene nanoplatelets electrodes in organic and aqueous electrolytes: effect of binders and scalable performance. J. Power. Sources **408**, 91–104 (2018). https://doi.org/10.1016/j.jpowsour.2018.10.072
22. C. Chang, X. Yang, S. Xiang, H. Que, M. Li, Layered MoS_2/PPy nanotube composites with enhanced performance for supercapacitors. J. Mater. Sci. Mater. Electron. **28**, 1777–1784 (2017). https://doi.org/10.1007/s10854-016-5725-5
23. C.-M. Chang, C.-J. Weng, C.-M. Chien, T.-L. Chuang, T.-Y. Lee, J.-M. Yeh, Y. Wei, Polyaniline/carbon nanotube nanocomposite electrodes with biomimetic hierarchical structure for supercapacitors. J Mater Chem A **1**(46), 14719–14728 (2013). https://doi.org/10.1039/C3TA13758A
24. T. Chau, R. Singhal, D. Lawrence, V. Kalra, Polyaniline-coated freestanding porous carbon nanofibers as efficient hybrid electrodes for supercapacitors. J. Power. Sources **293**, 373–379 (2015). https://doi.org/10.1016/j.jpowsour.2015.05.054
25. I.W.P. Chen, Y.C. Chou, P.Y. Wang, Integration of ultrathin MoS_2/PANI/CNT composite paper in producing all-solid-state flexible supercapacitors with exceptional volumetric energy density. J. Phys. Chem. C **123**(29), 17864–17872 (2019). https://doi.org/10.1021/acs.jpcc.9b04046

26. Y. Chen, W. Ma, K. Cai, X. Yang, C. Huang, In situ growth of polypyrrole onto three-dimensional tubular MoS_2 as an advanced negative electrode material for supercapacitor. Electrochim. Acta **246**, 615–624 (2017). https://doi.org/10.1016/j.electacta.2017.06.102
27. N.R. Chiou, A.J. Epstein, Polyaniline nanofibers prepared by dilute polymerization. Adv. Mater. **17**, 1679–1683 (2005). https://doi.org/10.1002/adma.200401000
28. S. Chu, A. Majumdar, Opportunities and challenges for a sustainable energy future. Nature **488**, 294–303 (2012). https://doi.org/10.1038/nature11475
29. H.D. Chul, R. Vinodh, C.V.V.M. Gopi, C. Deviprasath, H.J. Kim, M. Yi, Effect of the cobalt and zinc ratio on the preparation of zeolitic imidazole frameworks (ZIFs): synthesis, characterization and supercapacitor applications. Dalton Trans. **48**, 14808–14819 (2019). https://doi.org/10.1039/C9DT03306H
30. H.P. Cong, X.C. Ren, P. Wang, S.H. Yu, Flexible graphene–polyaniline composite paper for high-performance supercapacitor. Energy Environ. Sci. **6**, 1185–1191 (2013). https://doi.org/10.1039/C2EE24203F
31. J.M. D'Arcy, M.F. El-Kady, P.P. Khine, L. Zhang, S.H. Lee, N.R. Davis, D.S. Liu, M.T. Yeung, S.Y. Kim, C.L. Turner, A.T. Lech, P.T. Hammond, R.B. Kaner, Vapor-phase polymerization of nanofibrillar poly(3,4-ethylenedioxythiophene) for supercapacitors. ACS Nano **8**, 1500–1510 (2014). https://doi.org/10.1021/nn405595r
32. P. Damlin, C. Kvarnström, A. Ivaska, Electrochemical synthesis and in situ spectroelectrochemical characterization of poly(3,4-ethylenedioxythiophene) (PEDOT) in room temperature ionic liquids. J. Electroanal. Chem. **570**, 113–122 (2004). https://doi.org/10.1016/j.jelechem.2004.03.023
33. V.A. Davankov, A.V. Pastukhov, M.P. Tsyurupa, Unusual mobility of hypercrosslinked polystyrene networks: Swelling and dilatometric studies. J. Polym. Sci. Part B Polym. Phys. **38**, 1553–1563 (2000). https://doi.org/10.1002/(SICI)1099-0488(20000601)38:11%3c1553::AID-POLB160%3e3.0.CO;2-L
34. V.A. Davankov, S.V. Rogoshin, M.P. Tsyurupa, Macronet isoporous gels through crosslinking of dissolved polystyrene. J. Polym. Sci. Polym. Symp. **47**, 95–101 (1974). https://doi.org/10.1002/polc.5070470113
35. V.A. Davankov, S.V. Rogozhin, M.P. Tsyurupa, Macronet polystyrene structures for ionites and method of producing same. US Patent 3729457 (1973)
36. V.A. Davankov, G.I. Timofeeva, M.M. Ilyin, M.P. Tsyurupa, Formation of regular clusters through self-association of intramolecularly hypercrosslinked polystyrene-type nanosponges. J. Polym. Sci. Part A Polym. Chem. **35**, 3847–3852 (1997). https://doi.org/10.1002/(SICI)1099-0518(199712)35:17%3c3847::AID-POLA23%3e3.0.CO;2-C
37. V.A. Davankov, M.P. Tsyurupa, Structure and properties of hypercrosslinked polystyrene—the first representative of a new class of polymer networks. React. Polym. **13**, 27–42 (1990). https://doi.org/10.1016/0923-1137(90)90038-6
38. D. Dhawale, A. Vinu, C. Lokhande, Stable nanostructured polyaniline electrode for supercapacitor application. Electrochim. Acta **56**, 9482–9487 (2011). https://doi.org/10.1016/j.electacta.2011.08.042
39. R.S. Diggikar, J.D. Ambekar, M.V. Kulkarni, B.B. Kale, Nanocrystalline silver vanadium sulfide (SVS) anchored polyaniline (PANI): new nanocomposite system for supercapacitor. New J. Chem. **37**, 3236–3243 (2013). https://doi.org/10.1039/C3NJ00500C
40. D. Dong, Ternary composite MnO_2@MoS_2/Polypyrrole from in-situ synthesis for binder-free and flexible supercapacitor. J. Bioresour. Bioprod. **4**(4), 242–250 (2019). https://doi.org/10.12162/jbb.v4i4.010
41. D.P. Dubal, S.V. Patil, W.B. Kim, C.D. Lokhande, Supercapacitors based on electrochemically deposited polypyrrole nanobricks. Mater. Lett. **65**, 2628–2631 (2011). https://doi.org/10.1016/j.matlet.2011.05.114
42. A. Elschner, S. Kirchmeyer, W. Lovenich, U. Merker, K. Reuter, *PEDOT Principle and Applications of an Intrinsically Conductive Polymer*, 1st edn (CRC Press, 2010), pp. 167–264. https://doi.org/10.1201/b10318

43. L.Z. Fan, J. Maier, High-performance polypyrrole electrode materials for redox supercapacitors. Electrochem. Commun. **8**, 937–940 (2006). https://doi.org/10.1016/j.elecom.2006.03.035
44. K.D. Fong, T. Wang, S.K. Smoukov, Multidimensional performance optimization of conducting polymer-based supercapacitor electrodes. Sustain Energy Fuels **1**, 1857–1874 (2017). https://doi.org/10.1039/C7SE00339K
45. E. Frackowiak, V. Khomenko, K. Jurewicz, K. Lota, F. Béguin, Supercapacitors based on conducting polymers/nanotubes composites. J. Power. Sources **153**, 413–418 (2006). https://doi.org/10.1016/j.jpowsour.2005.05.030
46. C. Fu, H. Zhou, R. Liu, Z. Huang, J. Chen, Y. Kuang, Supercapacitor based on electropolymerized polythiophene and multi-walled carbon nanotubes composites. Mater. Chem. Phys. **132**, 596–600 (2012). https://doi.org/10.1016/j.matchemphys.2011.11.074
47. H. Gao, X. Wang, G. Wang, C. Hao, S. Zhou, C. Huang, An urchin-like $MgCo_2O_4$@PPy core–shell composite grown on Ni foam for a high-performance all-solid-state asymmetric supercapacitor. Nanoscale **10**, 10190–10202 (2018). https://doi.org/10.1039/C8NR02311E
48. X. Gao, B. Dong, S. Chen, B. Chen, X. Xiao, J. Zhou, X. Su, D. Zou, Facile preparation of poly(3,4-ethylenedioxythiophene)/MnO_2 composite electrodes for efficient supercapacitors. ChemElectroChem **3**(11), 1746–1752 (2016). https://doi.org/10.1002/celc.201600350
49. Y. Ge, R. Jalili, C. Wang, T. Zheng, Y. Chao, G.G. Wallace, A robust free-standing MoS_2/poly(3,4-ethylenedioxythiophene):poly(styrenesulfonate) film for supercapacitor applications. Electrochim. Acta **235**, 348–355 (2017). https://doi.org/10.1016/j.electacta.2017.03.069
50. H. Gomez, M.K. Ram, F. Alvi, P. Villalba, E. Stefanakos, A. Kumar, Graphene-conducting polymer nanocomposite as novel electrode for supercapacitors. J. Power. Sources **196**, 4102–4128 (2011). https://doi.org/10.1016/j.jpowsour.2010.11.002
51. J.B. Goodenough, *Basic Research Needs for Electrical Energy Storage* (2007). https://doi.org/10.2172/935429
52. C.V.V.M. Gopi, S. Sambasivam, K.V.G. Raghavendra, R. Vinodh, I.M. Obaidat, H.J. Kim, Facile synthesis of hierarchical flower-like $NiMoO_4$-$CoMoO_4$ nanosheet arrays on nickel foam as an efficient electrode for high rate hybrid supercapacitors. J Energy Storage **30**, 101550 (2020). https://doi.org/10.1016/j.est.2020.101550
53. H. Guan, L.Z. Fan, H. Zhang, X. Qua, Polyaniline nanofibers obtained by interfacial polymerization for high-rate supercapacitors. Electrochim. Acta **56**, 964–968 (2010). https://doi.org/10.1016/j.electacta.2010.09.078
54. G. Han, Y. Liu, L. Zhang, E. Kan, S. Zhang, J. Tang, W. Tang, Nanorods intercalating graphene oxide/polyaniline ternary composites for robust high-performance supercapacitors. Sci. Rep. **4**, 4824 (2014). https://doi.org/10.1038/srep04824
55. Y. Han, X. Qing, S. Ye, Y. Lu, Conducting polypyrrole with nanoscale hierarchical structure. Synth. Met. **160**, 1159–1166 (2010). https://doi.org/10.1016/j.synthmet.2010.03.002
56. C.C. Hu, C.H. Chu, Electrochemical impedance characterization of polyaniline-coated graphite electrodes for electrochemical capacitors-effects of film coverage/thickness and anions. J. Electroanal. Chem. **503**, 105–116 (2001). https://doi.org/10.1016/S0022-0728(01)00385-0
57. C. Huang, C. Hao, W. Zheng, S. Zhou, L. Yang, X. Wang, C. Jiang, L. Zhu, Synthesis of polyaniline/nickel oxide/sulfonated graphene ternary composite for all-solid-state asymmetric supercapacitor. Appl. Surf. Sci. **505**, 144589 (2020). https://doi.org/10.1016/j.apsusc.2019.144589
58. J.X. Huang, R.B. Kaner, Nanofiber formation in the chemical polymerization of aniline: a mechanistic study. Angew. Chem. **116**, 5817–5821 (2004). https://doi.org/10.1002/anie.200460616
59. J.X. Huang, R.B. Kaner, A general chemical route to polyaniline nanofibers. J. Am. Chem. Soc. **126**, 851–855 (2004). https://doi.org/10.1021/ja0371754
60. J.X. Huang, S. Virji, B.H. Weiller, R.B. Kaner, Polyaniline nanofibers: facile synthesis and chemical sensors. J. Am. Chem. Soc. **125**, 314–315 (2003). https://doi.org/10.1021/ja028371y

61. Y. Huang, H. Li, Z. Wang, M. Zhu, Z. Pei, Q. Xue, Y. Huang, C. Zhi, Nanostructured polypyrrole as a flexible electrode material of supercapacitor. Nano Energy **22**, 422–438 (2016). https://doi.org/10.1016/j.nanoen.2016.02.047
62. A. Imani, G. Farzi, Facile route for multi-walled carbon nanotube coating with polyaniline: tubular morphology nanocomposites for supercapacitor applications. J. Mater. Sci. Mater. Electron. **26**, 7438–7444 (2015). https://doi.org/10.1007/s10854-015-3377-5
63. M.D. Ingram, H. Staesche, K.S. Ryder, 'Activated' polypyrrole electrodes for high-power supercapacitor applications. Solid State Ion. **169**, 51–57 (2004). https://doi.org/10.1016/j.ssi.2002.12.003
64. D. Jacob, P.A. Mini, A. Balakrishnan, S. Nair, K. Subramanian, Electrochemical behaviour of graphene–poly (3,4-ethylenedioxythiophene) (PEDOT) composite electrodes for supercapacitor applications. Bull. Mater. Sci. **37**, 61–69 (2014). https://doi.org/10.1007/s12034-013-0610-9
65. Y. Jiao, G. Chen, D. Chen, J. Pei, Y. Hu, Bimetal–organic framework assisted polymerization of pyrrole involving air oxidant to prepare composite electrodes for portable energy storage. J. Mater. Chem. A **5**, 23744–23752 (2017). https://doi.org/10.1039/C7TA07464F
66. H.C. Jung, R. Vinodh, C.V.V.M. Gopi, M. Yi, H.J. Kim, Novel composite electrode material derived from hypercross-linked polymer of pyrene and polyaniline for symmetric supercapacitor. Mater. Lett. **257**, 126732 (2019). https://doi.org/10.1016/j.matlet.2019.126732
67. J.P. Jyothibasu, M.Z. Chen, R.H. Lee, Polypyrrole/carbon nanotube freestanding electrode with excellent electrochemical properties for high-performance all-solid-state supercapacitors. ACS Omega **5**(12), 6441–6451 (2020). https://doi.org/10.1021/acsomega.9b04029
68. J.P. Jyothibasu, D.W. Kuo, R.H. Lee, Flexible and freestanding electrodes based on polypyrrole/carbon nanotube/cellulose composites for supercapacitor application. Cellulose **26**, 4495–4513 (2019). https://doi.org/10.1007/s10570-019-02376-2
69. A. Karimi, I. Kazeminezhad, L. Naderi, S. Shahrokhian, Construction of a ternary nanocomposite, polypyrrole/Fe–Co sulfide-reduced graphene oxide/nickel foam, as a novel binder-free electrode for high-performance asymmetric supercapacitors. J. Phys. Chem. C **124**(8), 4393–4407 (2020). https://doi.org/10.1021/acs.jpcc.9b11010
70. S.N. Karri, P. Srinivasan, Synthesis of PEDOT:PSS using benzoyl peroxide as an alternative oxidizing agent for ESD coating and electro-active material in supercapacitor. Mater. Sci. Energy Technol. **2**, 208–215 (2019). https://doi.org/10.1016/j.mset.2019.01.008
71. A. Khosrozadeh, M. Xing, Q. Wang, A high-capacitance solid-state supercapacitor based on free-standing film of polyaniline and carbon particles. Appl. Energy **153**, 87–93 (2015). https://doi.org/10.1016/j.apenergy.2014.08.046
72. S.H. Kim, R. Vinodh, C.V.V.M. Gopi, V.G.R. Kummara, S. Sambasivam, I.M. Obaidat, H.J. Kim, Novel porous carbon material derived from hypercross-linked polymer of p-xylene for supercapacitors electrode. Mater. Lett. **263**, 127222 (2020). https://doi.org/10.1016/j.matlet.2019.127222
73. P.J. Kulesza, M. Skunik, B. Baranowska, K. Miecznikowski, M. Chojak, K. Karnicka, E. Frackowiak, F. Beguin, A. Kuhn, M.H. Delville, B. Starobrzynska, A. Ernst, Fabrication of network films of conducting polymer-linked polyoxometallate-stabilized carbon nanostructures. Electrochim. Acta **51**, 2373–2379 (2006). https://doi.org/10.1016/j.electacta.2005.06.041
74. M.S. Kumar, K.Y. Yasoda, S.K. Batabyal, N.K. Kothurkar, Carbon-polyaniline nanocomposites as supercapacitor materials. Mater. Res. Exp. **5**(4), 045505 (2018). https://doi.org/10.1088/2053-1591/aab911
75. N. Kurra, J. Park, H.N. Alshareef, A conducting polymer nucleation scheme for efficient solid-state supercapacitors on paper. J Mater Chem A **2**, 17058–17065 (2014). https://doi.org/10.1039/C4TA03603D
76. A. Laforgue, P. Simon, C. Sarrazin, J.F. Fauvarque, Polythiophene-based supercapacitors. J. Power. Sources **80**, 142–148 (1999). https://doi.org/10.1016/S0378-7753(98)00258-4

77. J.A. Lee, M.K. Shin, S.H. Kim, H.U. Cho, G.M. Spinks, G.G. Wallace, M.D. Lima, X. Lepró, M.E. Kozlov, R.H. Baughman, S.J. Kim, Ultrafast charge and discharge biscrolled yarn supercapacitors for textiles and microdevices. Nat. Commun. **4**, 1970 (2013). https://doi.org/10.1038/ncomms2970
78. J.A. Lee, M.K. Shin, S.H. Kim, S.J. Kim, G.M. Spinks, G.G. Wallace, R. Ovalle-Robles, M.D. Lima, M.E. Kozlov, R.H. Baughman, Hybrid nanomembranes for high power and high energy density supercapacitors and their yarn application. ACS Nano **6**, 327–334 (2012). https://doi.org/10.1021/nn203640a
79. J.S.M. Lee, M.E. Briggs, C.C. Hu, A.I. Cooper, Controlling electric double-layer capacitance and pseudocapacitance in heteroatom-doped carbons derived from hypercrosslinked microporous polymers. Nano Energy **46**, 277–289 (2018). https://doi.org/10.1016/j.nanoen.2018.01.042
80. D. Li, J.X. Huang, R.B. Kaner, Polyaniline nanofibers: a unique polymer nanostructure for versatile applications. Acc. Chem. Res. **42**, 135–145 (2009). https://doi.org/10.1021/ar800080n
81. H.L. Li, J.X. Wang, Q.X. Chu, Z. Wang, F.B. Zhang, S.C. Wang, Theoretical and experimental specific capacitance of polyaniline in sulfuric acid. J. Power. Sources **190**, 578–586 (2009). https://doi.org/10.1016/j.jpowsour.2009.01.052
82. Z.F. Li, H. Zhang, Q. Liu, L. Sun, L. Stanciu, J. Xie, Fabrication of high-surface-area graphene/polyaniline nanocomposites and their application in supercapacitors. ACS Appl. Mater. Interf. **5**, 2685–2691 (2013). https://doi.org/10.1021/am4001634
83. C. Liu, F. Li, L.P. Ma, H.M. Cheng, Advanced materials for energy storage. Adv. Mater. **22**, E28–E62 (2010). https://doi.org/10.1002/adma.200903328
84. Y. Liu, A.P.F. Turner, M. Zhao, W.C. Mak, Facile synthesis of highly processable and water dispersible polypyrrole and poly(3,4-ethylenedioxythiophene) microspheres for enhanced supercapacitive performance. Eur. Polym. J. **99**, 332–339 (2018). https://doi.org/10.1016/j.eurpolymj.2017.12.013
85. Y. Liu, N. Xu, W. Chen, X. Wang, C. Suna, Z. Su, Supercapacitor with high cycling stability through electrochemical deposition of metal–organic frameworks/polypyrrole positive electrode. Dalton Trans. **47**, 13472–13478 (2018). https://doi.org/10.1039/C8DT02740D
86. Y. Lu, J. Liang, S. Deng, Q. He, S. Deng, Y. Hu, D. Wang, Hypercrosslinked polymers enabled micropore-dominant N, S co-doped porous carbon for ultrafast electron/ion transport supercapacitors. Nano Energy **65**, 103993 (2019). https://doi.org/10.1016/j.nanoen.2019.103993
87. Y. Ma, C. Hou, H. Zhang, Q. Zhang, H. Liu, S. Wu, Z. Guo, Three-dimensional core-shell Fe_3O_4/polyaniline coaxial heterogeneous nanonets: preparation and high performance supercapacitor electrodes. Electrochim. Acta **315**, 114–123 (2019). https://doi.org/10.1016/j.electacta.2019.05.073
88. M. Mastragostino, C. Arbizzani, F. Sovai, Polymer-based supercapacitors. J. Power. Sources **97**, 812–815 (2001). https://doi.org/10.1016/S0378-7753(01)00613-9
89. C.Z. Meng, C.H. Liu, L.Z. Chen, C.H. Hu, S.S. Fan, Highly flexible and all-solid-state paper-like polymer supercapacitors. Nano Lett. **10**, 4025–4031 (2010). https://doi.org/10.1021/nl1019672
90. Q. Meng, K. Cai, Y. Chen, L. Chen, Review: research progress on conducting polymer based supercapacitor electrode materials. Nano Energy **36**, 268–285 (2017). https://doi.org/10.1016/j.nanoen.2017.04.040
91. H.Y. Mi, X.G. Zhang, S.Y. An, X.G. Ye, S.D. Yang, Microwave-assisted synthesis and electrochemical capacitance of polyaniline/multi-wall carbon nanotubes composite. Electrochem. Commun. **9**, 2859–2862 (2007). https://doi.org/10.1016/j.elecom.2007.10.013
92. S. Mu, C. Chen, J. Wang, The kinetic behavior for the electrochemical polymerization of aniline in aqueous solution. Synth. Met. **88**, 249–254 (1997). https://doi.org/10.1016/S0379-6779(97)03863-0

93. M.A. Nasalevich, M.G. Goesten, T.J. Savenije, F. Kapteijn, J. Gascon, Enhancing optical absorption of metal–organic frameworks for improved visible light photocatalysis. Chem. Commun. **49**, 10575–10577 (2013). https://doi.org/10.1039/C3CC46398B
94. X. Ning, W. Zhong, L. Wan, Ultrahigh specific surface area porous carbon nanospheres and its composite with polyaniline: preparation and application for supercapacitors. RSC Adv. **6**, 25519–25524 (2016). https://doi.org/10.1039/C6RA00596A
95. Z.Q. Niu, P.S. Luan, Q. Shao, H.B. Dong, J.Z. Li, J. Chen, D. Zhao, L. Cai, W.Y. Zhou, X.D. Chen, S. Xie, A "skeleton/skin" strategy for preparing ultrathin free-standing single-walled carbon nanotube/polyaniline films for high performance supercapacitor electrodes. Energy Environ. Sci. **5**, 8726–8733 (2012). https://doi.org/10.1039/C2EE22042C
96. B. Pal, S. Yang, S. Ramesh, V. Thangadurai, R. Jose, Electrolyte selection for supercapacitive devices: a critical review. Nanoscale Adv. **1**, 3807–3835 (2019). https://doi.org/10.1039/C9NA00374F
97. G.P. Pandey, A.C. Rastogi, C.R. Westgate, All-solid-state supercapacitors with poly(3,4-ethylenedioxythiophene)-coated carbon fiber paper electrodes and ionic liquid gel polymer electrolyte. J. Power. Sources **245**, 857–865 (2014). https://doi.org/10.1016/j.jpowsour.2013.07.017
98. A.D. Pasquier, A. Laforgue, P. Simon, G.G. Amatucci, J.F. Fauvarque, A nonaqueous asymmetric hybrid $Li_4Ti_5O_{12}$/poly(fluorophenylthiophene) energy storage device. J. Electrochem. Soc. **149**, A302–A306 (2002). https://doi.org/10.1149/1.1446081
99. A.V. Pastukhov, M.P. Tsyurupa, V.A. Davankov, Hypercrosslinked polystyrene: a polymer in a non-classical physical state. J. Polym. Sci. Part B Polym. Phys. **37**, 2324–2333 (1999). https://doi.org/10.1002/(SICI)1099-0488(19990901)37:17%3c2324::AID-POLB4%3e3.0.CO;2-B
100. B.H. Patil, A.D. Jagadale, C.D. Lokhande, Synthesis of polythiophene thin films by simple successive ionic layer adsorption and reaction (SILAR) method for supercapacitor application. Synth. Met. **162**, 1400–1405 (2012). https://doi.org/10.1016/j.synthmet.2012.05.023
101. B.H. Patil, S.J. Patil, C.D. Lokhande, Electrochemical characterization of chemically synthesized polythiophene thin films: performance of asymmetric supercapacitor device. Electroanalysis **26**, 2023–2032 (2014). https://doi.org/10.1002/elan.201400284
102. U. Patil, S.C. Lee, S. Kulkarni, J.S. Sohn, M.S. Nam, S. Han, S.C. Jun, Nanostructured pseudocapacitive materials decorated 3D graphene foam electrodes for next generation supercapacitors. Nanoscale **7**, 6999–7021 (2015). https://doi.org/10.1039/C5NR01135C
103. H. Peng, G. Ma, K. Sun, J. Mu, H. Wanga, Z. Lei, High-performance supercapacitor based on multi-structural CuS@polypyrrole composites prepared by in situ oxidative polymerization. J. Mater. Chem. A **2**, 3303–3307 (2014). https://doi.org/10.1039/C3TA13859C
104. S. Peng, L. Fan, C. Wei, H. Bao, H. Zhang, W. Xu, J. Xu, Polypyrrole/nickel sulfide/bacterial cellulose nanofibrous composite membranes for flexible supercapacitor electrodes. Cellulose **23**, 2639–2651 (2016). https://doi.org/10.1007/s10570-016-0981-3
105. S. Peng, Q. Xu, L. Fan, C. Wei, H. Bao, W. Xu, J. Xu, Flexible polypyrrole/cobalt sulfide/bacterial cellulose composite membranes for supercapacitor application. Synth. Met. **222**, 285–292 (2016). https://doi.org/10.1016/j.synthmet.2016.11.002
106. T.N. Phan, M.K. Gong, R. Thangavel, Y.S. Lee, C.H. Ko, Enhanced electrochemical performance for EDLC using ordered mesoporous carbons (CMK-3 and CMK-8): role of mesopores and mesopore structures. J. Alloys Compd. **780**, 90–97 (2019). https://doi.org/10.1016/j.jallcom.2018.11.348
107. W.J. Qian, J.Y. Zhu, Y. Zhang, X. Wu, F. Yan, Condiment-derived 3D architecture porous carbon for electrochemical supercapacitors. Small **11**, 4959–4969 (2015). https://doi.org/10.1002/smll.201500859
108. K.V.G. Raghavendra, C.V.V.M. Gopi, R. Vinodh, I.K. Durga, H.J. Kim, One-step facile synthesis of dense cloud-like tiny bundled nanoparticles of CuS nanostructures as an efficient electrode material for high-performance supercapacitors. J. Energy Storage **27**, 101148 (2020). https://doi.org/10.1016/j.est.2019.101148

109. K.V.G. Raghavendra, C.V.V.M. Gopi, R. Vinodh, S.S. Rao, I.M. Obaidat, H.J. Kim, Facile synthesis of nanoparticles anchored on honeycomb-like $MnCo_2S_4$ nanostructures as a binder-free electroactive material for supercapacitors. J Energy Storage **27**, 101159 (2020). https://doi.org/10.1016/j.est.2019.101159
110. K. Rajendra Prasad, N. Munichandraiah, Electrochemical studies of polyaniline in a gel polymer electrolyte. Electrochem. Solid-State Lett. **5**, A271–A274 (2002). https://doi.org/10.1149/1.1516910
111. L. Ren, G. Zhang, Z. Yan, L. Kang, H. Xu, F. Shi, Z. Lei, Z.H. Liu, Three-dimensional tubular MoS_2/PANI hybrid electrode for high rate performance supercapacitor. ACS Appl. Mater. Interf. **7**(51), 28294–28302 (2015). https://doi.org/10.1021/acsami.5b08474
112. A. Roy, A. Ray, P. Sadhukhan, S. Saha, S. Das, Morphological behaviour, electronic bond formation and electrochemical performance study of V_2O_5-polyaniline composite and its application in asymmetric supercapacitor. Mater. Res. Bull. **107**, 379–390 (2018). https://doi.org/10.1016/j.materresbull.2018.08.013
113. H.S. Roy, M.M. Islam, A. Yousuf, M.Y.A. Mollah, M.A.B.H. Susan, Polyaniline-MnO_2 composites prepared in-situ during oxidative polymerization of aniline for supercapacitor applications. Mater. Today Proc. **29**, 1013–1019 (2020). https://doi.org/10.1016/j.matpr.2020.04.635
114. K.S. Ryu, Y.G. Lee, Y.S. Hong, Y.J. Park, X. Wu, K.M. Kim, M.G. Kang, N.G., Park, S.H. Chang, Poly(ethylenedioxythiophene) (PEDOT) as polymer electrode in redox supercapacitor. Electrochim. Acta. **50**, 843–847 (2004). https://doi.org/10.1016/j.electacta.2004.02.055
115. R.R. Salunkhe, Y.H. Lee, K.H. Chang, J.M. Li, P. Simon, J. Tang, N.L. Torad, C.C. Hu, Y. Yamauchi, Nanoarchitectured graphene-based supercapacitors for next-generation energy-storage applications. Chem. Eur. J. **20**, 13838–13852 (2014). https://doi.org/10.1002/chem.201403649
116. D. Sarmah, A. Kumar, Ion beam modified molybdenum disulfide-reduced graphene oxide/polypyrrole nanotubes ternary nanocomposite for hybrid supercapacitor electrode. Electrochim. Acta **312**, 392–410 (2019). https://doi.org/10.1016/j.electacta.2019.04.174
117. M. Sarno, R. Castaldo, E. Ponticorvo, D. Scarpa, M. Cocca, V. Ambrogi, G. Gentile, Hyper-cross-linked polymer loaded with graphite nanoplatelets for supercapacitor application. Chem. Eng. Trans. **73**, 121–126 (2019). https://doi.org/10.3303/CET1973021
118. L. Shao, Q. Wang, Z. Ma, Z. Ji, X. Wang, D. Song, Y. Liu, N. Wang, A high-capacitance flexible solid-state supercapacitor based on polyaniline and metal-organic framework (UiO-66) composites. J. Power. Sources **379**, 350–361 (2018). https://doi.org/10.1016/j.jpowsour.2018.01.028
119. K.W. Shen, F. Ran, X.X. Zhang, C. Liu, N.J. Wang, X.Q. Niu, Y. Liu, D.J. Zhang, L.B. Kong, L. Kang, S.W. Chen, Supercapacitor electrodes based on nano-polyaniline deposited on hollow carbon spheres derived from cross-linked co-polymers. Synth. Met. **209**, 369–376 (2015). https://doi.org/10.1016/j.synthmet.2015.08.012
120. H. Sheng, M. Wei, A. D'Aloia, G. Wu, Heteroatom polymer-derived 3D high-surface-area and mesoporous graphene sheet-like carbon for supercapacitors. ACS Appl. Mater. Interf. **8**, 30212–30224 (2016). https://doi.org/10.1021/acsami.6b10099
121. S. Sivakkumar, W.J. Kim, J.A. Choi, D.R. Macfarlane, M. Forsyth, D.W. Kim, Electrochemical performance of polyaniline nanofibres and polyaniline/multi-walled carbon nanotube composite as an electrode material for aqueous redox supercapacitors. J. Power. Sources **171**, 1062–1068 (2007). https://doi.org/10.1016/j.jpowsour.2007.05.103
122. G.A. Snook, A.S. Best, Co-deposition of conducting polymers in a room temperature ionic liquid. J. Mater. Chem. **19**, 4248–4254 (2009). https://doi.org/10.1039/B905156B
123. G.A. Snook, P. Kao, A.S. Best, Conducting-polymer-based supercapacitor devices and electrodes. J. Power. Sources **196**, 1–12 (2011). https://doi.org/10.1016/j.jpowsour.2010.06.084
124. G.A. Snook, C. Peng, D.J. Fray, G.Z. Chen, Achieving high electrode specific capacitance with materials of low mass specific capacitance: potentiostatically grown thick micro-nanoporous

PEDOT films. Electrochem. Commun. **9**, 83–88 (2007). https://doi.org/10.1016/j.elecom.2006.08.037
125. A. Subramania, S.L. Devi, Polyaniline nanofibers by surfactant-assisted dilute polymerization for supercapacitor applications. Polym. Adv. Technol. **19**, 725–727 (2008). https://doi.org/10.1002/pat.1016
126. D. Sun, L. Jin, Y. Chen, J.R. Zhang, J.J. Zhu, Microwave-assisted in situ synthesis of graphene/PEDOT hybrid and its application in supercapacitors. ChemPlusChem **78**, 227–234 (2013). https://doi.org/10.1002/cplu.201200206
127. H. Sun, P. She, K.L. Xu, Y.X. Shang, S.Y. Yin, Z.N. Liu, A self-standing nanocomposite foam of polyaniline@reduced graphene oxide for flexible super-capacitors. Synth. Met. **209**, 68–73 (2015). https://doi.org/10.1016/j.synthmet.2015.07.001
128. L. Sun, T. Miyakai, S. Seki, M. Dinca, Mn_2(2,5-disulfhydrylbenzene-1,4-dicarboxylate): a microporous metal–organic framework with infinite $(-Mn-S-)_\infty$ chains and high intrinsic charge mobility. J. Am. Chem. Soc. **135**, 8185–8188 (2013). https://doi.org/10.1021/ja4037516
129. X. Sun, M. Gan, L. Ma, H. Wang, T. Zhou, S. Wang, W. Dai, H. Wang, Fabrication of PANI-coated honeycomb-like MnO_2 nanospheres with enhanced electrochemical performance for energy storage. Electrochim. Acta **180**, 977–982 (2015). https://doi.org/10.1016/j.electacta.2015.09.056
130. Y.S. Sung, L.Y. Lin, Systematic design of polypyrrole/carbon fiber electrodes for efficient flexible fiber-type solid-state supercapacitors. Nanomaterials **10**, 248 (2020). https://doi.org/10.3390/nano10020248
131. L. Tan, B. Tan, Research progress in hypercrosslinked microporous organic polymers. Acta Chim. Sin. **73**, 530–540 (2015). https://doi.org/10.6023/A15020096
132. L. Tan, B. Tan, Hypercrosslinked porous polymer materials: design, synthesis, and applications. Chem. Soc. Rev. **46**, 3322–3356 (2017). https://doi.org/10.1039/C6CS00851H
133. A.K. Thakur, R.B. Choudhary, M. Majumder, G. Gupta, M.V. Shelke, Enhanced electrochemical performance of polypyrrole coated MoS_2 nanocomposites as electrode material for supercapacitor application. J. Electroanal. Chem. **782**, 278–287 (2016). https://doi.org/10.1016/j.jelechem.2016.10.050
134. A.K. Thakur, M. Majumder, R.B. Choudhary, S.N. Pimpalkar, Supercapacitor based on electropolymerized polythiophene and multiwalled carbon nanotubes composites. Mater. Sci. Eng. **149**, 012166 (2016). https://doi.org/10.1088/1757-899X/149/1/012166
135. M.P. Tsyurupa, Z.K. Blinnikova, N.A. Proskurina, A.V. Pastukhov, L.A. Pavlova, V.A. Davankov, Hypercrosslinked polystyrene: the first nanoporous polymeric material. Nanotechnol. Russ. **4**, 665–675 (2009). https://doi.org/10.1134/S1995078009090109
136. M.P. Tsyurupa, V.A. Davankov, Hypercrosslinked polymers: basic principle of preparing the new class of polymeric materials. React. Funct. Polym. **53**, 193–203 (2002). https://doi.org/10.1016/S1381-5148(02)00173-6
137. M.P. Tsyurupa, V.A. Davankov, Porous structure of hypercrosslinked polystyrene: state-of-the-art mini-review. React. Funct. Polym. **66**, 768–779 (2006). https://doi.org/10.1016/j.reactfunctpolym.2005.11.004
138. R. Vinodh, A. Abidov, M.M. Peng, C.M. Babu, M. Palanichamy, W.S. Cha, H.T. Jang, A new strategy to synthesize hypercross-linked conjugated polystyrene and its application towards CO_2 sorption. Fiber Polym. **16**, 1458–1467 (2015). https://doi.org/10.1007/s12221-015-5151-y
139. R. Vinodh, A. Abidov, M.M. Peng, C.M. Babu, M. Palanichamy, W.S. Cha, H.T. Jang, Synthesis and characterization of semiconducting porous carbon for energy applications and CO_2 adsorption. J. Ind. Eng. Chem. **32**, 273–281 (2015). https://doi.org/10.1016/j.jiec.2015.09.002
140. R. Vinodh, C.M. Babu, A. Abidov, M.M. Peng, M. Palanichamy, W.S. Cha, H.T. Jang, Microporous spheres of tiny semiconducting graphene sheets from hypercross-linked polymers: absorption and CO_2 sorption characteristics. Adv. Polym. Technol. **37**, 714–723 (2018). https://doi.org/10.1002/adv.21713

141. R. Vinodh, R.S. Babu, C.V.V.M. Gopi, C. Deviprasath, R. Atchudan, L.M. Samyn, A.L.F. de Barros, H.J. Kim, M. Yi, Influence of annealing temperature in nitrogen doped porous carbon balls derived from hypercross-linked polymer of anthracene for supercapacitor applications. J. Ener. Storage **28**, 101196 (2020). https://doi.org/10.1016/j.est.2020.101196
142. R. Vinodh, M. Ganesh, M.M. Peng, A. Abidov, M. Palanichamy, W.S. Cha, H.T. Jang, Microporous hypercross-linked conjugated quinonoid chromophores of anthracene: novel polymers for CO_2 adsorption. Chin. J. Polym. Sci. **33**, 224–235 (2015). https://doi.org/10.1007/s10118-015-1573-7
143. R. Vinodh, P. Hemalatha, M. Ganesh, M.M. Peng, A. Abidov, M. Palanichamy, W.S. Cha, H.T. Jang, Novel microporous hypercross-linked conjugated quinonoid chromophores with broad light absorption and CO_2 sorption characteristics. RSC Adv. **4**, 3668–3674 (2014). https://doi.org/10.1039/C3RA45466E
144. R. Vinodh, E.M. Jung, M. Ganesh, M.M. Peng, A. Abidov, M. Palanichamy, W.S. Cha, H.T. Jang, Novel microporous hypercross-linked polymers as sorbent for volatile organic compounds and CO_2 adsorption. J. Ind. Eng. Chem. **21**, 1231–1238 (2015). https://doi.org/10.1016/j.jiec.2014.05.039
145. G.P. Wang, L. Zhang, J.J. Zhang, A review of electrode materials for electrochemical supercapacitors. Chem. Soc. Rev. **41**, 797–828 (2012). https://doi.org/10.1039/C1CS15060J
146. H.L. Wang, Q.L. Hao, X.J. Yang, L.D. Lu, X. Wang, Graphene oxide doped polyaniline for supercapacitors. Electrochem. Commun. **11**, 1158–1161 (2009). https://doi.org/10.1016/j.elecom.2009.03.036
147. J. Wang, Z. Wu, H. Yin, W. Li, Y. Jiang, Poly(3,4-ethylenedioxythiophene)/MoS_2 nanocomposites with enhanced electrochemical capacitance performance. RSC Adv. **4**, 56926–56932 (2014). https://doi.org/10.1039/C4RA12683A
148. J. Wang, Y. Xu, X. Chen, X. Du, Electrochemical supercapacitor electrode material based on poly(3,4-ethylenedioxythiophene)/polypyrrole composite. J. Power. Sources **163**, 1120–1125 (2007). https://doi.org/10.1016/j.jpowsour.2006.10.004
149. L. Wang, M. Bo, Z. Guo, H. Li, Z. Huang, H. Che, Z. Feng, Y. Wang, J. Mu, Construction of ultra-stable trinickel disulphide (Ni_3S_2)/polyaniline (PANI) electrodes based on carbon fibers for high performance flexible asymmetric supercapacitors. J. Colloid Interf. Sci. **577**, 29–37 (2020). https://doi.org/10.1016/j.jcis.2020.05.046
150. L. Wang, X. Feng, L.T. Ren, Q.H. Piao, J.Q. Zhong, Y.B. Wang, H.W. Li, Y.F. Chen, B. Wang, Flexible solid-state supercapacitor based on a metal-organic framework interwoven by electrochemically-deposited PANI. J. Am. Chem. Soc. **137**, 4920–4923 (2015). https://doi.org/10.1021/jacs.5b01613
151. M. Wang, L. Zhang, Y. Zhong, M. Huang, Z. Zhen, H. Zhu, In situ electrodeposition of polypyrrole onto $TaSe_2$ nanobelts quasi-arrays for high-capacitance supercapacitor. Nanoscale **10**, 17341–17346 (2018). https://doi.org/10.1039/C8NR05261A
152. S.Y. Wang, L. Ma, M.Y. Gan, S.N. Fu, W.Q. Dai, T. Zhou, X.W. Sun, H.H. Wang, H.N. Wang, Free-standing 3D graphene/polyaniline composite film electrodes for high-performance supercapacitors. J. Power. Sources **299**, 347–355 (2015). https://doi.org/10.1016/j.jpowsour.2015.09.018
153. X. Wang, J. Liu, Y. Wang, C. Zhao, W. Zheng, $Ni(OH)_2$ nanoflakes electrodeposited on Ni foam-supported vertically oriented graphene nanosheets for application in asymmetric supercapacitors. Mater. Res. Bull. **52**, 89–95 (2014). https://doi.org/10.1016/j.materresbull.2013.12.051
154. Y. Wang, C.C. Mayorga-Martinez, M. Pumera, Polyaniline/MoS_X supercapacitor by electrodeposition. Bull. Chem. Soc. Jpn **90**, 847–853 (2017). https://doi.org/10.1246/bcsj.20170076
155. Y. Wang, S. Tao, Y. An, S. Wu, C. Meng, Bio-inspired high performance electrochemical supercapacitors based on conducting polymer modified coral-like monolithic carbon. J. Mater. Chem. A **1**, 8876–8887 (2013). https://doi.org/10.1039/C3TA11348E

156. Z. Wang, P. Tammela, J. Huo, P. Zhang, M. Strømme, L. Nyholm, Solution-processed poly(3,4-ethylenedioxythiophene) nanocomposite paper electrodes for high-capacitance flexible supercapacitors. J. Mater. Chem. A **4**, 1714–1722 (2016). https://doi.org/10.1039/C5TA10122K
157. R. Warren, F. Sammoura, K.S. Teh, A. Kozinda, X. Zang, L. Lin, Electrochemically synthesized and vertically aligned carbon nanotube–polypyrrole nanolayers for high energy storage devices. Sens. Actuators A Phys. **231**, 65–73 (2015). https://doi.org/10.1016/j.sna.2014.07.010
158. M. Winter, R.J. Brodd, What are batteries, fuel cells, and supercapacitors? Chem. Rev. **104**, 4245–4270 (2004). https://doi.org/10.1021/cr020730k
159. W. Wu, L. Yang, S. Chen, Y. Shao, L. Jing, G. Zhao, H. Wei, Core–shell nanospherical polypyrrole/graphene oxide composites for high performance supercapacitors. RSC Adv. **5**, 91645–91653 (2015). https://doi.org/10.1039/C5RA17036B
160. Z.S. Wu, K. Parvez, S. Li, S. Yang, Z.Y. Liu, S.H. Liu, X.L. Feng, K. Müllen, Alternating stacked graphene-conducting polymer compact films with ultrahigh areal and volumetric capacitances for high-energy micro-supercapacitors. Adv. Mater. **27**, 4054–4061 (2015). https://doi.org/10.1002/adma.201501643
161. D.F. Xu, X. Xiao, J. Cai, J. Zhou, L.N. Zhang, Highly rate and cycling stable electrode materials constructed from polyaniline/cellulose nanoporous microspheres. J. Mater. Chem. A **3**, 16424–16429 (2015). https://doi.org/10.1039/C5TA03917G
162. J.J. Xu, K. Wang, S.Z. Zu, B.H. Han, Z.X. Wei, Hierarchical nanocomposites of polyaniline nanowire arrays on graphene oxide sheets with synergistic effect for energy storage. ACS Nano **4**, 5019–5026 (2010). https://doi.org/10.1021/nn1006539
163. R. Xu, J. Wei, F. Guo, X. Cui, T. Zhang, H. Zhu, K. Wang, D. Wu, Highly conductive, twistable and bendable polypyrrole–carbon nanotube fiber for efficient supercapacitor electrodes. RSC Adv. **5**, 22015–22021 (2015). https://doi.org/10.1039/C5RA01917F
164. X. Xu, J. Tang, H. Qian, S. Hou, Y. Bando, M.S.A. Hossain, L. Pan, Y. Yamauchi, Three-dimensional networked metal–organic frameworks with conductive polypyrrole tubes for flexible supercapacitors. ACS Appl. Mater. Interf. **9**(44), 38737–38744 (2017). https://doi.org/10.1021/acsami.7b09944
165. Y. Xu, S. Wu, S. Ren, J. Ji, Y. Yue, J. Shen, Nitrogen-doped porous carbon materials generated via conjugated microporous polymer precursors for CO_2 capture and energy storage. RSC Adv. **7**, 32496–32501 (2017). https://doi.org/10.1039/C7RA05551J
166. J. Yang, D.C. Martin, Microporous conducting polymers on neural microelectrode arrays: II. Physical characterization. Sens. Actuators A Phys. **113**, 204–211 (2004). https://doi.org/10.1016/j.sna.2004.02.029
167. S. Yang, L. Sun, X. An, X. Qian, Construction of flexible electrodes based on ternary polypyrrole@cobalt oxyhydroxide/cellulose fiber composite for supercapacitor. Carbohydr. Polym. **229**, 115455 (2020). https://doi.org/10.1016/j.carbpol.2019.115455
168. Y. Yang, L. Zhang, S. Li, W. Yang, J. Xu, Y. Jiang, J. Wen, Electrochemical performance of conducting polymer and its nanocomposites prepared by chemical vapor phase polymerization method. J. Mater. Sci. Mater. Electron. **24**, 2245–2253 (2013). https://doi.org/10.1007/s10854-013-1086-5
169. C. Yin, H. Zhou, J. Li, Influence of doped anions on PEDOT/Ni-Mn-Co-O for supercapacitor electrode material. Appl. Surf. Sci. **464**, 220–228 (2019). https://doi.org/10.1016/j.apsusc.2018.09.028
170. A. Yu, A. Davies, Z. Chen, in *Electrochemical Technologies for Energy Storage and Conversion* ed. by R.S. Liu, Z. Lei, X. Sun, H. Liu, J. Zhang. Electrochemical supercapacitors (Wiley-VCH Verlag & Co. KGaA, Weinheim, 2012)
171. L. Yu, M. Gan, L. Ma, H. Huang, H. Hu, Y. Li, Y. Tu, C. Ge, F. Yang, J. Yan, Facile synthesis of MnO_2/polyaniline nanorod arrays based on graphene and its electrochemical performance. Synth. Met. **198**, 167–174 (2014). https://doi.org/10.1016/j.synthmet.2014.10.010
172. B. Yue, C. Wang, X. Ding, G.G. Wallace, Electrochemically synthesized stretchable polypyrrole/fabric electrodes for supercapacitor. Electrochim. Acta **113**, 17–22 (2013). https://doi.org/10.1016/j.electacta.2013.09.024

173. D. Zhang, Q.Q. Dong, X. Wang, W. Yan, W. Deng, L.Y. Shi, Preparation of a three-dimensional ordered macroporous carbon nanotube/polypyrrole composite for supercapacitors and diffusion modeling. J. Phys. Chem. C **117**, 20446–20455 (2013). https://doi.org/10.1021/jp405850w
174. H. Zhang, Z. Hu, M. Li, L. Hu, S. Jiao, A high-performance supercapacitor based on a polythiophene/multiwalled carbon nanotube composite by electropolymerization in an ionic liquid microemulsion. J. Mater. Chem. A **2**, 17024–17030 (2014). https://doi.org/10.1039/C4TA03369H
175. J. Zhang, X.S. Zhao, Conducting polymers directly coated on reduced graphene oxide sheets as high-performance supercapacitor electrodes. J. Phys. Chem. C **116**, 5420–5426 (2012). https://doi.org/10.1021/jp211474e
176. K. Zhang, L.L. Zhang, X.S. Zhao, J. Wu, Graphene/polyaniline nanofiber composites as supercapacitor electrodes. Chem. Mater. **22**(4), 1392–1401 (2010). https://doi.org/10.1021/cm902876u
177. R. Zhang, Y. Liao, S. Ye, Z. Zhu, J. Qian, Novel ternary nanocomposites of MWCNTs/PANI/MoS_2: preparation, characterization and enhanced electrochemical capacitance. R. Soc. Open Sci. **5**(1), 171365 (2018). https://doi.org/10.1098/rsos.171365
178. P. Zhao, N. Wang, M. Yao, H. Ren, W. Hu, Hydrothermal electrodeposition incorporated with CVD-polymerisation to tune PPy@MnO_2 interlinked core-shell nanowires on carbon fabric for flexible solid-state asymmetric supercapacitors. Chem. Eng. J. **380**, 122488 (2020). https://doi.org/10.1016/j.cej.2019.122488
179. M. Zhou, Y. Li, Q. Gong, Z. Xia, Y. Yang, X. Liu, J. Wang, Q. Gao, Polythiophene grafted onto single-wall carbon nanotubes through oligo(ethylene oxide) linkages for supercapacitor devices with enhanced electrochemical performance. ChemElectroChem **6**, 4595–4607 (2019). https://doi.org/10.1002/celc.201901074
180. C. Zhu, Y. He, Y. Liu, N. Kazantseva, P. Saha, Q. Cheng, ZnO@MOF@PANI core-shell nanoarrays on carbon cloth for high-performance supercapacitor electrodes. J. Energy Chem. **35**, 124–131 (2019). https://doi.org/10.1016/j.jechem.2018.11.006
181. J. Zhu, Y. Xu, J. Wang, J. Wang, Y. Bai, X. Du, Morphology controllable nano-sheet polypyrrole–graphene composites for high-rate supercapacitor. Phys. Chem. Chem. Phys. **17**, 19885–19894 (2015). https://doi.org/10.1039/C5CP02710A

Chapter 8
Hybrid Nanocomposite Materials for Supercapacitor Applications

D. Karthigaimuthu, Abdel-Hamid I. Mourad, and Elangovan Thangavel

Abstract The attention toward supercapacitors, especially hybrid ones, has surged due to their distinctive electrochemical properties. The high performance in specific capacitance, stability, energy, and power density makes hybrid nanostructured materials, known for their cost-effectiveness and unique attributes, a promising renewable option for eco-friendly energy storage devices. This chapter explains the ideas, types, properties, and ways to make electrode materials for hybrid supercapacitors. This chapter emphasizes various materials like graphene-metal oxide hybrids, Li- and Na-ion hybrids, polymer-mixed hybrids, organic-inorganic hybrids, advanced nano-hybrid metal-organic frameworks, and flexible, wearable-based nano-hybrid materials, aiming to make hybrid supercapacitors better.

Keywords Nanostructures · Hybrid supercapacitors · Energy storage · Hybrid nanostructures · Polymer nanocomposites · Metal-organic framework

8.1 Introduction

Looking at energy from a scientific perspective, it has a long history of changing society with exciting innovations. Over the last thirty years, there has been a strong focus on improving how we collect and use energy [1, 2]. As technology advances, electronic devices are getting better at using both renewable and non-renewable energy. Now, there is a growing interest in finding alternative energy sources to meet the energy needs of people [3]. This has a big impact on our twenty-first-century

D. Karthigaimuthu · A.-H. I. Mourad
Department of Mechanical and Aerospace Engineering, College of Engineering, United Arab Emirates University, Al-Ain, United Arab Emirates

National Water and Energy Center, United Arab Emirates University, Al Ain, United Arab Emirates

D. Karthigaimuthu · E. Thangavel (✉)
Smart Energy Materials Research Laboratory (SEMRL), Department of Energy Science and Technology, Periyar University, Salem, India
e-mail: elangoes@periyaruniversity.ac.in

© The Author(s), under exclusive license to Springer Nature Switzerland AG 2026
K. Raju et al. (eds.), *Hybrid Nanostructures for Advanced Supercapacitors*, Modern Aspects of Electrochemistry 63, https://doi.org/10.1007/978-3-032-04593-5_8

electronic era, where devices like laptops, cameras, and aerospace equipment rely on a quick and steady supply of energy [4, 5]. To power these devices, we often use solid conductors, and in many cases, primary batteries, secondary batteries, fuel cells, and supercapacitors are popular choices. These are preferred because they have simpler charging circuits, can be used immediately, and are quick to supply energy in various fields [6].

Nanotechnology is a fascinating new technology that creates customized nanomaterials with the potential to make better-performing goods [5, 7]. Nanomaterials, with their unique properties like mechanical, optical, magnetic, electrical, and chemical features, are used in various fields such as medicine, energy production, electronics, and more. Recently, scientists have been exploring nanostructured materials with different shapes, including nanoparticles, nanorods, nanosheets, and porous frameworks (Fig. 8.1) [8, 9]. These nanostructures are important because they can be modified to create complex structures, leading to better and unique features. Their atomic-scale structures and morphologies, such as length and width dimensions, play a key role in determining physical and chemical features, influencing electronic and optical properties. These structures have been recognized as building blocks for sensors, electronics, bioelectronics, and energy storage, contributing to advancements in various technologies [10]. In the rapidly changing landscape of electronic energy scenarios, electrochemical energy storage devices with high-power densities and long lifetimes are crucial to meet energy demands. Through extensive research in materials science, energy storage devices have achieved an extended cyclic life, impacting portable electronic devices [11]. The field of electronics has been greatly influenced by battery technology, especially in the context of environmental protection and the need for sustainable energy sources like air, waves, and solar energy [12].

In the realm of energy storage, despite notable advancements, several challenges persist. These challenges include the need for efficient energy storage, improved performance, increased longevity, and prevention of damage during use. Researchers are currently directing their efforts toward the development of energy storage devices that can effectively address global energy demands. One key focus is on advancing the materials used in electrodes for systems such as batteries and electrochemical capacitors. Supercapacitors are emerging as a promising next-generation energy storage technology due to their higher energy density compared to traditional batteries and capacitors. The success of high-performance supercapacitors relies on the characteristics of electrode materials, such as their shape, surface area, and porosity.

The rapid evolution of portable electronic devices has motivated researchers to explore integrated energy storage systems with features like lightweight, flexibility, and even roll-up capabilities. To meet these requirements, extensive research has been dedicated to the development of symmetric and asymmetric supercapacitors. Researchers have experimented with various types of electrodes and designs across a range of electrolytes in recent years. A significant breakthrough has been the discovery of hybrid supercapacitors, which offer a wide range of energy storage possibilities. These hybrid supercapacitors combine different electrode types to

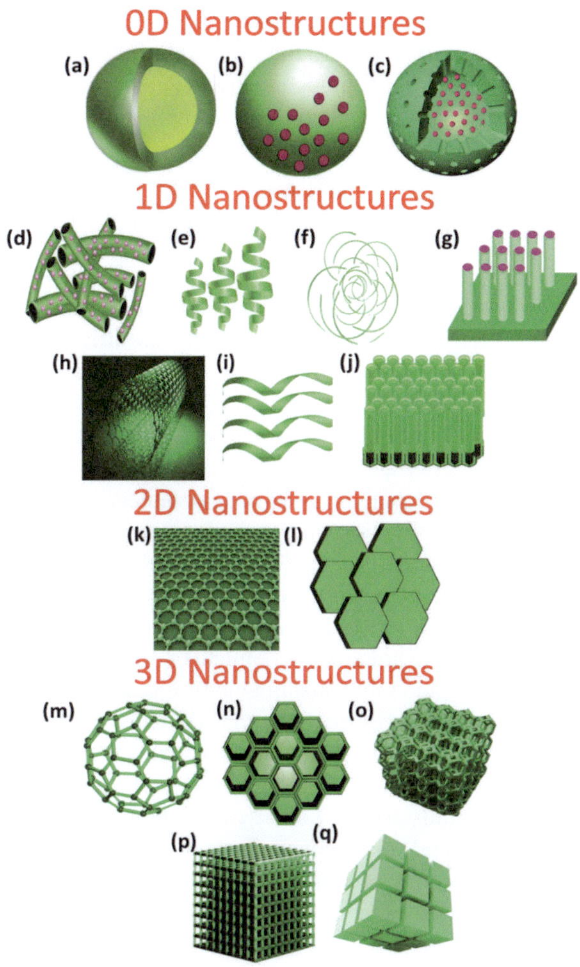

Fig. 8.1 Different types of nanostructures: zero dimensions **a** nano semisphere; **b** nanoparticles on nanosphere; **c** quantum dots on nanosphere. One dimension **d** nanotube; **e** nanoribbons; **f** nanoneedles; **g** nanorod; **h** nanotube; **i** nanowires; **j** nanobars. Two dimensions **k** nanosheets, **l** nanoflakes. Three dimensions: **m** porous polymeric microspheres; **n** nano mesoporous; **o** nano polycrystals **p** 3D microporous; **q** nanocubes

replace traditional asymmetric configurations. The technological exploration of hybrid supercapacitors emphasizes innovations in geometry, size, and composite materials.

Hybrid nanomaterials, composed of two or more components with at least one in the nano range, play a crucial role in enhancing the performance of supercapacitors. When combined, these hybrid nanostructures retain the unique properties of their components while also exhibiting novel features useful in practical applications. The chapter explores various types of hybrid supercapacitors, providing a summary of their electrodes based on recent works. The discussion covers the manufacturing processes, capacitance properties, and potential future applications. The analysis also considers the benefits and drawbacks of different-sized hybrid supercapacitors. Ultimately, the focus shifts to the promising future of hybrid supercapacitor technology in energy storage, accompanied by an exploration of the development challenges that lie ahead.

8.1.1 Electrochemical Energy Storage Devices

The limitations of both traditional and newer energy sources, coupled with the financial constraints faced by governments, are driving a shift toward alternative energy solutions and storage technologies. Recognizing this need, the global economy is now investing resources in the research and development of seemingly ordinary applications, such as internal battery backup and memory protection in supercapacitor technology. Batteries and supercapacitors stand out as promising technologies that could significantly impact the worldwide commercial electronic market. The electrochemical energy conversion and storage systems they employ are not only reliable but also environmentally friendly with simple operation. These devices (Fig. 8.2) play a crucial role in meeting peak power demands, particularly in portable electronic devices, energy harvesting, and transportation.

This shift toward electrochemical energy conversion and storage holds the potential for a greener, more sustainable future. These technologies have vast applications in the global energy market, thanks to their impressive energy density and suitability for micro-energy harvesting applications. The development of such systems could usher in a new era of green energy, contributing to a more sustainable and environmentally friendly energy landscape worldwide.

Batteries and supercapacitors share similar operating principles. Both involve two electrodes in contact with solid, liquid, or gel electrolytes, and the storage of energy occurs at the interface between the electrode and electrolyte. The movement of ions or electrons facilitates energy conversion or storage at this interface. However, there

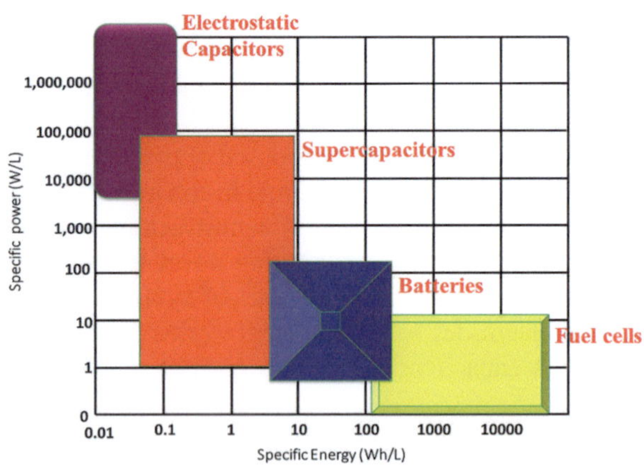

Fig. 8.2 Specific power and energy comparison of electrochemical energy conversion and storage devices. Supercapacitors exhibited very high-power density when compared to batteries and fuel cells

Table 8.1 Glance over batteries and supercapacitors

S. No.	Batteries	Supercapacitors
1	Ideally constant discharge and charge potential, excluding Li—intercalation batteries	Charge and discharge curves that are naturally sloping
2	The intrinsic state of the charge indicator is not very accurate. Charge time is 10–60 min	Intrinsic state of charge indication, charge time is 1–10 s
3	Based on the mass of active material and the potential window. The energy density is high. (100–200 Wh kg^{-1})	The poor energy density (5 Wh kg^{-1})
4	Lower power density, depending on the kinetics (1000–3000 W kg^{-1})	High-power density. (up to 10,000 W kg^{-1})
5	Due to irreversibility in redox and the phase difference of the electrode material, the cycle durability of the electrode is reduced. (500 and higher)	High cyclic durability, due to simple addition and extraction of charges on the surface of the electrode (one million or 30,000 h)
6	There is less internal IR because the surface area of the electrode/ electrolyte contacts increases	Internal IR drop due to high surface area of electrode/electrolyte contact
7	Lifespan is poor due to the degradation of electrode materials. (5–10 yrs)	Longer life span (10–15 yrs)

are notable distinctions between supercapacitors and batteries. In supercapacitors, charging comes from an external source, and there is no electron transfer at the electrode-electrolyte interface known as the Faradic process. On the other hand, batteries rely on the conversion of energy from chemical reactions like redox process involving the transfer of electrons in the outer circuit. For a quick reference, the key features of batteries and supercapacitors are summarized in Table 8.1 [13–15].

8.2 Principles and Mechanism of Supercapacitors

Since the 1980s, capacitor development has seen significant advancements, particularly in the realm of all-electrochemical capacitors or ultracapacitors, marking a departure from traditional capacitor designs. In earlier years, capacitors consisted of a single positive and negative electrode immersed in an electrolyte, separated by an insulator to facilitate charge transport [16]. Currently, a variety of capacitors are available for energy storage, utilizing the chemical bonds of different materials. The specific characteristics of each energy storage device depend on the electrical charges, influenced by the physical state of the prepared electrode constituents. These modern devices are commonly known as electric double-layer capacitors (EDLCs), pseudocapacitors, and hybrid supercapacitors (Fig. 8.3). They play a crucial role in

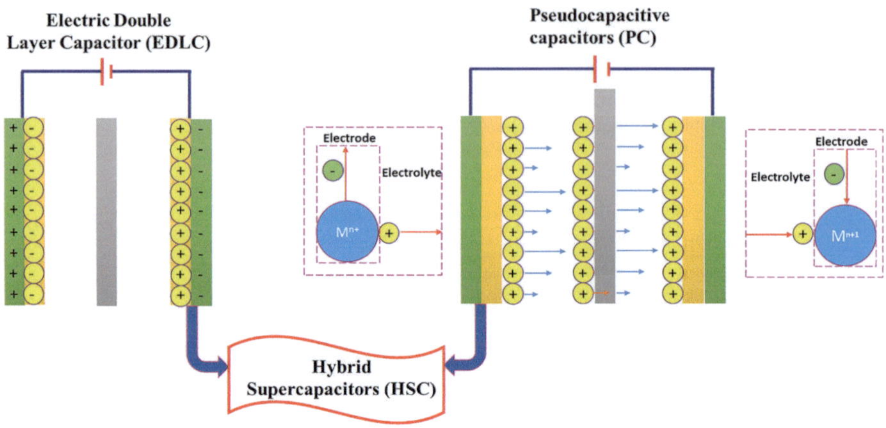

Fig. 8.3 Different types of supercapacitors. Electrochemical double-layer capacitors, i.e., the charge stored electrostatically whereas pseudocapacitors store the charge electrochemically, and the hybrid supercapacitor stores charge both electrochemically and electrostatically

storing and accumulating energy, particularly in the utilization of renewable energy sources [2].

Capacitors store energy by charging and discharging at the electrode-electrolyte interface. Supercapacitors operate on the same principle as regular capacitors but are specifically designed for rapid energy release and storage. Each type of capacitor has distinct applications: normal capacitors find use in small trimming in electronics, while supercapacitors are employed for high-voltage applications in electronics. The specific capacitance and energy density of supercapacitors can exceed those of standard capacitors by over 100,000 times. This improvement is achieved through the incorporation of active electrode materials with significantly larger surface areas, nanoscale dielectric distances, and rapid faradaic reactions [8.1]. As a result, conventional capacitors typically store energy in the micro-to-milli Farad range, whereas a single supercapacitor device can store several thousand Farads. Additionally, supercapacitors maintain high operational specific power with low equivalent series resistance, contributing to their efficiency [1, 2].

Research comparing batteries and supercapacitors reveals that supercapacitors excel in rapid charge and discharge rates, occurring within seconds or minutes, despite having lower specific energy density [17]. On the other hand, batteries boast high-power densities, along with several advantages over supercapacitors such as high efficiency, superior cycling life, and exceptional safety during operation, ensuring stable high performance. A key difference between supercapacitors and batteries lies in the voltage behavior during constant current charging. Unlike batteries, supercapacitors exhibit a linear increase and decrease in voltage. This distinction is significant in understanding the behavior of these energy storage devices.

While supercapacitors offer advantages, their specific power and specific energy can vary significantly. Through precise engineering, the range of specific energy and specific power in supercapacitors can be adjusted across multiple orders of magnitude. This flexibility makes supercapacitors a versatile storage option and allows for their categorization among capacitors. Energy storage in capacitors relies on two criteria: the coulomb charge, which is the attraction near the surface of the electrode to the electrolyte in the electric double-layer capacitor (EDLC); and the superficial redox reaction occurring at specific potentials, leading to pseudocapacitance. Additionally, hybrid supercapacitors, incorporating both coulomb charges and superficial redox reactions, represent a unified storage mechanism. These hybrid mechanisms are often discussed separately for clarity and convenience in understanding their operation.

8.2.1 General Classification and Mechanism of Capacitors

Supercapacitors operate through surface reactions involving cations and anions, as well as polarized or non-polarized interfaces between electrode materials and the electrode/electrolyte faradic redox reactions [18]. The composition of electrode materials, including carbon, conducting polymers, and metal oxides, contributes to their ability to generate high gravimetric energy density and volumetric energy density. This enables supercapacitors to deliver high currents efficiently.

There are three main classes of capacitor devices:

Electric Double-Layer Capacitors (EDLCs): These capacitors are non-Faradaic and rely on carbon materials. They have a lower discharge rate and charging efficiency.

Pseudocapacitive Capacitors: These capacitors are Faradaic and are based on conducting polymers or transition metal oxides. Their large specific surface area is advantageous for efficient energy storage.

Hybrid Supercapacitors: These capacitors combine both faradaic and non-faradaic reactions. They use a combination of carbon materials, metal oxides and polymers, like the operation and construction of lithium-ion batteries [19].

In this section, we will provide a brief overview of the characteristics of EDLCs, pseudocapacitive supercapacitors, and hybrid supercapacitors, along with the latest advancements in these technologies.

8.2.2 Electrochemical Double-Layer Capacitor (EDLC)

The electrochemical double-layer capacitor (EDLC), also known as a supercapacitor or ultracapacitor, was initially explored by von Helmholtz in 1853. EDLC energy storage devices, which operate based on the double-layer electrostatic effect, exhibit excellent cycle stability, lasting for millions of cycles, and durability [20]. The fundamental principle of an EDLC supercapacitor involves two electrodes immersed in an electrolyte interface and separated by an insulator. This setup provides a large operating potential window and high specific and areal capacity. During the polarization and depolarization process, ions at the electrode surface change. When a potential is applied to the device, there is no charge transfer across the electrode or electrolyte contact during the charging and discharging process. This characteristic results in high-power density, superior transient response, low weight, low internal resistance, and a compact volume. Notably, the energy delivery involves parallel movement of electrons in the external wire, and no chemical redox reactions occur. The capacitance of an EDLC can be calculated using the following formula [17].

$$C = \frac{\varepsilon_0 \varepsilon_r}{d} A \tag{8.1}$$

where ε_r, ε_0, A, and d represent, relative permittivity and permittivity in a vacuum, effective contact area between electrode and electrolyte and thickness of EDLC.

Key properties required for EDLC electrodes include:

Large Potential Range: EDLC electrodes need to support a wide potential range.

Fast Charge and Discharge Rate: The ability to quickly charge and discharge is crucial for optimal performance.

Large Operative Surface Area: A substantial surface area is necessary for efficient energy storage.

High Conductivity: Efficient energy transfer relies on high conductivity [21].

In applications involving mobile and stationary systems, especially those requiring high-power pulses, EDLCs play a vital role. Despite their ability to provide a longer cycle life with higher power, EDLCs face the challenge of relatively low energy density. Therefore, ongoing research focuses on enhancing the power and energy density of existing electrode materials and developing novel materials to improve energy storage performance [22, 23].

Porous carbons and carbon composites are widely used in the industrialization of EDLC as electrode materials due to their advantages, particularly a large specific surface area and improved pore-size distribution. Carbon, in various forms such as carbon itself [24], graphene [25] carbon nanotubes [26], and carbon nanofibers, serves as active materials in EDLC electrodes. These materials contribute to improved capacitive efficiency, often leveraging the principle of electrochemical double-layer storage devices. High-surface-area composite materials, such as carbon nanotubes and microporous carbon composites have been developed for EDLC electrodes. These materials offer an extended cycle life, rapid charge, and discharge intervals, and provide a safe and clean means of advanced energy storage, particularly in applications like hybrid electric vehicles [26]. EDLCs find applications across a spectrum of power and energy demands and serve as backup energy sources for devices like solar or wind energy storage systems.

8.2.3 Pseudocapacitors

Pseudocapacitance is a fascinating aspect of energy storage, distinguished by its thermodynamic nature and sensitivity to changes in potential and charge acceptance. Ruthenium dioxide (RuO_2) is notable as the first material to exhibit pseudocapacitance, despite its faradaic behavior in charge storage. A subsequent study demonstrated that a porous nanoscale structure of RuO_2 exhibited an impressive high capacitance of 700 F g^{-1} [27, 28]. The primary electrochemical principle behind pseudocapacitors is faradic, involving redox reactions during charge and discharge processes. The aim is to achieve high energy density, fast kinetics, and long cycle life [29, 30]. In this redox process, cations are adsorbed on the surface of the electrode material, enabling reversible and rapid charge transfer at the electrolyte/electrical interface. Pseudocapacitors offer advantages such as lower preparation costs, enhanced safety, and the ability to achieve higher capacitance than EDLC electrodes [31, 32].

Different electrode materials, including conducting polymers, and organic, inorganic, and transition metal oxides, can exhibit pseudocapacitance. These materials significantly improve pseudocapacitor properties. Four key factors contribute to enhancing the performance of these materials: (1) doping the contacting polymer (2) providing a large surface area (3) having a wide voltage window (4) achieving a high charge-discharge rate [18, 29, 33]. Nanostructured pseudocapacitive materials with a large surface-to-volume ratio are particularly emphasized for improved electrochemical performance [30]. Despite their advantages, pseudocapacitive materials are inherently non-electrostatic and involve electrochemical charge transfer with a finite number of active materials. Conducting polymers, a common pseudocapacitive material, undergo significant volumetric changes during charge-discharge processes. Nonetheless, pseudocapacitive electrode materials are gaining importance due to their quick charging-discharging rates, larger surface areas, and higher capacities compared to EDLCs.

Both EDLCs and pseudocapacitors share benefits and drawbacks, such as low energy density and poor cycle stability. To overcome these limitations, nanocomposite hybrid systems have been introduced. Hybrid supercapacitors, a new device type, combine the performance features of EDLCs and pseudocapacitive supercapacitors. Researchers view hybrid supercapacitors as an extreme of high-power devices, potentially bridging the gap between high-power and secondary batteries. Nanocomposite hybrid supercapacitors have the potential to fulfill the requirements of high energy density, fast kinetics, and long cycle life [30, 31].

8.2.4 Hybrid Supercapacitors

In recent years, there has been significant progress in the development of supercapacitors, particularly in the study of redox reaction architectures and materials for hybrid supercapacitors. These advancements aim to achieve extraordinary energy and power densities along with long life cycles, as compared to EDLCs and pseudocapacitive capacitors [34]. It has been confirmed that hybrid capacitors are more efficient in designing high-performance electrodes than both EDLCs and pseudocapacitors. Hybrid supercapacitors are designed with nanocomposite electrodes that combine the characteristics of both batteries and supercapacitors. This design enhances the energy storage capabilities of these devices [35, 36].

In a hybrid supercapacitor, there are both non-faradaic electrodes which involve electrostatic adsorption-desorption and faradaic electrodes which undergo reversible reaction phenomena. The hybrid electrode materials contribute to the efficient energy output of the device. Although the efficiency and cycling quality may be lower, the faradaic electrode, responsible for the power output, shows well-defined redox peaks. The non-faradaic electrode, on the other hand, does not involve Faraday charge transfer; instead, the charge storage occurs in the form of a double layer on the surface. This provides a quick response to changes, making it crucial for power output.

Additionally, batteries and capacitive electrodes exhibit different electrochemical dynamics, as expressed in the equation below [37]:

$$i = Cv^b \tag{8.2}$$

where i is the current and v is the scan rate of a cyclic voltammetry curve. The electrochemical behavior of the battery-type material with $b = 0.5$ is controlled by diffusion, whereas the surface controls the electrochemical behavior of the capacitance-type material.

Presently, commercially available lithium-ion batteries connect to the graphite anode through a micro-carbon capacitive electrode, a configuration also adopted by many widely used hybrid supercapacitors, often referred to as Li-ion capacitors. These Li-ion capacitors represent the state of the art in terms of performance, offering high energy density (> 20 Wh kg^{-1}) currently. However, incorporating lithium-ion battery negative electrodes in Li-ion capacitor systems compromises their power efficiency compared to EDLCs and introduces uncertainty in cycle lifespan, especially during high-rate cycles [38]. The distinction between symmetric and asymmetric supercapacitors is dependent on the configuration of their assembly. In the following discussion, we explore electrode materials by examining their structure, morphology, and fabrication features, and how these aspects relate to the electrochemical performance of hybrid supercapacitor devices.

8.3 Hybrid Nanocomposites Materials Synthesis Methods

The techniques used to produce electrode materials play a crucial role in determining the architectures and characteristics of hybrid nanocomposite materials. The quality of these hybrid nanocomposites heavily depends on the synthesis techniques and reagents employed. Over the past decade, various top-down synthesis

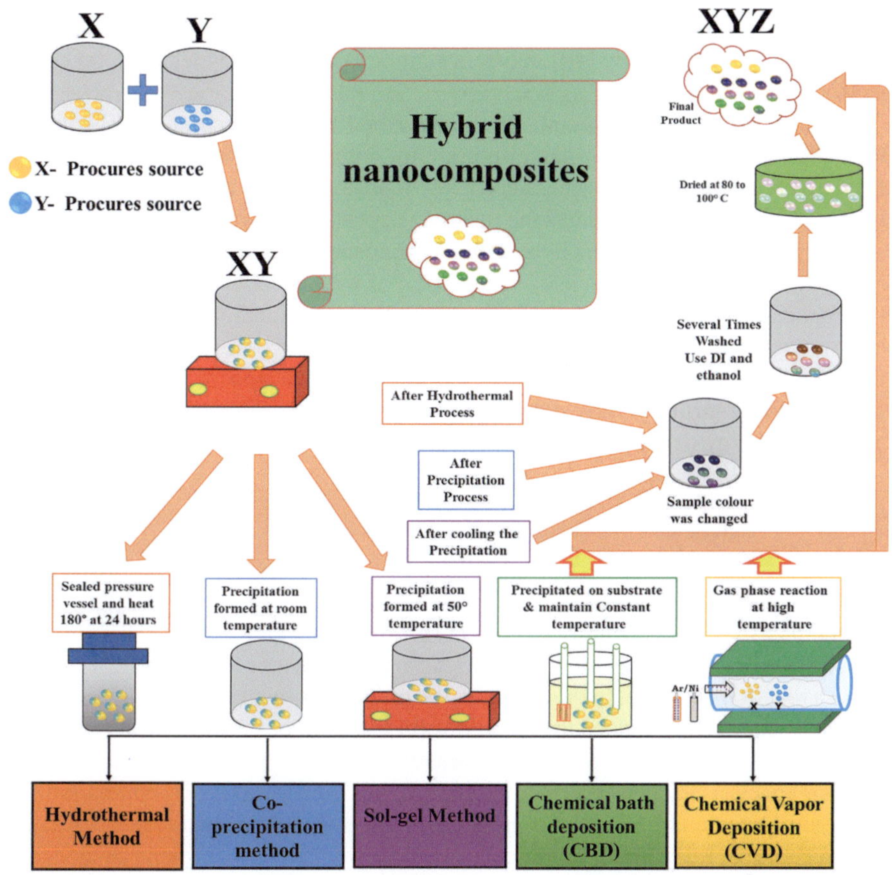

Fig. 8.4 Different types of synthesis methods to synthesize various nanostructured materials and hybrid nanocomposites with excellent physical and chemical properties for advanced supercapacitor applications

techniques for hybrid nanocomposites have been developed, demonstrating good scalability and a straightforward processing pathway to achieve desired features. Many of these synthesis methods have been thoroughly reviewed and discussed by experts, including electrochemical deposition, chemical bath deposition, chemical vapor deposition, and the sol-gel technique (Fig. 8.4).

8.3.1 Chemical Bath Deposition

Chemical bath deposition has been in use for a considerable time, especially in low-temperature thin film deposition [39]. This method is commonly chosen for thin film deposition due to its relatively low cost, suitability for large substrate areas, and low-temperature processing [40]. It offers simplicity, particularly when compared to electrode deposition.

In recent years, Zhang et al. 2016 described the production of 3D hierarchical porous quaternary zinc-nickel-aluminum-cobalt oxide nanosheets directly on nickel foam using a simple chemical bath deposition. The resulting zinc-nickel-aluminum-cobalt oxide served as the positive electrode, and activated carbon served as the negative electrode. This hybrid nanocomposite exhibited high energy density (72.4 Wh kg^{-1}) and power density (533 W kg^{-1}) with excellent cycling stability of 90% capacitance retention after 10,000 cycles [41]. Additionally, a hybrid nanocomposite asymmetric supercapacitor device, PB-MnO_2-PG, developed by Das et al., 2017 displayed notable features such as good specific capacitance (98 F g^{-1} at 1 A g^{-1}), long-term cycling stability of 93% retention after 4000 cycles and a substantial energy density of 16.5 Wh Kg^{-1} [42].

8.3.2 Chemical Vapor Deposition

The chemical vapor deposition procedure is a fundamental technique that involves exposing a substrate to appropriate precursors in the presence of heat or plasma energy within a reaction chamber. In this process, the precursors react and decompose on the substrate's surface to produce the desired solid thin film or powder. Chemical vapor deposition is commonly employed for generating carbon compounds in the context of supercapacitor devices and their manufacturing. For instance, carbon nanotubes were synthesized on highly porous carbon nanocups using the chemical vapor deposition technique, resulting in specific capacitances of 45 F/g with stable capacitance over 10,000 cycles [43]. Another application involved the use of N-doped porous carbon nanofibers and porous silver network hybrid film (NPCNFs-PSN) treated with ammonia gas in the chemical vapor deposition manufacturing of carbon nanofibers (CNFs). In a three-electrode system, the synthesized N-PCNFs-PSN supercapacitor electrode material increased the capacitive range to 245 Fg^{-1} at 1 Ag^{-1} and exhibited a rate capability of 90.8%. Additionally, the two-electrode device demonstrated a high energy density of 8.5 Wh/kg and power density of 250 W/kg with stable cycling stability of 103% after 10,000 cycles at 20 Ag^{-1} [44].

Plasma-enhanced chemical vapor deposition (PECVD) is a variation of chemical vapor deposition that enables the production of vertically dense, aligned carbon nanotubes at lower temperatures compared to traditional chemical vapor deposition. Recently, the PECVD technique was utilized to synthesize multi-walled carbon nanotubes on reduced graphene oxide for supercapacitor electrodes. The electrodes demonstrated impressive performance, with current densities of 20 Ag^{-1} resulting in a specific capacitance of around 120 F/g [45].

8.3.3 Sol-Gel Method

The sol-gel technique is a method used to create a structured network by connecting tiny particles in a liquid solution (sol) and organizing them regularly into a gel. The sol-gel process provides better control over the material's texture, purity, composition, and other features [46]. Scientists use this method to make hybrid nanomaterials for electrodes in supercapacitors. These hybrid materials have improved performance when it comes to storing and releasing electrical energy. The sol-gel method allows scientists to control various factors, such as the type of materials used, the chemicals involved, and the time it takes for the reaction to happen. For instance, researchers have used the sol-gel method to create hybrid aerogels made of vanadium pentoxide and multi-walled carbon nanotubes. These materials show excellent properties like a high ability to store electrical charge and a lot of energy stored per unit mass and they can endure many charging and discharging cycles without losing performance. Another example involves materials like carbon nanosheets and electrochemically exfoliated graphene. These materials have specific capacitance values, which indicate how much charge they can hold, even after many charging and discharging cycles. This shows that the sol-gel method is a powerful technique for creating advanced materials with improved electrical performance [47].

8.3.4 Co-precipitation Method

This technique is straightforward and involves creating large-scale powder samples. For success, the solubility limit must be lower than the concentration limit of the solvent, and the temperature needs to be high enough to dissolve the solvent rapidly [48]. This rapid method is mainly used in challenging morphology-based supercapacitor technology to make electrode materials. Numerous hybrid nanocomposite supercapacitors have been developed using this approach [49]. For instance, amorphous nickel borate (NiB_xO_y) hybrid material synthesized through this method exhibits impressive performance including a high specific capacity of 137.9 mAh g^-, high energy density of 42.4 Wh kg^{-1}, and power density of 800 W kg^{-1}. The device retains a remarkable 97% specific capacity retention over 10,000 cycles [50]. In another study, $La_2Mo_3O_{12}$ and $Cu_3Mo_2O_9$ were reported to have specific capacities of 195.85 and 71.94 Cg^{-1}, respectively. The asymmetric supercapacitor device derived from these materials demonstrates good energy densities of 21.84 and 2.92 Wh kg^{-1} and power densities of 1.61 and 10.86 kW kg^{-1} with stable cycles of 119.4 and 95.5% after 5000 cycles [51].

8.3.5 Hydrothermal Method

The hydrothermal method is an efficient technique, known for generating high pressure at elevated temperatures within a confined space, making it an environmentally friendly process with straightforward procedures [52, 53]. The high autogenous pressure created within the sealed reaction vessel regulates the settling of volatile products in an orderly manner. Consequently, this method is commonly employed in synthesizing nanocomposites with a well-organized particle arrangement. Patil et al. 2016 reported a straightforward hydrothermal synthesis of the rGO-$BiVO_4$ hybrid, demonstrating a specific capacitance of 151 Fg^{-1} at 0.15 mA cm^{-2} for the a prepared electrode [54]. Sarkar et al. 2018 disclosed the facile hydrothermal synthesis of cobalt disulfide and MWCNT hybrids, exhibiting a supercapacitor performance of 1486 Fg^{-1} at 1 Ag^{-1} [55].

8.4 Electrochemical Performance

8.4.1 Cyclic Voltammetry

The capacitance of electrode materials was estimated using Eq. (8.3).

$$C = \frac{i}{v} \tag{8.3}$$

where voltage-current density (A g^{-1}) is given by *i*, while scan rate (v) is given by v (V s^{-1}). Using the cyclic voltammetry measurement, an EDLC or a typical pseudocapacitive behavior with a rectangular cyclic voltammetry curve may be identified right away. Therefore, an electrode's average capacitance cannot be calculated just by measuring cyclic voltammetry because of redox peaks in the curves.

8.4.2 Galvanostatic Charge-Discharge

The specific capacitance of supercapacitors measured using the galvanostatic charge-discharge test is the most efficient in hybrid electrode materials. There is a relationship between capacitance (C) F g^{-1}, current density (i) A g^{-1}, charging-discharging time (t), and the electrode working potential window (V) that may be expressed as (8.4)

$$C = \frac{i * t}{V} \tag{8.4}$$

Also, the EDLC behavior indicates possible independent charge storage. In this case, the estimated capacitance is independent of the working potential window. Some pseudocapacitive or intercalation pseudocapacitive behaviors center on potential windows. The evaluated capacitances of these behaviors vary with the working potential windows used.

8.4.3 Electrochemical Impedance Spectroscopy

Supercapacitor impedance data is often collected at a given voltage of 5 or 10 mV throughout a wide frequency range of 0.01 Hz to 100 kHz during EIS testing. It has a Nyquist plot with three distinct regions: (1) semicircle at a high frequency, range larger than 104 Hz, (2) medium frequency, range 104 to 1 Hz, and (3) near-vertical line at low frequency, range less than 1 Hz. The corresponding frequency indicates the interface resistance, charge transfer resistance and the capacitance behavior of the device. Also, it is possible to establish the connection between imaginary impedance |Z| and frequency (f) using electrochemical impedance spectroscopy measurements.

Equation (8.5) may be used to estimate the capacitance (C):

$$C = \frac{1}{2\pi f\,|Z|} \tag{8.5}$$

The Bode plot is the linear component of a log |Z| versus log f curve and the capacitance reduces with increasing frequency range.

8.4.4 Energy and Power Density Assessment

The total performance of supercapacitors is assessed using the Ragone plot, which is constructed based on the outcomes of this evaluation. The Ragone plot provides a comprehensive representation of the electrochemical characteristics of the supercapacitor device. In contrast, the numerical calculation of energy density is determined by integrating discharge curves using a specific equation.

$$E = \int_{t_2}^{t_1} C(V_1 + V_2)(V_2 - V_1) \tag{8.6}$$

where E, C, V_1, and V_2 are represented by energy density Wh kg^{-1}, capacitance F g^{-1}, V_1, and V_2 denote the voltages of charge and discharge (V), respectively. The voltage window V_2–V_1 reflects the device's capacitive nature. P symbolizes power density (P) was calculated by using the following equation,

$$P = \frac{E}{t} \tag{8.7}$$

$$P = \frac{V_{max}^2}{4R_s} \tag{8.8}$$

where E and t denote energy density (Wh kg^{-1}) and time. The square of the maximum voltage (V_{max}) is defined by maximum power density (P_{max}) and equivalent series resistance denoted by R_S as mentioned in the below equation.

8.5 Design and Fabrication Methods

Several methods for fabricating supercapacitor devices with interdigitated patterns have been described thus far. Various manufacturing processes (Fig. 8.5) can be employed to construct the interdigitated structure of supercapacitors, which is subsequently paired with suitable electrodes and electrolytes. This section discusses processes such as ink printing, laser processing, screen printing, plasma etching, photolithography, stamping, 3D printing, and other techniques involved in the fabrication of supercapacitors.

8.5.1 Inkjet Printing

Printing enables digital printing by offering non-contact forming and flexibility with roll-to-roll manufacturing formats, which is advantageous for the supercapacitor fabrication process. Generally, suspensions or viscous pastes containing active materials are used in ink printing for supercapacitors [56, 57]. This method involves

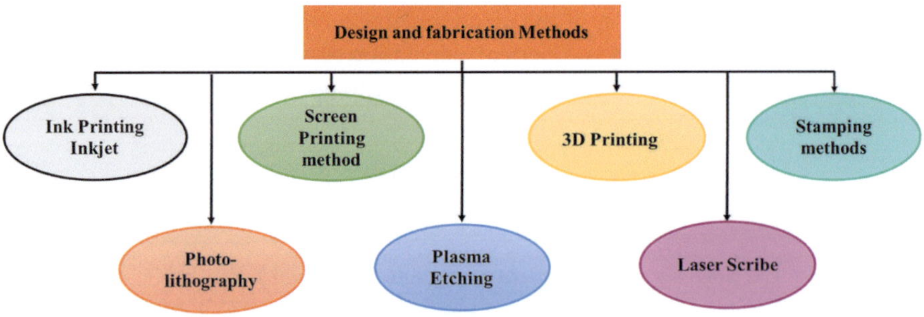

Fig. 8.5 Various manufacturing processes like printing, etching, lithography for the fabrication of interdigitated patterns of electrode materials for supercapacitors, especially hybrid capacitors mainly used in flexible and wearable electronics

employing a printing approach to generate interdigitated electrode patterns on the material surface. The distinctive structures of the electrode materials are achieved by configuring and fluidly printing the ink onto the substrate's surface. To avoid nozzle jamming during ink printing, despite its lower resolution, it is crucial to prepare suitable inks with excellent fluidity. This has proven to be an effective method for creating supercapacitors. Additionally, in the pursuit of efficient supercapacitor device fabrication, ink printing with appropriately prepared inks exhibiting good fluidity, despite the lower resolution, has been identified as a necessary approach [58, 59].

8.5.2 Photolithography

A widely used and well-established method for fabricating micro and nanoscale energy devices is photolithography technology. This method is preferred due to its high resolution, maturity, and compatibility with various electrode materials [59–63].

8.5.3 Screen-Printing Method

Screen printing is a simple and low-cost approach for rapidly fabricating energy devices using technology [64]. Achieving high precision in intermittent patterns is challenging with the screen-printing method. The combination of manufacturing processes, such as electrical deposition, may integrate ink printing and screen printing, where one approach complements the other. This involves the rational development of printable inks containing active electrode materials. The conversion of the electrode's active ingredient through electrical deposition incorporates both deposition techniques [65, 66].

8.5.4 Plasma Etching

The fabrication method involves a plasma source, wherein a high-speed glow discharge stream, composed of a cost-effective plasma source gas mixture, is directed onto the substrate [67, 68]. This plasma etching-based fabrication is primarily utilized to improve the surface properties of electrode pattern manufacturing [69, 70].

8.5.5 3D Printing

Additive manufacturing, also known as 3D printing, is the process of constructing 3D objects layer by layer. This technique offers several advantages for preparing hybrid composites, including excellent geometric precision and cost-effectiveness [71, 72]. Importantly, the 3D printing approach can enhance the surface area of the electrode material while maintaining high accuracy in creating electrode patterns within a confined space. As a result, the electrochemical characteristics of the fabricated supercapacitor devices were outstanding [73–77].

8.5.6 *Laser Scribe*

One of the most widely used and efficient methods for fabricating interdigitated electrodes is Laser scribe technology, which is becoming an increasingly attractive method for supercapacitor fabrication [78, 79]. In this method, polymer materials are primarily employed under the high energy of laser irradiation due to their chain fracture and gasification properties [80]. Laser processing involves an in-situ laser conversion of active electrode materials and electrode depositions of active materials [81–83].

8.5.7 *Stamping Methods*

The stamping method is considered the best and most cost-effective for printing electronics [84, 85]. Graphene and MXene ink stamps have been utilized to create diverse designs [84, 86, 87]. However, it is important to note that, before employing the imprint technique, a proper template must be created, requiring both time and effort [85, 88–90].

8.6 Physical and Chemical Characterizations

8.6.1 *Structural Characterization*

The crucial composition characterization of hybrid nanomaterials involves structural phase studies, which examine the presence of various compositions and functional groups. X-ray diffraction techniques are employed to identify the purity of molecular and atomic-level crystals or polycrystalline materials. The crystalline size is determined using the Debye-Scherrer formula [$D = k\lambda/(\beta \cos\theta)$] [91–93]. X-ray diffraction plays a pivotal role in the structural analysis of chemically prepared hybrid nanocomposites, including those involving metal ions, metal oxide, transition dichalcogenides, rare earth, and polymer-based supercapacitor electrodes [94–100]. X-ray diffraction peaks corresponding to Zn-ions [94], RE_2O_3 (RE = Nd, Gd, and Yb) [95], carbon nanotubes-$Li_4Ti_5O_{12}$ [96], graphene oxide-Fe_3O_4-polyaniline [97], $NiCo_2S_4$

[98], MoS_2-rGO-polyaniline [99], and Nb_2O_5-carbon [100] composites indicate the successful formation of hybrid materials. The X-ray diffraction patterns of these synthesized hybrid nanocomposites and their supercapacitor electrode materials demonstrate the growth of a completely ordered crystalline structure.

Fourier Transform Infrared (FTIR) spectral analysis is invaluable for detecting the structural functional groups and chemical bonding information of organic and inorganic hybrid nanocomposites [101] .The shifting of peaks due to chemical reactions provides insights into structure changes. FTIR studies on various hybrid materials, such as single-walled carbon nanotubes and poly(3, 4-ethylenedioxythiophene) [102], vanadium pentoxide (V_2O_5)-hybrid aerogel, V_2O_5-graphene hybrid aerogel [103], $CoMoO_4$, $MnMoO_4$, and $NiMoO_4$ [104], graphene-Co_3O_4, NiO, and Ni-Co [105], carbon dots co-doped by phosphorus, oxygen, and nitrogen, elements and hierarchical porous carbon (NPOCD-HPC) [106], γ-FeOOH and Ni–Mn hydroxide on carbon nanofoam paper [107] confirm the conformation of asymmetric and symmetric stretching vibrations in hybrid nanocomposites.

Raman spectroscopy is a well-known characterization method that distinguishes between different shapes of identical nanomaterials. In the case of hybrid nanocomposites, vibrational, rotational, chemical, physical, and structural properties are determined by Raman spectroscopy. Researchers have reported Raman characteristics of various hybrid nanocomposites, including graphene-polyaniline [108], graphitic carbon and $Ni_2P_2O_7$ [109], hybrid carbon nanotubes-$NiCo_2S_4$ composites [110], Fe^{3+} doped spinel $Li_4Ti_5O_{12}$-reduced graphene oxide [111], MOF-74 derived double hydroxide (MDH)–Ni–Co MDH (65Ni-MDH) [112], hollow carbon spheres (HCS) and solid carbon nanoparticles (SCN) [113], nickel-cobalt oxide-3D graphitic carbon (Co–Ni–O/3DG) and PTC-C [114]. These studies showcase the excellence of Raman spectroscopy in characterizing hybrid nanocomposites.

X-ray photoelectron spectroscopy is widely used to determine the elemental composition, structures, chemical electronic states, and orbital bonding of nanocomposites. Researchers have characterized hybrid nanocomposites using X-ray photoelectron spectroscopy, revealing features such as chemically reduced graphene sheets and graphene-polyaniline [115], α-MnS (α-manganese sulfide) nanoflakes-reduced graphene oxide sheets [116], functionalized carbon cloth-polyaniline array-rGO [117], hybrid MnO_2–$NiCo_2O_4$ nanosheets [118], and hybrid NiO–CuO [119]. These X-ray photoelectron spectroscopy results confirm the surface chemical composition of hybrid nanocomposites.

8.6.2 *Optical Characterization*

UV-visible spectroscopy is employed for the absorption measurement of hybrid nanocomposite materials and fabrication substrates (FTO-coated glass substrates). In UV-vis spectroscopy, previous studies have demonstrated distinctive absorption features for various hybrid nanocomposites. For example, polyaniline-graphene oxide exhibits identified absorption patterns at different pH values [120]. Carbon quantum dots-polypyrrole-titania nanotube shows absorption at 235 nm [121]. Graphene oxide-Fe_3O_4-polyaniline displays multiple distinct absorption peaks [97]. MoS_2 quantum dot-polyaniline exhibits absorption in the near-UV region (298 nm). Polyaniline hybrid and polyaniline hybrid-chloroform record an absorption peak at 348 nm [122]. The band gap of hybrid materials is determined by estimating band gap energies using Tauc's plot, as indicated by the following relation [123, 124].

$$(ahv)^2 = C(hv - E_g) \tag{8.9}$$

Where h, v, α, and C are Plank's constant, photon frequency, absorption coefficient and a constant, respectively. By extrapolating the linear region of $(ahv)^2$ versus the photon energy (E_g) axis the optical band gaps can be estimated. In addition to the optical studies mentioned above, photoluminescence spectroscopy has been employed in past research. Notable findings from photoluminescence studies include hybrid ZnO–ZnS nanoforests exhibiting photoluminescence spectra at 380 and 522 nm. MoS_2-graphene (GNS) hybrid shows photoluminescence excitation at 375 nm and emission spectra at 660 nm [125]. Co_3O_4–Ni–PTA (p-benzenedicarboxylic acid)-TEOA (triethanolamine) has photoluminescence excitation at 514 nm and displays a broad band at 800–1000 nm [126]. α–MnO_2–C-dot green hybrid Nanorods achieve photoluminescence intensity with excitation at 400 and 750 nm emission maxima at 495 nm [127]. Furthermore, the optical characteristics of hybrid nanocomposites are comprehensively understood through UV-visible and photoluminescence spectroscopy.

8.6.3 *Morphological Characterization*

Nanocomposites exhibit diverse shapes, and their morphology depends on the synthesis procedures employed. Various morphologies, such as nanospheres, nanoparticles, nanoclusters, nanosheets, nanotubes, nanofibers, nanowires, and nanoflowers, can be observed in hybrid nanocomposite materials. Specific morphology plays a crucial role in determining the properties such as optical, electrical, electronic, magnetic, and thermal and further technological applications of these materials [123, 128, 129]. Structural characterizations provide insights into morphology, including phase composition, crystallographic defects, different phases, and chemical composition. Several morphological characterization techniques are available, with a focus on scanning electron microscopy and transmission electron microscopy.

Scanning electron microscopy and transmission electron microscopy are powerful tools for characterizing nanocomposite materials, offering high-resolution and detailed structural information through virtual images at high magnification. Energy dispersive X-ray spectroscopy (EDX) integrated into these microscopes allows the identification of the geochemical composition of elements. Noteworthy findings in various morphologies include:

(i) Hybrid nanoparticles have gained attention in recent decades [130–136].
(ii) Nanospheres are typical in hybrid nanocomposite materials [137–141].
(iii) Nanotubes have been widely studied for hybrid conductive electrodes [142–147].
(iv) Nanowire structures have been extensively explored for energy storage applications [148–155].
(v) Hybrid nanoflower structures effectively enhance the electrochemical properties of nanocomposites [156–164].
(vi) Nanoplate structures serve as excellent electrode materials in supercapacitors [165–173].
(vii) Nanofiber structures are commonly used in electrode materials [125, 174–186].
(viii) Nanorod structures exhibit good electrochemical properties and outstanding rate capability in hybrid nanocomposites [187–196].

As an example, the $NiCo_2S_4$ hybrid structure prepared using the sequential ion exchange method forms onion-like particles with a hollow shell nanostructure. This material, utilized as an electrode for hybrid supercapacitors, demonstrates an electrochemical performance of 1016 F g^{-1} at current densities of 2 A g^{-1} [197].

8.7 Recent Hybrid Nanostructured Materials

8.7.1 Graphene-Metal Oxide Hybrid Nanostructures for High-Performance Supercapacitors

Graphene stands out as a leading and extensively investigated hybrid nanostructured material, particularly for high-performance applications in energy storage. It is characterized by a monolayer and hexagonal lattice structure formed by tightly packed carbon atoms bonded in a sp^2 configuration [198]. Graphene possesses remarkable intrinsic properties, including unique mechanical, electrical, and thermal features that make it highly attractive for various potential applications [199]. Noteworthy characteristics of graphene include large carrier mobility at room temperature (200,000 cm^2V^{-1} s^{-1}), a substantial specific surface area (2630 m^2g^{-1}), high thermal conductivity ($\sim$ 5000 Wm^{-1} K^{-1}), excellent optical transmittance ($\sim$ 97.7%), and a high young's modulus ($\sim$ 1.0 TPa) [200]. To enhance the functionalities of graphene, surface modification approaches involving graphene oxide or reduced graphene oxide are employed. These derivatives introduce a variety of oxygen functional groups, playing a crucial role in the substantial surface growth of metal oxide nanostructures [199, 201]. Additionally, the combination of transition metal oxides with graphene results in the formation of graphene-metal oxide hybrid nanocomposites. This approach is well-established in hybrid nanocomposites, leading to improved electrical conductivity and a large specific surface area, facilitating the rapid movement of electrolyte ions in various energy storage devices [202, 203]. The hybrid nanostructures of metal oxides formed are strongly bonded to the GO-rGO surfaces, significantly enhancing electrochemical properties. For instance, Zhang et al. 2020 developed a reduced graphene oxide and NiO (rGO-NiO) based supercapacitor electrode through the plasma electrolysis method, exhibiting a maximum specific capacitance of 1093 F g^{-1} at 1 A g^{-1}, with excellent cycle stability [204]. Similarly, Zhu et al. 2017 reported reduced graphene oxide and thumb-ring-like α–Fe_2O_3-based nanocomposites α–Fe_2O_3-rGO, synthesized by a facile hydrothermal method, displaying good electrochemical performance with only a 10% decay after 11,000 cycles of galvanostatic charge–discharge tests [205].

Another example involves the rational design of carbon nanotubes-rGO-$MnMoO_4$ hybrid nanocomposites by Mu, et al., 2017 via hydrothermal growth on three-dimensional nickel foam. The binder-free supercapacitor cathode fabricated from this material exhibited the highest specific capacitance of 2374.9 F g^{-1} at a scan rate of 2 mV s^{-1} [206]. Additionally, Zhang et al. 2021 successfully synthesized carbonate hydroxide ($Ni_{0.3}Co_{0.6}Cu_{0.1}(CO_3)_{0.5}(OH)$) by co-substituting Co with Ni and Cu, resulting in an urchin-like active material with an excellent specific capacitance of 1480 F g^{-1} at 1 A g^{-1} and 72.6% capacitance retention at high current density [207].

8.7.2 Hybrid Nanostructures in Li- and Na-Ion Supercapacitors

Recently, there has been significant progress in the development of novel electrochemical energy storage devices capable of providing high energy and power densities [208–210]. Among these, the lithium and sodium-ion hybrid supercapacitor has emerged as a promising energy storage device, effectively bridging the gap between lithium-ion and sodium-ion capacitors [211, 212]. Like Li- and Na-ion capacitors, the fabrication of Li- and Na-ion hybrid supercapacitors involves combining the favorable characteristics of both batteries and capacitors. This design allows Li- and Na-ion hybrid capacitors to deliver higher energy and power densities, with the anode behaving like a battery offering a high specific capacity and the cathode resembling a capacitor providing high-power density. This makes Li- and Na-ion hybrid capacitors suitable for high-performance electrochemical energy storage applications that demand both high energy and power densities [213, 214]. Zhan et al. 2018 developed a hybrid supercapacitor based on Li/Na-ion hybrid supercapacitors with an interlayer expansion between MoS_2-rGO composite, demonstrating ultrahigh energy density of 1600 mAh g^{-1}, 580 mAh g^{-1} at 100 mAh g^{-1} corresponding to Li- and Na-ion hybrid supercapacitors and power density of 320 mAh g^{-1} at 10 A g^{-1} along with excellent cycle stability of over 99% capacitance retention after 10,000 cycles [215].

Chen et al. 2018 reported experimental and theoretical studies on rechargeable sodium-ion batteries and symmetric lithium-ion batteries using hybrid composites in the form of polyoxovanadate $Li_7V_{15}O_{36}(CO_3)$. The sodium and lithium-ion battery system exhibited an ultrahigh specific power of 52 kW kg^{-1}, specific energy of > 125 W h kg^{-1} and a long cycling life of > 500 cycles. These rechargeable Na-ion batteries demonstrated a high specific capacity of 240 mA h g^{-1} and a high specific energy of 390 Wh kg^{-1}, in line with theoretical predictions [216]. Yuan et al. 2018 prepared sodium-ion hybrid capacitors using porous NaBi as an anode, activated carbon as a cathode, and a 1.5 M $NaPF_6$ sodium-ion hybrid capacitors electrolyte. The sodium-ion hybrid capacitors exhibited a specific capacity of 298 mA h g^{-1} with a capacity retention of 98.6% after 1000 cycles [217]. The development of Li- and Na-ion hybrid supercapacitors represents a new and highly sought-after class of energy storage devices. Sodium-based technologies, especially sodium-ion batteries, are experiencing rapid growth and are expected to become commercialized soon.

8.7.3 Polymer Nanostructures-Based Hybrid Supercapacitors

Polymer-based electrochemical supercapacitors involve conjugated polymers bonded to the polymer chain, conducting electricity throughout the entire device. Over the past two decades, significant research has been conducted on hybrid polymer supercapacitors due to their high charge density, lower cost, and high redox nature compared to expensive metal oxides [218, 219]. The electrostatic charge is stored at the interface between the electrode and electrolyte, and this process is particularly fast in polymer-based supercapacitors. Notably, polyaniline, polypyrrole, and polythiophene and their derivatives exhibit superior performance among various polymer electrodes for supercapacitors [220–223].

In recent years, considerable progress has been made in the synthesis of polymer-based hybrid supercapacitors in various forms, such as bulk powder, nanowalls, nanorods, nanosheets, and nanowires, leveraging surface interactions and achieving high surface-to-volume ratios with impressive electrochemical characteristics [224, 225]. For example, Fan et al. 2017 reported the simple fabrication of a graphene-polypyrrole film, demonstrating a high-powered hybrid electrode with a capacitance of 416 F cm^{-3} at 1.0 A g^{-1} in a 6 M KOH electrolyte [226]. Yu et al. 2021 developed a 3D hierarchical nanosheet-structured carbon cloth-nickel cobaltite-polypyrrole through a two-step electro-deposition process. The resulting hybrid electrode exhibited a high specific capacitance of 1687.2 F g^{-1} at 1 A g^{-1} and robust cycling stability, retaining 91.9% capacitance after 10,000 cycles at 10 A g^{-1}. Additionally, asymmetric supercapacitors made of carbon cloth-nickel cobaltite-polypyrrole as anode and activated carbon as a cathode in 6 M KOH aqueous electrolyte demonstrated excellent cycling stability over 10,000 cycles at 4 A g^{-1} [227]. Another straightforward method for creating a high-performance electrode material using ternary-doped polyaniline with cetyl trimethyl ammonium bromide (CTAB) surfactant-wrapped environmentally friendly activated carbon. The composites exhibited a specific capacitance of 1250 F g^{-1} and good cyclic stability, retaining 87% capacity after 1000 cycles [228].

8.7.4 *Organic-Inorganic Hybrid Nanomaterials*

Hybrid nanocomposites consist of two or more components, typically combining inorganic components such as metal ions, sulfides, oxides, clusters, particles, non-metallic elements, salts, and their composites with organic components such as organic ligands, groups, surfactants, and polymers [229–231]. Generally, organic materials offer structural flexibility, convenient processing, tunable electronic properties, efficient luminescence, photoconductivity, and the potential for semiconducting behavior. On the other hand, inorganic compounds provide potential bandgap tunability, high carrier motilities, excellent dielectric, and magnetic properties, as well as thermal and mechanical stability [232, 233]. The synergy between these materials results in hybrid nanocomposites with outstanding properties.

Recently, Nagaraju et al. 2017 reported an environmentally friendly and simple approach for creating flexible battery electrodes using binder-free double hydroxide nanosheets on nickel fabric, LDHs-NFAs-NSs in Ni fabric. The electrochemical performance of the core-shell-like nickel-cobalt layered double hydroxide nanoflake arrays on nickel fabric was evaluated in 1 M KOH electrolyte, showing a current density of 2 mA/cm^2, capacity retention of 78.3% at 30 $mAcm^{-2}$ (420.5 Ah cm^{-2}), and 87% capability after 1000 cycles [234]. A controllable and efficient annealing of Fe_3O_4 nanoparticles was achieved using a dip-coating approach to generate ultrafine Fe_3O_4 nanoparticles-graphene on carbon fabric, serving as a supercapacitor electrode. This electrode exhibited a specific capacitance of 406 F g^{-1} at 1 A g^{-1}, high-rate capability of 56.9% retention, energy density of 800.2 W kg^{-1}, power density of 8614.7 W kg^{-1}, and outstanding cyclic stability [235]. Huang et al. 2020 produced a hierarchical nickel-cobalt sulfide-nickel-copper-based layered hydroxides nanotube-nanosheet hybrid electrode directly on carbon cloth using a multi-step solution-based method. The optimized electrode showed a high capacity of 632.0 Ah cm^{-2} at 2 mA cm^{-2}, strong rate capability due to rational nanostructure engineering and atomic replacement of nickel by copper. Additionally, the flexible solid-state hybrid supercapacitor device using $NiCo_2S_4$–NiCu-layer double hydroxide exhibited a maximum volumetric energy density of 2.7 mWh cm^{-3} at a power density of 21.3 mWh cm^{-3}, excellent long-term cycling stability over 2000 cycles, and remarkable flexibility under repeated bending tests [236].

8.7.5 Advanced Nano-Hybrid Metal-Organic Frameworks in Supercapacitors

Over the past two decades, metal-organic frameworks have garnered significant attention for their construction involving strong chemical bonds between inorganic metal ions and organic linkers, resulting in structures with dimensions such as one dimension (1D), two dimensions (2D), or three dimensions (3D) [237, 238]. Researchers have identified metal–organic frameworks as a promising class of nanomaterials due to their electrical, optical, magnetic, and structural features [239, 240]. Metal-organic frameworks-based nanocomposites showcase excellent electrochemical properties and find applications in various energy storage devices, benefiting from their large surface areas, flexibility, uniformly ordered structures, modular nature, and high porosities [239, 240]. The next generation of metal-organic frameworks-based energy storage devices is expected to demonstrate higher power densities, increased energy densities, and longer cycling stability [241]. In a study by Yang Jiao et al., an efficient technique for enhancing the conductance of metal-organic frameworks involved partially replacing Ni^{2+} in the Ni-metal-organic frameworks with Co^{2+} or Zn^{2+}. The resulting hybrid supercapacitors, metal–organic frameworks -carbon nanotubes-COOH showed improved performance due to the elimination of phase transitions during electrochemical cycling tests. Notably, the Co/Ni-metal–organic frameworks-carbon nanotubes-COOH (Co/Ni-MOF-CNT-COOH) exhibited energy and power densities of 49.5 $Whkg^{-1}$ and 1450 Wkg^{-1}, respectively [242]. Saraf et al. 2016 described the fabrication of a Cu-metal-organic framework---rGO hybrid through easy ultrasonication. This hybrid demonstrated excellent charge storage efficiency, high energy, and power density of 685.33 F g^{-1}, 137.066 Wh kg^{-1}, and 4800.04 W kg^{-1}, respectively along with great stability on the electrode surface, as evidenced by its extended cycle life of 91.91% after 1000 cycles [243]. Jayaramulu et al. 2021 reported on the electrochemical characterizations of the hybrid GA-UiO-66-NH_2, which serves as an effective energy-storing material with high capacitance at 651 F g^{-1}, surpassing other graphene-based materials [244]. Li et al. 2019 proposed a two-step approach for regulating the growth of Ni-Ni-metal-organic framework arrays on $NiCo_2O_4$ nanowire surfaces. This hybrid structure, combining the unique properties of the $NiCo_2O_4$ core and Ni-metal-organic framework shell, exhibited a high specific capacity of 208.8 mAh/g at 2 mA/cm^2 [245]. Using metal-organic frameworks as a template, Liu et al., 2020 created hybrid nanosheets with phosphorus-doped Co_3O_4 nanoparticles in a P–N co-doped carbon matrix (P–Co_3O_4–PNC). The P-Co_3O_4-PNC hybrid nanosheets demonstrated a specific capacity of 614 mC cm^{-2} at 1 mA cm^{-2} and excellent cycle stability [246].

8.7.6 *Nano-Hybrid Materials in Flexible and Wearable Supercapacitors*

Despite the rapid development of wearable electronics, the next-generation trend focuses on thinner thickness, lighter weight, increased conformability, and enhanced adaptability to the human body [247, 248]. The rising demand for flexible and wearable electronics is driven by their stable cycling life, rapid charge-discharge capability, and high power and energy density, making them increasingly popular [249, 250]. Devices employing self-supported P-Co_3O_4–PNC and P–N co-doped carbon matrix materials in asymmetric supercapacitors have demonstrated high energy density and remarkable cycling stability, achieving 750 W/kg in flexible solid-state asymmetric supercapacitors [246]. The solvothermal preparation of nickel-based metal-organic frameworks (Ni–MOFs) has been streamlined, significantly enhancing capacitance performance. Ni-MOFs-GO, with a 3% weight concentration of graphene oxide and modulated with HCl, exhibited pseudocapacitance of 2192.4 F g^{-1} at a current density of 1 A g^{-1}, surpassing the performance of Ni-MOFs with 1457.7 F g^{-1} at 180 °C [251].

Recent research on nano-hybrid supercapacitors has shifted toward solid-state devices that are flexible and wearable while maintaining high electrochemical performance [252, 253]. Polymers emerge as promising materials for flexible and wearable supercapacitors due to their flexibility, low mechanical stiffness, and various soft properties [254]. However, challenges remain in the fabrication of flexible and wearable supercapacitor products [255]. In the fields of electric vehicles, industry, and consumer electronics, hybrid supercapacitors play a significant role in reducing costs, size, and lifespan. They offer substantial output, charging devices quickly within minutes or seconds, providing a distinct advantage over batteries. Novel hybrid supercapacitors, offering purely organic, printable, flexible, and wearable energy storage solutions, utilize a variety of materials, including metals, metal oxides, metal chalcogenides, polymers, and carbonaceous materials.

8.8 Conclusions

In conclusion, this chapter delved into the critical aspects of energy storage, elucidating the mechanisms, categorizations, unique attributes, and limitations of supercapacitors. The focus then shifted toward introducing the cutting-edge technology of hybrid supercapacitors, providing insights into recent advancements, the assembly of hybrid nanocomposites, and the creation of their distinctive features. Electrode material design, manufacturing procedures, and the resulting electrochemical performance were thoroughly explored. Significant emphasis was placed on conceivable intercalation methods, material preparation, structural characterization, and electrode fabrication, providing a comprehensive understanding of the electrochemical reactions within hybrid supercapacitors. The exploration extended to techniques such as tuning morphology, adjusting phase structure, and enhancing crystalline purity all contributing to improved electrochemical characteristics. The chapter concluded by discussing the electrochemical properties of various current hybrid nanostructured materials, encompassing graphene-metal oxide hybrids, Li- and Na-ion supercapacitors, polymer-based hybrid supercapacitors, organic-inorganic hybrids, advanced nano-hybrid, metal-organic frameworks, and nano-hybrid materials in flexible and wearable applications. Experimental studies consistently indicated that interfacial alterations and structural controls have the potential to significantly enhance the specific capacity, cycle lifespan, and rate performance of hybrid supercapacitors. It is anticipated that ongoing basic research on hybrid supercapacitors will address existing disadvantages and unstable characteristics, paving the way for broad applications in energy storage. With continued advancements, hybrid supercapacitors are poised to overcome current challenges, guiding in a new era of efficient and versatile energy storage solutions.

Acknowledgement The co-author Prof. Abdel-Hamid I. Mourad would like to acknowledge the support of National Water and Energy Center, United Arab Emirates University, UAE for the financial support through research grants 12R227 and 12R304.

References

1. A.C. Forse, J.M. Gri, C.P. Grey, New perspectives on the charging mechanisms of supercapacitors. J. Am. Ceram. Soc. **138**, 5731–5744 (2016). https://doi.org/10.1021/jacs.6b02115
2. C. Soc, G. Wang, J. Zhang, A review of electrode materials for electrochemical supercapacitors. Chem. Soc. Rev. Crit. Rev., 797–828 (2012). https://doi.org/10.1039/c1cs15060j
3. H. Mikulčić, J. Baleta, J.J. Klemeš, X. Wang, Energy transition and the role of system integration of the energy, water and environmental systems. J. Clean. Prod. **292** (2021). https://doi.org/10.1016/j.jclepro.2021.126027
4. T.H. Jeon, M.S. Koo, H. Kim, W. Choi, Dual-functional photocatalytic and Photoelectrocatalytic Systems for Energy- and Resource-Recovering Water Treatment. ACS Catal. **8**, 11542–11563 (2018). https://doi.org/10.1021/acscatal.8b03521
5. J. Chen, P.S. Lee, Electrochemical supercapacitors: from mechanism understanding to multifunctional applications. Adv. Energy Mater. **11** (2021). https://doi.org/10.1002/aenm.202003311
6. J. Zhao, A.F. Burke, Review on supercapacitors: technologies and performance evaluation. J. Energy Chem. **59**, 276–291 (2021). https://doi.org/10.1016/j.jechem.2020.11.013
7. E. Barry, R. Burns, W. Chen, G.X. De Hoe, J.M.M. De Oca, J.J. De Pablo, J. Dombrowski, J.W. Elam, A.M. Felts, G. Galli, J. Hack, Q. He, X. He, E. Hoenig, A. Iscen, B. Kash, H.H. Kung, N.H.C. Lewis, C. Liu, X. Ma, A. Mane, A.B.F. Martinson, K.L. Mulfort, J. Murphy, K. Mølhave, P. Nealey, Y. Qiao, V. Rozyyev, G.C. Schatz, S.J. Sibener, D. Talapin, D.M. Tiede, M.V. Tirrell, A. Tokmakoff, G.A. Voth, Z. Wang, Z. Ye, M. Yesibolati, N.J. Zaluzec, S.B. Darling, Advanced materials for energy-water systems: the central role of water/solid interfaces in adsorption, reactivity, and transport. Chem. Rev. **121**, 9450–9501 (2021). https://doi.org/10.1021/acs.chemrev.1c00069

8. H. Zhao, Y. Lei, 3D nanostructures for the next generation of high-performance nanodevices for electrochemical energy conversion and storage. Adv. Energy Mater. **10**, 1–8 (2020). https://doi.org/10.1002/aenm.202001460
9. K.S. Kumar, N. Choudhary, Y. Jung, J. Thomas, Recent advances in two-dimensional nanomaterials for supercapacitor electrode applications. ACS Energy Lett. **3**, 482–495 (2018). https://doi.org/10.1021/acsenergylett.7b01169
10. S. Chen, W. Xing, J. Duan, X. Hu, S.Z. Qiao, Nanostructured morphology control for efficient supercapacitor electrodes. J Mater. Chem. A **1**, 2941–2954 (2013). https://doi.org/10.1039/c2ta00627h
11. Y. Sun, S. Guo, H. Zhou, Exploration of advanced electrode materials for rechargeable sodium-ion batteries. Adv. Energy Mater. **9**, 1–18 (2019). https://doi.org/10.1002/aenm.201800212
12. G. Song, H. Bin Son, D.Y. Han, M. Je, S. Nam, S. Park, A renewable future: a comprehensive perspective from materials to systems for next-generation batteries. Mater. Chem. Front. **5**, 3344–3377 (2021). https://doi.org/10.1039/d1qm00071c
13. J. Cherusseri, D. Pandey, J. Thomas, Symmetric, asymmetric, and battery-type supercapacitors using two-dimensional nanomaterials and composites. Batter. Supercaps. **3**, 860–875 (2020). https://doi.org/10.1002/batt.201900230
14. E. Pomerantseva, F. Bonaccorso, X. Feng, Y. Cui, Y. Gogotsi, Energy storage: the future enabled by nanomaterials. Science **366**, (2019). https://doi.org/10.1126/science.aan8285
15. H. Gu, Y.E. Zhu, J. Yang, J. Wei, Z. Zhou, Nanomaterials and technologies for lithium-ion hybrid supercapacitors. ChemNanoMat. **2**, 578–587 (2016). https://doi.org/10.1002/cnma.201600068
16. S. Wu, Y. Zhu, Highly densified carbon electrode materials towards practical supercapacitor devices. Sci. China Mater., 1–14 (2016). https://doi.org/10.1007/s40843-016-5109-4
17. L.L. Zhang, X.S. Zhao, Carbon-based materials as supercapacitor electrodes. Chem. Soc. Rev., 2520–2531 (2009). https://doi.org/10.1039/b813846j
18. P. Simon, Y. Gogotsi, Materials for electrochemical capacitors electrochemical. Nat. Mater. **7**, 845–854 (2008). https://doi.org/10.1038/nmat2297
19. C. Liu, F. Li, M. Lai-Peng, H.M. Cheng, Advanced materials for energy storage. Adv. Mater. **22**, 28–62 (2010). https://doi.org/10.1002/adma.200903328
20. S. Faraji, F.N. Ani, The development supercapacitor from activated carbon by electroless plating—a review. Renew. Sust. Energ. Rev. **42**, 823–834 (2015). https://doi.org/10.1016/j.rser.2014.10.068
21. J. Zhang, X. Zhang, Y. Zhou, S. Guo, K. Wang, Z. Liang, Q. Xu, Nitrogen-doped hierarchical porous carbon Nanowhisker ensembles on carbon Nano fiber for high-performance supercapacitors. ACS Sustain. Chem. Eng. **2**, 1525–1533 (2014) 10.1021/sc500221s
22. X. Huang, S. Kim, M.S. Heo, J.E. Kim, H. Suh, I. Kim, Easy synthesis of hierarchical carbon spheres with superior capacitive performance in supercapacitors. Langmuir **29**, 12266–12274 (2013). https://doi.org/10.1021/la4026969
23. X. Chen, H. Wang, H. Yi, X. Wang, Anthraquinone on porous carbon nanotubes with improved supercapacitor performance. J. Phys. Chem. **118**, 8262–8270 (2014). https://doi.org/10.1021/jp5009626
24. F.W. Richey, C. Tran, V. Kalra, Y.A. Elabd, Ionic liquid dynamics in Nanoporous carbon nano fibers in supercapacitors measured with in operando infrared spectroelectrochemistry. J. Phys. Chem. **118**, 21846–21855 (2014). https://doi.org/10.1021/jp506903m
25. X. Yang, F. Zhang, L. Zhang, T. Zhang, Y. Huang, Y. Chen, A high-performance graphene oxide-doped ion gel as gel polymer electrolyte for all-solid-state supercapacitor applications. Adv. Funct. Mater. **23**, 3353–3360 (2013). https://doi.org/10.1002/adfm.201203556
26. A.B. Fuertes, M. Sevilla, Hierarchical microporous/mesoporous carbon nanosheets for high-performance supercapacitors. ACS Appl. Mater. Interfaces (2015). https://doi.org/10.1021/am508794f
27. W. Dmowski, T. Egami, K.E. Swider-Lyons, C.T. Love, D.R. Rolison, Local atomic structure and conduction mechanism of nanocrystalline hydrous RuO_2 from X-ray scattering. J. Phys. Chem. B **106**, 12677–12683 (2002). https://doi.org/10.1021/jp026228l

28. J.W. Long, K.E. Swider, C.I. Merzbacher, D.R. Rolison, Voltammetric characterization of ruthenium oxide-based aerogels and other RuO_2 solids: The nature of capacitance in nanostructured materials. Langmuir **15**, 780–784 (1999). https://doi.org/10.1021/la980785a
29. X. Yu, S. Yun, J.S. Yeon, P. Bhattacharya, L. Wang, S.W. Lee, X. Hu, H.S. Park, Emergent Pseudocapacitance of 2D nanomaterials. Adv. Energy Mater. **8**, 1–33 (2018). https://doi.org/10.1002/aenm.201702930
30. S. Fleischmann, J.B. Mitchell, R. Wang, C. Zhan, D.E. Jiang, V. Presser, V. Augustyn, Pseudocapacitance: From fundamental understanding to high power energy storage materials. Chem. Rev. **120**, 6738–6782 (2020). https://doi.org/10.1021/acs.chemrev.0c00170
31. V. Augustyn, P. Simon, B. Dunn, Pseudocapacitive oxide materials for high-rate electrochemical energy storage. Energy Environ. Sci. **7**, 1597–1614 (2014). https://doi.org/10.1039/c3ee44164d
32. T. Brousse, D. Bélanger, K. Chiba, M. Egashira, F. Favier, J. Long, J.R. Miller, M. Morita, K. Naoi, P. Simon, W. Sugimoto, *Materials for Electrochemical Capacitors* (Springer, 2017), pp. 495–561. https://doi.org/10.1007/978-3-662-46657-5_16
33. K. Chen, Y. Yang, K. Li, Z. Ma, Y. Zhou, D. Xue, $CoCl_2$ designed as excellent pseudocapacitor electrode materials. ACS Sustain. Chem. Eng. **2**, 440–444 (2014). https://doi.org/10.1021/sc400338c
34. C. Guan, Z. Zeng, X. Li, X. Cao, Y. Fan, X. Xia, G. Pan, H. Zhang, H.J. Fan, Atomic-layer-deposition-assisted formation of carbon nanoflakes on metal oxides and energy storage application. Small **10**, 300–307 (2014). https://doi.org/10.1002/smll.201301009
35. K. Naoi, Y. Nagano, Li-ion-based hybrid supercapacitors in organic medium, supercapacitors mater. Syst. Appl., 239–256 (2013). https://doi.org/10.1002/9783527646661.ch7
36. S. Yang, X. Song, P. Zhang, J. Sun, L. Gao, Self-assembled α-Fe_2O_3 mesocrystals/graphene nanohybrid for enhanced electrochemical capacitors. Small **10**, 2270–2279 (2014). https://doi.org/10.1002/smll.201303922
37. J. Wang, J. Polleux, J. Lim, B. Dunn, Pseudocapacitive contributions to electrochemical energy storage in TiO_2 (anatase) nanoparticles. J. Phys. Chem. C **111**, 14925–14931 (2007). https://doi.org/10.1021/jp074464w
38. P. Han, G. Xu, X. Han, J. Zhao, X. Zhou, G. Cui, Lithium ion capacitors in organic electrolyte system: scientific problems, material development, and key technologies. Adv. Energy Mater. **8**, 1–30 (2018). https://doi.org/10.1002/aenm.201801243
39. H. Khallaf, I.O. Oladeji, L. Chow, Optimization of chemical bath deposited CdS thin films using nitrilotriacetic acid as a complexing agent. Thin Solid Films **516**, 5967–5973 (2008). https://doi.org/10.1016/j.tsf.2007.10.079
40. S.A. Pawar, R.S. Devan, D.S. Patil, A.V. Moholkar, M.G. Gang, Y.R. Ma, J.H. Kim, P.S. Patil, Improved solar cell performance of chemosynthesized cadmium selenide pebbles. Electrochim. Acta **98**, 244–254 (2013). https://doi.org/10.1016/j.electacta.2013.03.016
41. Q. Zhang, B. Zhao, J. Wang, C. Qu, H. Sun, K. Zhang, Nano energy high-performance hybrid supercapacitors based on self-supported 3D ultrathin porous quaternary Zn–Ni–Al–co oxide nanosheets. Nano Energy **28**, 475–485 (2016). https://doi.org/10.1016/j.nanoen.2016.08.049
42. A.K. Das, et al., Fabrication of an advanced asymmetric supercapacitor based on microcubical PB@MnO_2 hybrid and PANI/GNP composite with excellent electrochemical behaviour. J. Mater. Chem. A (2017). https://doi.org/10.1039/C7TA05134D
43. M.G. Hahm, A.L.M. Reddy, D.P. Cole, M. Rivera, J.A. Vento, J. Nam, H.Y. Jung, Y.L. Kim, N.T. Narayanan, D.P. Hashim, C. Galande, Y.J. Jung, M. Bundy, S. Karna, P.M. Ajayan, R. Vajtai, Carbon nanotube-nanocup hybrid structures for high power supercapacitor applications. Nano Lett. **12**, 5616–5621 (2012). https://doi.org/10.1021/nl3027372
44. Q. Meng, K. Qin, L. Ma, C. He, E. Liu, F. He, C. Shi, Q. Li, J. Li, N. Zhao, N-doped porous carbon nanofibers/porous silver network hybrid for high-rate supercapacitor electrode. ACS Appl. Mater. Interfaces **9**, 30832–30839 (2017). https://doi.org/10.1021/acsami.7b08610
45. S. Pansri, S. Noothongkaew, MWCNTs/r-GO hybrid films fabricated by layer by layer assembly for supercapacitor electrodes. J Energy Storage **22**, 153–156 (2019). https://doi.org/10.1016/j.est.2019.02.009

46. S. Pandey, S.B. Mishra, Sol-gel derived organic-inorganic hybrid materials: synthesis, characterizations and applications. J. Sol-Gel Sci. Technol. **59**, 73–94 (2011). https://doi.org/10.1007/s10971-011-2465-0
47. Y. Liu, X. Qiu, X. Liu, Y. Liu, L.Z. Fan, 3D porous binary-heteroatom doped carbon nanosheet/electrochemically exfoliated graphene hybrids for high performance flexible solid-state supercapacitors. J. Mater. Chem. A **6**, 8750–8756 (2018). https://doi.org/10.1039/c8ta01148f
48. K. Dharmalingam, V. Gurudevan, G. Dhanasekaran, D. Sekar, R. Gopal, D. Alshamsi, E. Thangavel, S. Sambasivam, Synthesis and characterization of lamellar-like $Cu_2(OH)_3NO_3$ nanosheets integrated with $Mg(OH)_2$ nanoparticles heterojunction for photocatalytic activity. J. Mater. Res. **2** (2023). https://doi.org/10.1557/s43578-023-01218-z
49. H. Dong, G.M. Koenig, A review on synthesis and engineering of crystal precursors produced: via coprecipitation for multicomponent lithium-ion battery cathode materials. CrystEngComm **22**, 1514–1530 (2020). https://doi.org/10.1039/c9ce00679f
50. W. Qin, Y. Liu, X. Liu, G. Yang, Facile and scalable production of amorphous nickel borate for high performance hybrid supercapacitors. J Mater Chem A **6**, 19689–19695 (2018). https://doi.org/10.1039/c8ta07385f
51. V. Gajraj, C.R. Mariappan, Electrochemical performances of asymmetric aqueous supercapacitor based on porous $Cu_3Mo_2O_9$ petals and $La_2Mo_3O_{12}$ nanoparticles fabricated through a simple co-precipitation method. Appl. Surf. Sci. **512**, 145648 (2020). https://doi.org/10.1016/j.apsusc.2020.145648
52. D. Karthigaimuthu, P. Saradhi, B.A. Kumar, G. Ramalingam, T. Elangovan, S. Sangaraju, Hydrothermal synthesis of MoS_2-$Mg(OH)_2$-$BiVO_4$ ternary hierarchical heterostructures for dye-sensitized solar cell application. Mater. Lett. **359**, 135890 (2024). https://doi.org/10.1016/j.matlet.2024.135890
53. D. Karthigaimuthu, A.K. Bojarajan, E. Thangavel, P.S. Maram, S. Venkidusamy, S. Sangaraju, A.H.I. Mourad, Synergistic effects in $MoS_2/Co_3O_4/Cu_2O$ nanocomposites for superior solar cell and photodegradation efficiency. J. Alloys Compd. **1010** 177672 (2025) https://doi.org/10.1016/j.jallcom.2024.177672
54. S.S. Patil, D.P. Dubal, V.G. Deonikar, M.S. Tamboli, J.D. Ambekar, P. Gomez-Romero, S.S. Kolekar, B.B. Kale, D.R. Patil, Fern-like rGO/$BiVO_4$ hybrid nanostructures for high-energy symmetric supercapacitor. ACS Appl. Mater. Interfaces **8**, 31602–31610 (2016). https://doi.org/10.1021/acsami.6b08165
55. A. Sarkar, A.K. Chakraborty, S. Bera, S. Krishnamurthy, Novel hydrothermal synthesis of CoS_2/MWCNT Nanohybrid electrode for supercapacitor: a systematic investigation on the influence of MWCNT. J. Phys. Chem. C **122**, 18237–18246 (2018). https://doi.org/10.1021/acs.jpcc.8b04137
56. H. Li, J. Liang, Recent development of printed micro-supercapacitors: printable materials, printing technologies, and perspectives. Adv. Mater. **32**, 1–19 (2020). https://doi.org/10.1002/adma.201805864
57. T.S. Tran, N.K. Dutta, N.R. Choudhury, Graphene-based inks for printing of planar micro-supercapacitors: a review. Materials (Basel). **12** (2019). https://doi.org/10.3390/ma12060978
58. L. Nayak, S. Mohanty, S.K. Nayak, A. Ramadoss, A review on inkjet printing of nanoparticle inks for flexible electronics. J. Mater. Chem. C **7**, 8771–8795 (2019). https://doi.org/10.1039/c9tc01630a
59. A. Sajedi-Moghaddam, E. Rahmanian, N. Naseri, Inkjet-printing technology for supercapacitor application: current state and perspectives. ACS Appl. Mater. Interfaces **12**, 34487–34504 (2020). https://doi.org/10.1021/acsami.0c07689
60. J.H. Sung, S.J. Kim, S.H. Jeong, E.H. Kim, K.H. Lee, Flexible micro-supercapacitors. J. Power Sources **162**, 1467–1470 (2006). https://doi.org/10.1016/j.jpowsour.2006.07.073
61. Z.S. Wu, K. Parvez, X. Feng, K. Müllen, Photolithographic fabrication of high-performance all-solid-state graphene-based planar micro-supercapacitors with different interdigital fingers. J. Mater. Chem. A **2**, 8288–8293 (2014). https://doi.org/10.1039/c4ta00958d

62. D. Kim, G. Lee, D. Kim, J.S. Ha, Air-stable, high-performance, flexible microsupercapacitor with patterned ionogel electrolyte. ACS Appl. Mater. Interfaces **7**, 4608–4615 (2015). https://doi.org/10.1021/am5077843
63. X. Pu, M. Liu, L. Li, S. Han, X. Li, C. Jiang, C. Du, J. Luo, W. Hu, Z.L. Wang, Wearable textile-based in-plane microsupercapacitors. Adv. Energy Mater. **6** (2016). https://doi.org/10.1002/aenm.201601254
64. L. Liu, Q. Lu, S. Yang, J. Guo, Q. Tian, W. Yao, Z. Guo, V.A.L. Roy, W. Wu, All-printed solid-state microsupercapacitors derived from self-template synthesis of ag@PPy nanocomposites. Adv. Mater. Technol. **3**, 1–9 (2018). https://doi.org/10.1002/admt.201700206
65. S. Bellani, E. Petroni, A.E. Del Rio Castillo, N. Curreli, B. Martín-García, R. Oropesa-Nuñez, M. Prato, F. Bonaccorso, Scalable production of graphene inks via wet-jet milling exfoliation for screen-printed micro-supercapacitors. Adv. Funct. Mater. **29**, 1–14 (2019). https://doi.org/10.1002/adfm.201807659
66. X. Shi, S. Pei, F. Zhou, W. Ren, H.M. Cheng, Z.S. Wu, X. Bao, Ultrahigh-voltage integrated micro-supercapacitors with designable shapes and superior flexibility. Energy Environ. Sci. **12**, 1534–1541 (2019). https://doi.org/10.1039/c8ee02924e
67. L. Liu, D. Ye, Y. Yu, L. Liu, Y. Wu, Carbon-based flexible micro-supercapacitor fabrication via mask-free ambient micro-plasma-jet etching. Carbon N. Y. **111**, 121–127 (2017). https://doi.org/10.1016/j.carbon.2016.09.037
68. L. Liu, et al., Silver nanowires as the current collector for flexible in-plane micro-supercapacitor via a one-step mask-free patterning strategy, Nanotechnology (2017). https://doi.org/10.1088/1361-6528/aa9aaf
69. Z.S. Wu, K. Parvez, A. Winter, H. Vieker, X. Liu, S. Han, A. Turchanin, X. Feng, K. Müllen, Layer-by-layer assembled heteroatom-doped graphene films with ultrahigh volumetric capacitance and rate capability for micro-supercapacitors. Adv. Mater. **26**, 4552–4558 (2014). https://doi.org/10.1002/adma.201401228
70. Z.S. Wu, K. Parvez, X. Feng, K. Müllen, Graphene-based in-plane micro-supercapacitors with high power and energy densities. Nat. Commun. **4** (2013). https://doi.org/10.1038/ncomms3487
71. W. Yang, J. Yang, J.J. Byun, F.P. Moissinac, J. Xu, S.J. Haigh, M. Domingos, M.A. Bissett, R.A.W. Dryfe, S. Barg, 3D printing of freestanding MXene architectures for current-collector-free supercapacitors. Adv. Mater. **31**, 1–8 (2019). https://doi.org/10.1002/adma.201902725
72. J.P. Mensing, T. Lomas, A. Tuantranont, 2D and 3D printing for graphene based supercapacitors and batteries: a review. Sustain. Mater. Technol. **25**, e00190 (2020). https://doi.org/10.1016/j.susmat.2020.e00190
73. J. Orangi, F. Hamade, V.A. Davis, M. Beidaghi, 3D printing of additive-free 2D Ti_3C_2Tx (MXene) ink for fabrication of micro-supercapacitors with ultra-high energy densities. ACS Nano **14**, 640–650 (2020). https://doi.org/10.1021/acsnano.9b07325
74. P. Chang, H. Mei, S. Zhou, K.G. Dassios, L. Cheng, 3D printed electrochemical energy storage devices. J Mater Chem A **7**, 4230–4258 (2019). https://doi.org/10.1039/c8ta11860d
75. K. Shen, J. Ding, S. Yang, 3D printing quasi-solid-state asymmetric micro-supercapacitors with ultrahigh areal energy density. Adv. Energy Mater. **8**, 1–7 (2018). https://doi.org/10.1002/aenm.201800408
76. J. Zhao, Y. Zhang, Y. Huang, J. Xie, X. Zhao, C. Li, J. Qu, Q. Zhang, J. Sun, B. He, Q. Li, C. Lu, X. Xu, W. Lu, L. Li, Y. Yao, 3D printing Fiber electrodes for an all-Fiber integrated electronic device via hybridization of an asymmetric supercapacitor and a temperature sensor. Adv. Sci. **5** (2018). https://doi.org/10.1002/advs.201801114
77. L. Li, Z. Lou, W. Han, D. Chen, K. Jiang, G. Shen, Highly stretchable micro-supercapacitor arrays with hybrid MWCNT/PANI electrodes. Adv. Mater. Technol. **2** (2017). https://doi.org/10.1002/admt.201600282
78. W. Gao, N. Singh, L. Song, Z. Liu, A.L.M. Reddy, L. Ci, R. Vajtai, Q. Zhang, B. Wei, P.M. Ajayan, Direct laser writing of micro-supercapacitors on hydrated graphite oxide films. Nat. Nanotechnol. **6**, 496–500 (2011). https://doi.org/10.1038/nnano.2011.110

79. Z. Peng, R. Ye, J.A. Mann, D. Zakhidov, Y. Li, P.R. Smalley, J. Lin, J.M. Tour, Flexible boron-doped laser-induced graphene microsupercapacitors. ACS Nano **9**, 5868–5875 (2015). https://doi.org/10.1021/acsnano.5b00436
80. J. Lee, J.Y. Seok, S. Son, M. Yang, B. Kang, High-energy, flexible micro-supercapacitors by one-step laser fabrication of a self-generated nanoporous metal/oxide electrode. J. Mater. Chem. A **5**, 24585–24593 (2017). https://doi.org/10.1039/c7ta07960e
81. S. Wang, Y. Yu, S. Luo, X. Cheng, G. Feng, Y. Zhang, Z. Wu, G. Compagnini, J. Pooran, A. Hu, All-solid-state supercapacitors from natural lignin-based composite film by laser direct writing. Appl. Phys. Lett. **115** (2019). https://doi.org/10.1063/1.5118340
82. Z. Zhang, M. Song, J. Hao, K. Wu, C. Li, C. Hu, Visible light laser-induced graphene from phenolic resin: A new approach for directly writing graphene-based electrochemical devices on various substrates. Carbon N. Y. **127**, 287–296 (2018). https://doi.org/10.1016/j.carbon.2017.11.014
83. S. Wang, Y. Yu, R. Li, G. Feng, Z. Wu, G. Compagnini, A. Gulino, Z. Feng, A. Hu, High-performance stacked in-plane supercapacitors and supercapacitor array fabricated by femtosecond laser 3D direct writing on polyimide sheets. Electrochim. Acta **241**, 153–161 (2017). https://doi.org/10.1016/j.electacta.2017.04.138
84. F. Li, J. Qu, Y. Li, J. Wang, M. Zhu, L. Liu, J. Ge, S. Duan, T. Li, V.K. Bandari, M. Huang, F. Zhu, O.G. Schmidt, Stamping fabrication of flexible planar micro-supercapacitors using porous graphene inks. Adv. Sci. **7**, 1–9 (2020). https://doi.org/10.1002/advs.202001561
85. B. Nagar, D.P. Dubal, L. Pires, A. Merkoçi, P. Gómez-Romero, Design and fabrication of printed paper-based hybrid micro-supercapacitor by using graphene and redox-active electrolyte. ChemSusChem **11**, 1849–1856 (2018). https://doi.org/10.1002/cssc.201800426
86. E. Quain, T.S. Mathis, N. Kurra, K. Maleski, K.L. Van Aken, M. Alhabeb, H.N. Alshareef, Y. Gogotsi, Direct writing of additive-free MXene-in-water ink for electronics and energy storage. Adv. Mater. Technol. **4**, 1–7 (2019). https://doi.org/10.1002/admt.201800256
87. C.J. Zhang, M.P. Kremer, A. Seral-Ascaso, S.H. Park, N. McEvoy, B. Anasori, Y. Gogotsi, V. Nicolosi, Stamping of flexible, coplanar micro-supercapacitors using MXene inks. Adv. Funct. Mater. **28**, 1–10 (2018). https://doi.org/10.1002/adfm.201705506
88. L. Jia, Y. Shi, Q. Zhang, X. Xu, Green synthesis of ultrafine methyl-cellulose-derived porous carbon/MnO_2 nanowires for asymmetric supercapacitors and flexible pattern stamping. Appl. Surf. Sci. **462**, 923–931 (2018). https://doi.org/10.1016/j.apsusc.2018.08.213
89. Z. Song, W. Li, Y. Bao, Z. Sun, L. Gao, M.H. Nawaz, D. Han, L. Niu, Enhanced pseudocapacitance and electrolyte-wettability of graphene hydrogels to tailor high mass loading all-solid-state supercapacitor with ultra-high volumetric energy density. Carbon N. Y. **136**, 46–53 (2018). https://doi.org/10.1016/j.carbon.2018.04.036
90. A. Lamberti, F. Clerici, M. Fontana, L. Scaltrito, A highly stretchable supercapacitor using laser-induced graphene electrodes onto elastomeric substrate. Adv. Energy Mater. **6**, 1–6 (2016). https://doi.org/10.1002/aenm.201600050
91. C.F. Holder, R.E. Schaak, Tutorial on powder X-ray diffraction for characterizing nanoscale materials. ACS Nano **13**, 7359–7365 (2019). https://doi.org/10.1021/acsnano.9b05157
92. K. Dharamalingam, B.A. Kumar, G. Ramalingam, S.S. Florence, K. Raju, P.S. Kumar, S. Govindaraju, E. Thangavel, The role of sodium dodecyl sulfate mediated hydrothermal synthesis of MoS_2 nanosheets for photocatalytic dye degradation and dye-sensitized solar cell application. Chemosphere **294**, 133725 (2022). https://doi.org/10.1016/j.chemosphere.2022.133725
93. D. Karthigaimuthu, M. Alsawalha, A.A.H. Mourad, A.A. Yadav, Y.M. Hunge, E. Thangavel, A. Hamid, I. Mourad, Construction of g-C_3N_4/MoS_2/SnO_2 hybrid as 2D/2D/1D architecture for counter electrode of dye-sensitized solar cells and photodegradation of pharmaceutical drugs from wastewater. Appl. Surf. Sci. Adv. **27**, 100771 (2025). https://doi.org/10.1016/j.apsadv.2025.100771
94. P. Zhang, Y. Li, G. Wang, F. Wang, S. Yang, F. Zhu, X. Zhuang, O.G. Schmidt, X. Feng, Zn-ion hybrid micro-supercapacitors with ultrahigh areal energy density and Long-term durability. Adv. Mater. **31**, 1–6 (2019). https://doi.org/10.1002/adma.201806005

95. M. Majumder, R.B. Choudhary, A.K. Thakur, C.S. Rout, G. Gupta, Rare earth metal oxide (RE_2O_3; RE = Nd, Gd, and Yb) incorporated polyindole composites: Gravimetric and volumetric capacitive performance for supercapacitor applications. New J. Chem. **42**, 5295–5308 (2018). https://doi.org/10.1039/c8nj00221e
96. H. Huang, X. Wang, E. Tervoort, G. Zeng, T. Liu, X. Chen, A. Sologubenko, M. Niederberger, Nano-sized structurally disordered metal oxide composite aerogels as high-power anodes in hybrid supercapacitors. ACS Nano **12**, 2753–2763 (2018). https://doi.org/10.1021/acsnano.7b09062
97. S. Mondal, U. Rana, S. Malik, Reduced graphene oxide/Fe_3O_4/polyaniline nanostructures as electrode materials for an all-solid-state hybrid supercapacitor. J. Phys. Chem. C **121**, 7573–7583 (2017). https://doi.org/10.1021/acs.jpcc.6b10978
98. B.Y. Guan, L. Yu, X. Wang, S. Song, X.W.D. Lou, Formation of onion-like $NiCo_2S_4$ particles via sequential ion-exchange for hybrid supercapacitors. Adv. Mater. **29**, 3–7 (2017). https://doi.org/10.1002/adma.201605051
99. X. Li, C. Zhang, S. Xin, Z. Yang, Y. Li, D. Zhang, P. Yao, Facile synthesis of MoS_2/reduced graphene oxide@polyaniline for high-performance supercapacitors. ACS Appl. Mater. Interfaces **8**, 21373–21380 (2016). https://doi.org/10.1021/acsami.6b06762
100. E. Lim, H. Kim, C. Jo, J. Chun, K. Ku, S. Kim, H.I. Lee, I.S. Nam, S. Yoon, K. Kang, J. Lee, Advanced hybrid supercapacitor based on a mesoporous niobium pentoxide/carbon as high-performance anode. ACS Nano **8**, 8968–8978 (2014). https://doi.org/10.1021/nn501972w
101. D. Karthigaimuthu, K. Raju, S. Chakrabortty, S. Ghosh, B. Arjunkumar, T. Elangovan, S. Sambasivam, Rational design of $Mg(OH)_2/Cu_2(OH)_3(NO_3)$ binary heterostructure electrodes for enriched supercapacitors performance. Ionics (Kiel). **2** (2023). https://doi.org/10.1007/s11581-023-05304-4
102. N. Zhang, W. Zhou, Q. Zhang, P. Luan, L. Cai, F. Yang, X. Zhang, Q. Fan, W. Zhou, Z. Xiao, X. Gu, H. Chen, K. Li, S. Xiao, Y. Wang, H. Liu, S. Xie, Biaxially stretchable supercapacitors based on the buckled hybrid fiber electrode array. Nanoscale **7**, 12492–12497 (2015). https://doi.org/10.1039/c5nr03027g
103. Y. Wu, G. Gao, G. Wu, Self-assembled three-dimensional hierarchical porous V_2O_5/graphene hybrid aerogels for supercapacitors with high energy density and long cycle life. J. Mater. Chem. A **3**, 1828–1832 (2015). https://doi.org/10.1039/c4ta05537c
104. M. Zhou, F. Lu, X. Shen, W. Xia, H. He, X. Zeng, One-pot construction of three dimensional $CoMoO_4/Co_3O_4$ hybrid nanostructures and their application in supercapacitors. J. Mater. Chem. A **3**, 21201–21210 (2015). https://doi.org/10.1039/c5ta05658f
105. B.Y. Guan, A. Kushima, L. Yu, S. Li, J. Li, X.W.D. Lou, Coordination polymers derived general synthesis of multishelled mixed metal-oxide particles for hybrid supercapacitors. Adv. Mater. **29**, 1–8 (2017). https://doi.org/10.1002/adma.201605902
106. J.S. Wei, C. Ding, P. Zhang, H. Ding, X.Q. Niu, Y.Y. Ma, C. Li, Y.G. Wang, H.M. Xiong, Robust negative electrode materials derived from carbon dots and porous hydrogels for high-performance hybrid supercapacitors. Adv. Mater. **31**, 1–7 (2019). https://doi.org/10.1002/adma.201806197
107. T. Nguyen, M., Fátima Montemor, γ-FeOOH and amorphous Ni–Mn hydroxide on carbon nanofoam paper electrodes for hybrid supercapacitors. J. Mater. Chem. A **6**, 2612–2624 (2018). https://doi.org/10.1039/c7ta05582j
108. H. Wang, Q. Hao, X. Yang, L. Lu, X. Wang, A nanostructured graphene / polyaniline hybrid material for supercapacitors. Nanoscale, 2164–2170 (2010). https://doi.org/10.1039/c0nr00224k
109. B. Senthilkumar, Z. Khan, S. Park, K. Kim, H. Ko, Y. Kim, Highly porous graphitic carbon and $Ni_2P_2O_7$ for a high performance aqueous hybrid supercapacitor. J Mater Chem A **3**, 21553–21561 (2015). https://doi.org/10.1039/c5ta04737d
110. D. Li, Y. Gong, C. Pan, Facile synthesis of hybrid CNTs / $NiCo_2S_4$ composite for high performance supercapacitors. Nat. Publ. Gr., 1–7 (2016). https://doi.org/10.1038/srep29788

111. S. Repp, E. Harputlu, S. Gurgen, M. Castellano, N. Kremer, N. Pompe, J. Wörner, A. Hoffmann, R. Thomann, F.M. Emen, S. Weber, K. Ocakoglu, E. Erdem, Synergetic effects of Fe^{3+} doped spinel $Li_4Ti_5O_{12}$ nanoparticles on reduced graphene oxide for high surface electrode hybrid supercapacitors. Nanoscale **10**, 1877–1884 (2018). https://doi.org/10.1039/c7nr08190a
112. C. Qu, B. Zhao, Y. Jiao, D. Chen, S. Dai, Y. Chen, K.S. Walton, R. Zou, M. Liu, Functionalized bimetallic hydroxides derived from metal-organic frameworks for high performance hybrid supercapacitor with ex-ceptional cycling stability. ACS Energy Lett. **2**, 1263–1269 (2017). https://doi.org/10.1021/acsenergylett.7b00265
113. S. Chen, L. Ma, K. Zhang, M. Kamruzzaman, C. Zhi, J.A. Zapien, A flexible solid-state zinc ion hybrid supercapacitor based on co-polymer derived hollow carbon spheres. J Mater Chem A **7**, 7784–7790 (2019). https://doi.org/10.1039/C9TA00733D
114. C. Xiang, Y. Liu, Y. Yin, P. Huang, Y. Zou, M. Fehse, Z. She, F. Xu, D. Banerjee, D.H. Merino, A. Longo, H. Kraatz, D.F. Brougham, B. Wu, L. Sun, Facile green route to Ni/co oxide nanoparticle embedded 3D graphitic carbon Nanosheets for high performance hybrid supercapacitor devices. ACS Appl. Energy Mater. **2**, 3389–3399 (2019). https://doi.org/10.1021/acsaem.9b00202
115. L. Jianhua, A. Junwei, Z. Yecheng, M. Yuxiao, L. Mengliu, Y. Mei, L. Songmei, Preparation of an amide group-connected graphene–polyaniline nanofiber hybrid and its application in supercapacitors. ACS Appl. Mater. Interfaces **4**, 2870–2876 (2012) dx.doi.org/10.1021/am300640y
116. S.G. Mohamed, S.Y. Attia, Y.F. Barakat, H.H. Hassan, W. Al Zoubi, Hydrothermal synthesis of α-MnS Nanoflakes@nitrogen and Sulfur co-doped rGO for high-performance hybrid supercapacitor. ChemistrySelect. **3**, 6061–6072 (2018). https://doi.org/10.1002/slct.201801042
117. P. Du, Y. Dong, H. Kang, X. Yang, Q. Wang, J. Niu, S. Wang, P. Liu, Graphene-wrapped polyaniline nanowire Array modified functionalized of carbon cloth for high-performance flexible solid-state supercapacitor. ACS Sustain. Chem. Eng. **6**, 14723–14733 (2018). https://doi.org/10.1021/acssuschemeng.8b03278
118. D. Zhao, X. Wu, C. Guo, Hybrid MnO_2@$NiCo_2O_4$ nanosheets for high performance asymmetric supercapacitors. Inorg. Chem. Front. **5**, 1378–1385 (2018). https://doi.org/10.1039/c8qi00170g
119. Z. Fang, S. ur Rehman, M. Sun, Y. Yuan, S. Jin, H. Bi, Hybrid NiO-CuO mesoporous nanowire array with abundant oxygen vacancies and a hollow structure as a high-performance asymmetric supercapacitor. J. Mater. Chem. A **6**, 21131–21142 (2018). https://doi.org/10.1039/C8TA08262F
120. T. Lee, T. Yun, B. Park, B. Sharma, H.K. Song, B.S. Kim, Hybrid multilayer thin film supercapacitor of graphene nanosheets with polyaniline: Importance of establishing intimate electronic contact through nanoscale blending. J. Mater. Chem. **22**, 21092–21099 (2012). https://doi.org/10.1039/c2jm33111j
121. Y. Xie, H. Du, Electrochemical capacitance of a carbon quantum dots-polypyrrole/titania nanotube hybrid. RSC Adv. **5**, 89689–89697 (2015). https://doi.org/10.1039/c5ra16538e
122. G. Konwar, S.C. Sarma, S.C. Sarma, D. Mahanta, S.C. Peter, S.C. Peter, S.C. Peter, Polyaniline hybrid nanofibers via green interfacial polymerization for all-solid-state symmetric supercapacitors. ACS Omega., 7 (2020). https://doi.org/10.1021/acsomega.0c01158
123. D. Karthigaimuthu, S. Ramasundaram, P. Nisha, B. Arjun Kumar, J. Sriram, G. Ramalingam, P. Vijaibharathy, T.H. Oh, T. Elangovan, Synthesis of MoS_2/$Mg(OH)_2$/$BiVO_4$ hybrid photocatalyst by ultrasonic homogenization assisted hydrothermal methods and its application as sunlight active photocatalyst for water decontamination. Chemosphere **308**, 136406 (2022). https://doi.org/10.1016/j.chemosphere.2022.136406
124. K. Dharamalingam, S. Ramasundaram, V. Kumar, Facile synthesis of polymer—based magnesium hydroxide nanocomposites for photocatalytic degradation for methylene blue dye and antibacterial application, biomass convers. Biorefinery., 1–14 (2022). https://doi.org/10.1007/s13399-022-02770-0

125. F. Paquin, J. Rivnay, A. Salleo, N. Stingelin, C. Silva, Multi-phase semicrystalline microstructures drive exciton dissociation in neat plastic semiconductors. J. Mater. Chem. C **3**, 10715–10722 (2015). https://doi.org/10.1039/b000000x
126. L. Zhang, Y. Zhang, S. Huang, Y. Yuan, H. Li, Z. Jin, J. Wu, Q. Liao, L. Hu, J. Lu, S. Ruan, Y.J. Zeng, Co_3O_4/Ni-based MOFs on carbon cloth for flexible alkaline battery-supercapacitor hybrid devices and near-infrared photocatalytic hydrogen evolution. Electrochim. Acta **281**, 189–197 (2018). https://doi.org/10.1016/j.electacta.2018.05.162
127. R. Genc, M.O. Alas, E. Harputlu, S. Repp, N. Kremer, M. Castellano, S.G. Colak, K. Ocakoglu, E. Erdem, High-capacitance hybrid supercapacitor based on multi-colored fluorescent carbon-dots. Sci. Rep. **7**, 1–13 (2017). https://doi.org/10.1038/s41598-017-11347-1
128. K. Dharmalingam, A.K. Bojarajan, R. Gopal, E. Thangavel, S.A.B. Al Omari, S. Sangaraju, Direct Z-scheme heterojunction impregnated MoS_2–NiO–CuO nanohybrid for efficient photocatalyst and dye-sensitized solar cell. Sci. Rep. **14**(14518), 1–20 (2024). https://doi.org/10.1038/s41598-024-65163-5
129. K. Dharmalingam, E. Thangavel, P.C. Tsai, P.V. Pham, K. Prakasham, G. Andaluri, K.B. Manjappa, Y.C. Lin, V.K. Ponnusamy, Novel MoS_2-In_2O_3-WS_2 (2D/3D/2D) ternary heterostructure nanocomposite material: Efficient photocatalytic degradation of antimicrobial agents under visible-light. Environ. Res. **261**, 119759 (2024). https://doi.org/10.1016/j.envres.2024.119759
130. X. Liu, C. Wang, Y. Dou, A. Zhou, T. Pan, J. Han, M. Wei, A NiAl layered double hydroxide@carbon nanoparticles hybrid electrode for high-performance asymmetric supercapacitors. J. Mater. Chem. A **2**, 1682–1685 (2014). https://doi.org/10.1039/c3ta14350c
131. Q. Liao, N. Li, S. Jin, G. Yang, C. Wang, All-solid-state symmetric supercapacitor based on Co_3O_4 nanoparticles on vertically aligned graphene. ACS Nano **9**, 5310–5317 (2015). https://doi.org/10.1021/acsnano.5b00821
132. Y. Hu, C. Guan, Q. Ke, Z.F. Yow, C. Cheng, J. Wang, Hybrid Fe_2O_3 nanoparticle clusters/rGO paper as an effective negative electrode for flexible supercapacitors. Chem. Mater. **28**, 7296–7303 (2016). https://doi.org/10.1021/acs.chemmater.6b02585
133. E. Lim, W.G. Lim, C. Jo, J. Chun, M.H. Kim, K.C. Roh, J. Lee, Rational design of Li_3VO_4@carbon core-shell nanoparticles as Li-ion hybrid supercapacitor anode materials. J. Mater. Chem. A **5**, 20969–20977 (2017). https://doi.org/10.1039/c7ta05863b
134. L. Ye, Y. Zhou, Z. Bao, Y. Zhao, Y. Zou, L. Zhao, Q. Jiang, Sheet-membrane Mn-doped nickel hydroxide encapsulated via heterogeneous Ni_3S_2 nanoparticles for efficient alkaline battery-supercapacitor hybrid devices. J. Mater. Chem. A **6**, 19020–19029 (2018). https://doi.org/10.1039/c8ta04613a
135. E. Kim, H. Kim, B.J. Park, Y.H. Han, J.H. Park, J. Cho, S.S. Lee, J.G. Son, Etching-assisted crumpled graphene wrapped spiky iron oxide particles for high-performance Li-ion hybrid supercapacitor. Small **14**, 1–9 (2018). https://doi.org/10.1002/smll.201704209
136. R. Liu, A. Zhang, J. Tang, J. Tian, W. Huang, J. Cai, C. Barrow, W. Yang, J. Liu, Fabrication of Cobaltosic oxide nanoparticle-doped 3 D MXene/graphene hybrid porous aerogels for all-solid-state supercapacitors. Chem. A Eur. J. **25**, 5547–5554 (2019). https://doi.org/10.1002/chem.201806342
137. J. Zhu, J. He, Facile synthesis of graphene-wrapped honeycomb MnO_2 nanospheres and their application in supercapacitors. ACS Appl. Mater. Interfaces **4**, 1770–1776 (2012). https://doi.org/10.1021/am3000165
138. W. Fan, Y.Y. Xia, W.W. Tjiu, P.K. Pallathadka, C. He, T. Liu, Nitrogen-doped graphene hollow nanospheres as novel electrode materials for supercapacitor applications. J. Power Sources **243**, 973–981 (2013). https://doi.org/10.1016/j.jpowsour.2013.05.184
139. G. Ma, K. Li, Y. Li, B. Gao, T. Ding, Q. Zhong, J. Su, A high-performance hybrid supercapacitor based on graphene-wrapped mesoporous T-Nb_2O_5 Nanospheres anode and mesoporous carbon-coated graphene cathode. ChemElectroChem (2016). https://doi.org/10.1002/celc.201600181
140. M. Guo, J. Balamurugan, X. Li, N.H. Kim, J.H. Lee, Hierarchical 3D cobalt-doped Fe_3O_4 Nanospheres@NG hybrid as an advanced anode material for high-performance asymmetric supercapacitors. Small **13** (2017). https://doi.org/10.1002/smll.201701275

141. P. Yu, X. Liu, L. Wang, C. Tian, H. Yu, H. Fu, Urchin-like V_2O_3/C hollow Nanosphere hybrid for high-capacity and long-cycle-life lithium storage. ACS Sustain. Chem. Eng. **5**, 11238–11245 (2017). https://doi.org/10.1021/acssuschemeng.7b01640
142. X. Zhao, C. Johnston, P.S. Grant, A novel hybrid supercapacitor with a carbon nanotube cathode and an iron oxide/carbon nanotube composite anode. J. Mater. Chem. **19**, 8755–8760 (2009). https://doi.org/10.1039/b909779a
143. P. Chen, G. Shen, Y. Shi, H. Chen, C. Zhou, Preparation and characterization of flexible asymmetric supercapacitors. ACS Nano **4**, 4403–4411 (2010). https://doi.org/10.1021/nn100856y
144. A.L.M. Reddy, M.M. Shaijumon, S.R. Gowda, P.M. Ajayan, Multisegmented au-MnO_2/carbon nanotube hybrid coaxial arrays for high-power supercapacitor applications. J. Phys. Chem. C **114**, 658–663 (2010). https://doi.org/10.1021/jp908739q
145. W. Jiang, D. Yu, Q. Zhang, K. Goh, L. Wei, Y. Yong, R. Jiang, J. Wei, Y. Chen, Ternary hybrids of amorphous nickel hydroxide-carbon nanotube-conducting polymer for supercapacitors with high energy density, excellent rate capability, and long cycle life. Adv. Funct. Mater. **25**, 1063–1073 (2015). https://doi.org/10.1002/adfm.201403354
146. T. Yeo, D. Shin, J. Shin, H. Hwang, B. Seo, J. Lee, W. Choi, DC-field-driven combustion waves for one-step fabrication of reduced manganese oxide/multi-walled carbon nanotube hybrid nanostructures as high-performance supercapacitor electrodes. J Mater Chem A **5**, 24707–24719 (2017). https://doi.org/10.1039/c7ta07812a
147. S.S. Raut, B.R. Sankapal, M.S.A. Hossain, S. Pradhan, R.R. Salunkhe, Y. Yamauchi, Zinc ferrite anchored multiwalled carbon nanotubes for high-performance supercapacitor applications. Eur. J. Inorg. Chem. **2018**, 137–142 (2018). https://doi.org/10.1002/ejic.201700836
148. C. Guan, J. Liu, C. Cheng, H. Li, X. Li, W. Zhou, H. Zhang, H.J. Fan, Hybrid structure of cobalt monoxide nanowire @ nickel hydroxidenitrate nanoflake aligned on nickel foam for high-rate supercapacitor. Energy Environ. Sci. **4**, 4496–4499 (2011). https://doi.org/10.1039/c1ee01685g
149. X. Wang, A. Sumboja, M. Lin, J. Yan, P.S. Lee, Enhancing electrochemical reaction sites in nickel-cobalt layered double hydroxides on zinc tin oxide nanowires: A hybrid material for an asymmetric supercapacitor device. Nanoscale **4**, 7266–7272 (2012). https://doi.org/10.1039/c2nr31590d
150. Y. Ai, X. Geng, Z. Lou, Z.M. Wang, G. Shen, Rational synthesis of branched $CoMoO_4$@$CoNiO_2$ Core/Shell nanowire arrays for all-solid-state supercapacitors with improved performance. ACS Appl. Mater. Interfaces **7**, 24204–24211 (2015). https://doi.org/10.1021/acsami.5b07599
151. Y. Li, F. Tang, R. Wang, C. Wang, J. Liu, Novel dual-ion hybrid supercapacitor based on a $NiCo_2O_4$ nanowire cathode and MoO_2-C Nanofilm anode. ACS Appl. Mater. Interfaces **8**, 30232–30238 (2016). https://doi.org/10.1021/acsami.6b10249
152. J. Lin, H. Liang, H. Jia, S. Chen, Y. Cai, J. Qi, J. Cao, W. Fei, J. Feng, Hierarchical $CuCo_2O_4$@$NiMoO_4$ core-shell hybrid arrays as a battery-like electrode for supercapacitors. Inorg. Chem. Front. **4**, 1575–1581 (2017). https://doi.org/10.1039/c7qi00361g
153. K.R. Shrestha, S. Kandula, G. Rajeshkhanna, M. Srivastava, N.H. Kim, J.H. Lee, An advanced sandwich-type architecture of $MnCo_2O_4$@N-C@MnO_2 as an efficient electrode material for a high-energy density hybrid asymmetric solid-state supercapacitor. J Mater Chem A **6**, 24509–24522 (2018). https://doi.org/10.1039/c8ta08976k
154. X. Liu, G. Sheng, M. Zhong, X. Zhou, Hybrid nanowires and nanoparticles of WO_3 in a carbon aerogel for supercapacitor applications. Nanoscale **10**, 4209–4217 (2018). https://doi.org/10.1039/c7nr07191d
155. Y. Fan, S. Liu, X. Han, R. Xiang, Y. Gong, Ni-co-based nanowire arrays with hierarchical core-shell structure electrodes for high-performance supercapacitors. ACS Appl Energy Mater **3**, 7580–7587 (2020). https://doi.org/10.1021/acsaem.0c01004
156. M. Ahmad, S. Yingying, A. Nisar, H. Sun, W. Shen, M. Wei, J. Zhu, Synthesis of hierarchical flower-like ZnO nanostructures and their functionalization by au nanoparticles for improved

photocatalytic and high performance Li-ion battery anodes. J. Mater. Chem. **21**, 7723–7729 (2011). https://doi.org/10.1039/c1jm10720h

157. Y. Wang, J. Park, B. Sun, H. Ahn, G. Wang, Wintersweet-flower-like $CoFe_2O_4$/MWCNTs hybrid material for high-capacity reversible lithium storage. Chem. Asian J. **7**, 1940–1946 (2012). https://doi.org/10.1002/asia.201200257
158. R. Wang, X. Yan, J. Lang, Z. Zheng, P. Zhang, A hybrid supercapacitor based on flower-like $co(OH)_2$ and urchin-like VN electrode materials. J. Mater. Chem. A **2**, 12724–12732 (2014). https://doi.org/10.1039/c4ta01296h
159. N. Padmanathan, S. Selladurai, K.M. Razeeb, Ultra-fast rate capability of a symmetric supercapacitor with a hierarchical Co_3O_4 nanowire/nanoflower hybrid structure in non-aqueous electrolyte. RSC Adv. **5**, 12700–12709 (2015). https://doi.org/10.1039/c4ra13327g
160. R. Wang, A. Jayakumar, C. Xu, J.M. Lee, $Ni(OH)_2$ nanoflowers/graphene hydrogels: A new assembly for supercapacitors. ACS Sustain. Chem. Eng. **4**, 3736–3742 (2016). https://doi.org/10.1021/acssuschemeng.6b00362
161. Y. Zhao, L. Hu, S. Zhao, L. Wu, Preparation of $MnCo_2O_4$@$Ni(OH)_2$ Core–Shell flowers for asymmetric supercapacitor materials with ultrahigh specific capacitance. Adv. Funct. Mater. **26**, 4085–4093 (2016). https://doi.org/10.1002/adfm.201600494
162. L. Wu, L. Hao, B. Pang, G. Wang, Y. Zhang, X. Li, MnO_2 nanoflowers and polyaniline nanoribbons grown on hybrid graphene/Ni 3D scaffolds by in situ electrochemical techniques for high-performance asymmetric supercapacitors. J Mater Chem A **5**, 4629–4637 (2017). https://doi.org/10.1039/c6ta10757e
163. S. Wang, Z.S. Wu, F. Zhou, X. Shi, S. Zheng, J. Qin, H. Xiao, C. Sun, X. Bao, All-solid-state high-energy planar hybrid micro-supercapacitors based on 2D VN nanosheets and $co(OH)_2$ nanoflowers. Npj 2D Mater. Appl. **2**, 1–8 (2018). https://doi.org/10.1038/s41699-018-0052-8
164. Y.L. Liu, C. Yan, G.G. Wang, H.Y. Zhang, L.Y. Dang, B.W. Wu, Z.Q. Lin, X.S. An, J.C. Han, Achieving ultrahigh capacity with self-assembled $Ni(OH)_2$ Nanosheet-decorated hierarchical flower-like $MnCo_2O_{4.5}$ Nanoneedles as advanced electrodes of battery-supercapacitor hybrid devices. ACS Appl. Mater. Interfaces **11**, 9984–9993 (2019). https://doi.org/10.1021/acsami.8b21803
165. B. Luo, Y. Fang, B. Wang, J. Zhou, H. Song, L. Zhi, Two dimensional graphene-SnS_2 hybrids with superior rate capability for lithium ion storage. Energy Environ. Sci. **5**, 5226–5230 (2012). https://doi.org/10.1039/c1ee02800f
166. Q. Fan, L. Lei, G. Yin, Y. Sun, Self-weaving CNT–$LiNbO_3$ nanoplate–polypyrrole hybrid as a flexible anode for Li-ion batteries. Chem. Commun. **50**, 2370–2373 (2014). https://doi.org/10.1039/c3cc48367c
167. S. Maiti, A.K. Das, S.K. Karan, B.B. Khatua, Carbon nanohorn-graphene nanoplate hybrid : An excellent electrode material for supercapacitor application. J. Appl. Polym. Sci. **42118**, 1–6 (2015). https://doi.org/10.1002/app.42118
168. Z. Long, L. Wei, L. Shuang, W. Jifei, W. Huanlei, C. Jiaxin, Fe_3O_4 nanoplates/carbon network synthesized by in situ pyrolysis of an organic-inorganic layered hybrid as a high-performance lithium-ion battery anode. J. Mater. Chem. A **3**, 14210–14216 (2015). https://doi.org/10.1039/c5ta01926e
169. W. Xu, Z. Xie, X. Cui, K. Zhao, L. Zhang, L. Mai, Y. Wang, Direct growth of an economic green energy storage material: A monocrystalline jarosite-$KFe_3(SO_4)_2(OH)_6$-nanoplates@rGO hybrid as a superior lithium-ion battery cathode. J. Mater. Chem. A **4**, 3735–3742 (2016). https://doi.org/10.1039/c5ta10622b
170. W. Li, K. Wang, S. Cheng, K. Jiang, A two-dimensional hybrid of SbO_x nanoplates encapsulated by carbon flakes as a high performance sodium storage anode. J. Mater. Chem. A **5**, 1160–1167 (2017). https://doi.org/10.1039/c6ta09265a
171. P. Sivakumar, M. Jana, M.G. Jung, A. Gedanken, H.S. Park, Hexagonal plate-like Ni-co-Mn hydroxide nanostructures to achieve high energy density of hybrid supercapacitors. J. Mater. Chem. A **7**, 11362–11369 (2019). https://doi.org/10.1039/c9ta02583a
172. T. Quan, N. Goubard-Bretesché, E. Härk, Z. Kochovski, S. Mei, N. Pinna, M. Ballauff, Y. Lu, Highly dispersible hexagonal carbon–MoS_2 –carbon Nanoplates with hollow Sandwich

structures for supercapacitors. Chem. A Eur. J. **25**, 4757–4766 (2019). https://doi.org/10.1002/chem.201806060
173. S. Wu, Y. Chen, T. Jiao, J. Zhou, J. Cheng, B. Liu, S. Yang, K. Zhang, W. Zhang, An aqueous Zn-ion hybrid supercapacitor with high energy density and ultrastability up to 80,000 cycles. Adv. Energy Mater. **9**, 1–7 (2019). https://doi.org/10.1002/aenm.201902915
174. J. Liu, J. Essner, J. Li, Hybrid supercapacitor based on coaxially coated manganese oxide on vertically aligned carbon nanofiber arrays. Chem. Mater. **22**, 5022–5030 (2010). https://doi.org/10.1021/cm101591p
175. M.B. Gholivand, A. Azadbakht, A. Pashabadi, An electrochemical sensor based on carbon nanotube bimetallic au-Pt inorganic-organic nanofiber hybrid nanocomposite electrode applied for detection of guaifenesin. Electroanalysis **23**, 2771–2779 (2011). https://doi.org/10.1002/elan.201100381
176. M.X. Jing, Z.C. Pi, H.A. Zhai, J.Q. Li, L.L. Chen, X.Q. Shen, X.M. Xi, K.S. Xiao, Three-dimensional $Li_3V_2(PO_4)_3$/C nanowire and nanofiber hybrid membrane as a self-standing, binder-free cathode for lithium ion batteries. RSC Adv. **6**, 71574–71580 (2016). https://doi.org/10.1039/c6ra13686a
177. H. Yin, M.L. Cao, X.X. Yu, H. Zhao, Y. Shen, C. Li, M.Q. Zhu, Self-standing Bi_2O_3 nanoparticles/carbon nanofiber hybrid films as a binder-free anode for flexible sodium-ion batteries. Mater. Chem. Front. **1**, 1615–1621 (2017). https://doi.org/10.1039/c7qm00128b
178. H. Xie, X. Liu, R. Wu, J. Liu, J. Wu, L. Li, High-performance supercapacitor with faster energy storage and Long cyclic life based on CuO@MnO_2 Nano-Core-Shell Array on carbon Fiber surface. ACS Appl Energy Mater **3**, 7325–7334 (2020). https://doi.org/10.1021/acsaem.0c00590
179. G. Zhang, X.W.D. Lou, Controlled growth of $NiCo_2O_4$ nanorods and ultrathin nanosheets on carbon nanofibers for high-performance supercapacitors. Sci. Rep. **3**, 2–7 (2013). https://doi.org/10.1038/srep01470
180. X. Zhou, L.J. Wan, Y.G. Guo, Electrospun silicon nanoparticle/porous carbon hybrid nanofibers for lithium-ion batteries. Small **9**, 2684–2688 (2013). https://doi.org/10.1002/smll.201202071
181. L. Qiao, X. Wang, X. Sun, X. Li, Y. Zheng, D. He, Single electrospun porous NiO-ZnO hybrid nanofibers as anode materials for advanced lithium-ion batteries. Nanoscale **5**, 3037–3042 (2013). https://doi.org/10.1039/c3nr34103h
182. Q. Niu, K. Gao, Z. Shao, Cellulose nanofiber/single-walled carbon nanotube hybrid non-woven macrofiber mats as novel wearable supercapacitors with excellent stability, tailorability and reliability. Nanoscale **6**, 4083–4088 (2014). https://doi.org/10.1039/c3nr05929d
183. Z. Zhang, Q. Li, Y. Lai, J. Li, Confine sulfur in polyaniline-decorated hollow carbon nanofiber hybrid nanostructure for lithium-sulfur batteries. J. Phys. Chem. C **118**, 13369–13376 (2014). https://doi.org/10.1021/jp5005117
184. Z. Zhang, Q. Li, S. Jiang, K. Zhang, Y. Lai, J. Li, Sulfur encapsulated in a TiO_2-anchored hollow carbon nanofiber hybrid nanostructure for lithium-sulfur batteries. Chem. A Eur. J. **21**, 1343–1349 (2015). https://doi.org/10.1002/chem.201404686
185. Y. Wang, H. Wei, J. Wang, J. Liu, J. Guo, X. Zhang, B.L. Weeks, T.D. Shen, S. Wei, Z. Guo, Electropolymerized polyaniline/manganese iron oxide hybrids with an enhanced color switching response and electrochemical energy storage. J. Mater. Chem. A **3**, 20778–20790 (2015). https://doi.org/10.1039/c5ta04439a
186. L. Zhang, Q. Ding, Y. Huang, H. Gu, Y.E. Miao, T. Liu, Flexible hybrid membranes with $Ni(OH)_2$ Nanoplatelets vertically grown on electrospun carbon nanofibers for high-performance supercapacitors. ACS Appl. Mater. Interfaces **7**, 22669–22677 (2015). https://doi.org/10.1021/acsami.5b07528
187. L. Yuan, X. Lu, X. Xiao, T. Zhai, J. Dai, F. Zhang, B. Hu, X. Wang, Flexible solid-state supercapacitors based on carbon nanoparticles/MnO_2 Nanorods hybrid structure. ACS Nano **6**, 656–661 (2012). https://doi.org/10.1021/nn2041279
188. L. Ruizhi, X. Ren, F. Zhang, C. Du, J. Liu, Synthesis of Fe_3O_4@SnO_2 core–shell nanorod film and its application as a thin-film supercapacitor electrode. Chem. Commun. **48**, 5010–5012 (2012). https://doi.org/10.1039/c2cc31786a

189. J. Xu, Q. Wang, X. Wang, Q. Xiang, B. Liang, D. Chen, G. Shen, Flexible asymmetric supercapacitors based upon Co_9S_8 nanorod//Co_3O_4@RuO_2 nanosheet arrays on carbon cloth. ACS Nano **7**, 5453–5462 (2013). https://doi.org/10.1021/nn401450s
190. Y. Shao, H. Wang, Q. Zhang, Y. Li, High-performance flexible asymmetric supercapacitors based on 3D porous graphene/MnO_2 nanorod and graphene/ag hybrid thin-film electrodes. J. Mater. Chem. C **1**, 1245–1251 (2013). https://doi.org/10.1039/c2tc00235c
191. J. Rodríguez-Moreno, E. Navarrete-Astorga, E.A. Dalchiele, R. Schrebler, J.R. Ramos-Barrado, F. Martín, Vertically aligned ZnO@CuS@PEDOT core@shell nanorod arrays decorated with MnO_2 nanoparticles for a high-performance and semi-transparent supercapacitor electrode. Chem. Commun. **50**, 5652–5655 (2014). https://doi.org/10.1039/c4cc01984a
192. R. Li, Y. Wang, C. Zhou, C. Wang, X. Ba, Y. Li, X. Huang, J. Liu, Carbon-stabilized high-capacity Ferroferric oxide Nanorod Array for flexible solid-state alkaline battery-supercapacitor hybrid device with high environmental suitability. Adv. Funct. Mater. **25**, 5384–5394 (2015). https://doi.org/10.1002/adfm.201502265
193. M. Sethi, U.S. Shenoy, D.K. Bhat, Porous graphene-$NiCo_2O_4$ nanorod hybrid composite as a high performance supercapacitor electrode material. New J. Chem. **44**, 4033–4041 (2020). https://doi.org/10.1039/c9nj05725k
194. A. De Adhikari, R. Oraon, S.K. Tiwari, J.H. Lee, N.H. Kim, G.C. Nayak, A V_2O_5 nanorod decorated graphene/polypyrrole hybrid electrode: A potential candidate for supercapacitors. New J. Chem. **41**, 1704–1713 (2017). https://doi.org/10.1039/c6nj03580a
195. Y. Ruan, L. Lv, Z. Li, C. Wang, J. Jiang, Ni nanoparticles@Ni-Mo nitride nanorod arrays: A novel 3D-network hierarchical structure for high areal capacitance hybrid supercapacitors. Nanoscale **9**, 18032–18041 (2017). https://doi.org/10.1039/c7nr05560a
196. K.V. Sankar, Y. Seo, S.C. Lee, S. Chan Jun, Redox additive-improved electrochemically and structurally robust binder-free nickel pyrophosphate Nanorods as superior cathode for hybrid supercapacitors. ACS Appl. Mater. Interfaces **10**, 8045–8056 (2018). https://doi.org/10.1021/acsami.7b19357
197. B.Y. Guan, L. Yu, X. Wang, S. Song, X.W.D. Lou, Formation of onion-like $NiCo_2S_4$ particles via sequential ion-exchange for hybrid supercapacitors. Adv. Mater. **29**, 2–6 (2017). https://doi.org/10.1002/adma.201605051
198. X. Huang, X. Qi, F. Boey, H. Zhang, Graphene-based composites. Chem. Soc. Rev. **41**, 666–686 (2012). https://doi.org/10.1039/c1cs15078b
199. P. Cataldi, A. Athanassiou, I.S. Bayer, Graphene nanoplatelets-based advanced materials and recent progress in sustainable applications. Appl. Sci. **8** (2018). https://doi.org/10.3390/app8091438
200. S. Pang, Y. Hernandez, X. Feng, K. Müllen, Graphene as transparent electrode material for organic electronics. Adv. Mater. **23**, 2779–2795 (2011). https://doi.org/10.1002/adma.201100304
201. M. Srivastava, J. Singh, T. Kuila, R.K. Layek, N.H. Kim, J.H. Lee, Recent advances in graphene and its metal-oxide hybrid nanostructures for lithium-ion batteries. Nanoscale **7**, 4820–4868 (2015). https://doi.org/10.1039/c4nr07068b
202. V. Chabot, D. Higgins, A. Yu, X. Xiao, Z. Chen, J. Zhang, A review of graphene and graphene oxide sponge: Material synthesis and applications to energy and the environment. Energy Environ. Sci. **7**, 1564–1596 (2014). https://doi.org/10.1039/c3ee43385d
203. Y. Ito, Y. Tanabe, K. Sugawara, M. Koshino, T. Takahashi, K. Tanigaki, H. Aoki, M. Chen, Three-dimensional porous graphene networks expand graphene-based electronic device applications. Phys. Chem. Chem. Phys. **20**, 6024–6033 (2018). https://doi.org/10.1039/c7cp07667c
204. Y. Zhang, Y. Shen, X. Xie, W. Du, L. Kang, Y. Wang, X. Sun, Z. Li, B. Wang, One-step synthesis of the reduced graphene oxide @ NiO composites for supercapacitor electrodes by electrode-assisted plasma electrolysis. Mater. Des. **196**, 109111 (2020). https://doi.org/10.1016/j.matdes.2020.109111

205. Y. Zhu, S. Cheng, W. Zhou, J. Jia, L. Yang, M. Yao, M. Wang, J. Zhou, P. Wu, M. Liu, Construction and performance characterization of α-Fe_2O_3/rGO composite for long-cycling-life supercapacitor anode. ACS Sustain. Chem. Eng. **5**, 5067–5074 (2017). https://doi.org/10.1021/acssuschemeng.7b00445
206. M. Xuemei et al., Construction of hierarchical CNT/rGO supported MnMoO nanosheets on Ni foam for high performance aqueous hybrid supercapacitors. Appl. Mater. Interfaces (2017). https://doi.org/10.1021/acsami.7b09005
207. M. Zhang, X. Liu, M. Ma, G. Ni, Z. Sun, J. Liu, Y. Wu, 3D hierarchical urchin-like $Ni_{0.3}Co_{0.6}Cu_{0.1}(CO3)_{0.5}(OH)$ microspheres for supercapacitors with high specific capacitance. Energy Fuels. **35**, 20358–20366 (2021). https://doi.org/10.1021/acs.energyfuels.1c03299
208. P. Sengodu, A.D. Deshmukh, Conducting polymers and their inorganic composites for advanced Li-ion batteries: A review. RSC Adv. **5**, 42109–42130 (2015). https://doi.org/10.1039/c4ra17254j
209. V. Aravindan, M. Ulaganathan, S. Madhavi, Research progress in Na-ion capacitors. J Mater. Chem. A **4**, 7538–7548 (2016). https://doi.org/10.1039/c6ta02478e
210. W. Zuo, R. Li, C. Zhou, Y. Li, J. Xia, J. Liu, Battery-supercapacitor hybrid devices: recent progress and future prospects. Adv. Sci. **4** (2017). https://doi.org/10.1002/advs.201600539
211. X. Yin, H. Li, H. Wang, Z. Zhang, R. Yuan, J. Lu, Q. Song, J.G. Wang, L. Zhang, Q. Fu, Self-templating synthesis of cobalt Hexacyanoferrate hollow structures with superior performance for Na-ion hybrid supercapacitors. ACS Appl. Mater. Interfaces **10**, 29496–29504 (2018). https://doi.org/10.1021/acsami.8b08455
212. K. Leng, F. Zhang, L. Zhang, T. Zhang, Y. Wu, Y. Lu, Y. Huang, Y. Chen, Graphene-based Li-ion hybrid supercapacitors with ultrahigh performance. Nano Res **6**, 581–592 (2013). https://doi.org/10.1007/s12274-013-0334-6
213. W. Liu, J. Li, K. Feng, A. Sy, Y. Liu, L. Lim, G. Lui, R. Tjandra, L. Rasenthiram, G. Chiu, A. Yu, Advanced Li-ion hybrid supercapacitors based on 3D graphene-foam composites. ACS Appl. Mater. Interfaces **8**, 25941–25953 (2016). https://doi.org/10.1021/acsami.6b07365
214. J.G. Wang, Z. Zhang, X. Zhang, X. Yin, X. Li, X. Liu, F. Kang, B. Wei, Cation exchange formation of prussian blue analogue submicroboxes for high-performance Na-ion hybrid supercapacitors. Nano Energy **39**, 647–653 (2017). https://doi.org/10.1016/j.nanoen.2017.07.055
215. C. Zhan, W. Liu, M. Hu, Q. Liang, X. Yu, Y. Shen, R. Lv, F. Kang, Z.H. Huang, High-performance sodium-ion hybrid capacitors based on an interlayer-expanded MoS2/rGO composite: Surpassing the performance of lithium-ion capacitors in a uniform system. NPG Asia Mater. **10**, 775–787 (2018). https://doi.org/10.1038/s41427-018-0073-y
216. J.J. Chen, J.C. Ye, X.G. Zhang, M.D. Symes, S.C. Fan, D.L. Long, M. Sen Zheng, D.Y. Wu, L. Cronin, Q.F. Dong, Design and performance of rechargeable sodium ion batteries, and symmetrical Li-ion batteries with supercapacitor-like power density based upon Polyoxovanadates. Adv. Energy Mater. **8**, 1–6 (2018). https://doi.org/10.1002/aenm.201701021
217. Y. Yuan, C. Wang, K. Lei, H. Li, F. Li, J. Chen, Sodium-ion hybrid capacitor of high power and energy density. ACS Cent. Sci. **4**, 1261–1265 (2018). https://doi.org/10.1021/acscentsci.8b00437
218. S. Ghosh, T. Maiyalagan, R.N. Basu, Nanostructured conducting polymers for energy applications: towards a sustainable platform. Nanoscale **8**, 6921–6947 (2016). https://doi.org/10.1039/c5nr08803h
219. A. Noori, M.F. El-Kady, M.S. Rahmanifar, R.B. Kaner, M.F. Mousavi, Towards establishing standard performance metrics for batteries, supercapacitors and beyond. Chem. Soc. Rev. **48**, 1272–1341 (2019). https://doi.org/10.1039/c8cs00581h
220. H. Hu, K. Zhang, S. Li, S. Ji, C. Ye, Flexible, in-plane, and all-solid-state micro-supercapacitors based on printed interdigital au/polyaniline network hybrid electrodes on a chip. J. Mater. Chem. A **2**, 20916–20922 (2014). https://doi.org/10.1039/c4ta05345a

221. T. Chen, Y. Fan, G. Wang, Q. Yang, R. Yang, Rationally designed hierarchical $ZnCo_2O_4$/polypyrrole nanostructures for high-performance supercapacitor electrodes. RSC Adv. **5**, 74523–74530 (2015). https://doi.org/10.1039/c5ra14808a
222. C.O. Baker, X. Huang, W. Nelson, R.B. Kaner, Polyaniline nanofibers: broadening applications for conducting polymers. Chem. Soc. Rev. **46**, 1510–1525 (2017). https://doi.org/10.1039/c6cs00555a
223. C. Yu, P. Ma, X. Zhou, A. Wang, T. Qian, S. Wu, Q. Chen, All-solid-state flexible supercapacitors based on highly dispersed polypyrrole nanowire and reduced graphene oxide composites. ACS Appl. Mater. Interfaces **6**, 17937–17943 (2014). https://doi.org/10.1021/am5059603
224. I. Shown, A. Ganguly, L.C. Chen, K.H. Chen, Conducting polymer-based flexible supercapacitor. Energy Sci. Eng. **3**, 1–25 (2015). https://doi.org/10.1002/ese3.50
225. K. Wang, K. Amin, Z. An, Z. Cai, H. Chen, H. Chen, Y. Dong, X. Feng, W. Fu, J. Gu, Y. Han, D. Hu, R. Hu, D. Huang, F. Huang, F. Huang, Y. Huang, J. Jin, X. Jin, Q. Li, T. Li, Z. Li, Z. Li, J. Liu, J. Liu, S. Liu, H. Peng, A. Qin, X. Qing, Y. Shen, J. Shi, X. Sun, B. Tong, B. Wang, H. Wang, L. Wang, S. Wang, Z. Wei, T. Xie, C. Xu, H. Xu, Z.-K. Xu, B. Yang, Y. Yu, X. Zeng, X. Zhan, G. Zhang, J. Zhang, M.Q. Zhang, X.-Z. Zhang, X. Zhang, Y. Zhang, Y. Zhang, C. Zhao, W. Zhao, Y. Zhou, Z. Zhou, J. Zhu, X. Zhu, B.Z. Tang, Advanced functional polymer materials. Mater. Chem. Front. (2020). https://doi.org/10.1039/d0qm00025f
226. Z. Fan, J. Zhu, X. Sun, Z. Cheng, Y. Liu, Y. Wang, High density of free-standing holey graphene/PPy films for superior volumetric capacitance of supercapacitors. ACS Appl. Mater. Interfaces **9**, 21763–21772 (2017). https://doi.org/10.1021/acsami.7b03477
227. J. Yu, D. Yao, Z. Wu, G. Li, J. Song, H. Shen, X. Yang, W. Lei, F. Wu, Q. Hao, Construction of a high-performance three-dimensional structured $NiCo_2O_4$@PPy Nanosheet Array free-standing electrode for a hybrid supercapacitor. ACS Appl Energy Mater **4**, 3093–3100 (2021). https://doi.org/10.1021/acsaem.0c02827
228. A. Chonat, S. Palaty, Enhanced electrochemical performance of a hybrid Supercapacitive material based on ternary doped polyaniline/activated carbon composite. Energy Fuel **34** (2020). https://doi.org/10.1021/acs.energyfuels.0c01555
229. S.H. Mir, L.A. Nagahara, T. Thundat, P. Mokarian-Tabari, H. Furukawa, A. Khosla, Review—organic-inorganic hybrid functional materials: An integrated platform for applied technologies. J. Electrochem. Soc. **165**, B3137–B3156 (2018). https://doi.org/10.1149/2.0191808jes
230. D. Singh, R. Tao, G. Lubineau, A synergetic layered inorganic–organic hybrid film for conductive, flexible, and transparent electrodes. NPJ Flex. Electron. **3**, 1–8 (2019). https://doi.org/10.1038/s41528-019-0054-4
231. B.W. Park, S. Il Seok, Intrinsic Instability of inorganic–organic hybrid halide perovskite materials. Adv. Mater. **31**, 1–17 (2019). https://doi.org/10.1002/adma.201805337
232. M. Hong, Inorganic-organic hybrid coordination polymers: a new frontier for materials research. Cryst. Growth Des. **7**, 10–14 (2007). https://doi.org/10.1021/cg068015h
233. M. Karg, T. Hellweg, Smart inorganic/organic hybrid microgels: synthesis and characterisation. J. Mater. Chem. **19**, 8714–8727 (2009). https://doi.org/10.1039/b820292n
234. G. Nagaraju, S. Chandra Sekhar, L. Krishna Bharat, J.S. Yu, Wearable fabrics with self-branched bimetallic layered double hydroxide coaxial nanostructures for hybrid supercapacitors. ACS Nano **11**, 10860–10874 (2017). https://doi.org/10.1021/acsnano.7b04368
235. S. Su, L. Lai, R. Li, Y. Lin, H. Dai, X. Zhu, Annealing-assisted dip-coating synthesis of ultrafine Fe_3O_4 nanoparticles/graphene on carbon cloth for flexible quasi-solid-state symmetric supercapacitors. ACS Appl Energy Mater **3**, 9379–9389 (2020). https://doi.org/10.1021/acsaem.0c01745
236. J. Huang, J. Xie, L. Wang, J. Zhang, P. Wang, P. Sun, Z. Yao, Y. Yang, Boosted electrochemical performance of honeycomb-like NiCu-LDH Nanosheets anchoring on $NiCo_2S_4$ nanotube arrays for flexible solid-state hybrid supercapacitors. Energy and Fuels. **34**, 13157–13166 (2020). https://doi.org/10.1021/acs.energyfuels.0c03135
237. H. Furukawa, K.E. Cordova, M. O'Keeffe, O.M. Yaghi, The chemistry and applications of metal-organic frameworks. Science **341**, 6149 (2013). https://doi.org/10.1126/science.1230444

238. S. Kuyuldar, D.T. Genna, C. Burda, On the potential for nanoscale metal-organic frameworks for energy applications. J. Mater. Chem. A **7**, 21545–21576 (2019). https://doi.org/10.1039/c9ta09896h
239. K. Zhang, K.O. Kirlikovali, Q. Van Le, Z. Jin, R.S. Varma, H.W. Jang, O.K. Farha, M. Shokouhimehr, Extended metal-organic frameworks on diverse supports as electrode nanomaterials for electrochemical energy storage. ACS Appl. Nano Mater. **3**, 3964–3990 (2020). https://doi.org/10.1021/acsanm.0c00702
240. S. Tajik, H. Beitollahi, F.G. Nejad, K.O. Kirlikovali, Q. Van Le, H.W. Jang, R.S. Varma, O.K. Farha, M. Shokouhimehr, Recent electrochemical applications of metal–organic framework-based materials. Cryst. Growth Des. (2020). https://doi.org/10.1021/acs.cgd.0c00601
241. X. Cao, C. Tan, M. Sindoro, H. Zhang, Hybrid micro−/nano-structures derived from metal-organic frameworks: preparation and applications in energy storage and conversion. Chem. Soc. Rev. **46**, 2660–2677 (2017). https://doi.org/10.1039/c6cs00426a
242. Y. Jiao, J. Pei, D. Chen, C. Yan, Y. Hu, Q. Zhang, G. Chen, Mixed-metallic MOF based electrode materials for high performance hybrid supercapacitors. J. Mater. Chem. A **5**, 1094–1102 (2017). https://doi.org/10.1039/C6TA09805C
243. M. Saraf, R. Rajak, S.M. Mobin, A fascinating multitasking Cu-MOF/rGO hybrid for high performance supercapacitors and highly sensitive and selective electrochemical nitrite sensors. J. Mater. Chem. A **4**, 16432–16445 (2016). https://doi.org/10.1039/c6ta06470a
244. K. Jayaramulu, M. Horn, A. Schneemann, H. Saini, A. Bakandritsos, V. Ranc, M. Petr, V. Stavila, C. Narayana, B. Scheibe, Š. Kment, M. Otyepka, N. Motta, D. Dubal, R. Zbořil, R.A. Fischer, Covalent graphene-MOF hybrids for high-performance asymmetric supercapacitors. Adv. Mater. **33** (2021). https://doi.org/10.1002/adma.202004560
245. G. Li, H. Cai, X. Li, J. Zhang, D. Zhang, Y. Yang, J. Xiong, Construction of hierarchical $NiCo_2O_4$@Ni-MOF hybrid arrays on carbon cloth as superior battery-type electrode for flexible solid-state hybrid supercapacitor. ACS Appl. Mater. Interfaces (2019). https://doi.org/10.1021/acsami.9b11994
246. S. Liu, L. Kang, J. Zhang, E. Jung, S. Lee, S.C. Jun, Structural engineering and surface modification of MOF-derived cobalt-based hybrid nanosheets for flexible solid-state supercapacitors. Energy Storage Mater **32**, 167–177 (2020). https://doi.org/10.1016/j.ensm.2020.07.017
247. F. Yi, H. Ren, J. Shan, X. Sun, D. Wei, Z. Liu, Wearable energy sources based on 2D materials. Chem. Soc. Rev. **47**, 3152–3188 (2018). https://doi.org/10.1039/c7cs00849j
248. L. Wang, D. Chen, K. Jiang, G. Shen, New insights and perspectives into biological materials for flexible electronics. Chem. Soc. Rev. **46**, 6764–6815 (2017). https://doi.org/10.1039/c7cs00278e
249. J.V. Vaghasiya, C.C. Mayorga-martinez, M. Pumera, Flexible energy generation and storage devices: focus on key role of heterocyclic solid-state organic ionic conductors. Chem. Soc. Rev. (2020). https://doi.org/10.1039/d0cs00698j
250. T. Chen, L. Dai, Flexible supercapacitors based on carbon nanomaterials. J. Mater. Chem. A **2**, 10756–10775 (2014). https://doi.org/10.1039/c4ta00567h
251. Y. Zhou, Z. Mao, W. Wang, Z. Yang, X. Liu, In-situ fabrication of graphene oxide hybrid Ni-based metal-organic framework (Ni-MOFs@GO) with ultrahigh capacitance as electrochemical Pseudocapacitor materials. ACS Appl. Mater. Interfaces **8**, 28904–28916 (2016). https://doi.org/10.1021/acsami.6b10640
252. D.G. Mackanic, T.H. Chang, Z. Huang, Y. Cui, Z. Bao, Stretchable electrochemical energy storage devices. Chem. Soc. Rev. **49**, 4466–4495 (2020). https://doi.org/10.1039/d0cs00035c
253. L. Dong, C. Xu, Y. Li, Z.H. Huang, F. Kang, Q.H. Yang, X. Zhao, Flexible electrodes and supercapacitors for wearable energy storage: a review by category. J. Mater. Chem. A **4**, 4659–4685 (2016). https://doi.org/10.1039/c5ta10582j
254. C. Zhao, X. Jia, K. Shu, C. Yu, G.G. Wallace, C. Wang, Conducting polymer composites for unconventional solid-state supercapacitors. J. Mater. Chem. A **8**, 4677–4699 (2020). https://doi.org/10.1039/c9ta13432h

255. S. Shrivastava, T.Q. Trung, N.E. Lee, Recent progress, challenges, and prospects of fully integrated mobile and wearable point-of-care testing systems for self-testing. Chem. Soc. Rev. **49**, 1812–1866 (2020). https://doi.org/10.1039/c9cs00319c

Chapter 9
Advanced Nanostructured Materials for Hybrid Lithium- and Sodium-Ion Capacitors

Ostar A. Seerane, Daniel M. Teffu, Mpho S. Ratsoma, Amanda-Lee E. Manicum, Shehla Honey, Kumar Raju, Kabir O. Oyedotun, Kwena D. Modibane, and Katlego Makgopa

Abstract To boost the energy efficiency of advanced energy storage technologies, various innovative strategies, i.e. smart electrode materials, hybrid mechanisms, and progressive ionic media, are currently being explored to overcome the escalating energy demand as well as minimize the negative consequences resulting from the overuse of fossil fuels. In this regard, hybrid metal-ion capacitors like Li^+ and Na^+ capacitors are experiencing great attention due to their ability to deliver high energy density, faster energy delivery, and long cycle life. The advantages of these hybrid metal-ion capacitors arise from the synergetic properties derived from both conventional batteries and supercapacitors. Hence, lithium-ion capacitors and sodium-ion capacitors serve as a bridge between the metal-ion batteries and supercapacitors capabilities. Recently, extensive efforts have been dedicated to the development of nanostructured materials with the primary focus on structural and compositional optimization, incorporation of heteroatoms, and pore-size manipulation to improve the performance of hybrid capacitor applications. The objective of this chapter is to provide the recent developments on the use of nanostructured materials as potential electrodes for lithium- and sodium-ion capacitors. This is channelled through by first

O. A. Seerane · M. S. Ratsoma · A.-L. E. Manicum · K. Makgopa (✉)
Department of Chemistry, Faculty of Science, Tshwane University of Technology (Arcadia Campus), Pretoria, South Africa
e-mail: MakgopaK@tut.ac.za

D. M. Teffu · K. D. Modibane
Nanotechnology Research Lab, Department of Chemistry, School of Physical and Mineral Sciences, University of Limpopo (Turfloop), Polokwane, South Africa

S. Honey
Center for Nanosciences, University of Okara, Okara, Pakistan

K. Raju
Electrochemical Energy Technologies (EET), Energy Centre, CSIR, Pretoria, South Africa

K. O. Oyedotun
College of Science, Engineering and Technology (CSET), University of South Africa, Florida Campus, Johannesburg, South Africa

© The Author(s), under exclusive license to Springer Nature Switzerland AG 2026
K. Raju et al. (eds.), *Hybrid Nanostructures for Advanced Supercapacitors*, Modern Aspects of Electrochemistry 63, https://doi.org/10.1007/978-3-032-04593-5_9

examining the background as well as the design and mechanism of these technologies. Finally, the concluding prospects of both lithium- and sodium-ion capacitor technologies are described.

Keywords Lithium-ion capacitors · Sodium-ion capacitors · Hybrid metal-ion capacitors · Energy storage · Nanostructured materials

9.1 Introduction

Energy crises, environmental degradation, and climate change are all being fought with ever-more-potent weapons: renewable energy systems like solar and wind energy [1, 2]. Nonetheless, there is a great demand for grid storage to combine solar and wind energy due to the intermittent and unpredictable nature of electricity supply and consumption. However, it is more geographically diffuse and offers a modular solution, and electrochemical energy storage technology will present a sizable market opportunity [2]. Supercapacitors and secondary batteries are the two main topologies used in electrochemical energy storage. The ultra-long-life cycles offer great advantages to supercapacitors, as they often charge and discharge for a million cycles resulting in high-power performance of greater than 10,000 W kg^{-1} [3, 4]. The drawbacks of supercapacitors are numerous. The most important one concerns the energy density, which is extremely low in supercapacitors of less than 10 Wh kg^{-1}. They also self-discharge quickly, especially the aqueous supercapacitors [5]. In contrast, the majority of commercial lithium-ion batteries deliver a high energy density of up to 200 Wh kg^{-1} and a low self-discharge rate but only have a 350 W kg^{-1} and 1000 cycle cycling lifespan, respectively [1, 5]. Supercapacitors and lithium-ion batteries have extremely similar electrochemical performance characteristics. As a result, they are frequently utilized together in situations where high energy-power output is required. For instance, supercapacitors can be utilized for accelerating power and regenerative braking energy harvesting in electric vehicles (EVs), but lithium-ion batteries are frequently employed for driving power [6, 7].

To fulfil the objectives of high-degree lightweight and integration design, delivering both high energy and high power in a single device is an increasing target for advanced electrochemical energy storage systems. There are also hybrid energy storage systems known as hybrid capacitors, where one electrode employs electrical double-layer storage while the other runs on faradaic processes [8]. Hybrid capacitors can be divided into three categories based on the movement of charge carriers during the electrochemical process, namely (i) Accordion mechanism, where the typical configuration includes electrochemical intercalation negative electrode and physical adsorption positive electrode. The electrolyte's cations and anions are separated into the negative electrode and the positive electrode, respectively, in this type of hybrid capacitor. (ii) Rocking chair mechanisms, where the typical configuration is a negative electrode with physical adsorption and a positive electrode with electrochemical intercalation. Similar to lithium-ion batteries, this type of hybrid capacitor charges by having cations deintercalated from the positive electrode and adsorbed on the surface of the negative electrode, and vice versa. (iii) Hybrid mechanism, in which one or both of the electrodes contain both materials that are used in batteries and capacitors. The usual design of this hybrid technology involves the electrochemical intercalation of ionic species and physical adsorption of charges at the electrodes. During charging, cations are removed from the capacitor-type positive electrode and together with the cations from the electrolyte, intercalate into the battery-type negative electrode. Simultaneously, anions from the electrolyte are adsorbed by the positive electrode [1].

In terms of energy-power cycling, hybrid capacitors fall in between lithium-ion batteries or sodium-ion batteries and supercapacitors [9]. Importantly, a positive electrode and a negative electrode are more scientific names than cathode and anode for the two electrodes of hybrid capacitors since most capacitive electrodes in hybrid capacitors perform adsorption–desorption processes rather than redox reactions. The lithium-ion capacitors and sodium-ion capacitors complete cells have undergone tremendous efforts to design electrode materials and build them over the past few decades [10]. Combining both high-energy components like lithium-ion batteries with high-power and long-life components like EDLCs is the ultimate goal of hybrid capacitors. Regrettably, the existing hybrid capacitors cannot compete with lithium-ion batteries in terms of energy density. As a result, there is still much work to be done in order to create high-performance hybrid capacitors. Hybrid system research is still in its early stages. To make significant advancements in sophisticated electrode materials, appropriate electrolytes, and device layouts, more straining efforts are still needed. High-performance hybrid devices are anticipated to be available soon as more researchers concentrate on the discovery and assessment of hybrid capacitors [1, 10, 11]. The kinetic gap between the electrodes used in capacitor- and battery-type batteries must be closed in order to produce high-performance hybrid capacitors. In general, the kinetics of electrodes of the battery type coupled with a Faradaic reaction process are substantially worse than those of electrodes of the capacitor type coupled with a physical desorption–adsorption process. To close such a kinetic gap, it is necessary to design battery-type materials with amazing power properties.

The development of hybrid Li^+ and Na^+ capacitors, which combine Faradaic and double-layer techniques, has recently attracted a lot of attention due to their increased electrochemical performance in terms of both energy and power densities. Due to the high quantity of sodium in the Earth's crust, its low cost, and its similar intercalation chemistry to lithium, studies on SIBs have recently gained a lot of attention as one of the most promising alternatives to lithium-ion batteries [12–14]. Although activated porous carbons have been used as the cathode in several papers, creating effective sodium storage materials for the creation of high-performance sodium-ion capacitor devices is still a difficult task [13, 14]. Rechargeable Li^+ and Na^+ batteries that deliver higher energy and power densities are highly essential to meet the growing energy demand from portable electronics and automotive sectors. In this regard, the trade-off between specific energy and power densities in these systems needs to be addressed with newer storage mechanisms. Here, we have discussed current developments in nanostructured materials for making electrodes for hybrid metal-ion capacitors. In addition, a discussion of the design principles for fabricating such materials and cells is done.

9.2 Design and Mechanism of Lithium- and Sodium-Ion Capacitor

9.2.1 Lithium-Ion Capacitor

The main challenge for lithium-ion batteries and electric double-layer capacitors (EDLCs) is to meet the fast-growing demands for energy storage [15]. To address these challenges, lithium-ion capacitors have been introduced. As illustrated in Fig. 9.1a a new design suggests using a pseudocapacitor electrode instead of the battery or EDLC-type electrode to enhance power performance, and these three types of lithium-ion capacitors have different materials that result in unique power and energy performances [16]. The galvanostatic charge–discharge and cyclic voltammetry curves in Fig. 9.1b illustrate these differences. EDLC-type materials show a straight galvanostatic charge–discharge curve and a rectangular cyclic voltammetry curve in ideal situations, indicating pure physical changes during charge–discharge with surface ion adsorption–desorption. On the other hand, battery-type materials exhibit a galvanostatic charge–discharge curve with redox plateaus and a cyclic voltammetry curve with redox peaks. However, pseudocapacitor-type material are a special kind of Faradaic material that also present nonlinear galvanostatic charge–discharge with and without obvious redox voltage plateaus and a nonideal rectangular cyclic voltammetry curve [17].

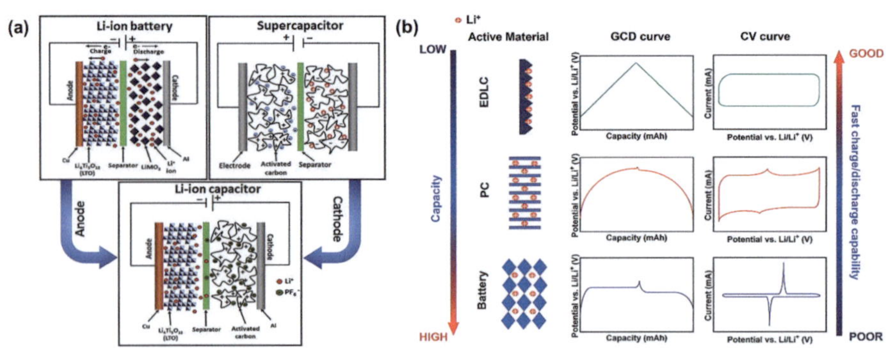

Fig. 9.1 **a** Different types of Li^+-based energy storage devices with supercapacitors and **b** their electrochemical signatures, especially the galvanostatic charge–discharge cycles and rectangular cyclic voltammograms. Reprinted with permission of Wiley from [16]

Due to the low power density, the battery materials show slow kinetics, whereas the pseudocapacitor and EDLCs do fast charge and discharge; however, the capacity is in the reverse order where the EDLCs have less capacity than the battery materials [18, 19]. Based on different energy storage kinetics and performance, there is a need to decipher ways to configure a lithium-ion capacitor with advanced materials that incorporate both pseudocapacitive and double-layer capacitive mechanisms to improve the electrochemical performances like high-rate capability, energy density and cycle stability of the device. To design a better lithium-ion capacitor, the prerequisites are the optimal mass loading and cell balance (N:P ratio), thickness of the active materials, pre-lithiation strategies, and density of the electrodes [20]. Too thick electrodes will trade-off between the diffusion issue and the energy density whereas the thinner loading affects the energy density. It is necessary to find a suitable range of electrode mass, thickness, and active materials for maximizing the energy density of lithium-ion capacitors.

Lithium-ion capacitors represent a cutting-edge hybrid energy storage solution, incorporating a lithium insertion–desertion-type anode and an electric EDLC-type cathode [1]. The remarkable energy performance is attained by leveraging the broad operating potential of the non-aqueous electrolyte, the substantial specific capacity of the battery-type anode, and the rapid charging-discharging capabilities of the EDLC-type cathode. With superior energy and power densities, lithium-ion capacitors effectively bridge the divide between supercapacitors and lithium-ion batteries, as illustrated in Fig. 9.1a. A lithium-ion capacitor is essentially a fusion of high-power EDLC-type positive electrodes and high-energy lithium insertion–desertion-type negative electrodes, utilizing a Li-based organic electrolyte. To evaluate or calculate the performance of the lithium-ion capacitor, the following equations are to be followed.

$$\text{Capacitance(C)} : C = \frac{it}{\Delta V}, \tag{9.1}$$

$$\text{Specific capacitance } \mathrm{C_{SP}} : C_{SP} = \frac{4C}{m}, \tag{9.2}$$

where i is the applied current, t is the discharge time, m is the total mass of the two electrodes, and ΔV is the potential difference.

$$\text{Power density (P)} = \frac{\Delta V i}{m}, \tag{9.3}$$

$$\text{Energy density}(E) = P * t, \tag{9.4}$$

where $\Delta V = (V_{\max} + V_{\min})/2$ and $V_{\min}$ and $V_{\max}$ are the lower and upper limits of the voltage window at different current densities, respectively.

In instances where the galvanostatic charge–discharge profiles show a nonlinear curve, as depicted in Fig. 9.2, the determination of energy density and power density

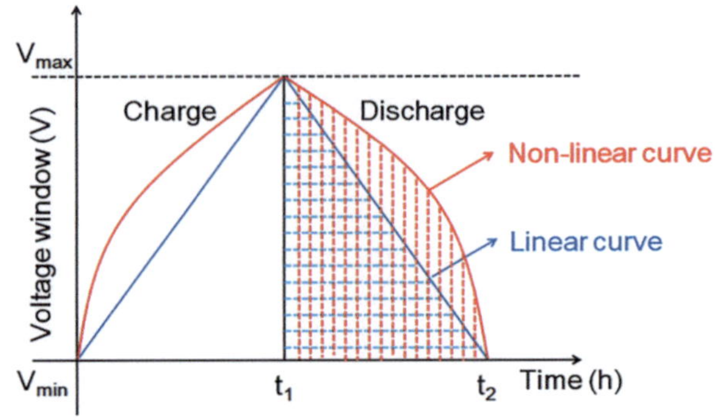

Fig. 9.2 Electrochemical profile of a typical lithium-ion capacitor with the linear (blue line) and nonlinear (red line) charge–discharge curve. Reprinted with permission of Wiley from [21]

for lithium-ion capacitors involves a direct integration of the area underneath the galvanostatic discharge curve. The relevant equations are delineated as follows:

$$E_{\max} = \int \frac{UdQ}{m} = \int \frac{UIdt}{m} = I \int \frac{Udt}{m}, \tag{9.5}$$

$$P \max = \frac{U^2}{4mRs}, \tag{9.6}$$

$$Rs = \frac{U_{IR}}{2I}, \tag{9.7}$$

where I represents the constant discharge current, and U denotes the maximum operational voltage of the cell, dictated by the thermodynamic stability of the electrolyte. Meanwhile, m signifies the aggregate mass of active materials within a lithium-ion capacitor, Rs corresponds to the equivalent series resistance (ESR) of the cell, and UIR refers to the initial iR drop. The presence of a factor of 2 in Eq. 9.7 is attributed to the alteration in current from I to − I at the zenith of the voltage profile.

9.2.2 *Sodium-Ion Capacitor*

Sodium-ion capacitors engineered to achieve high energy density, swift energy delivery, and prolonged lifespan have garnered considerable interest due to their performance comparable to lithium-ion capacitors and the abundance of sodium resources [22]. The conventional design of sodium-ion capacitors involves utilizing battery-like anodes and capacitive cathodes. The anode materials undergo various reactions such as insertion, alloying, and conversion reactions, while the cathode materials typically rely on activated carbon [23]. Nevertheless, recent research endeavours have explored alternative configurations, such as sodium-ion capacitors with battery-like cathodes and capacitive anodes and a combination of both [24]. Amatucci et al. (2001) introduced the first lithium-ion capacitor in 2001, utilizing

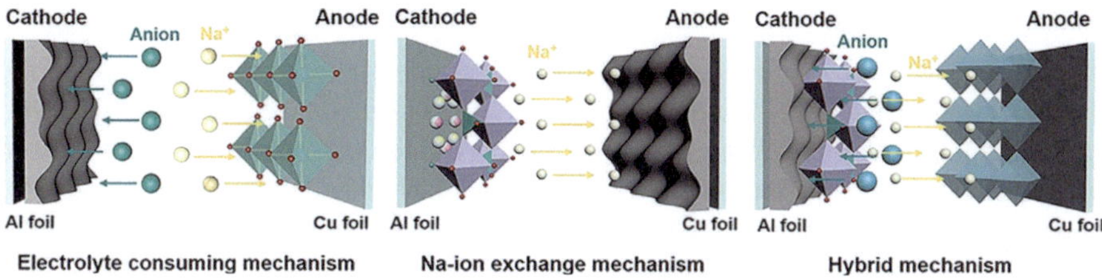

Fig. 9.3 Three kinds of energy storage mechanisms, i.e. electrolyte consuming, Na-ion exchange, and hybrid of sodium-ion capacitors. Reprinted with permission of Wiley from [25]

nanostructured $Li_4Ti_5O_{12}$ (LTO) as the anode and activated carbon as the cathode. Due to the escalating cost of lithium compounds from limited resources, there has been a shift towards sodium-ion capacitors since 2012. The initial sodium-ion capacitor employed a V_2O_5-carbon nanotube composite anode and activated carbon cathode, achieving a high cell voltage of 2.80 V and a maximum energy density of approximately 40 Wh kg^{-1}.

There are three kinds of energy storage mechanisms employed in sodium-ion capacitors i.e. electrolyte consumption, Na-ion exchange, and hybrid as shown in Fig. 9.3.

Electrolyte Consumption Mechanism: In this arrangement, battery materials, encompassing Na-ion deintercalation compounds, metal oxides, and carbonaceous substances, function as the anode. Meanwhile, capacitor-type materials, such as activated carbon derived from biomass or metal–organic frameworks, graphene, and carbon nanotubes, typically operate as the cathode. During the charging process, cations and anions undergo separation, moving towards the anode and cathode, respectively, driven by voltage. This process mirrors the mechanism of supercapacitors but with a distinction: Na-ion intercalation occurs in Na-ion inclusive compounds, or a reduction reaction transpires in the anode, as opposed to simple physical Na-ion adsorption. In the discharging phase, Na-ions return from the anode to the electrolyte, accompanied by the desorption of anions from the cathode to maintain charge balance.

Na-Ion Exchange Mechanism: In this system, the cathode comprises a battery-type material that furnishes Na-ions, while the anode is made of a capacitor-type material. Throughout the charge–discharge cycles, the electrolyte concentration remains constant and primarily facilitates the transfer of Na-ions, resembling the concept of "Rocking-Chair" Na-ion batteries. In contrast to conventional batteries, Na-ions deintercalate from the cathode and adsorb onto the surface of the anode during the charging process in sodium-ion capacitors, and the reverse occurs during discharging.

Hybrid Mechanism: This sodium-ion capacitor exhibits a distinctive characteristic wherein one or both electrodes incorporate a combination of battery-type materials and capacitor-type materials. In the charging phase, Na-ions are deintercalated from the cathode, entering the electrolyte. Simultaneously, activated carbon in the cathode absorbs free anions from the electrolyte, while all Na ions deintercalated from the cathode and provided by the electrolyte intercalated into the anode. During discharging, activated carbon releases the absorbed anions into the electrolyte to aid in balancing the Na ions deintercalated from the anode. Additionally, a portion of the Na-ions intercalates into the cathode to restore the Na-inclusive state. The galvanostatic charge–discharge curve of the sodium-ion capacitor is nonlinear (Fig. 9.2, red line) because of the combination of battery behaviour and capacitive behaviour and, therefore, the energy density (E) is normally estimated by directly integrating the area below the charge–discharge curve by using Eq. (9.5).

9.3 Nanostructured Materials for Lithium- and Sodium-Ion Capacitors

As explained above, the performance of both the lithium-ion capacitors and sodium-ion capacitors extremely depends on the choice of electrode materials. The explosion of nanotechnology and nanosciences greatly benefitted the engineering of advanced battery technologies in terms of their electrochemical performances [20, 26]. Nanomaterials are widely defined as any developed materials that have a size ranging from 1 to 100 nm (unit size < 100 nm) [23, 27]. There are various electrode materials used for the fabrication of electrodes such as cathode and anode in the lithium-ion capacitor and sodium-ion capacitor devices. Figure 9.4 shows various nanostructured electrode materials that have been used in hybrid metal-ion capacitors including carbonaceous materials, metal oxides, and crystalline porous materials [28, 29].

The development of anodic electrode materials focused on the improvement of electrical conductivity and fabrication of nanostructured materials in order to achieve high rate capability. In contrast, cathodic electrode materials have been prepared with

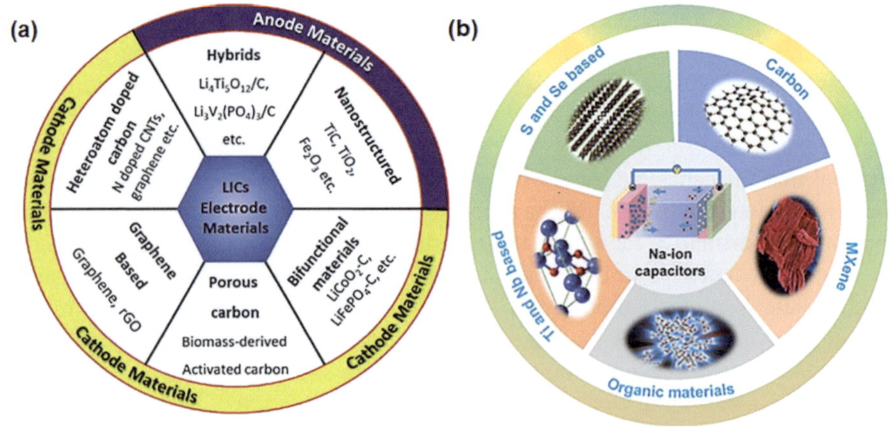

Fig. 9.4 Diagrams showing electrode materials used for **a** lithium-ion capacitors, Reprinted with permission of Elsevier from [28], and **b** sodium-ion capacitors, Reprinted with permission of Wiley from [29]

the intention of improving the capacitive performance of the capacitor by doping heteroatom or fabricating high surface area carbon materials [30, 31]. Although some developments on the anodic and cathodic electrodes have been done, the charging-discharging rate of capacitor-type cathodes supersedes that of battery-type anodes, while the capacity achieved by cathodes is much lower than that of anodes. Therefore, optimizing the matching of cathodes and anodes is important in obtaining maximum performance in lithium-ion capacitor and sodium-ion capacitor devices. The doping of heteroatoms into the electrode materials is utilized to improve the conductivity whereas constructing the hybrid structure using nanostructured carbon allotropes and other materials to form nanocomposite improves the transmission rate of Li-ion and electrons. These are some of the methods that are mainly used to optimize the electrochemical performance [32].

9.3.1 Nanostructured Carbon Materials

The carbon atom has four valence electrons which allow it to bond in a variety of configurations, which span from single sp^3 bonds to double sp^2 and triple sp bonds with all elements [27]. Nanostructured carbon materials are popular electrode materials in electrochemical energy storage i.e. supercapacitors research and are mainly used as capacitor-type utilizing physical adsorption–desorption behaviour electrode materials in hybrid metal-ion capacitors [33]. Nonetheless, graphite is a popular battery-type electrode material for lithium-ion capacitors and they possess active sites for lithium-ion intercalation–deintercalation [34], while hard carbon is a favoured battery-type electrode for sodium-ion capacitors because of its durability paired with high capacity and an operating voltage range that is closer to that of alkali metals [35]. Carbon materials range from hard carbon, graphite, activated carbon, and porous carbon, which includes carbon nanotubes (CNTs), graphene, and graphene-related materials: graphene oxide (GO) [36, 37] and reduced graphene oxide (rGO) [38]. The use of carbon materials in lithium-ion capacitors and sodium-ion capacitors is influenced by their excellent properties such as large surface area, good conductivity, good electrolyte diffusion, and large capacitance from the electrochemical double-layer [28, 33, 39, 40]. Porous carbon materials have outstanding electrolyte accessibility, which is notably rendered by the presence of macro-, micro-, and mesopores. The morphological make-up of the available surface area of porous carbons constitutes the extent of electrode–electrolyte interaction thus affecting their performance. The porosity limits the retardation of the reaction kinetics of the electrode by shortening the distance for the transportation of charges during charging and discharging [41]. Nanoengineering carbon materials into nanostructures allow for improved rate capabilities by shortened pathways for electrolyte diffusion attributed to better accessibility of the surface area [39]. One other effort used to increase the rate capability of carbon materials is doping with hetero atoms (N, S,

O, P) [38]. The incorporation of heteroatoms in the carbon structures not only introduces extra electrons which enhance wettability and electronic conductivity, but it also induces physical defects on the carbon surface inherently creating reactive sites.

Using graphite as an anode for lithium-ion capacitors, Zeng et al. [42] synthesized high-energy microcrystalline graphite from denim material. The activation method employed in this synthesis was sodium nitrate which introduced nitrogen doping to the graphite microcrystalline, which increased reactive sites for faster kinetics. Illustrated in Fig. 9.5a the cyclic voltammograms of graphite microcrystalline carbon (GMC)//GMC depict good charge separation at different scan rates, with a quasi-rectangular shape testament to great rate capability; (b) galvanostatic charge–discharge cycles with the largest discharge time above 2500 s at 0.10 A g^{-1}; and (c) stability analysis shows capacitance retention of 82% promoted by the abundance of nitrogen vacancies on the increased surface area with adequate pores for electrolyte immersion and storage [42]. The graphene microcrystalline anode electrode delivered a high reversible capacity of 1195 mAh g^{-1} at 0.10 A g^{-1} for the GMC//GMC lithium-ion capacitor device with a high energy density of 190.63 Wh kg^{-1} at 225 W kg^{-1} [42].

In the instance of hard carbon being utilized as electrode material for sodium-ion capacitors, Wang et al. [43] synthesized porous nanospheres of nitrogen-doped hard carbon. The electrochemical analysis of the nitrogen-doped hard carbon as the

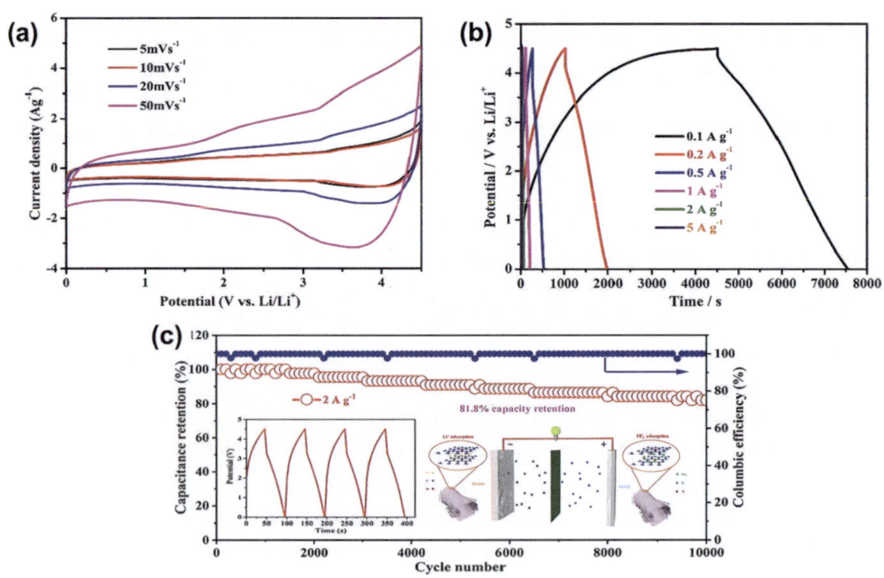

Fig. 9.5 Electrochemical performance of graphite microcrystalline carbon//graphite microcrystalline carbon (GMC//GMC) in lithium-ion capacitor indicating **a** cyclic voltammograms at various scan rates, **b** galvanostatic charge–discharge cycles at different current densities, at the potential range of 4.5 V, and **c** cycling capability at 2 A g^{-1}. Reprinted with permission of Elsevier from [42]

symmetric electrode is shown in Fig. 9.6, where Fig. 9.6a shows the cyclic voltammograms of the nitrogen-doped hard carbon employed as an anode electrode in a sodium-ion capacitor. For the first three cycles at 0.01–3 V, there are no redox peaks observed signifying that Na^+ is stored predominantly by physical and chemical adsorption. From the galvanostatic charge–discharge cycles (Fig. 9.6b) the first charge and discharge capacities were calculated and found to be 188 and 258 mAh g^{-1}, respectively. Additionally, to fully understand the effect of nitrogen doping, the performance of nitrogen-doped hard carbon was compared to pristine hard carbon (HC). In Fig. 9.6c, the cycling performance of nitrogen-doped hard carbon to hard carbon shows higher capacity retention of 122 mAh g^{-1} at 1000 mA g^{-1} after 8000 cycles, with a Coulombic efficiency of $\approx$ 100%. This is further reinforced by the apparent capacity decay observed for hard carbon in similar conditions. The nitrogen-doped hard carbon still showed superior rate performance in comparison to hard carbon (Fig. 9.6d). The introduction of nitrogen atoms in hard carbon propagated the production of defects in its carbon structure, which facilitated easier ion adsorption considering opened channels for electrolyte penetration, and the increased conductivity was facilitated by the extra electrons pairs from the nitrogen atoms that enhanced the conductivity of nitrogen-doped hard carbon consequently improving the cycling and rate performance of nitrogen-doped hard carbon [43].

9.3.2 Titanium Oxides and Derivatives

Titanium dioxide (TiO_2) has a variety of three kinds of three-dimensional (3D) crystalline polymorphs, and it also can form a series of two-dimensional (2D) layered titanates. The three dominant crystalline polymorphs are rutile (tetragonal structure), anatase (tetragonal structure), and brookite (orthorhombic structure) [44]. TiO_2 is a semiconductor that has long-term cycling stability, and high safety, it is inexpensive and poses minimum impact on the environment [45]. However, TiO_2 suffers from poor electronic conductivity and high impedance at the electrode–electrolyte interface. Consequently, the impression is that lowering charge transport pathways through nanostructuring TiO_2 can increase its efficiency [46]. Titanate nanostructures are mainly used in hybrid metal-ion capacitors (metal: Li or Na) because of their huge ion exchange capacity, fast ion diffusion and intercalation, and high surface charge density [47]. There are three main titanates found in literature, monoclinic tri-titanate ($H_2Ti_3O_7$), orthorhombic titanates ($H_2Ti_4O_9$), and monoclinic TiO_2 (B) [48, 49]. A variety of lithium and sodium-ion titanates $A_2Ti_nO_{2n+1}$, where A = Li or Na, have gained a lot of research attention as anode materials for hybrid metal-ion capacitors [29, 49]. A classic example of Na-ion-based titanate oxides includes $Na_2Ti_3O_7$ and $Na_2Ti_6O_{13}$, while a typical lithium-ion-based titanate oxide would be $Li_4Ti_5O_{12}$ and $Li_2Ti_3O_7$ [49–51]. These ternary alkali titanates normally possess the TiO_6 octahedral corners that are negatively charged and share oxygen atoms. They are promising anode materials with the advantage of low potentials (0.30 V vs. Na^+/Na for $Na_2Ti_3O_7$ and 1.55 V vs. Li^+/Li for $Li_4Ti_5O_{12}$), natural abundance, and low

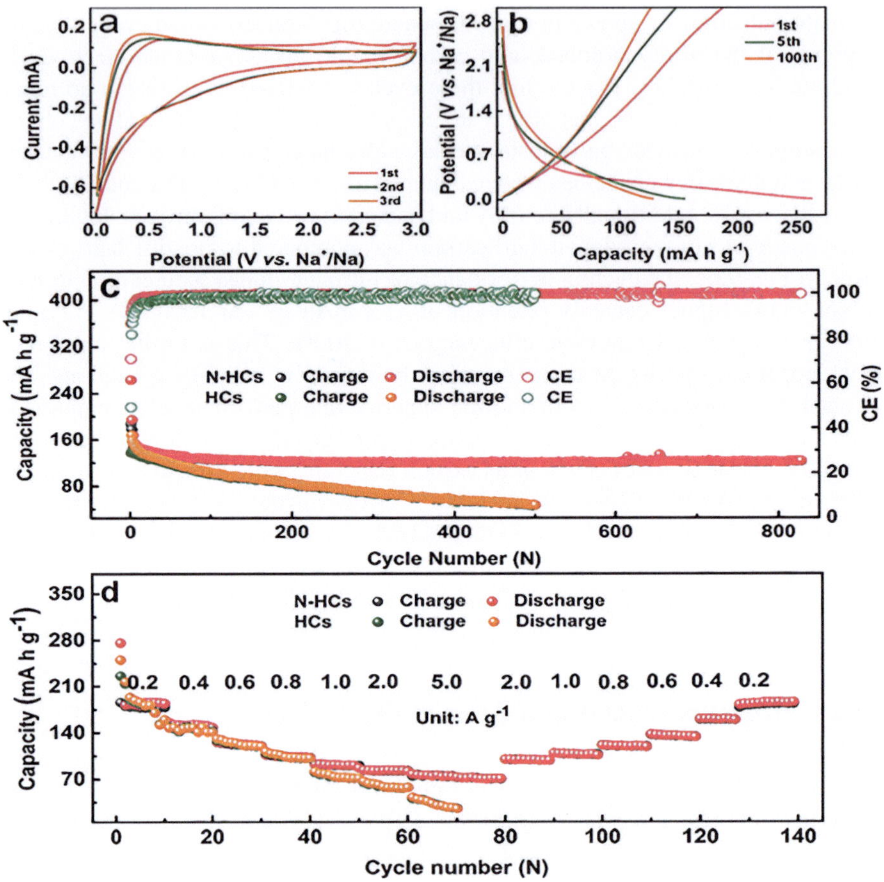

Fig. 9.6 **a** Cyclic voltammograms, **b** galvanostatic charge–discharge cycles, of nitrogen-doped hard carbon, **c** capacity retention and coulombic efficiency at 1 A g^{-1}, and **d** rate performances, of nitrogen-doped hard carbon and hard carbons. Reprinted with permission of the American Chemical Society from [43]

environmental toxicity [52–54]. The disadvantages of titanates are also common to TiO_2, which include sluggish kinetics and poor electronic conductivity, remediation includes (i) nanostructuring materials to shorten charge transport pathways through increasing the surface-to-volume ratio, (ii) carbon coating to improve conductivity and (iii) doping with hetero atoms (N, P, Sb) to facilitate charge transport [22, 23, 55].

Growing TiO_2 nanoparticles on a single-walled nanotube matrix were investigated by Luo et.al [56] for Na-ion insertion. Figure 9.7a shows the electrochemical impedance spectroscopy analysis of TiO_2-SCNT nanocomposite in comparison to pristine TiO_2 where the polarization resistance and charge transfer resistance are significantly lower for TiO_2-SCNT in comparison to TiO_2. The composite also showed better stability at a high current in contrast to the pristine materials, shown in

Fig. 9.7b, and the capacity retention of the composite ($\approx$ 100%) also surpassed that of the pristine material after being subjected to 1000 cycles. The presence of the SCNT matrix plays a pivotal role in easing the transport of Na ion and charge transfer in the composite giving good cycle performance which is a testament to the synergistic effect induced by adding carbon support [56]. The composite was also analysed as an anode electrode against an activated carbon cathode for the sodium-ion capacitor, the cell configuration denoted as TiO_2-SCNT//activated carbon. To significantly lower the irreversible capacity loss experienced on the initial first cycles, both electrodes were predicted. Depicted in Fig. 9.7c the cyclic voltammograms at scan rates between 1 and 20 mV s^{-1} have a near-rectangular shape suggesting the dominance of non-faradaic charge transfer in comparison to Na-ion insertion; furthermore, a slight deviation from the pseudocapacitive linear profile is displayed on the galvanostatic charge–discharge cycles (Fig. 9.7d) confirming the finite contribution of Na-ion insertion to the full cells charge storage mechanism [56]. The rate performance of the full cell was also analysed with the Coulombic efficiency from 1–100 C as shown in Fig. 9.7e, in the display the current rate was calculated to deliver an average capacity of 70 mAh g^{-1} at 1C, and it only experienced a 50% drop at 100 C. This occurred whilst the cell was maintaining $\approx$ 100% Coulombic efficiency with a $\approx$ 90% capacity retention at 10C. Additionally, capacity drops recorded after 20 C are negligible in comparison to the initial rates [56]. The sodium-ion capacitor performance of the TiO_2-SCNT//activated carbon showed the full capability to be employed as one of the future materials for energy storage applications.

Liu et al. [49] compared the effect of carbon nanotubes used as carbon support for porous lithium-ion titanates ($Li_2Ti_3O_7$), denoted as pLTO (CNT-pLTO) and dense

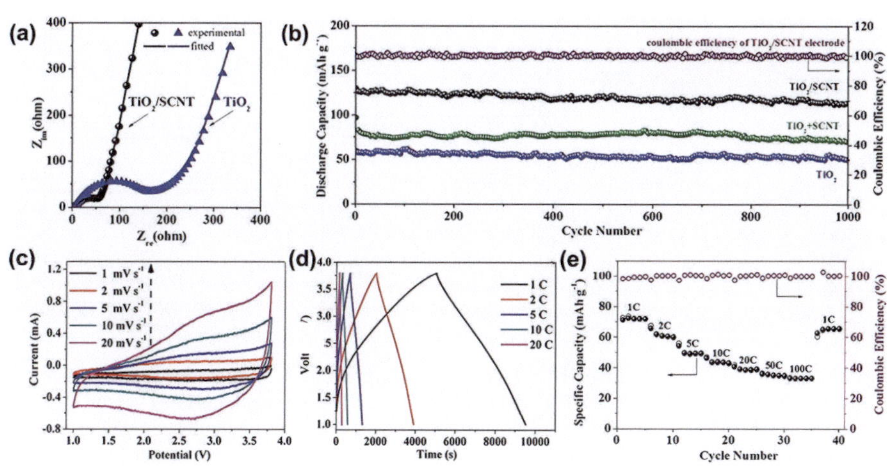

Fig. 9.7 **a** Electrochemical impedance profile of TiO_2-SCNT and TiO_2 electrodes, **b** Long cycle performances of TiO_2-SCNT composite; TiO_2 + SCNT mixture; and TiO_2 electrodes, **c** cyclic voltammograms at various scan rates, **d** galvanostatic charge–discharge cycles at different current densities, and **e** rate capability and Coulombic efficiency at various current densities, of TiO_2-SCNT//activated carbon. Reprinted with permission of Wiley from [56]

lithium titanates (CNT-LTO). The nanosized porous particles of pLTO are anticipated to deliver high magnitudes of charge transfer owing to the shortened diffusion paths and ease of access of the electrolyte to the electrode, thus facilitating high-energy performance. A schematic depiction of the lithium-ion capacitor system is shown in Fig. 9.8a. From Fig. 9.8b, the cyclic voltammograms performed at a scan rate of 1 mV s^{-1} demonstrate the redox couple (i.e. the anodic and cathodic peaks) observed for the CNT-LTO and CNT-pLTO. This is due to faradaic reactions which are attributed to a greater mass of the LTO in the composite in comparison to the CNT (0.75:0.15) [49]. The fitted log (peak current) versus log (scan rate) of CNT-pLTO was calculated to have a slope that is 66% greater than that of CNT-LTO (Fig. 9.8c), which correlates accordingly with the rate performance of the nanocomposites calculated from the galvanostatic charge–discharge of the half cell (Fig. 9.8d). The trend of improved performance from the CNT-pLTO nanocomposite is further confirmed by having a diffusion coefficient higher than that of CNT-LTO as depicted in Fig. 9.8e. The positive outcome from this work is illustrated by the intense similarity of the cycling performance, shown in Fig. 9.8f, with a coulombic efficiency of $\approx$ 100% for both nanocomposites. Indeed, (i) nanoarchietcturing creates a better channel in the materials to efficiently interact with the electrolyte, (ii) the incorporation of carbon materials for structural support and coating increases the conductive electronic network whilst facilitating increased electron transfer, and (iii) observing similar capacity retention proves that the integrity of the existing lithium titanate channels was preserved even as porous nanoparticles of lithium titanate [49, 57].

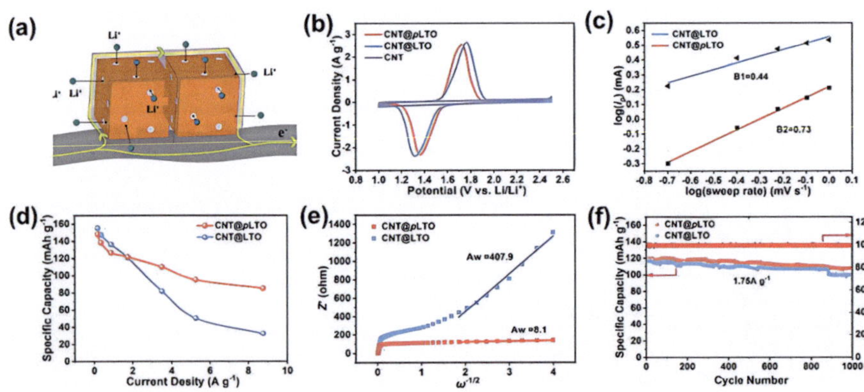

Fig. 9.8 **a** Lithium-ion capacitor cell displaying Li-ion migrations, **b** comparative cyclic voltammograms of carbon nanotubes, CNT-LTO, and CNT-pLTO, **c** fitted log(peak current) versus log(scan rate) curves, **d** specific capacity at various current densities, **e** electrochemical impedance spectroscopy analysis with calculated diffusion coefficient values, and **f** cycling performance, for CNT-LTO and CNT-pLTO. Reprinted with permission of Elsevier from [49]

9.3.3 Niobium and Vanadium Oxides

There are stoichiometric forms of niobium oxides: Nb_2O_5, NbO, NbO_2, $Nb_{10}O_{25}$, and non-stoichiometric forms: $Nb_{12}O_{29}$ and $Nb_{22}O_{54}$. These oxides have varying properties that can exist in several oxidation states (i.e. Nb_2O_5) allowing redox charge transfer and efficient ion intercalation–deintercalation abilities. However, these oxides also demonstrate poor electric conductivity of $\sim 3 \times 10^{-6}$ S cm^{-1} [44, 58, 59]. Nanoengineering Nb_2O_5 gained traction for ion capacitors seeing as it is an intrinsic pseudocapacitive material that is challenged heavily by rapid sodiation–desodiation due to the larger radius of the Na ion. Orthorhombic phase Nb_2O_5 (T-Nb_2O_5) theoretically showed better rate capabilities with a calculated capacity of at least 200 mAh g^{-1}, high cycling stability, and high-power characteristics, nonetheless, experienced severe pulverization during cycling and the capacity is much less than the theoretical value even when the material is in the nanosize with high porosity. Researchers have now embarked on using carbon materials with good electron conductivity like graphene, carbon nanotubes, and carbon nanospheres as support materials to grow T-Nb_2O_5 nanostructures to shield them from pulverization and opening.

The use of nanoengineering to control the morphology of nanostructures is portrayed well in the work of Hemmati et al. [60], where zero-dimensional (0D) nitrogen-doped T-Nb_2O_5 (N-NbOC). In Fig. 9.9, the synthesis process of N-NbOC is mapped out, and the NbO is reduced to T-Nb_2O_5 concurrently with the calcination of polyaniline (nitrogen stock).

From the distinguished electrochemical performance of the N-NbOC, it was proven that (i) the 0D porous nanostructures of the T-Nb_2O_5 in conjunction with the hollow design of the carbon microtubes facilitated easier transport of the electrolyte significantly shortening the path for Li-ion intercalation–deintercalation and (ii) the doping of nitrogen through polyaniline was a success and the confirmed presence of nitrogen on the N-NbOC enhanced the electron conductivity and rate capability [60]. The cyclic voltammograms show good charge separation with reversible peaks that serve as evidence of a high Li^+ storage mechanism in Fig. 9.10a, while displayed in Fig. 9.10b is the log (current) versus log(potential scan rate) to analyse the surface processes present in the nanocomposite. The slope from 0.10–10 mV s^{-1} is 0.92, which is characteristic of capacitive and pseudocapacitive charge storage contribution, while the slope between 15–50 mV s^{-1} decreased to 0.56 suggesting dominant ion-diffusion charge storage. Figure 9.10c shows from the cyclic voltammograms at 5 mV s^{-1} that the dominating charge storage process is capacitive (shaded area) and the redox peaks are a demonstration of pseudocapacitive charge storage performance. The verified doping of nitrogen to the metal oxide lattice was proven by the dominant capacitive and pseudocapacitive charge storage process and can be pinned to the presence of fast faradaic reaction because of the presence of electron-rich nitrogen-containing functional groups and an increase in the electronic conductivity attributed to increased ease on the electron donor properties of the nanocomposite [60, 61]. Furthermore, the rate capability of N-NbOC, Nb_2O_5, and carbon from

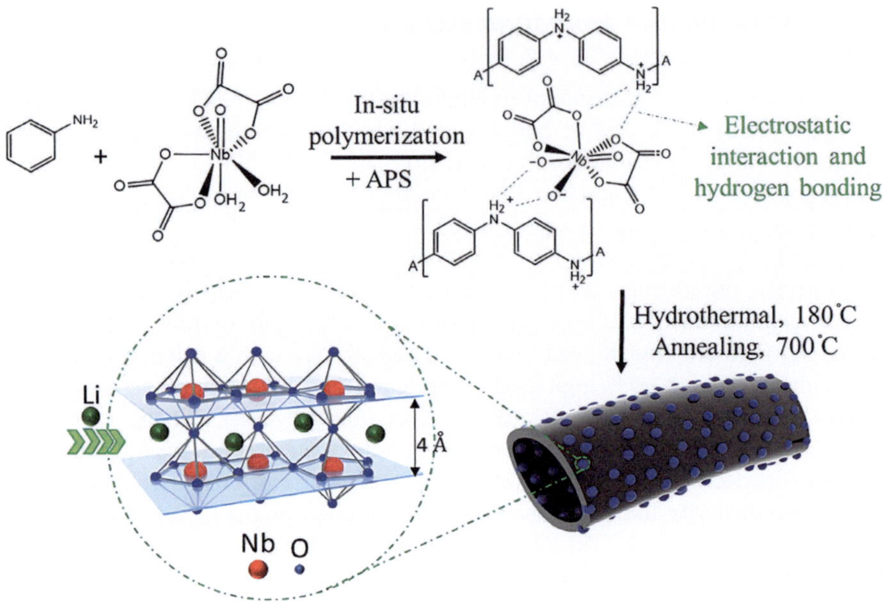

Fig. 9.9 Synthesis of 0D nitrogen-doped T-Nb_2O_5 embedded on carbon microtubes fabricated as an anode electrode in lithium-ion capacitor application. Reprinted with permission of Elsevier from [60]

PANI is shown in Fig. 9.10d confirming the higher specific capacitance of N-NbOC, which easily arises from nitrogen doping and increased porosity. The electrochemical impedance spectroscopy plot in Fig. 9.10e was recorded after the 2nd and 100th cycles and what was deduced from these plots is the lower polarization and charge transfer resistance experienced by N-NbOC in comparison to Nb_2O_5, which is also corroborated by the displayed superior cyclability in Fig. 9.10f [60]. This work proved the effect of carbon coating and heteroatom doping on the electrochemical performance of Nb_2O_5 from the increased wettability which correlates with increased ionic conductivity during cycling characteristics, thus resulting in good electrochemical performance.

The dominant pseudocapacitive behaviour rendered by nitrogen doping is confirmed by She et al. [62] through the anchoring of T-Nb_2O_s on nitrogen-doped graphene (Nb_2O_5-NG). In Fig. 9.11a, the CV curves of the first three cycles of Nb_2O_5-NG at 0.10 mV s^{-1} in the voltage range of 0.01–3 V versus Na/Na^+ depict a cathodic peak that is attributed to the formation of the solid electrolyte interphase (SEI). However, the cathodic peak is observed to be diminished after the first cycle due to capacitive kinetics becoming more dominant. The galvanostatic charge–discharge cycles of the 1st, 2nd, and 3rd charge–discharge cycles (Fig. 9.11b) were used to calculate the charge capacity of the 1st and 2nd cycles which were 352 and 126

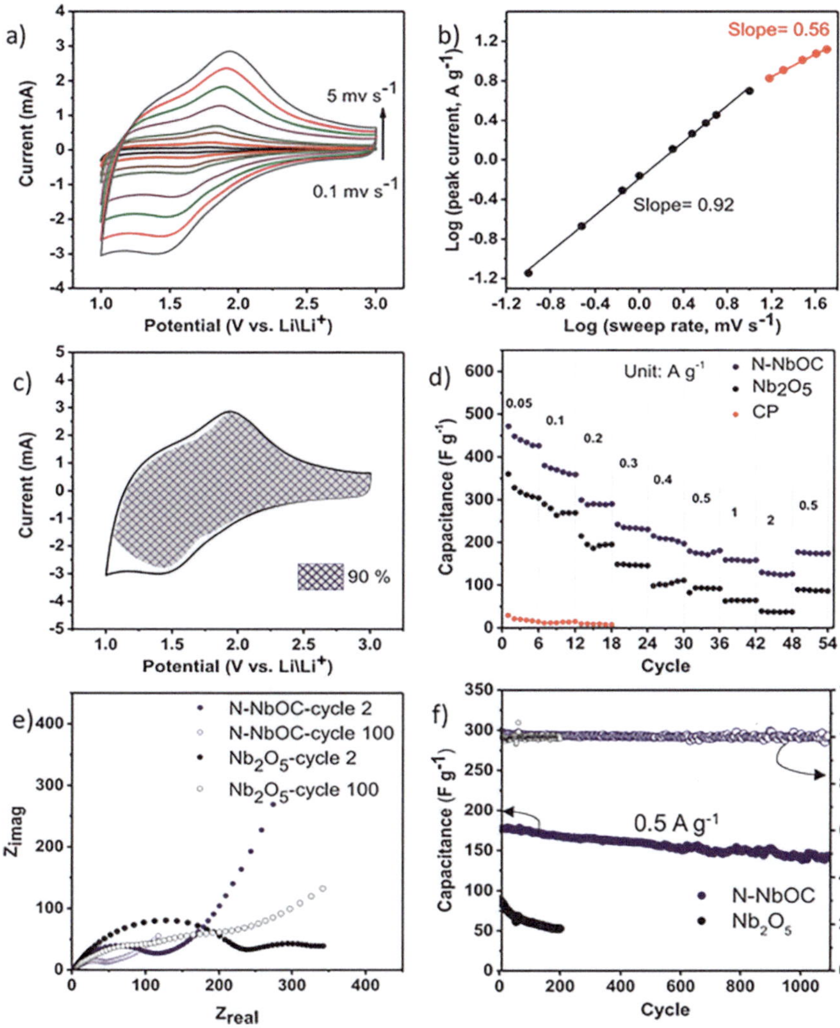

Fig. 9.10 **a** Cyclic voltammograms at various scan rates, **b** fitted log (peak current) versus log (scan rate) curves, **c** cyclic voltammogram indicating capacitance contribution (shaded area), of N-NbOC, **d** overlaid specific capacitance at various current densities of N-NbOC, Nb_2O_5, and conducting polymer, **e** electrochemical impedance spectroscopy for the 2nd and 100th cycles, and **f** cycling performance, of N-NbOC and Nb_2O_5. Reprinted with permission of Elsevier from [60]

mAh g^{-1}, respectively. The rate capability of Nb_2O_5-NG at varied current densities is distinctly higher than that of Nb_2O_5-graphene (Fig. 9.11c), and the same trend is apparent with higher cycling performance at 1 A g^{-1} in Fig. 9.11(d). The Nb_2O_5-NG electrode managed to deliver a high specific capacity of 76 mAh g^{-1} after 1000 cycles and showed an excellent Coulombic efficiency of $\approx$ 100%, as shown in Fig. 9.11e. Moreover, the lower charge transfer resistance depicted by

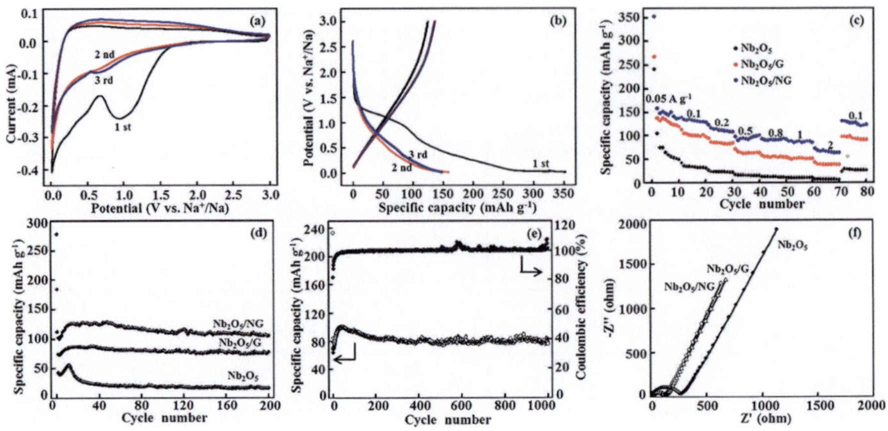

Fig. 9.11 **a** Cyclic voltammograms, **b** galvanostatic charge–discharge, of the first three cycles, of Nb_2O_5-NG, **c** overlaid cycling performance and rate capabilities at different current densities, **d** cycling performance at 0.20 A g^{-1}, of Nb_2O_5, Nb_2O_5-G, and Nb_2O_5-NG, **e** cycling performance at 1 A g^{-1} and the Coulombic efficiency, and **f** Nyquist plots of Nb_2O_5, Nb_2O_5-G, and Nb_2O_5-NG after 100 cycles at 1 A g^{-1}. Reprinted with permission of the American Chemical Society from [62]

the Nyquist plot of Nb_2O_5-NG Nyquist plot in Fig. 9.11f suggests an increased electronic conductivity that is induced by the introduction of graphene and nitrogen doping on Nb_2O_5 in Fig. 9.11f [62]. Nb_2O_5-NG showed excellent sodium-ion capacitor performance confirming the synergistic effects of doping heteroatoms on highly conductive nanostructured carbon materials decorated with metal oxides.

9.3.4 Conversion Compounds

Conversion compounds use reversible redox reactions to accommodate Li^+ and Na^+ through the formation of Li and Na compounds during charging. These materials are typically crystalline or amorphous binary and ternary oxide, sulphide, nitride, fluoride, and phosphide compounds of V, Cr, Mn, Fe, Co, Ni, Cu, Mo, W, and Ru [63, 64]. Upon charging, the metal cations in these compounds are displaced by Li^+ and Na^+ cations to form the electrochemically inert Li_2X-Na_2X compounds, as outlined by Eqs. 9.8 and 9.9 for binary and ternary compounds, respectively:

$$M_aX_b + cLi^+/Na^+ \leftrightarrow dM + mLi_2X/Na_2X \tag{9.8}$$

$$M'_aM''_aX_c + dLi^+/Na^+ \leftrightarrow M' + mM'' + nLi_2X/Na_2X \tag{9.9}$$

The Li_2X-Na_2X compounds then decompose to reform the original M_aX_b or $M'_aM''_bX_c$ compounds upon discharging. As such, conversion compounds inherently display greater energy storage capability due to the occurrence of redox reactions comprising multi-electron transfers compared to the intercalation-based materials [29, 63, 64]. Examples of the conversion compounds used in lithium-ion capacitors and sodium-ion capacitors include $NiCo_2O_4$, MoS_2, MnO, $CoWO_4$, and Sb_2O_3 [31, 63, 65]. Conversion compounds generally offer higher energy densities, cost-efficient synthesis procedures, and non-toxic or minimized toxicity towards the environment [31, 66]. On the other hand, these materials suffer from poor cycling lives and low Coulombic efficiencies that are brought on by poor electronic conductivities and higher fluctuations of the electrode volume during insertion-extraction of charges [31, 67]. These setbacks are usually circumvented by replacing the bulk materials with nanostructured materials and fabricating composites with other conductive materials.

For example, [68] encapsulated cubic $NiCo_2O_4$ particles with N-rGO ($NiCo_2O_4$-N-rGO) for application as an anode in sodium-ion capacitors. A half-cell analysis of the $NiCo_2O_4$-N-rGO composite revealed a reversible capacity of 452 mAh g^{-1}, whilst a $NiCo_2O_4$-N-rGO//activated carbon sodium-ion capacitor displayed a high energy density of 92.80 Wh kg^{-1}, a maximum power density of 9750 W kg^{-1}, and 84% retention of the initial energy density after 4000 cycles. Zhao et al. [69] fabricated a composite of $NiCo_2O_4$ nanowires with carbon tubular bundles (CTBs) for sodium-ion capacitor applications. The $NiCo_2O_4$-CTBs nanocomposite exhibited a high specific capacity of 298 mAh g^{-1} at 1 A g^{-1}, as well as 92% capacity retention after 500 cycles. A sodium-ion capacitor built using the $NiCo_2O_4$-CTBs nanocomposite and commercial activated carbon was able to achieve a high energy density of 99 Wh kg^{-1}, a maximum power density of 15,000 W kg^{-1}, and 85% capacity retention after 5000 cycles. The structural and electrochemical analyses of the study are shown in Fig. 9.12.

Wang et al. [71] increased the interlayer spacing of MoS_2 by interlacing its layers with those of the amorphous carbon layer (MoS_2-C) and supporting it on reduced graphene oxide (MoS_2-C/rGO). The group then assembled a lithium-ion capacitor using the MoS_2-C/rGO composite as an anode and polyaniline-derived porous carbon (PDPC) as a cathode, which delivered an energy density of approximately 188 Wh kg^{-1}, a maximum power density of 40,000 W kg^{-1}, and 80% retention of the initial capacity after 10,000 cycles at 2 A g^{-1}. Zhan et al. [72] followed a similar approach by employing an interlayer-expanded MoS_2-reduced graphene oxide nanocomposite as a sodium-ion capacitor anode. The nanocomposite was coupled with a nitrogen-doped 3D graphene (N-3DG) cathode in a sodium-ion capacitor that delivered a high energy density of 140 Wh kg^{-1} and a maximum power density of 103,000 W kg^{-1}.

Tang et al. [73] fabricated a new lithium-ion capacitor comprising a composite of Fe_2O_3 nanospheres wrapped in aerogel-like MXene ($Ti_3C_2T_x$-Fe_2O_3) as an anode and 3D N, S co-doped porous carbon (NS-DPC) as a cathode. The $Ti_3C_2T_x$-Fe_2O_3//NS-DPC device displayed a high specific capacity of 109 mAh g^{-1}, a high energy density of 216 Wh kg^{-1}, a maximum power density of 20,000 W kg^{-1}, and 87% retention of the initial capacity after 5000 cycles at 5 A g^{-1}. Considerable research

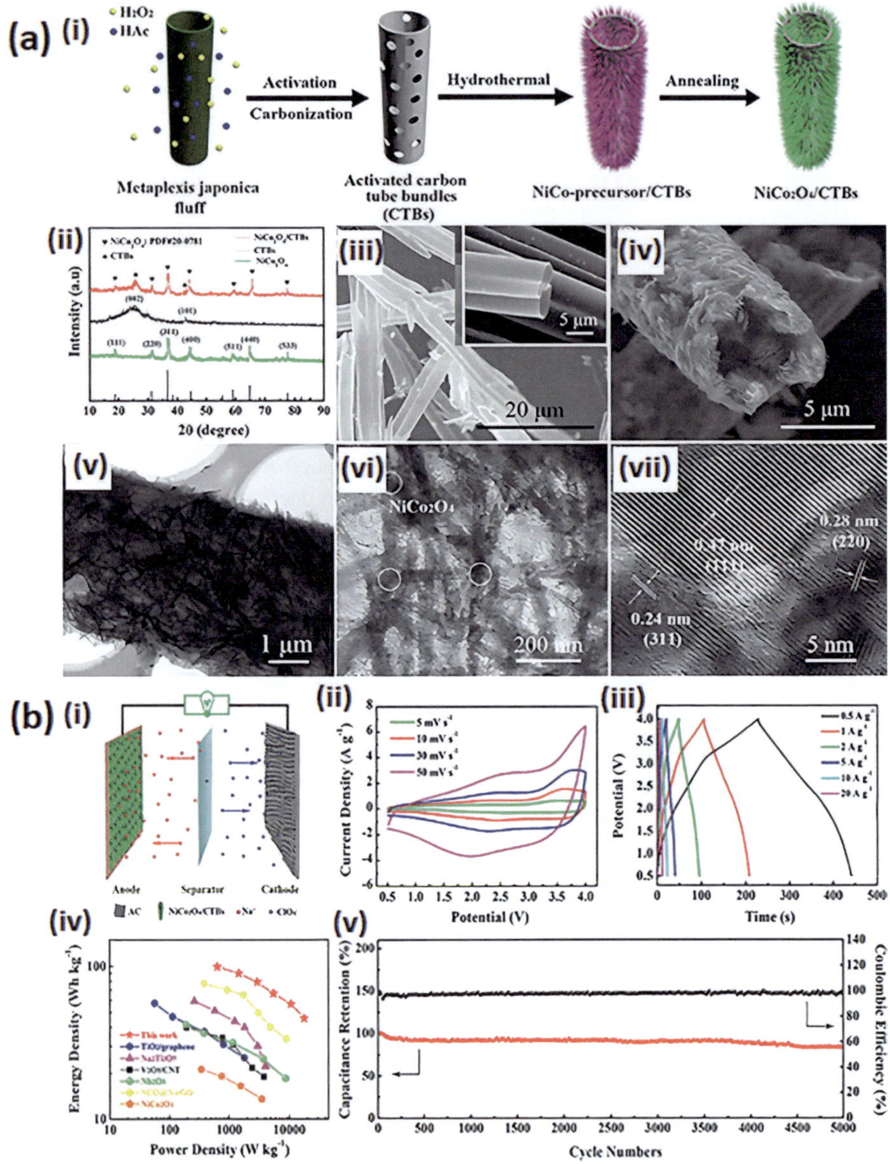

Fig. 9.12 **a** (i) Synthesis procedure, (ii–vii) structural and morphological characterization of $NiCo_2O_4$-CTBs composite, and **b** electrochemical performance of the $NiCo_2O_4$-CTBs// activated carbon sodium-ion capacitor. Reprinted with permission of Wiley from [70]

Table 9.1 Conversion compounds in lithium-ion capacitors and sodium-ion capacitors

Configuration	HIC	V_{max} (V)	Specific capacity (mAh g^{-1})	E_{max} (Wh kg^{-1})	P_{max} (kW kg^{-1})	References
δ-MnO_2@rGO/CC//AC	SIC	1.6	77.8	–	–	[69]
1 T-MoS_2/d-$Ti_3C_2T_x$//GNC	LIC	4.0	112	188	13	[74]
NCO//VACNFs	LIC	3.2	–	136.9	40	[75]
$ZnMn_2O_4$//NCN	LIC	4.0	–	202.8	21	[76]
MnO//AC	LIC	4.0	(270.3 F g^{-1})[a]	227	2.952	[77]
HT-MoO_3/CNTs//AC	LIC	4.3	(50 F g^{-1})[a]	90	2	[78]
GFeF//pG	LIC	3.5	(83.2 F g^{-1})[a]	129.6	1.87	[79]
NiF_2//AC	LIC	4.3	–	35	11.4	[80]
CoF_2//AC	LIC	4.3	–	36.9	11.4	[80]
NiF_2//AC	SIC	4.3	–	14.3	8.6	[80]
CoF_2//AC	SIC	4.3	–	14.3	8.6	[80]

[a]Specific capacitance (F g^{-1}), LIC—Lithium-ion capacitors, SIC—Sodium-ion capacitors, AC—Activated carbon, GNC—Graphene nanocomposite, VACNTs—Vertically aligned carbon nanotubes, NCN—Nitrogen-doped carbon nanosheets

efforts have been invested in the application of conversion compounds in lithium-ion capacitors and sodium-ion capacitors. Moreover, novel materials are still being discovered by researchers to add to the already vast number of materials that are known to date. Table 9.1 gives a summary of some of the literature reported on the use of conversion compounds in lithium-ion capacitors and sodium-ion capacitors.

9.3.5 *Natrium Super-Ionic Conductor (NASICON)-Type Compounds*

The NASICON-type compounds have found favour for application as lithium- and sodium-ion capacitor electrode materials due to their highly stable and open 3D crystal structures (see Fig. 9.13a) that allow rapid ion migration through the inherent channels and minimized volume changes [29, 31, 81]. These materials present a general chemical formula of $A_xM'_yM''_{2-y}(XO_4)_3$ (i.e. A = alkali (Li, Na, K) or alkaline earth metal (Mg, Ca), M' and M" = V, Cr, Mn, Fe, Co, and Ni, and X = P, S) [64]. The NASICON-type compounds generally store energy through intercalation and deintercalation of the active ions. Materials such as $Li_3V_2(PO_4)_3$ [82–84], $Na_3V_2(PO_4)_3$ [84, 85], $NaTi_2(PO_4)_3$ [86–88, 71], $Na_2CrTi(PO_4)_3$ [89] have been respectively employed as electrode materials in lithium-ion capacitors and sodium-ion capacitors. The analogous lithium vanadium phosphate and sodium vanadium phosphate materials have found the most research interest for application in lithium-ion capacitors and sodium-ion capacitors, respectively, because of the ability of vanadium to intercalate ions at various potentials as a result of its multiple oxidation states [63].

However, the NASICON-type compounds generally suffer from poor electronic conductivities, which result in sluggish intercalation kinetics and restrict their lifespan. This challenge is usually overcome by coating the materials with electrically conductive carbon materials, developing nanostructured particles, and integrating the compounds into composites with other conducive materials [31, 83, 88]. Böckenfeld and Balducci [90] reported the use of two lithium vanadium phosphate materials containing varied carbon residues from the synthesis procedures as electrode material for lithium-ion capacitors. Lithium-ion capacitors assembled with activated carbon as an anode and $Li_3V_2(PO_4)_3$-citric acid obtained from using citric acid (CA) as a reducing agent as a cathode (activated carbon// $Li_3V_2(PO_4)_3$ -CA) achieved an initial discharge capacity of 23 mAh g^{-1} and an energy density of 22.10 Wh kg^{-1}. On the other hand, a lithium-ion capacitor assembled with activated carbon as a cathode and $Li_3V_2(PO_4)_3$ -CA + OA obtained from using a mixture of citric acid and oxalic acid (CA + OA) as a reducing agent as a cathode (Activated carbon// $Li_3V_2(PO_4)_3$ -CA) achieved an initial discharge capacity of 55 mAh g^{-1} and an energy density of 76.30 Wh kg^{-1}. [83] followed suit by demonstrating the application of carbon-coated $Li_3V_2(PO_4)_3$ in lithium-ion capacitors. The group built two lithium-ion capacitors using $Li_3V_2(PO_4)_3$-C as an anode against activated carbon ($Li_3V_2(PO_4)_3$ -C//AC) and as a cathode against activated carbon (Activated carbon// $Li_3V_2(PO_4)_3$ -C), which displayed maximum energy densities of about 27 and 25 Wh kg^{-1}, respectively.

Wang et al. [46] reported the electrochemical performance of $Na_3V_2(PO_4)_3$ and $Na_3V_2(PO_4)_3$ -C composite as electrode materials for sodium-ion capacitors. The $Na_3V_2(PO_4)_3$ //activated carbon and $Na_3V_2(PO_4)_3$-C//activated carbon demonstrated specific capacitances of 59.53 and 50.98 F g^{-1}, high energy densities of 15.48 and 15.93 Wh kg^{-1}, and maximum power densities of 734.40 and 750 Wh kg^{-1}, respectively. Thangavel et al. [87] demonstrated the Na-ion storage capability of a carbon-coated $Na_3V_2(PO_4)_3$ anode coupled with a cinnamon stick-derived carbon (CDC). The C-$Na_3V_2(PO_4)_3$//CDC sodium-ion capacitor showed a specific capacitance of 105 F g^{-1} and a high energy density of 118 Wh kg^{-1} both at 0.12 mA cm^{-2}, as well as a maximum power density of 850 Wh kg^{-1} and 95% retention of the initial capacity after 10,000 cycles both at 1 mA cm^{-2}. The authors attributed the results to various the structural stability of C-$Na_3V_2(PO_4)_3$ that enabled it to retain its polyanion framework even at high current densities, and to the uniform carbon nanomaterial-coating that aided with enhancing the cyclability by increasing the flow of current on the electrode surface. Moreover, the carbon coating was observed to also aid in preventing side reactions with the electrolyte which also contributed to the retention of the polyanion framework.

Roh et al. [88] reported on the development of a $NaTi_2(PO4)_3$-rGO microsphere composite for application as a high-power sodium-ion capacitor anode. The $NaTi_2(PO4)_3$-rGO microsphere composite showed a specific capacity of 133 mAh g^{-1} at 0.10 C in a half-cell analysis. A sodium-ion capacitor constructed using a

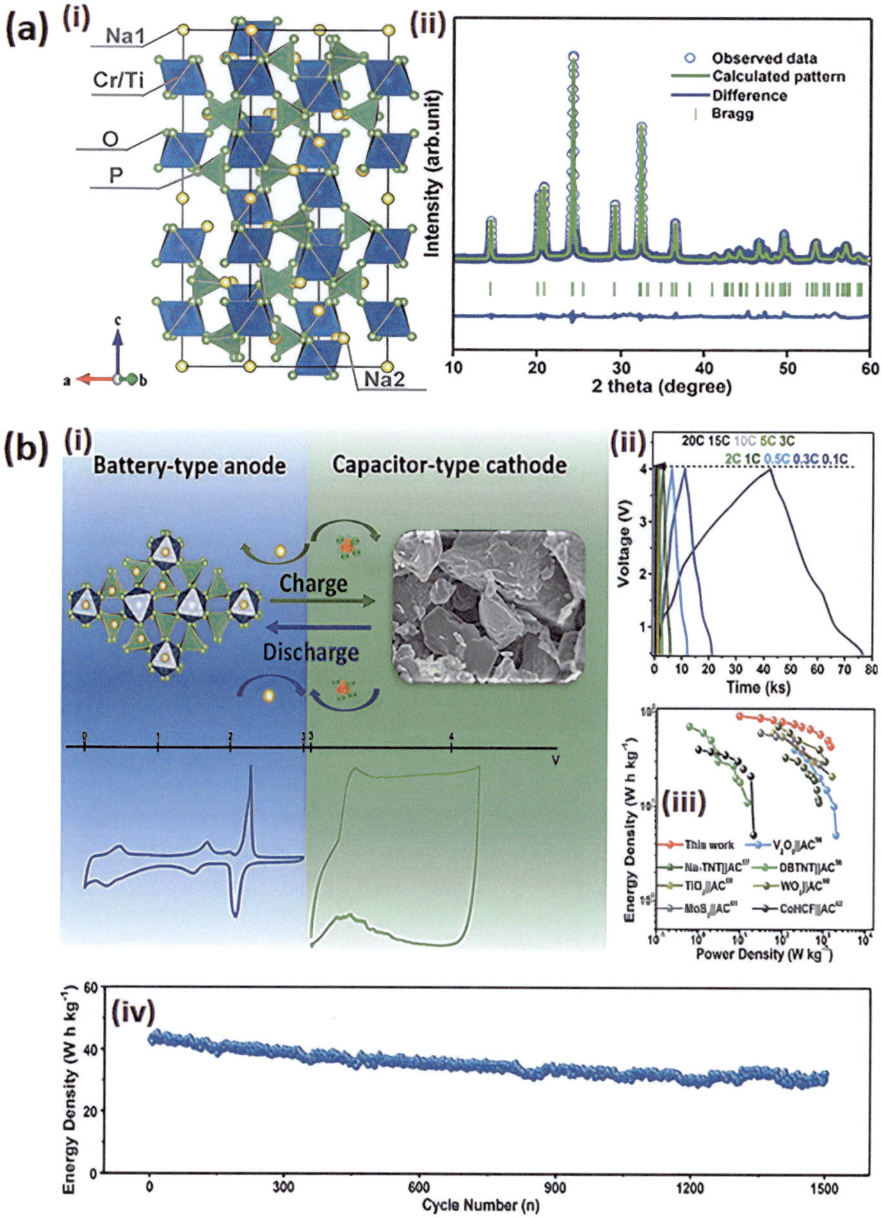

Fig. 9.13 **a** Structural properties of $Na_2CrTi(PO_4)_3$, and **b** the electrochemical performance of the $Na_2CrTi(PO_4)_3$//activated carbon fabricated for sodium-ion capacitor application. Reprinted with permission of the Royal Society of Chemistry from [89]

Table 9.2 NASICON-type compounds in lithium-ion capacitors and sodium-ion capacitors

Configuration	HIC	V_{max} (V)	Specific capacity (mAh g^{-1})	E_{max} (Wh kg^{-1})	P_{max} (kW kg^{-1})	References
G-NTP//GNS	SIC	3.0	–	31	8	[86]
PSC-NTP-C//NC	SIC	1.2	90	–	–	[91]
LVP-T2//AC	LIC	2.7	–	44.28	3.11	[84]
NTP/C-NFs//AC	SIC	1.5	–	56	1.162	[92]
NTP//AC	SIC	1.6	27.5	31.6	0.42	[93]
NTP//AC	SIC	2.5	–	56	4.096	[71]

LIC—Lithium-ion capacitors, SIC—Sodium-ion capacitors, AC —Activated carbon, GNS—Graphene nanosheets, NFs—Nanofibers, NC—Nanocomposite, PSC—Porous sulphur doped carbon, LVP— $Li_3V_2(PO_4)_3$, NVP-$Na_3V_2(PO_4)_3$, NTP - $NaTi_2(PO4)_3$.

$NaTi_2(PO_4)_3$-rGO anode and an activated carbon cathode achieved a high energy density of 53 Wh kg^{-1} and a maximum power density of 6700 W kg^{-1}. Wang et al. [34] fabricated a $Na_2CrTi(PO_4)_3$-C composite anodic application in sodium-ion capacitors. The composite displayed a reversible capacity of 220 at 0.1 C, an outstanding rate performance with a specific capacity of 110 mA h g^{-1} at 20 C, and exceptional electrochemical stability that reaches up to 20,000 cycles in a half-cell analysis. A sodium-ion capacitor of the composite and activated carbon ($Na_2CrTi(PO_4)_3$-C//Activated carbon) was able to achieve a high energy density of 92 Wh kg^{-1} and a maximum power density of 1700 W kg^{-1}. The electrochemical performance of the sodium-ion capacitor is shown in Fig. 9.13b. Table 9.2 lists some of the literature that reported on the application of NASICON-type compounds in lithium-ion capacitors and sodium-ion capacitors.

9.3.6 Metal–Organic Frameworks

Crystalline porous materials including metal–organic frameworks have been applied in various energy storage applications due to their precious properties. Metal–organic frameworks are constructed by combining metal ions (clusters) as central ions with organic ligands through chemical bonding. Different metal centres, organic linkers, and synthesis methods may be varied to obtain desired properties for different applications including batteries and supercapacitors [17, 94]. Recently, several metal–organic frameworks have been studied as electrodes for lithium-ion capacitor and sodium-ion capacitor storage and have shown potential to be good electrode materials. However, the practical application of traditional metal–organic frameworks in the field is still challenging due to poor conductivities and weak electrochemical stability which results in ineffective utilization of active sites and compromised cycling stability [95, 96]. Conductive metal–organic frameworks have been demonstrated to deliver encouraging electrochemical properties due to their good conductivities both for fast ion and electron transport. Conductive metal–organic frameworks can be contracted from active organic linkers, good choice of metal centres, and the introduction of other compounds including metal oxides, metal nanoparticles, carbon materials, and transitional metals [97]. For example, Chen et al. [1] developed an anode material from Ti-based metal–organic frameworks (TiO_xN_y) for a Na-ion capacitor. The anode half cell exhibited a high reversible capacity of 275 mAh g^{-1} at a current density of 50 mA g^{-1}. An N-doped hierarchical porous carbon (NHPC) produced from the TiO_2-based metal–organic frameworks (Fig. 9.14) exhibited an EDLC energy storage mechanism and was used as a cathode material for the sodium-ion capacitor. After coupling the TiO_xN_y anode and the NHPC cathode, the capacitor delivered a maximum energy density of 80 Wh kg^{-1} with a power density of 100 W kg^{-1}. The sodium-ion capacitor retained 90% of its capacity after 1200 cycles [1].

Sun et al. synthesised cobalt-based metal–organic frameworks-nanowires through a facile solvothermal synthesis for application as a battery-type anode in the lithium-ion capacitors. Figure 9.15a demonstrates the coordination of the Co^{2+} to its adjacent HHTP linkers resulting in an alternating 2D P-conjunction framework. The scanning electron microscopic images in Fig. 9.15b confirm the formation of the nanowires, and their lengths can be determined from the images. Furthermore, the electrochemical performance of metal–organic frameworks-nanowires was tested in a half-cell configuration. The cyclic voltammograms (Fig. 9.15c) and the charge–discharge (Fig. 9.15d) confirm the battery-type behaviour of the metal–organic frameworks-nanowire due to the redox activity of the lithium insertion-extraction. The electrode material was able to deliver a high specific capacity of 715 mAh g^{-1} at a current density of 0.10 A g^{-1}. To investigate the application of the nanowires in lithium-ion

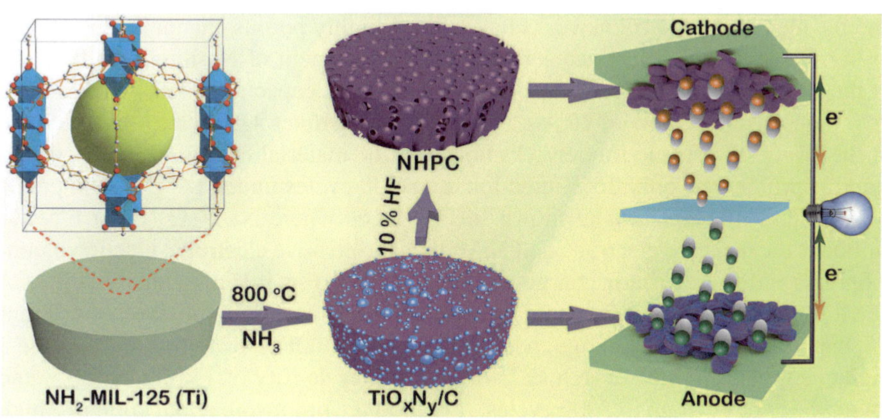

Fig. 9.14 Synthesis of the electrode materials and the relevant cathode and anode charge storage mechanisms in the sodium-ion capacitor. Reprinted with permission of Wiley from [1]

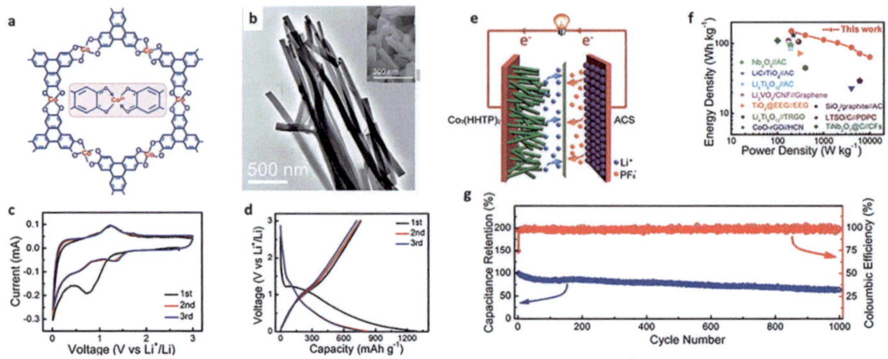

Fig. 9.15 **a** Chemical structure, **b** transmission electron microscopic and field-emission scanning electron microscopic image, **c** cyclic voltammograms, **d** galvanostatic charge–discharge cycle of Co-metal–organic frameworks nanowires, **e** demonstration of working principle of lithium-ion capacitor, **f** Ragone plot of the device and comparison with other reported lithium-ion capacitors, and **g** cycling stability at 0.10 A g^{-1}. Reprinted with permission of Wiley from [98]

capacitors, the nanowires were used as the anode and the cathode was constructed from activated carbon as shown in Fig. 9.15e. The $Co_3(HHTP)_2$//ACS hybrid cell exhibited a high specific capacitance of 67 F g^{-1} at a current density of 0.10 A g^{-1} and a high energy density of 150Wh kg^{-1} at a power density of 10,000 W kg^{-1} as indicated in Fig. 9.15d. Besides the high performance, the metal–organic frameworks—the nanowire-based device still managed to show a retention of 65% after cycling 1000 times [98].

Liang et al. [99] demonstrated conductive Ni-metal–organic frameworks that exhibited a high-efficiency ion and electronic transport path which facilitates the quick reversible function of Na-ion storage. The conductive Ni-metal–organic frameworks were prepared by deprotonation and oxidation processes in basic conditions, and the metal–organic frameworks have shown highly porous channels as indicated in Fig. 9.16a–b which can accelerate the free movement of Na-ions. The half cell of the metal–organic frameworks produced a specific capacity of 107 mAh g^{-1} at a very high current density of 10 A g^{-1}, which demonstrates a good rate characteristic of the Ni-metal–organic frameworks however, the material has shown poor cycling stability which was only 26% retention after 400 cycles under 1A g^{-1}. The prototype sodium-ion capacitor is shown in Fig. 9.16c where $Na_3V_2O_2(PO_4)_2F$/active carbon is the negative electrode and NiMOF is the positive electrode. Cyclic voltammograms and galvanostatic charge–discharge cycles (Fig. 9.16d–e) of NVOPF/AC// NiMOF cells show quasi-rectangular shapes which show the combination of sodium-ion batteries and EDLC storage mechanisms. The NVOPF/AC//NiMOF delivered a high energy density of 127 Wh kg^{-1}, a high-power density of 17,309 W kg^{-1}, and improved cycling of up to 5000 cycles, revealing the promising application in a high energy and power output and long cycling stability [99].

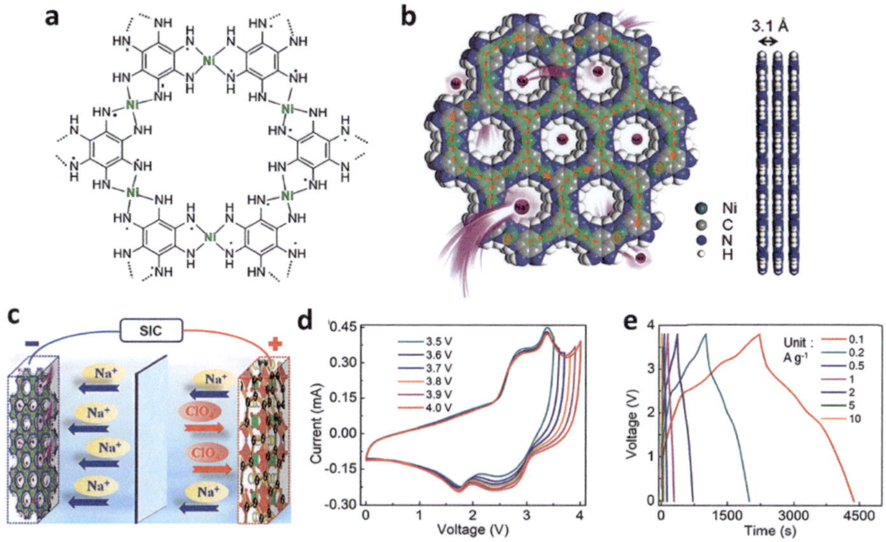

Fig. 9.16 **a** Chemical structure of Ni-metal–organic frameworks. **b** Space-filling model. Green: Ni, grey: C, blue: N, white: H. Electrochemical performance of the NVOPF/AC//Ni-metal–organic frameworks. **a** Diagrammatic configuration of the sodium-ion capacitor. **b** Cyclic voltammograms at fixed 1 mV s^{-1} under various voltage windows. **c** Charge–discharge curves at various current densities between 0 and 3.80 V. Reprinted with permission of Wiley from [99]

Recently, metal–organic frameworks have been used as sacrificial templates to produce porous carbon frameworks, which have been shown to improve the electrochemical nature by incorporating properties of both metal–organic frameworks and carbon materials [100]. Li et al. [101] developed carbon frameworks using Co-ZIF as the effective precursor and template. The conversion of the ZIF-67 was done by heating the ZIF-67 at 600 °C at a rate of 5 °C min^{-1} for 5 h under an argon atmosphere as indicated in Fig. 9.17a. Figure 9.17b–c shows the scanning electron microscope images with different magnifications of the nanocarbon material where rhombic dodecahedral shapes of the ZIF-67-Co are still preserved even after carbonization [101]. Dimas Chandra Permana [102] carbonized Fe-metal–organic frameworks (Basolite F300) to produce high surface area onion-like-carbons. Metal–organic frameworks-based onion-like carbons were used as the anode and activated carbon as the cathode material to construct the onion-like carbon//activated carbon device which achieved an outstanding maximum energy density of 14,436 W kg^{-1} and a maximum power density of 224 Wh kg^{-1}. [103] also developed a well-organized interleaved composite of graphene-like nanosheets embedded with MnO_2 nanoparticles (MnO_2-C-NS) using a manganese-based metal–organic framework and employed it as a promising anode material for the lithium-ion capacitor. The MnO_2-C-NS anode exhibited an impressive specific capacity of 1054 mAh g^{-1} at a current density of 0.10 A g^{-1}. Figure 9.17d(i) shows the assemblage of MnO_2-C-NS as an anode with a nanoporous carbon sheet (NPCS) as a cathode to fabricate

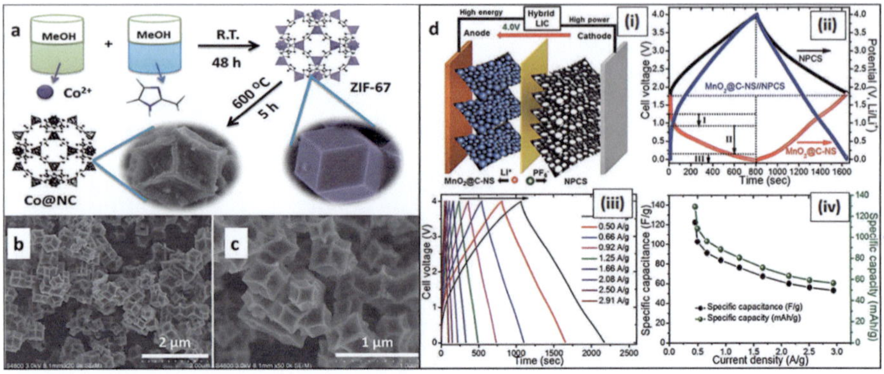

Fig. 9.17 **a** Synthesis of Co-NC derived from ZIF-67-Co, **b–c** Scanning electron microscopic images of the metal–organic frameworks -derived nanocarbon material, **d**(i) schematic diagram of the design of lithium-ion capacitor **d**(ii) galvanostatic charge–discharge cycle of MnO_2-C-NS//NPCS with corresponding potential distribution across MnO_2-C-NS anode and NPCS cathode versus Li-foil reference electrode, **d**(iii) galvanostatic charge–discharge cycles of MnO_2-C-NS//NPCS at different current densities, and **d**(iv) variation of specific capacitance and capacity with current density for MnO_2-C-NS//NPCS cell. Reprinted with permission of Wiley from [102]

a full-cell system (MnO_2-C-NS//NPCS). As shown, the Li-ions rapidly move in and out of the MnO_2-C-NS through surface faradaic reactions, while PF_6^- anions are adsorbed–desorbed on the porous NPCS. The charge–discharge profiles of the anode and cathode are shown in Fig. 9.17d(ii), and it is observed that the total amount of energy stored in the MnO_2-C-NS hybrid system is the sum of energy stored in the MnO_2 nanocrystals and carbon nanosheets. The shapes of galvanostatic charge–discharge cycles at different current densities exhibit a quasi-symmetric triangular shape, suggesting a good combination between the hybrid battery-type anode and EDLC cathode (Fig. 9.17d(iii). The maximum specific capacity of the device was 129 mAh g^{-1} at a current density of 0.50 A g^{-1} as observed in Fig. 9.17d(iv) [102].

9.3.7 MXenes

MXenes are an emerging family of 2D transition metal carbides, nitrides, and carbonitrides that are produced by etching out layers of the A element (usually a group III or IV element) or A-based compounds from MAX phase ceramic precursors ($M_{n+1}AX_n$). The remaining accordion-like layers (see Fig. 9.18a(i) and a(ii)) have a general chemical formula of $M_{n+1}X_nT_x$, where M is a transition metal, X represents either C, N, or CN, T_x denotes surface functional or termination groups, and n = 1, 2, or 3 [29, 63, 81]. As an example of pseudocapacitive materials, MXenes accommodate ions through intercalation reactions that are independent of diffusion-regulated faradaic processes. MXenes have found growing interest as ideal electrode materials in lithium-ion capacitor and sodium-ion capacitor applications due to their multi-layered structure, excellent metal-like conductivity of over 2.0×10^4 S cm^{-1}, and higher volumetric capacities [63]. Examples of MXenes studied in the literature include $Ti_3C_2T_x$ [104–109], $Zr_3C_2T_x$ [11, 109, 110], $MoTi_2C_2T_x$ [111], Nb_2CT_x [112, 113] with $Ti_3C_2T_x$ being the most studied Mxene in a variety of applications such as in hybrid metal-ion capacitors [105, 114], supercapacitors [11, 104, 107, 115], lithium-ion batteries [116], sodium-ion batteries [108, 109], electrochemical water-splitting (EWS) [111].

However, it has been shown in the literature that delaminated MXenes offer more charge storage compared to multi-layered MXenes [105, 116]. On the other hand, like many 2D materials, delaminated MXenes are prone to self-stacking that inhibits easy access of ions to active sites due to attenuation of the surface area. It has therefore become essential to improve the electrochemical performance of MXenes by integrating them with "pillar" materials that will restrict self-stacking. For example, [117] employed $Ti_3C_2T_x$ sheets pillared with Na ions as anode material for sodium-ion capacitors. A half-cell of the Na-$Ti_3C_2T_x$ nanosheets exhibited a high reversible capacity of 175 mAh g^{-1} at 0.10 A g^{-1}. A sodium-ion capacitor device built using the Na-$Ti_3C_2T_x$ nanosheets and activated carbon as anode and cathode, respectively, was able to achieve a high energy density of 80.20 Wh kg^{-1}, a maximum power density of 6172 W kg^{-1}, and over 78% retention of the initial capacity after 115,000 cycles at 2 A g^{-1}.

Xia et al. [105] demonstrated the pillaring of $Ti_3C_2T_x$ with Sn^{4+} that led to an in situ formation of amorphous SnO_2 layers between the MXene layers for lithium-ion capacitor applications. The modified MXene (AS-$Ti_3C_2T_x$) displayed a reversible capacity of 320 mAh g^{-1} at 0.05 A g^{-1} and 92% capacity retention after 1500 cycles at 2 A g^{-1} in a Li-ion half-cell analysis. As shown in Fig. 9.18b, alithium-ion capacitor employing AS-$Ti_3C_2T_x$ as the anode and activated carbon as the cathode delivered a good electrochemical performance with a high energy density of 105.70 Wh kg^{-1}, a maximum power density of 8000 W kg^{-1},, and a long cycle life that retains 82% of the initial capacity after 4000 cycles at 2 A g^{-1}. The authors attributed the excellent electrochemical performance of the AS-$Ti_3C_2T_x$ MXene to synergistic effects of faster ion-diffusion kinetics brought on by the intercalation of the Sn^{4+} and a lowered charge transfer barrier due to the presence of amorphous SnO_2-modified layers.

Yi et al. [114] devised an electrostatic self-assembly strategy to interlace $Ti_3C_2T_x$ with the negatively charged rGO sheets by modifying the surfaces of the MXene layers with positively charged tetrabutylammonium ion (TBA^+) for application as an anode in lithium-ion capacitors. A half-cell analysis of the Ti_3C_2Tx-rGO nanohybrid delivered an outstanding reversible capacity of 1394 mAh g^{-1} at 0.05 A g^{-1}. The authors further studied the practicality of the nanohybrid in a lithium-ion capacitor comprising the Ti_3C_2Tx-rGO nanohybrid as an anode and activated carbon as a cathode. The device demonstrated a high specific capacity of 70.70 F g^{-1} and an impressively high energy density of 147 Wh kg^{-1} at 0.10 A g^{-1}, as well as a maximum power density of 3125 W kg^{-1} at 5 A g^{-1}. The Ti_3C_2Tx-rGO//AC lithium-ion capacitor also exhibited excellent electrochemical stability, with over 80% retention of the initial capacity and almost 100% coulombic efficiency after 5000 cycles at 2 A g^{-1}.

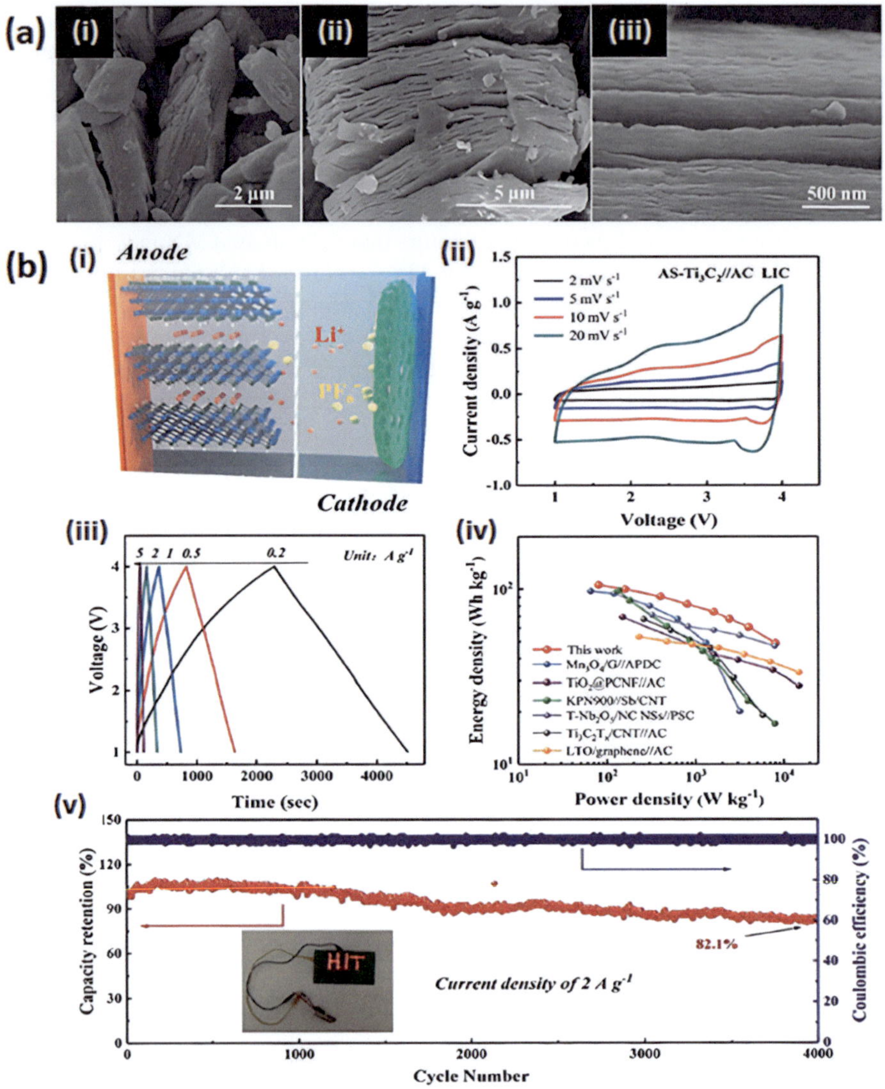

Fig. 9.18 **a** Scanning electron microscopic images of (i) Ti_3AlC_2, (ii) $Ti_3C_2T_x$, and (iii) AS-$Ti_3C_2T_x$, and **b** the electrochemical performance of the AS-$Ti_3C_2T_x$ //AC lithium-ion capacitor. Reprinted with permission of Elsevier from [105]

As MXenes are still an emerging group of materials and more novel variations are still being discovered, research into their applications as lithium-ion capacitors and sodium-ion capacitor electrode materials will increase. Future studies should be focused on advancing the electrochemical performance of MXenes in lithium-ion capacitors and sodium-ion capacitors with regard to extending their cycling life span.

Table 9.3 MXenes in lithium-ion capacitors and sodium-ion capacitors

Configuration	HIC	V_{max} (V)	Specific capacity (mAh g^{-1})	E_{max} (Wh kg^{-1})	P_{max} (kW kg^{-1})	References
Ti_2CT_x// $Na_2Fe_2(SO_4)_3$	SIC	2.4	90	320	1.44	[118]
$Ti_3C_2T_x$-CNT// $Na_{0.44}MnO_2$	SIC	4.0	(242 mAh cm^{-3})^a	–	–	[119]
CTAB – Sn(IV)@ $Ti_3C_2T_x$//AC	LIC	4.0	(51 F g^{-1})^b	105.56	10.8	[8]
$Ti_3C_2T_x$//AC	SIC	3.4	–	39	1.140	[120]
$Ti_3C_2T_x@Fe_2O_3$// NS-DPC	LIC	4.0	109	216	20	[73]
Nb_2CT_x-CNT// Nb_2CT_x-CNT	LIC	3.0	43	38	–	[113]
Nb_2CT_x-CNT// $LiFePO_4$	LIC	3.0	36	43	–	[113]
Lithiated graphite/ Nb_2CT_x-CNT	LIC	3.0	72	49	0.546	[113]
1 T-MoS_2/d-$Ti_3C_2T_x$/ /GNC	LIC	4.0	112	188	13	[74]
Ti_2CT_x// $LiNi_{1/3}Co_{1/3}Mn_{1/3}O_2$	LIC	2.5	206	160	–	[121]
HC//V_2CT_x	SIC	3.5	50	–	–	[122]
$Ti_3C_2T_x$-CNT//AC	LIC	3.0	(32.1 F g^{-1})^b	67	5.797	[123]

^{a}Volumetric capacity (mAh cm^{-3}); ^{b}Specific capacitance (F g^{-1}), LIC—Lithium-ion capacitors, SIC—Sodium-ion capacitors, AC—Activated carbon, GNC—Graphene nanocomposite, CNT—Carbon nanotubes

Moreover, their capacities could be enhanced by exploring a hybridization of their charge storage mechanism to include faradaic processes [63]. Nonetheless, considerable research efforts have been given to studying the performance of MXenes in lithium-ion capacitors and sodium-ion capacitors. Table 9.3 lists some of the literature reporting the electrochemical performance of MXenes in lithium-ion capacitors and sodium-ion capacitors devices.

9.4 Conclusion

Up to now, it is quite clear that the innovative strategies that involve the utilization of hybrid metal-ion such as Na and Li when combined with supercapacitive materials are essential for the advancement of energy storage technologies. Hence, the chapter extensively provided the current progress and understanding of mechanisms for the hybrid metal-ion capacitors such as Li^+ and Na^+ capacitors in order to identify the gaps that would have to be considered for further improvement of energy storage devices necessary for high energy and power outputs. The need to decipher ways to develop and design lithium-ion capacitors and sodium-ion capacitor devices requires advanced nanostructured electrode materials that already display the potential to increase their high-rate capability, energy density and cycle stability. The importance of the nanostructured architecture of the electrode materials ensures vast opportunities to avoid the usage of conductive additives, which deters adequate charge transfer, boosting the electronic properties. The incorporation of heteroatoms either through doping or surface functionalization in the backbone of nanostructured materials greatly enhances the performance amplifying the kinetics of the devices. From the literature, it is evident that sodium-ion capacitors exhibit comparable energy storage performance to that of lithium-ion capacitors in terms of energy and power densities. This is an added advantage as the overuse of Li ions is not sustainable since Li is scarce while Na is abundant in nature. From the performance of these lithium-ion capacitors and sodium-ion capacitors, it will be necessary to explore other hybrid metal-ion capacitors derived from the newly established battery technologies such as Al-ion, Mg-ion, K-ion, F-ion, and Zn-ion. The strategic development of metal–organic frameworks-derived porous carbon frameworks and the discovery of MXenes in electrochemical energy storage technologies have provided exciting opportunities, especially in hybrid metal-ion capacitors. Furthermore, extensive research is still required on the role that novel nanostructured materials such as metal–organic frameworks and MXenes play in the development of high-performing lithium-ion capacitors and sodium-ion capacitors.

References

1. S. Dong et al., Lithium-ion and sodium-ion hybrid capacitors: from insertion-type materials design to devices construction. Adv. Func. Mater. **31**(21), 2100455 (2021)
2. L. Trahey et al., Energy storage emerging: a perspective from the joint center for energy storage research. Proc. Natl. Acad. Sci. **117**(23), 12550–12557 (2020)
3. P. Simon, Y. Gogotsi, Perspectives for electrochemical capacitors and related devices. Nat. Mater. **19**(11), 1151–1163 (2020)
4. K.O. Oyedotun et al., Synthesis of ternary NiCo-MnO_2 nanocomposite and its application as a novel high energy supercapattery device. Chem. Eng. J. **335**, 416–433 (2018)
5. J. Zheng, Theoretical energy density for electrochemical capacitors with intercalation electrodes. J. Electrochem. Soc. **152**, A1864–A1869 (2005)
6. K.O. Oyedotun et al., Comparison of ionic liquid electrolyte to aqueous electrolytes on carbon nanofibres supercapacitor electrode derived from oxygen-functionalized graphene. Chem. Eng. J. **375**, 121906 (2019)
7. M.S. Ratsoma et al., Prospective cathode materials for all-solid-state batteries, in *Solid State Batteries: Design, Challenges and Market Demands*. ed. by N. Palaniyandy, K.P. Abhilash, B. Nalini (Springer International Publishing, Cham, 2022), pp.83–125
8. J. Luo et al., Pillared structure design of MXene with ultralarge interlayer spacing for high-performance lithium-ion capacitors. ACS Nano **11**(3), 2459–2469 (2017)
9. P. Jeżowski et al., Safe and recyclable lithium-ion capacitors using sacrificial organic lithium salt. Nat. Mater. **17**(2), 167–173 (2018)

10. J. Zheng, The limitations of energy density of battery/double-layer capacitor asymmetric cells. J. Electrochem. Soc. **150**, A484–A492 (2003)
11. K.O. Oyedotun et al., Electrochemical performance of two-dimensional Ti_3C_2-Mn_3O_4 nanocomposites and carbonized iron cations for hybrid supercapacitor electrodes. Electrochim. Acta **301**, 487–499 (2019)
12. S.S.M. Bhat et al., Nanostructured Na(2)Ti(9)O(19) for hybrid sodium-ion capacitors with excellent rate capability. ACS Appl. Mater. Interfaces **10**(1), 437–447 (2018)
13. V. Palomares et al., Na-ion batteries, recent advances and present challenges to become low cost energy storage systems. Energy Environ. Sci. **5**(3), 5884–5901 (2012)
14. D. Kundu et al., The emerging chemistry of sodium ion batteries for electrochemical energy storage. Angew. Chem. Int. Ed. **54**(11), 3431–3448 (2015)
15. K. Raju et al., Rational design of 2D manganese phosphate hydrate nanosheets as pseudocapacitive electrodes. ACS Energy Lett. **5**, 23–30 (2020)
16. J. Liang, D.-W. Wang, Design rationale and device configuration of lithium-ion capacitors. Adv. Energy Mater. **12**(25), 2200920 (2022)
17. K. Makgopa, M.S. Ratsoma, K.D. Modibane, Intrinsic properties of metal–organic frameworks (MOFs) in supercapacitor applications. Curr. Opin. Electrochem. **36**, 101112 (2022)
18. F.M. Auxilia et al., Au@TiO_2/reduced graphene oxide nanocomposites for lithium-ion capacitors. Chem. Eng. J. **362**, 136–143 (2019)
19. K.O. Oyedotun et al., Electrochemical properties of asymmetric supercapacitor based on optimized carbon-based nickel-cobalt-manganese ternary hydroxide and sulphur-doped carbonized iron-polyaniline electrodes. Electrochim. Acta **334**, 135610 (2020)
20. J.J. Lamb, O.S. Burheim, Lithium-ion capacitors: a review of design and active materials. Energies **14**(4), 979 (2021)
21. H. Wang et al., Nonaqueous hybrid lithium-ion and sodium-ion capacitors. Adv. Mater. **29**(46), 1702093 (2017)
22. B. Babu, M.M. Shaijumon, High performance sodium-ion hybrid capacitor based on $Na_2Ti_2O_4(OH)_2$ nanostructures. J. Power. Sources **353**, 85–94 (2017)
23. J. Ding et al., Review of hybrid ion capacitors: from aqueous to lithium to sodium. Chem. Rev. **118**, 6457–6498 (2018)
24. T. Huang et al., Boosting capacitive sodium-ion storage in electrochemically exfoliated graphite for sodium-ion capacitors. ACS Appl. Mater. Interfaces **12**, 52635–52642 (2020)
25. Z. Zhang et al., Sodium-ion capacitors: recent development in electrode materials. Batter. Supercaps **4**(11), 1680–1700 (2021)
26. Makgopa, K., P.M. Ejikeme, K.I. Ozoemena, *Nanostructured Manganese Oxides in Supercapacitors*, in *Nanomaterials in Advanced Batteries and Supercapacitors* (2016), pp. 345–376
27. K. Makgopa et al., Nanostructured carbon-based electrode materials for supercapacitor applications, in *Carbon Related Materials: Commemoration for Nobel Laureate Professor Suzuki Special Symposium at IUMRS-ICAM2017*. ed. by S. Kaneko et al. (Springer Singapore, Singapore, 2021), pp.317–355
28. A. Jagadale et al., Lithium ion capacitors (LICs): development of the materials. Energy Storage Mater. **19**, 314–329 (2019)
29. Y. Zhang et al., Sodium-ion capacitors: materials, mechanism, and challenges. Chemsuschem **13**(10), 2522–2539 (2020)
30. K. Makgopa et al., Nanostructured metal oxides for supercapacitor applications. in *Emerging Nanostructured Materials for Energy and Environmental Science*, ed. by N.M.R.K.B.R.S. Rajendran (Springer, 2019), pp. 247–303
31. P. Naskar et al., Frontiers in hybrid ion capacitors: a review on advanced materials and emerging devices. ChemElectroChem **8**(8), 1393–1429 (2021)
32. J. Xu et al., Recent progress in electrode materials for nonaqueous lithium-ion capacitors. J. Nanosci. Nanotechnol. **20**(5), 2652–2667 (2020)
33. B. Li et al., Electrode materials, electrolytes, and challenges in nonaqueous lithium-ion capacitors. Adv. Mater. **30**(17), 1705670 (2018)

34. X. Wang, L. Liu, Z. Niu, Carbon-based materials for lithium-ion capacitors. Mater. Chem. Front. **3**(7), 1265–1279 (2019)
35. M. Shirpour, J. Cabana, M. Doeff, New materials based on a layered sodium titanate for dual electrochemical Na and Li intercalation systems. Energy Environ. Sci. **6**(8), 2538–2547 (2013)
36. K. Makgopa, P.M. Ejikeme, K.I. Ozoemena, Graphene oxide-modified nickel (II) tetra-aminophthalocyanine nanocomposites for high-power symmetric pseudocapacitor. Electrochim. Acta **212**, 876–882 (2016)
37. M.J. Madito et al., Graphene-based electrode materials for supercapacitor applications. in *Electrochemical Devices for Energy Storage Applications*, ed. by M.A.E.F.I. Kebede (Taylor & Francis Group, CRC Press, 2019), pp. 101–130
38. K. Makgopa et al., One-step hydrothermal synthesis of nitrogen-doped reduced graphene oxide/hausmannite manganese oxide for symmetric and asymmetric pseudocapacitors. ACS Omega **6**, 31421–31434 (2021)
39. K. Zou et al., Carbon materials for high-performance lithium-ion capacitor. Curr. Opin. Electrochem. **21**, 31–39 (2020)
40. K. Zou et al., A kinetically well-matched full-carbon sodium-ion capacitor. J. Mater. Chem. A **7**(22), 13540–13549 (2019)
41. L. Shen et al., Peapod-like Li_3VO_4/N-doped carbon nanowires with pseudocapacitive properties as advanced materials for high-energy lithium-ion capacitors. Adv. Mater. **29**(27), 1700142 (2017)
42. D. Zeng et al., High-energy graphite microcrystalline carbon for high-performance lithium-ion capacitor: diffusion kinetics and lithium-storage mechanism. J. Colloid Interface Sci. **623**, 1190–1199 (2022)
43. X. Wang et al., Nitrogen-doped hard carbon as symmetric electrodes for sodium-ion capacitor. Energy Fuels **34**, 13144–13148 (2020)
44. R.S. Devan et al., One-dimensional metal-oxide nanostructures: recent developments in synthesis, characterization, and applications. Adv. Func. Mater. **22**(16), 3326–3370 (2012)
45. P. Voepel, B.M. Smarsly, Synthesis of titanium oxide nanostructures in ionic liquids. Z. Anorg. Allg. Chem. **643**, 3–13 (2017)
46. H. Wang et al., A high energy and power li-ion capacitor based on a TiO_2 nanobelt array anode and a graphene hydrogel cathode. Small **11**, 1470–1477 (2015)
47. Y. Zhang et al., Titanate and titania nanostructured materials for environmental and energy applications: a review. RSC Adv. **5**(97), 79479–79510 (2015)
48. A. Mukherjee et al., Surface modified sodium titanate as low voltage anode for sodium rechargeable cell with superior electrochemical properties. Mater. Lett. **301**, 130219 (2021)
49. Y. Liu et al., Fabrication of porous lithium titanate self-supporting anode for high performance lithium-ion capacitor. J. Energy Chem. **50**, 344–350 (2020)
50. S. Liu et al., TiO_2 nanorods grown on carbon fiber cloth as binder-free electrode for sodium-ion batteries and flexible sodium-ion capacitors. J. Power. Sources **363**, 284–290 (2017)
51. Z. Liu et al., Confined seeds derived sodium titanate/graphene composite with synergistic storage ability toward high performance sodium ion capacitors. Chem. Eng. J. **379**, 122418 (2020)
52. J. Chen et al., Improving the sodiation performance of $Na_2Ti_3O_7$ through Nb-doping. Electrochim. Acta **224**, 446–451 (2017)
53. H.-Y. Wang et al., Long life anode material sodium titanate synthesized by a moderate method. Mater. Lett. **186**, 326–329 (2017)
54. X. Li et al., Self-supporting lithium titanate nanorod/carbon nanotube/reduced graphene oxide flexible electrode for high performance hybrid lithium-ion capacitor. J. Alloy. Compd. **790**, 1157–1166 (2019)
55. B.F. Cottam et al., Accelerated synthesis of titanium oxide nanostructures using microfluidic chips. Lab Chip **7**, 167–169 (2007)
56. S. Luo et al., Enhanced ionic/electronic transport in nano-TiO_2/sheared CNT composite electrode for Na+ insertion-based hybrid ion-capacitors. Adv. Func. Mater. **30**, 1908309 (2020)

57. Y. Liu, S.P. Jiang, Z. Shao, Intercalation pseudocapacitance in electrochemical energy storage: recent advances in fundamental understanding and materials development. Mater. Today Adv. **7**, 100072–100103 (2020)
58. J. Zhang et al., High intercalation pseudocapacitance of free-standing T-Nb_2O_5 Nanowires@carbon cloth hybrid supercapacitor electrodes. J. Electrochem. Soc. **164**(4), A820 (2017)
59. G.G. Gorokh, A.N. Pligovka, A.A. Lozovenko, Columnar niobium oxide nanostructures: mechanism of formation, microstructure, and electrophysical properties. Tech. Phys. **64**(11), 1657–1665 (2019)
60. S. Hemmati et al., 3D N-doped hybrid architectures assembled from 0D T-Nb_2O_5 embedded in carbon microtubes toward high-rate Li-ion capacitors. Nano Energy **56**, 118–126 (2019)
61. K. Hernández-Burgos, Z.J. Barton, J. Rodríguez-López, Finding harmony between ions and electrons: new tools and concepts for emerging energy storage materials. Chem. Mater. **29**(21), 8918–8931 (2017)
62. L. She et al., Nb_2O_5 nanoparticles anchored on an N-doped graphene hybrid anode for a sodium-ion capacitor with high energy density. ACS Omega **3**, 15943–15951 (2018)
63. J. Ding et al., Review of hybrid ion capacitors: from aqueous to lithium to sodium. Chem. Rev. **118**(14), 6457–6498 (2018)
64. C. Julien et al., *Lithium Batteries: Science and Technology* 1–619 (2016)
65. H. Wang et al., Nonaqueous hybrid lithium-ion and sodium-ion capacitors. Adv. Mater. **29**(46) (2017)
66. J. Zhao et al., The application of nanostructured transition metal sulfides as anodes for lithium ion batteries. J. Energy Chem. **27**(6), 1536–1554 (2018)
67. N. Nitta et al., Li-ion battery materials: present and future. Mater. Today **18**(5), 252–264 (2015)
68. D. Yang et al., Encapsulation of $NiCo_2O_4$ in nitrogen-doped reduced graphene oxide for sodium ion capacitors. J. Mater. Chem. A **6**(29), 14146–14154 (2018)
69. J. Wang et al., Freestanding MnO_2 composite electrode via an in situ growth method for asymmetric sodium-ion capacitor and hybrid capacitive electrodialysis. J. Solid State Electrochem. **26** (2022)
70. J. Zhao et al., Nickel cobalt oxide nanowires-modified hollow carbon tubular bundles for high-performance sodium-ion hybrid capacitors. Int. J. Energy Res. **44**(5), 3883–3892 (2020)
71. T. Wei, G. Yang, C. Wang, Iso-oriented $NaTi_2(PO_4)_3$ mesocrystals as anode material for high-energy and long-durability sodium-ion capacitor. ACS Appl. Mater. Interfaces **9**(37), 31861–31870 (2017)
72. C. Zhan et al., High-performance sodium-ion hybrid capacitors based on an interlayer-expanded MoS2/rGO composite: surpassing the performance of lithium-ion capacitors in a uniform system. NPG Asia Mater. **10**(8), 775–787 (2018)
73. X. Tang et al., A novel lithium-ion hybrid capacitor based on an aerogel-like MXene wrapped Fe_2O_3 nanosphere anode and a 3D nitrogen sulphur dual-doped porous carbon cathode. Mater. Chem. Front. **2**(10), 1811–1821 (2018)
74. X. Zhang et al., Tetrabutylammonium-intercalated 1T-MoS2 nanosheets with expanded interlayer spacing vertically coupled on 2D delaminated MXene for High-performance lithium-ion capacitors. Adv. Funct. Mater. **31** (2021)
75. C.-F. Cheng et al., A high-performance lithium-ion capacitor with carbonized $NiCo_2O_4$ anode and vertically-aligned carbon nanoflakes cathode. Energy Storage Mater. **22**, 265–274 (2019)
76. S. Li et al., A high-performance lithium-ion capacitor based on 2D nanosheet materials. Small **13**(6) (2017)
77. C. Liu et al., Mesocrystal MnO cubes as anode for Li-ion capacitors. Nano Energy **22**, 290–300 (2016)
78. D. Han et al., Controlling MoO2 and MoO3 phases in MoOx/CNTs nanocomposites and their application to anode materials for lithium-ion batteries and capacitors. Electrochim. Acta **388**, 138635 (2021)

79. M. Li et al., Designed construction of a graphene and iron oxide freestanding electrode with enhanced flexible energy-storage performance. ACS Appl. Mater. Interfaces **8**(11), 6972–6981 (2016)
80. Y. Huang et al., Tetragonal MF2 (M=Ni, Co) micro/nanocrystals anodes for lithium/sodium-ion capacitors. Electrochim. Acta **329**, 135138 (2020)
81. H. Zhang et al., Advanced materials for sodium-ion capacitors with superior energy-power properties: progress and perspectives. Small **16**(15), 1902843 (2020)
82. B. Zhuang et al., Mesoporous carbon film inlaid with $Li_3V_2(PO_4)_3$ nanoclusters through delaying sol-gel method for high performance lithium-ion hybrid supercapacitors. Electrochim. Acta **283**, 1589–1599 (2018)
83. R. Satish et al., Carbon-coated $Li_3V_2(PO_4)_3$ as insertion type electrode for lithium-ion hybrid electrochemical capacitors: an evaluation of anode and cathodic performance. J. Power. Sources **281**, 310–317 (2015)
84. B. Zhuang et al., High-performance lithium-ion supercapatteries constructed using $Li_3V_2(PO_4)_3$/C mesoporous nanosheets. ChemistrySelect **4**(33), 9822–9828 (2019)
85. R. Thangavel et al., Going beyond lithium hybrid capacitors: proposing a new high-performing sodium hybrid capacitor system for next-generation hybrid vehicles made with bio-inspired activated carbon. Adv. Energy Mater. **6**(7), 1502199 (2016)
86. R. Thangavel et al., Pushing the energy output and cyclability of sodium hybrid capacitors at high power to new limits. Adv. Energy Mater. **7**(14), 1602654 (2017)
87. M. Wu et al., NASICON-structured NaTi(2)(PO(4))(3) for sustainable energy storage. Nanomicro Lett **11**(1), 44 (2019)
88. H.-K. Roh et al., A chemically bonded $NaTi_2(PO_4)_3$/rGO microsphere composite as a high-rate insertion anode for sodium-ion capacitors. J. Mater. Chem. A **5**(33), 17506–17516 (2017)
89. D. Wang et al., Intercalation pseudocapacitance in a NASICON-structured $Na_2CrTi(PO_4)_3$@carbon nanocomposite: towards high-rate and long-lifespan sodium-ion-based energy storage. J. Mater. Chem. A **7**(36), 20604–20613 (2019)
90. N. Schirmer, A. Balducci, On the use of lithium vanadium phosphate in high power devices. J. Power. Sources **235**, 265–273 (2013)
91. Q. Yang et al., Porous single-crystal $NaTi_2(PO_4)_3$ via liquid transformation of TiO_2 nanosheets for flexible aqueous Na-ion capacitor. Nano Energy **50**, 623–631 (2018)
92. P. Wei et al., Porous $NaTi_2(PO_4)$3/C hierarchical nanofibers for ultrafast electrochemical energy storage. ACS Appl. Mater. Interfaces **10**(32), 27039–27046 (2018)
93. S. Zhang et al., Development and characterization of aqueous sodium-ion hybrid supercapacitor based on $NaTi_2(PO_4)_3$//activated carbon. J. Alloy. Compd. **729**, 850–857 (2017)
94. D.M. Teffu et al., High-performance supercabattery based on reduced graphene oxide/metal organic framework nanocomposite decorated with palladium nanoparticles. Electrochim. Acta **412**, 140136 (2022)
95. Z. Cao et al., Metal-organic framework materials for electrochemical supercapacitors. Nano-Micro Letters **14**(1), 181 (2022)
96. V. Anh Tran et al., Metal-organic framework for lithium and sodium-ion batteries: progress and perspectivez. Fuel **319**, 123856 (2022)
97. W.-H. Li et al., Conductive MOFs. EnergyChem **2**(2), 100029 (2020)
98. J. Sun, L. Guo, X. Sun, J. Zhang, Y. Liu, L. Hou, C. Yuan, Conductive Co-based metal–organic framework nanowires: a competitive high-rate anode towards advanced Li-ion capacitors. J. Mater. Chem. A **7**(43), 24788–24791 (2019). https://doi.org/10.1039/C9TA08788E
99. Z. Liang et al., Pristine metal-organic frameworks and their composites for energy storage and conversion. Adv. Mater. **30**(37), 1702891 (2018)
100. D. Liu et al., Recent advances in MOF-derived carbon-based nanomaterials for environmental applications in adsorption and catalytic degradation. Chem. Eng. J. **427**, 131503 (2022)
101. X. Li et al., Magnetic cobalt nanoparticles embedded in hierarchically porous nitrogen-doped carbon frameworks for highly efficient and well-recyclable catalysis. J. Mater. Chem. A **4**(19), 7476–7482 (2016)

102. A. Dimas Chandra Permana et al., MOF-derived onion-like carbon with superior surface area and porosity for high performance lithium-ion capacitors. Batteries Supercaps **5**(5), e202100353 (2022)
103. D.P. Dubal et al., Metal-organic framework (MOF) derived electrodes with robust and fast lithium storage for li-ion hybrid capacitors. Adv. Func. Mater. **29**(19), 1900532 (2019)
104. J. Yan et al., Flexible MXene/graphene films for ultrafast supercapacitors with outstanding volumetric capacitance. Adv. Func. Mater. **27**, 1701264 (2017)
105. Y. Xia et al., Boosting ion/e—transfer of Ti_3C_2 via interlayered and interfacial co-modification for high-performance Li-ion capacitors. Chem. Eng. J. **404**, 127116 (2021)
106. H. Yu et al., Surface modified MXene-based nanocomposites for electrochemical energy conversion and storage. Small **15**, 1901503 (2019)
107. S. Yang et al., Fluoride-free synthesis of two-dimensional titanium carbide (MXene) using a binary aqueous system. Angew. Chem. Int. Ed. **57**(47), 15491–15495 (2018)
108. Y. Wu et al., 2D MXene/SnS2 composites as high-performance anodes for sodium ion batteries. Chem. Eng. J. **334**, 932–938 (2018)
109. Q. Meng et al., Theoretical investigation of zirconium carbide MXenes as prospective high capacity anode materials for Na-ion batteries. J. Mater. Chem. A **6**(28), 13652–13660 (2018)
110. J. Zhou et al., A two-dimensional zirconium carbide by selective etching of Al_3C_3 from nanolaminated $Zr_3Al_3C_5$. Angew. Chem. Int. Ed. **55**(16), 5008–5013 (2016)
111. J. Wang et al., FeNi nanoparticles on Mo2TiC2Tx MXene@nickel foam as robust electrocatalysts for overall water splitting. Nano Res. **14**(10), 3474–3481 (2021)
112. O. Mashtalir et al., Amine-assisted delamination of Nb2C MXene for li-ion energy storage devices. Adv. Mater. **27**(23), 3501–3506 (2015)
113. A. Byeon et al., Lithium-ion capacitors with 2D Nb2CTx (MXene)—carbon nanotube electrodes. J. Power. Sources **326**, 686–694 (2016)
114. S. Yi et al., Cationic intermediates assisted self-assembly two-dimensional Ti(3)C(2)T(x)/ rGO hybrid nanoflakes for advanced lithium-ion capacitors. Sci Bull (Beijing) **66**(9), 914–924 (2021)
115. S. Xu et al., Binder-free Ti_3C_2Tx MXene electrode film for supercapacitor produced by electrophoretic deposition method. Chem. Eng. J. **317**, 1026–1036 (2017)
116. J. Zhu et al., S-functionalized MXenes as electrode materials for Li-ion batteries. Appl. Mater. Today **5**, 19–24 (2016)
117. J. Luo et al., Tunable pseudocapacitance storage of MXene by cation pillaring for high performance sodium-ion capacitors. J. Mater. Chem. A **6**(17), 7794–7806 (2018)
118. X. Wang et al., Pseudocapacitance of MXene nanosheets for high-power sodium-ion hybrid capacitors. Nat. Commun. **6**(1), 6544 (2015)
119. X. Xie et al., Porous heterostructured MXene/carbon nanotube composite paper with high volumetric capacity for sodium-based energy storage devices. Nano Energy **26**, 513–523 (2016)
120. N. Kurra et al., Bistacked titanium carbide (MXene) anodes for hybrid sodium-ion capacitors. ACS Energy Lett. **3**(9), 2094–2100 (2018)
121. S. Kajiyama et al., Enhanced li-ion accessibility in mxene titanium carbide by steric chloride termination. Adv. Energy Mater. **7**(9), 1601873 (2017)
122. Y. Dall'Agnese et al., Two-dimensional vanadium carbide (MXene) as positive electrode for sodium-ion capacitors. J. Phys. Chem. Lett. **6**(12), 2305–2309 (2015)
123. P. Yu et al., Binder-free 2D titanium carbide (MXene)/carbon nanotube composites for high-performance lithium-ion capacitors. Nanoscale **10**(13), 5906–5913 (2018)

Chapter 10
Computational Modeling of Nanomaterials for Advanced Supercapacitors

Sheng Bi

Abstract Computational modeling has become essential for understanding and optimizing the nanoscale processes that are critical to supercapacitor performance. This review highlights recent molecular simulations of advanced nanomaterials for supercapacitors, emphasizing nanoporous carbons, conductive metal-organic frameworks (cMOFs), and two-dimensional layered materials such as MXenes and transition metal dichalcogenides. Key findings include the discovery that anomalous capacitance enhancement in nanoporous carbons arises from optimal ion confinement and structural disorder, as demonstrated through constant potential molecular dynamics. For cMOFs, high-throughput density functional theory screening has identified frameworks with high electronic conductivity, while molecular simulations have clarified the relationship between pore structure, ionic mobility, and interfacial capacitance, with values in good agreement with experimental observations. Sophisticated constant potential models for heterogeneous electrodes, such as constant chemical potential method, revealed unique charge-storage mechanisms in layered 2D materials, including dominant co-ion desorption in metallic 1T-MoS_2. The review also identifies challenges, including limitations in accurately modeling redox reactions, scalability of quantum-mechanical methods, and the need for diverse machine-learning training datasets. Overcoming these challenges requires developing multiscale modeling and continued synergy between simulation and experiment, paving the way for the rational design of next-generation supercapacitor materials.

Keywords Supercapacitors · Molecular simulations · Nanomaterials · Porous materials · 2D Materials · Double layer · Machine-learning

S. Bi (✉)
College of Materials, Xiamen University, Xiamen, Fujian, China
e-mail: sheng.bi@xmu.edu.cn

© The Author(s), under exclusive license to Springer Nature Switzerland AG 2026

K. Raju et al. (eds.), *Hybrid Nanostructures for Advanced Supercapacitors*, Modern Aspects of Electrochemistry 63, https://doi.org/10.1007/978-3-032-04593-5_10

10.1 Introduction

Supercapacitors are distinguished by their high power density, reliability, long cycle life, and inherent safety, complementing the performance of batteries. They are commonly classified into two main categories: *electrical double-layer capacitors (EDLCs)* and *pseudocapacitors* [1]. The key distinction between these categories lies in their charge-storage mechanisms: while EDLCs store charge via non-Faradaic ion adsorption at the electrode surface, pseudocapacitors depend on Faradaic (redox) reactions at the electrode-electrolyte interface [2]. As both processes occur at the electrode-electrolyte boundary, this nanometer-thick interface emerges as the pivotal point for supercapacitor performance.

From a materials perspective, achieving higher energy and power densities in supercapacitors hinges on optimizing both electrodes and electrolytes. Yet, these two components have traditionally been developed with limited cross-consideration. For instance, research on nanoporous activated carbons has often focused on increasing the specific surface area (SSA) and improving electrical conductivity, whereas electrolyte development has centered on expanding the electrochemical stability window and enhancing ion transport properties. These examples highlight the largely independent paths that electrode and electrolyte research have taken, with less attention given to their interplay.

However, in recent years, two fascinating opportunities have emerged that broaden the design space for supercapacitors. The first revolves around the advent of novel nanomaterials, including two-dimensional (2D) compounds (e.g., MXenes [3] and transition metal dichalcogenides [4]) as well as porous crystalline frameworks (e.g., metal-organic frameworks (MOFs) [5] and covalent organic frameworks (COFs) [6]). These materials often feature highly tunable surface properties and ordered nanoporous architectures, and many exhibit a blend of non-Faradaic and Faradaic (pseudocapacitive) charge-storage mechanisms [7]. Indeed, their multifaceted charge-storage behavior has prompted efforts to unify different underlying models [1, 8].

The second opportunity arises from recent advances in experimental and modeling techniques, which now permit in situ or atomistic-level investigation of charge-storage phenomena. A landmark discovery in this realm was the unexpected increase in capacitance when pore sizes closely match electrolyte-ion dimensions [9], spurring both theoretical [10, 11] and experimental [12, 13] investigations aimed at elucidating the fundamental interactions governing supercapacitor performance.

Despite the remarkable progress in in situ and operando techniques, such as X-ray scattering, advanced electron microscopy, and nuclear magnetic resonance, experimental approaches still face notable constraints when probing the fast and complex processes occurring at electrode-electrolyte interfaces. Limited spatial and temporal resolutions, difficulties in replicating extreme operating conditions, and challenges in directly observing ion transport or interfacial reactions all hinder a comprehensive understanding of charge-storage mechanisms. In this context, computational modeling, particularly molecular simulations, has emerged as a powerful tool for bridging these gaps [14–16]. By offering atomistic insights that are often inaccessible experimentally, simulations can shed light on ion adsorption, structural rearrangements, and the subtleties of electron transfer. Consequently, molecular modeling not only complements experimental findings but also propels supercapacitor research toward more predictive design strategies.

Advances in force-field development, quantum-based methods, and machine-learning accelerations are enabling researchers to tackle the intricate chemical and structural complexity of emerging electrode materials in unprecedented detail. High-throughput screening workflows with automated simulation pipelines, allowing for the rapid assessment of vast material libraries. This paradigm shift not only accelerates the discovery of innovative electrode and electrolyte candidates but also fosters a more predictive approach to materials design. Consequently, supercapacitor research is being reshaped by data-driven modeling, in which targeted experimental validation can focus on the most promising computational leads.

In the following, we focus on computational modeling of advanced supercapacitors, with a particular emphasis on those with cutting-edge electrode nanomaterials. First, we present a spectrum of simulation methods, including continuum theory, molecular-mechanical methods, quantum-mechanical methods and machine-learning-augmented frameworks. We also discuss the recent methodological advances, including improved electrode models, multiscale simulation strategies, and machine-learning-based methods. By highlighting specific studies on porous carbons, conductive metal-organic frameworks (cMOFs), and 2D layered electrodes such as MoS_2, and MXenes, we illustrate how accurate modeling can reveal structure-property relationships, especially in systems where charge transfer, redox reactions, and interfacial processes are paramount. Taken together, this chapter underscores the fast-evolving landscape of computational research in supercapacitors and its indispensable role in unraveling the mechanisms that will define the next generation of energy-storage technologies (Fig. 10.1).

10.2 Theory

In this section, we provide a brief overview of the primary approaches used to model supercapacitors. In the first subsection, we begin with continuum-based methods. Despite their limited atomistic fidelity, some of these earliest yet still evolving models remain cost-effective and readily integrable into larger-scale device simulations. We then examine atomistic modeling, including molecular mechanics techniques (classical Molecular Dynamics, Monte Carlo) and quantum-mechanical methods such as density functional theory (DFT) and *ab initio* molecular dynamics (AIMD). These advanced methods provide crucial molecular insights that go beyond purely continuum treatments [18]. Cutting-edge methods like hybrid approaches and machine-learning approaches are also discussed, which offer new methods to tackle the multiscale complexity in the simulation of supercapacitors. A comparative summary of these modeling methods, highlighting their scaling behavior, accessible system sizes,

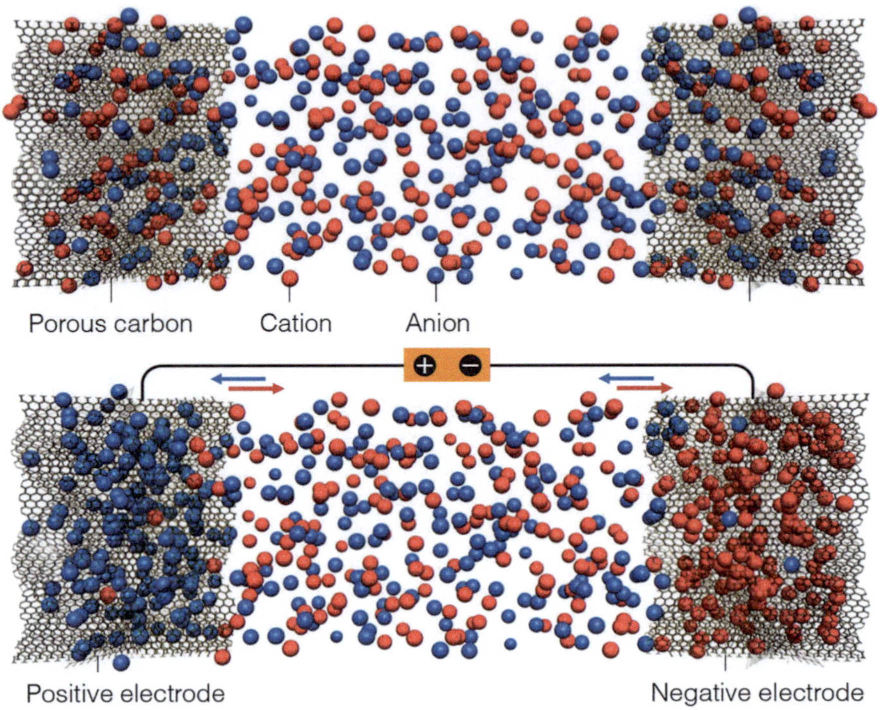

Fig. 10.1 Typical molecular modeling of a supercapacitor. Under null polarization, ions from the electrolyte reside either within the pores of the electrodes or in the bulk region (*top panel*). When an external voltage is applied, the ions reorganize, with cations accumulating near the negatively charged electrode, and anions gathering near the positively charged electrode (bottom panel). Adapted with permission from [17]

and inherent limitations, is presented in Table 10.1 and Fig. 10.2. The subsequent subsection delves into these techniques in greater detail, highlighting how each contributes to our understanding of charge-storage mechanism in supercapacitors across varying spatial and temporal scales.

10.2.1 Theoretical Frameworks

10.2.1.1 Continuum Theory

Continuum theory provides a macroscopic framework for modeling electrified solid/liquid interfaces, efficiently describing ion distributions, electrostatics, and

Table 10.1 Comparison among different modeling methods

Method	Scaling order	System size	Time scale	Limitation
Continuum model	$O(N)$	Macroscopic	μs–ms	No atomistic detail
DFTMD	$O(N_e^3)$	100 to 1000	10–100 ps	Time/length scale
Joint DFT	$O(N_{elec}^3)$	100 to 1000	–	Solvation model
QM/MM	$O(N_{elec}^3)$	1,000 to 10,000	10–100 ps	Selection of reactive region
cMD	$O(N \log N)$	>10,000	10–100 ns	Non-reactive
MLMD	$O(N \log N)$	>10,000	1–10 ns	Descriptor and training data

For continuum models, N refers to the number of discretization points; for other methods, N_e, N_{elec}, and N respectively denote the total number of electrons, electrons in the electrode, and total atoms

Adapted with permission from [16]. DFT: Density-functional-theory; DFTMD: DFT-based molecular dynamics; QM/MM: quantum mechanics/molecular mechanics; cMD: classical molecular dynamics; MLMD: machine-learning-based molecular dynamics

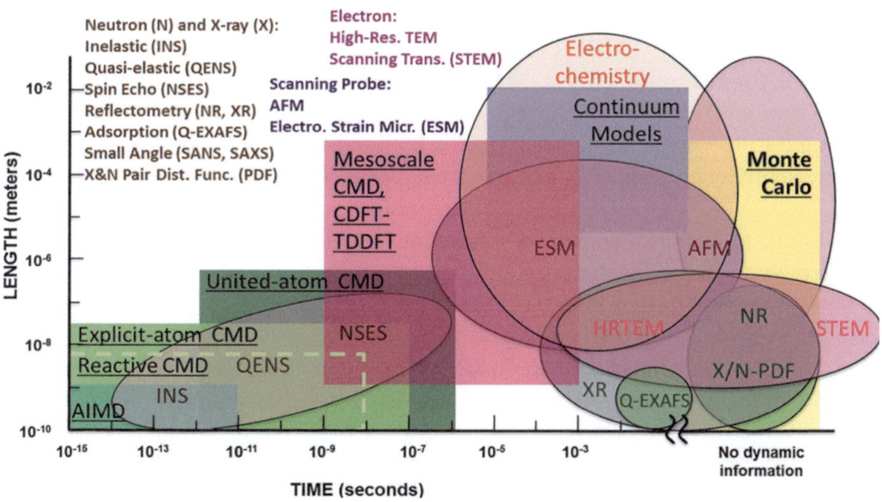

Fig. 10.2 Comparison of length and time scales accessible by various simulation and experimental techniques. CMD: classical molecular dynamics; AIMD: *ab initio* molecular dynamics; CDFT: classical density functional theory; TDDFT: time-dependent density functional theory. The latter two methods are not discussed in this chapter, as we focus specifically on molecular simulation techniques. Adapted with permission from [19]

charge transport without atomic scale resolution. Early models, such as the Poisson–Boltzmann (PB) equation (Gouy 1910; Chapman 1913), capture equilibrium ion distributions, while the Gouy–Chapman–Stern (GCS) model (Stern 1924) accounts for ion adsorption and finite ion size. For dynamic processes, the Poisson–Nernst–Planck (PNP) equations extend these principles to charge transport under applied fields.

Beyond electrostatic models, electrical circuit models, including resistor–capacitor (RC) and constant phase element (CPE) representations, provide simplified descriptions of supercapacitor behavior, widely used in experimental analysis. While continuum models lack atomistic detail, they remain essential for large-scale simulations and device-level studies, complementing molecular-scale approaches.

10.2.1.2 Molecular Mechanics and Quantum Mechanics

In contrast to continuum models, which describe supercapacitors at a macroscopic level, atomistic modeling provides a more detailed perspective by explicitly representing atomic interactions and molecular structures. Two primary approaches dominate this field: molecular mechanics (MM) and quantum mechanics (QM), each offering distinct advantages in terms of accuracy, system size, and computational efficiency. MM simulations rely on empirical force fields, such as OPLS [20], CHARMM [21], AMBER [22] to model atomic interactions without explicitly accounting for electronic effects. MM can handle relatively large-scale simulations ($>$10,000 atoms) over long timescales ($>$10 ns), making it particularly valuable for studying the electrical double layer and ion dynamics in supercapacitors [15]. However, it usually cannot capture charge transfer, oxidation-state changes, or chemical reactions.

In contrast, quantum mechanics simulations explicitly solve for electronic structure, providing deeper insights into charge redistribution and polarization effects. Methods such as Hartree-Fock [23], Density Functional Theory (DFT) [24], and wavefunction-based approaches allow for a fundamental understanding of electrode materials and interfacial electrochemistry. Quantum mechanics simulations are crucial for studying redox reactions and charge-storage mechanisms at the electrode-electrolyte interface. However, they are often computationally expensive, limiting their application to small system sizes ($\sim$100 atoms) and short timescales ($\sim$10 ps). Moreover, applying an electric potential constraint in quantum mechanics methods remains a challenge, despite recent advances in potential-dependent DFT formulations and finite-field approaches [25, 26].

10.2.1.3 Hybrid Approaches and Machine-Learning Acceleration

As discussed, quantum-mechanical methods, such as Density Functional Theory, can offer deep insights into charge transfer, redox reactions, and interfacial electrochemistry. However, their high computational cost limits their application to small system sizes and short timescales. On the other hand, molecular mechanics methods are computationally efficient and well-suited for large-scale simulations, but they lack explicit electronic structure information. Given these trade-offs, a natural strategy is to combine the strengths of both, leveraging the efficiency of MM while preserving essential electronic effects where necessary.

A well-established approach to this strategy is QM/MM, where the chemically active region (e.g., the electrode surface) is treated at the quantum mechanics level, while the surrounding electrolyte and other less critical regions are modeled classically [27]. This partitioning allows for accurate treatment of interfacial charge transfer and reaction mechanisms while significantly reducing computational cost. Another method, joint DFT, integrates an implicit classical description of the electrolyte within a DFT framework, thus improving computational efficiency while maintaining the precision required for interfacial electrochemistry [28]. These hybrid approaches have been widely used to study charge storage and transport in supercapacitors, particularly when oxidation-state changes and strong electrode-electrolyte interactions are involved.

In recent years, machine-learning (ML) acceleration has emerged as a powerful alternative to traditional QM/MM methods [29, 30]. ML models trained on high-accuracy QM reference data can learn complex many-body interactions, enabling near-DFT accuracy at a fraction of the computational cost. Unlike fixed classical force fields, machine-learning-based potentials can dynamically capture electronic polarization and charge-transfer effects, allowing for more flexible and accurate simulations of electrode-electrolyte interfaces. Machine-learning methods also facilitate large-scale simulations by predicting local charges, charge-response kernels, and ion transport behavior with high fidelity.

10.2.2 Simulation Techniques

10.2.2.1 Continuum Models

In the most macroscopic description, the interface between an electrode and an electrolyte can be treated as a continuous medium, where charged species are characterized by bulk properties such as permittivity, ion concentrations, and activity coefficients rather than explicit atomistic details. Within this framework, the formation of an electrical double layer emerges naturally from the balance of electrostatic forces, thermal motion, and ion size effects.

A seminal approach is the Gouy–Chapman (GC) model, originally formulated in the early 20th century by Gouy (1910) and Chapman (1913). It treats ions as point charges within a mean-field Poisson–Boltzmann (PB) framework, whereby the surface potential and ionic distributions are solved self-consistently. This approach yields key parameters such as the Debye screening length, which characterizes how quickly the electric field diminishes into the bulk electrolyte. Recognizing that real ions have finite size and can adsorb onto the electrode, Stern refined this theory in 1924 by dividing the electrical double layer (EDL) into a compact layer adjacent to the surface (the Stern layer) and a diffuse layer farther into the solution. This integrated Gouy–Chapman–Stern (GCS) model remains one of the most influential theoretical constructs in describing the EDL, and it continues to serve as a foundational basis for modern treatments of electrode–electrolyte interfaces.

While GCS theory captures many experimental trends, it can be insufficient for highly concentrated electrolytes such as ionic liquids and water-in-salt systems, where strong ion–ion correlations and excluded-volume effects become pronounced. In these regimes, extended continuum theories—notably those pioneered by Kornyshev and co-workers—introduce lattice-based or modified Poisson–Boltzmann formalisms to account for ionic packing and complex correlation phenomena [31, 32]. These refinements have shown success in describing voltage-dependent capacitance and ion layering in supercapacitor applications, offering valuable insights even though explicit molecular details are intentionally coarse-grained. Ultimately, continuum models remain powerful for obtaining quick, system-wide predictions of interface behavior and serve as useful boundary conditions or initial guesses for more detailed molecular-level simulations.

10.2.2.2 Classical Molecular Simulations

Classical molecular simulations, including molecular dynamics (MD) and Monte Carlo (MC), are widely employed to explore the vast configurational space of supercapacitors. Typically conducted in the canonical (NVT) ensemble, Monte Carlo can also operate in the grand canonical (μVT) ensemble to model ion exchange phenomena. While molecular dynamics deterministically integrates Newton's equations of motion, Monte Carlo uses stochastic moves accepted or rejected based on Boltzmann criteria. Both approaches have proven indispensable for studying the structure and dynamics of EDLCs with carbon-based electrodes [15].

The interactions among atoms in these simulations are described by empirical force fields, which do not explicitly treat electronic degrees of freedom and thus define the simulations as *classical*. Underpinning such simulations are classical force fields, which generally ignore explicit electronic polarization by fixing partial charges. In their simplest form, these force fields use fixed partial charges. More advanced variants incorporate polarization either via induced dipoles [33] or through reactive frameworks such as ReaxFF [34], enabling bond formation and breaking. Specialized force fields tailored to ionic liquids (e.g., CL&P) [35] and coarse-grained approaches further extend the scope to larger, more complex systems. Beyond these, polarizable and machine-learning-based force fields (see Sect. 10.2.2.5) enhance both accuracy and flexibility, making classical simulations an ever more powerful tool for uncovering the atomistic details of EC performance.

Constant Potential Method

Modeling a metallic electrode is crucial for accurately describing electrified interfaces, since the free electrons within a metal instantaneously screen any changing ionic configuration in the electrolyte. The constant potential method (CPM) addresses this need by enforcing a fixed electrode potential, V_{elec}, at each simulation timestep

[36, 37]. In contrast to constant-charge methods that hold the total electrode charge static, CPM dynamically adjusts the atomic charges on the electrode to maintain the same local potential on all electrode atoms (Fig. 10.3).

Consider a system composed of N atoms, where the first M atoms belong to the electrode, and the remaining $N - M$ atoms belong to the electrolyte. In fluctuating charge methods, each electrode atom i is assigned a fluctuating charge q_i, which dynamically responds to changes in electrolyte configuration. In contrast, the electrolyte atoms possess fixed charges. The electrostatic energy of the system can be expressed in a quadratic form as:

$$U_{\mathrm{el}}(\mathbf{r}^N, \mathbf{q}) = \frac{1}{2}\mathbf{q}^{\mathrm{T}}\mathbf{A}\mathbf{q} - \mathbf{q}^{\mathrm{T}}\mathbf{B}(\mathbf{r}^N) + C(\mathbf{r}^N)$$

Here, $\mathbf{r}^N$ denotes the set of coordinates of all atoms, the symmetric matrix $\mathbf{A}$ encodes the electrostatic interactions between electrode atoms, including both self and mutual Coulombic interactions. The vector $\mathbf{B}(\mathbf{r}^N)$ represents the electrostatic potentials induced by the electrolyte at each electrode atom, which explicitly couples electrode charges to electrolyte configurations. The scalar $C(\mathbf{r}^N)$ accounts for the electrostatic interactions within the electrolyte itself.

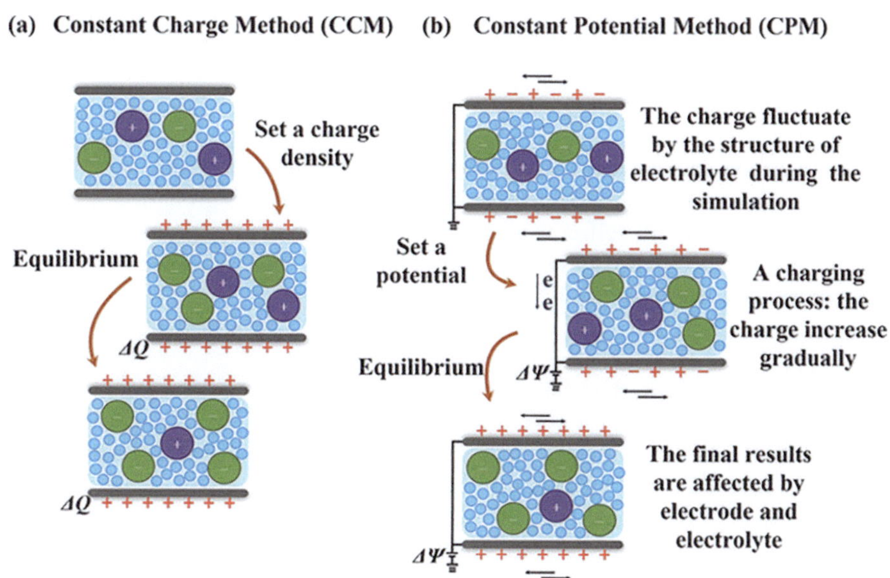

Fig. 10.3 The evolution of electrode charges and electrolyte structures with time by using **a** constant-charge method and **b** constant potential method. Adapted with permission from [38]

In the constant potential method, the charges q_i at electrode atoms are determined by minimizing this electrostatic energy subject to the constraint of maintaining each electrode atom at a fixed potential V_{elec}. This yields a set of linear equations:

$$\frac{\partial U_{\text{el}}(\mathbf{r}^N, \mathbf{q})}{\partial \mathbf{q}^T} = \mathbf{A}\mathbf{q} - \mathbf{B}(\mathbf{r}^N) = \mathbf{V}_{\text{elec}},$$

where solving for $\mathbf{q}$ gives the instantaneous charge distribution:

$$\mathbf{q} = \mathbf{A}^{-1}(\mathbf{B} + \mathbf{V}_{\text{elec}}).$$

Thus, fluctuating charge methods provide a direct and explicit link between electrode charges and electrolyte configurations at given electrode potential, enabling realistic modeling of electrode polarization and interfacial electrochemical processes.

Several open-source software packages implement the constant potential method, including METALWALLS [39], LAMMPS [40], and GROMACS [41], enabling large-scale simulations of porous electrodes. For instance, CPM has been applied to carbon-based systems [11] to reveal how electrode porosity and ion confinement jointly determine capacitance. Recently, Zeng et al. [42] compared CPM with simpler constant-charge approximations across diverse electrode/electrolyte conditions, illustrating the crucial role of dynamic polarization for accurate predictions. Furthermore, advanced variants of CPM, including constant chemical potential method (CPM_χ) [43] and heteroatomic constant potential method (HCPM) [44], can differentiate the response of distinct atomic species in multi-component electrodes, capturing effects such as dopants or heterogeneous electronegativity.

Overall, constant potential method provides a physically grounded boundary condition for metals by self-consistently matching the electrode potential to any external bias or specified voltage, thus improving the predictive power of atomistic simulations for supercapacitor interfaces.

10.2.2.3 Density Functional Theory

Although not originally devised for supercapacitors, Density Functional Theory (DFT) has become a crucial tool for understanding the electronic properties that underpin high-performance electrode and electrolyte materials. DFT calculations can elucidate band structures, surface energetics, and adsorption sites, thereby guiding the rational design of next-generation supercapacitor materials. Among various DFT formulations, the Kohn–Sham approach is the most widely adopted, as it recasts the many-electron problem into an equivalent set of non-interacting electrons moving in an effective potential. Mathematically, it involves solving single-particle orbitals:

$$\left[-\frac{\hbar^2}{2m}\nabla^2 + V_{\text{eff}}(\mathbf{r})\right]\psi_i(\mathbf{r}) \;=\; \varepsilon_i\,\psi_i(\mathbf{r}), \tag{10.1}$$

where $\psi_i(\mathbf{r})$ are Kohn–Sham orbitals, ε_i are their energies, and $V_{\text{eff}}(\mathbf{r})$ encompasses external potentials, classical Coulomb interactions, and exchange–correlation effects. Different exchange–correlation functionals—such as LDA [24], GGA [45], and meta-GGA [46]—offer various trade-offs in accuracy and computational cost.

Ab Initio Molecular Dynamics

Ab initio molecular dynamics (AIMD) is a powerful simulation technique in which interatomic forces are computed directly from quantum-mechanical electronic structure calculations rather than from empirical force fields. By integrating Newton's equations of motion with forces determined from first principles, *Ab initio* molecular dynamics can accurately model complex physical and chemical processes at the atomic scale without relying on empirical parameters. Among the different *Ab initio* molecular dynamics approaches, DFT-based molecular dynamics (DFTMD) has become particularly prominent, owing to its optimal balance between computational efficiency and predictive accuracy. DFTMD leverages density functional theory to calculate electronic structure and forces at each simulation step, enabling it to capture intricate phenomena such as ion insertion, electronic polarization, structural rearrangements, and interfacial charge transfer at electrified interfaces [47].

However, the computational expense typically restricts these simulations to smaller systems over short timescales. Recent innovations like finite-field DFTMD [48], which applies a constant electric displacement field, have helped accelerate polarization convergence, enabling direct computation of interfacial properties such as Helmholtz capacitance. In this way, DFT and DFTMD serve as powerful complements to classical models, linking electronic structure details with macroscopic electrochemical performance.

10.2.2.4 Hybrid QM/MM and Joint DFT

For large electrochemical systems, a fully quantum-mechanical treatment rapidly becomes computationally intractable, especially if one aims to capture extended solvent environments and complex pore geometries in supercapacitors. Two major strategies–QM/MM and joint DFT–have emerged to address this challenge by combining quantum detail where it matters most with more approximate treatments elsewhere.

Quantum Mechanics/Molecular Mechanics

In the quantum mechanics/molecular mechanics (QM/MM) scheme, the physically or chemically active region of the system (e.g., the electrode surface or a reactive site) is treated at the quantum mechanics level, while the surrounding solvent or less critical regions are described by a classical force field. This partitioning limits the computationally demanding quantum mechanics calculations to a fraction of the total system, thus extending the feasible size and timescale of simulations. For supercapacitor applications, QM/MM approaches can illuminate how local defects, dopants, or functional groups on the electrode surface influence charge-transfer events, while still allowing for the simulation of a realistic electrolyte environment. The main challenges involve defining a robust QM–MM boundary, avoiding artifacts such as electron leakage at the interface, and consistently representing polarization across both regions (Fig. 10.4).

Joint DFT

An alternative method is joint DFT, which employs a quantum-mechanical description of the electrode but couples the electrolyte to a continuum classical model [28, 50]. In essence, joint DFT treats the electrode's electron density *ab initio* while describing the ions and solvent through an electrostatic free-energy functional. This setup captures key interfacial properties—such as differential capacitance and local electric fields—without the overhead of simulating thousands of solvent molecules. Joint DFT has proven effective in evaluating thermodynamic properties like adsorption energies or redox potentials for pseudocapacitive electrodes, but its continuum description of the electrolyte can miss subtleties such as specific ion–electrode interactions, ion pairing, or solvent structuring near the surface.

10.2.2.5 Machine-Learning-Based Approaches

Both QM/MM and joint DFT lie between purely classical simulations and fully *ab initio* molecular dynamics, balancing accuracy with computational efficiency. They can be particularly powerful for supercapacitor studies where redox reactions,

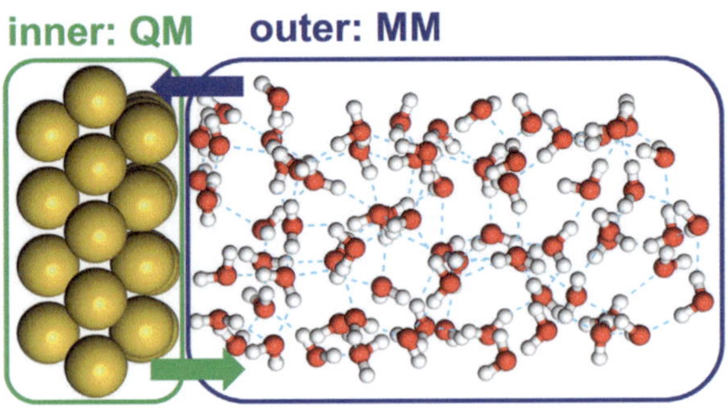

Fig. 10.4 The embedded inner quantum mechanics (QM) part (green box) and the outer molecular mechanics (MM) region (blue box) exchange forces and energies using a coupling scheme that ensures seamless interaction between the quantum and classical regions (e.g., electrostatic embedding). Adapted with permission from [49]

bond rearrangements, or complex charge distributions occur at specific sites, while the remainder of the electrolyte primarily serves as a reservoir. In recent years, however, a complementary strategy has emerged to achieve near-DFT accuracy at speeds approaching classical force fields: *machine-learning force fields*. Trained on datasets from high-level quantum mechanics calculations, these models learn complex many-body interactions that can capture bond formation, charge transfer, and polarization with greater fidelity than fixed-charge force fields [29, 30].

Machine-Learning Force Fields

A key development in the quest for fast yet accurate simulations is the advent of machine-learning force fields (MLFFs). Unlike purely classical models that rely on pre-defined functional forms and fixed charges, MLFFs learn interatomic potentials from high-level quantum-mechanical data, promising near-DFT accuracy at a fraction of the computational cost. Early *first-generation* models, such as the Behler–Parrinello neural network potential [51], relied on carefully chosen symmetry functions to encode the local atomic environment. Subsequent *second-generation* approaches, exemplified by Gaussian Approximation Potentials (GAP) [52], introduced kernel-based learning and the Smooth Overlap of Atomic Positions (SOAP) descriptor to capture many-body correlations more flexibly.

Recent MLFFs incorporate message-passing neural networks (MPNNs) [53] and advanced graph-based representations [54], often benefiting from active learning strategies that iteratively refine training datasets to cover diverse configurational space. Finally, *fourth-generation* models introduce further sophistication by including equivariant neural networks and physically inspired constraints, pushing MLFFs closer to ab initio fidelity while preserving scalability [29]. Across all generations, these force fields have demonstrated marked improvements in describing complex phenomena such as bond breaking, polarization, and charge transfer. Their emergence stands to radically expand the scope of molecular simulations for supercapacitors, from predicting electrode/electrolyte interfaces under realistic working conditions to guiding experimental efforts with rapid, high-throughput computational screening (Fig. 10.5).

Machine-learning Polarizable Electrode Models

Despite constant potential method allowing constant potential, incorporating chemical complexity and heterogeneity of electrodes remains challenging. Machine-learning models, such as those in PiNNwall [55], use neural networks (e.g., PiNet-χ, PiNet-dipole) to predict electrode response (charge-response kernels or atomic charges) in heterogeneous materials. This effectively integrates Machine-learning-based polarization with molecular dynamics engines like MetalWalls [39, 56], enabling realistic simulations of multi-element, doped, or defective electrodes.

Other emerging Machine-learning models aim to predict the electron density or electronegativity variation in electrodes as a function of applied fields, potentially allowing QM/MM-level simulations of supercapacitors at reduced cost [57, 58].

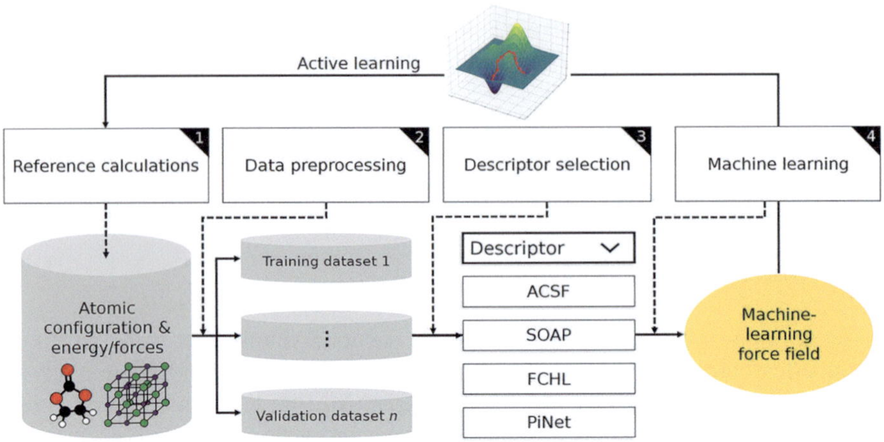

Fig. 10.5 A typical workflow for generating machine-learning force fields (MLFFs). Initial reference data are labeled and prepossessed, which usually come from *ab-inito* molecular dynamics simulations. This provides atomic configurations as well as the energy and/or forces labels. Then, before applying any machine-learning method (e.g., neural network), atomic configurations need to be encoded using a chosen descriptor (e.g., ACSFs [51], SOAP [52]. More labeled data with unique configurations as compared to the initial reference data are usually needed to produce a well-functioning MLFF and this can be done using so-called active learning schemes. Reproduced with permission from [16]

10.3 Modeling of Nanomaterials for Advanced Supercapacitors

In this section, we focus on the computational modeling of emerging nanomaterials for supercapacitors, emphasizing both their intrinsic properties and their interactions with electrolytes. We begin by discussing how these nanomaterials can be modeled at the atomistic level in isolation, shedding light on factors such as structure, electronic states, and property predictions. We then move to the more complex task of modeling the electrode–electrolyte interface, where ion adsorption, solvent structuring, and charge transfer all play critical roles.

The following subsections illustrate these approaches by examining specific types of electrode materials: porous carbons, conductive metal–organic frameworks (MOFs), transition metal dichalcogenides (TMDs), and MXenes. In each case, we summarize the computational methods and the obtained charge storage mechanisms. Note that the key challenges and modeling objectives differ across these materials, requiring specialized computational strategies to address each system's unique requirements.

10.3.1 Porous Carbons

Porous carbons have long been the primary choice in commercial supercapacitors, thanks to their high specific surface area (SSA) and favorable electrical conductivity. However, capturing the realistic morphology of these materials in simulations remains challenging: they frequently exhibit significant disorder across multiple length scales. Consequently, various computational strategies—from reverse Monte Carlo (RMC) to machine-learning-based structure generation—have been employed to better represent their amorphous nature and reconcile it with experimental scattering data.

10.3.1.1 Atomistic Models of Porous Carbons

Constructing an accurate atomistic model of porous carbons typically involves methods such as reverse Monte Carlo (RMC) or machine-learning-driven structure generation, both of which incorporate experimental data (e.g., from X-ray or neutron scattering) to refine initial atomistic configurations until key structural descriptors—such as radial distribution functions and pore-size distributions—match the measured values.

A notable example is the approach described by Thomson and Gubbins [59], who illustrated how iterative adjustment of atomic positions in disordered materials can be guided by scattering data until the simulated partial structure factors converge toward experimental values (Fig. 10.6). In their workflow, the microporous carbon structures are reconstructed by iteratively refining atomistic configurations to align with experimental data. The process begins with a disordered assembly of rigid carbon basal planes, which are then adjusted using a Monte Carlo algorithm that applies translation, rotation, or displacement moves within a 3D simulation box. These moves are accepted or rejected based on their ability to improve agreement with experimentally derived radial distribution functions (RDFs), which capture both short-range graphitic order and longer-range porous connectivity. By systematically minimizing the deviation between computed and experimental RDFs (Fig. 10.6a–b), the reverse Monte Carlo algorithm ensures that the resulting structures accurately reflect the target microporous carbons (Fig. 10.6c–d). Further validation is performed by comparing nitrogen adsorption isotherms, pore-size distributions, and surface areas with experimental data, confirming that the generated models not only exhibit realistic porosity but also capture capillary condensation effects, which are often absent in traditional slit-pore representations.

More recently, machine-learned potentials such as the Atomic Cluster Expansion (ACE) model has been proposed to generate amorphous carbon structures with controlled levels of disorder. The ACE model can systematically expand atomic interactions using complete basis functions, ensuring both high transferability and computational efficiency. In specific, the model is trained on a diverse dataset spanning graphene, graphite, diamond, fullerenes, amorphous carbons, and even liquid carbon

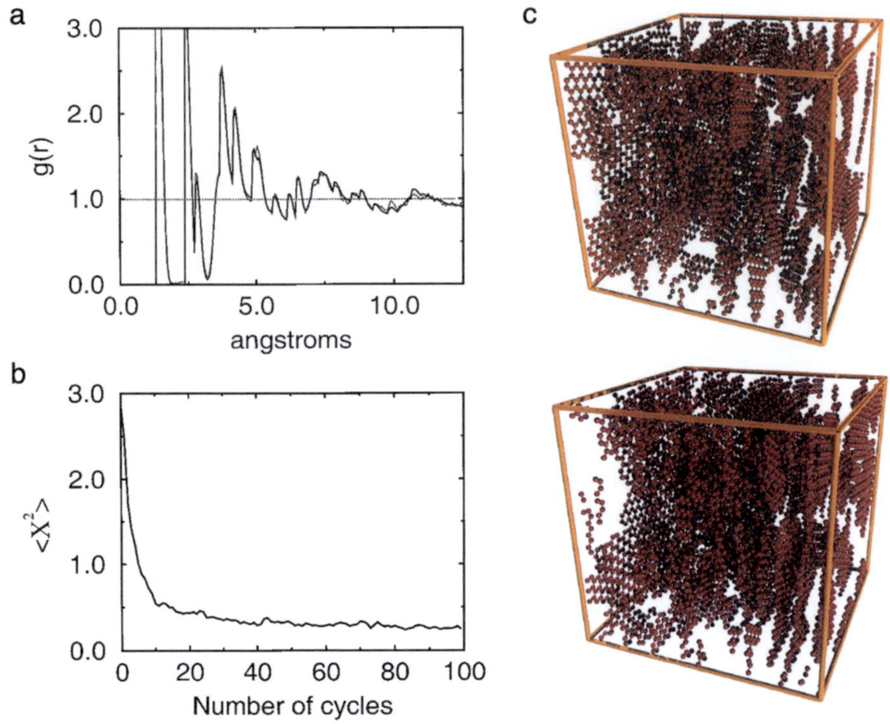

Fig. 10.6 Atomistic porous carbon model generated by reverse Monte Carlo (RMC) method. **a** C–C radial distribution function for the *target* structure (solid line) and the converged *model* structure (gray dotted line). **b** Convergence of the square deviation function χ^2 as a function of RMC simulation cycles for the generation of the *model* structure. **c** Structural representation of the structure *target* (top panel) and the converged structure model (bottom panel). Reproduced with permission from [59]

phases [60], making it well-suited for modeling porous carbons in energy storage applications (Fig. 10.7a–b). Compared to the previously discussed RMC method, ACE is trained directly on high-fidelity DFT datasets, allowing it to capture both structural disorder and electronic properties more accurately (Fig. 10.7c–d). Furthermore, unlike traditional neural network potentials, ACE employs a hierarchical polynomial expansion that improves generalization, making it robust for simulating nanoconfined carbon electrodes where quantum capacitance effects are critical.

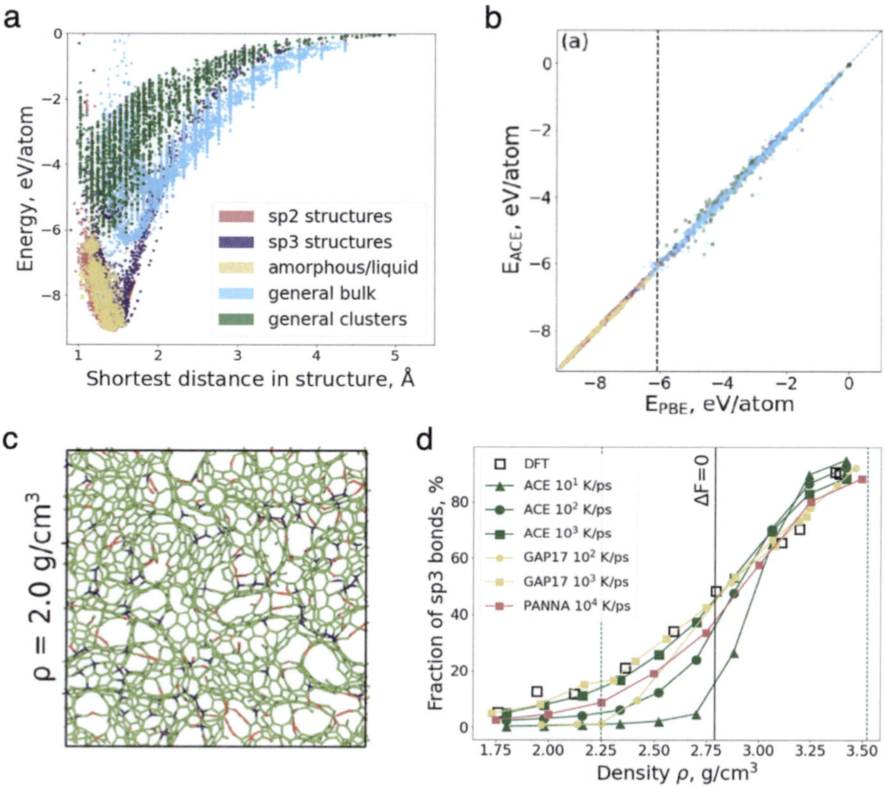

Fig. 10.7 Amorphous carbon structures generated by atomic cluster expansion (ACE) model. **a** Training dataset visualized as a scatter plot of energy per atom *w.r.t* the nearest interatomic distance with in each structure. **b** Energy and force errors for a split test data set consisting of 1912 structures. **c** Representative amorphous carbon structures at a given density using 1000 K/ps quench rate. **d** Variation of *sp*3 bond fraction in amorphous carbons with density for different quench rates and methods. Adapted with permission from [60]

10.3.1.2 Nanoconfinement and Anomalous Capacitance

Although increasing SSA has historically guided electrode design, a landmark discovery by Chmiola et al. [9] revealed that subnanometer pores can give rise to *anomalous* increases in capacitance. Recent research by Forse et al. [13], however, demonstrated a lack of direct correlation between pore size and capacitance, identifying structural disorder as the dominant factor influencing capacitance (Fig. 10.8). Nevertheless, these findings shifted the focus from mere surface area to the effects of ion confinement and local structural disorder.

For instance, Feng and Cummings [10] explored this phenomenon using molecular dynamics simulations, uncovering that capacitance exhibits an *oscillatory* dependence on pore size rather than following a monotonic increase (Fig. 10.9a–b). Their

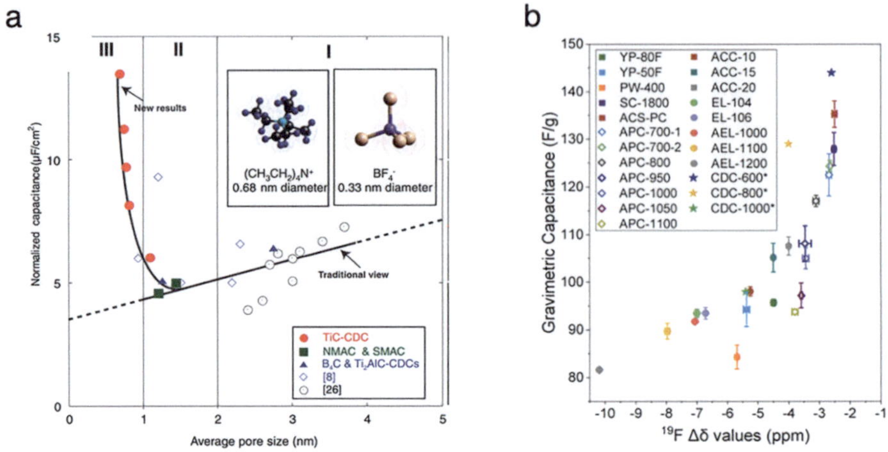

Fig. 10.8 Anomalous capacitance increase in porous carbon electrodes. **a** Correlation between areal capacitance for the carbons and average pore sizes [9]. **b** Correlation between gravimetric capacitance and ^{19}F $\Delta\delta$ values for commercial carbons and thermally annealed carbons [13]. Adapted with permission from [9, 13]

study introduced the concept of electrical double-layer interference, where ion layering within slit-shaped nanopores results from constructive or destructive overlap of EDLs originating from opposing pore walls. They observed two capacitance peaks at 0.7 nm and 1.4 nm, suggesting that optimal charge storage occurs when constructive interference amplifies the local charge density. This insight challenges the conventional wisdom that capacitance scales straightforwardly with SSA, emphasizing that pore-size distribution is crucial for designing high-performance supercapacitor electrodes.

Simulations using the constant potential method combined with realistic porous carbon models provide a highly accurate molecular-level depiction of carbon-based supercapacitors. A landmark study by Merlet et al. [11] leveraged this approach to elucidate the molecular origins of anomalous capacitance in nanoporous carbon electrodes. Unlike earlier studies that employed simplified slit-pore models, their work incorporated an atomistically realistic representation of carbide-derived carbon (CDC) structures and explicitly accounted for electrode polarization effects using constant potential method (Fig. 10.9c). The study demonstrated that the unusually high capacitance observed in subnanometer pores arises from a unique charge-storage mechanism. By tracking ion dynamics in these highly confined environments, they revealed that charge separation occurs without the traditional overscreening effects that are observed at planar interfaces, leading to more efficient storage. Their simulations achieved capacitance values ($\sim$125 F/g) in close agreement with experimental results, suggesting the importance of accounting for realistic pore morphologies and electrostatic screening effects.

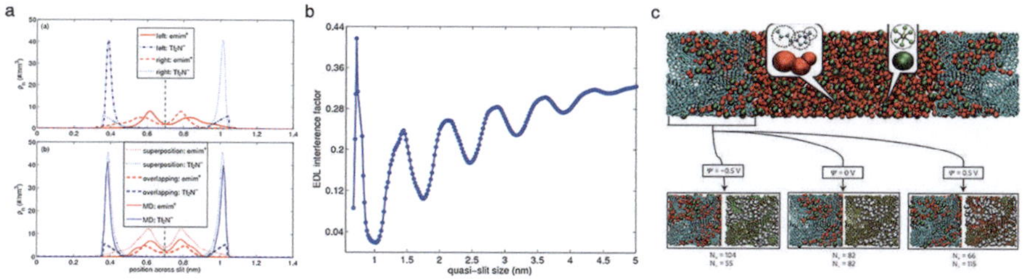

Fig. 10.9 Modeling aspect of anomalous capacitance increase in porous carbons. **a** Superposition (top panel) and over-lapping of ion number densities (bottom panel) in a quasi-slit with size of 1.4 nm [10]. **b** EDL interference factor as a function of quasi-slit size [10]. **c** The simulation cell consists of an ionic liquid electrolyte surrounded by two porous carbon electrodes held at constant electrical potentials (top panel); Structure of the ions in the electrode for various voltages (bottom panel) [11]. Reproduced with permission from [10, 11]

10.3.1.3 Partial Electronic Screening and Quantum Capacitance

Despite the relatively good electrical conductivity of carbons, perfect metallic screening may not always hold, especially in ultrathin or defect-rich regions. This can induce a quantum or incomplete screening effect, often referred to as *quantum capacitance*, which limits the total charge-storage capacity [61].

To address this issue, a Thomas–Fermi (TF) screening model has been proposed which refines the conventional constant potential simulations by incorporating finite electronic screening effects, bridging the gap between classical and quantum descriptions of electrode polarization [62]. Unlike perfect metallic electrodes, where charge redistributes instantly, the TF approach penalizes charge fluctuations via an additional energy term proportional to q^2, leading to a more gradual electrostatic potential decay inside the electrode (Fig. 10.10a). This model was applied to assess how varying degrees of electronic screening impact supercapacitor performance. Their results show that increasing the TF screening length l_{TF} (which corresponds to the strength of electronic screening) lowers the integral capacitance by allowing charge to spread into the electrode bulk (Fig. 10.10b), reducing the effective surface charge density. The altered charge distribution modifies the electric field at the interface (Fig. 10.10c), affecting ion adsorption and solvent orientation. Moreover, electrolyte dynamics differs from those observed in perfect-metal models, with changes in ion layering and water structuring near the surface. This approach provides a computationally efficient way to preserve key electronic effects of realistic electrodes.

In parallel, a recent study introduces the spillover constant potential method (sCPM), which refines the conventional CPM framework by allowing fluctuating Gaussian charge distributions that extend beyond atomic sites [63]. This modification accounts for electron spillover, where charge redistributes beyond the nuclear positions due to the delocalized nature of electronic states in metallic materials. In sCPM, the potential constraint is not imposed at the nuclei but rather at the center of mass of the fluctuating charge distributions (Fig. 10.11a), ensuring a more accurate

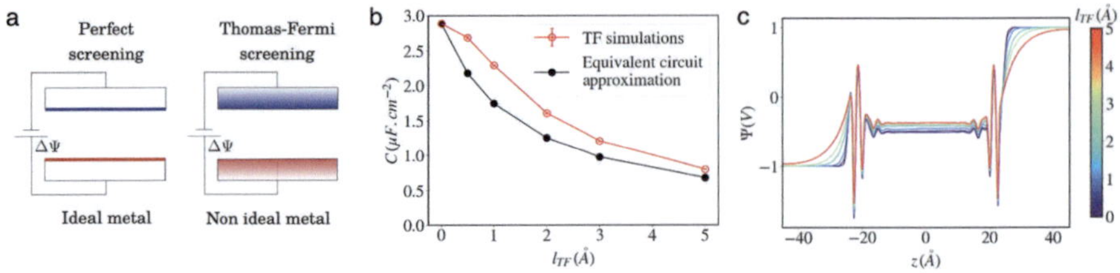

Fig. 10.10 The Thomas-Fermi electrode model. **a** Constant potential simulations (left) correspond to a perfect screening of the charges, whereas the Thomas-Fermi model introduces a screening length to account for the imperfect screening of the charge in a non-ideal metal. **b** Variation of the capacitance with TF screening length (l_{TF}). **c** Poisson potential across the simulation cell for a system made of two (100) gold-like electrodes in contact with a NaCl aqueous solution. Reproduced with permission from [62]

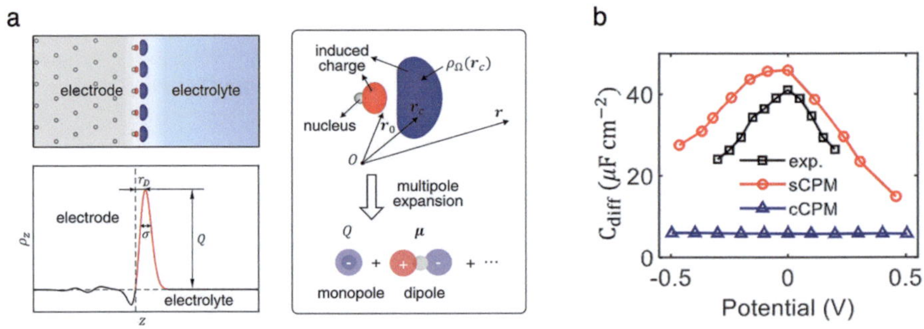

Fig. 10.11 The spillover constant potential method for modeling metallic electrodes. **a** Schematics of the spillover constant potential method (sCPM): (Top-left panel) Induced charge in electrode-electrolyte interface. (Right panel) Multi-pole expansion of induced charge. Red and blue iso-surfaces refer to the positive and negative induced charges, respectively. (Bottom-left panel) Schematic of induced charge distribution in metals. **b** Potential dependence of the capacitance of Au(111) in 2 M $NaClO_4$ solution derived from experiments, sCPM, and conventional constant potential method (CPM). Adapted with permission from [63]

description of electrostatic screening at the electrode-electrolyte interface. Using this improved method, molecular dynamics simulations of aqueous $NaClO_4$ electrolytes in contact with Au(111) electrodes reveal significantly more accurate differential capacitance profiles compared to classical CPM. By comparison, the sCPM method can recover the experimentally observed bell-shaped capacitance-potential curve, in contrast to the flat and underestimated capacitance predicted by cCPM (Fig. 10.11b).

10.3.2 Conductive Metal–Organic Frameworks

Since their inception in the 1990 s MOFs have garnered considerable attention for applications ranging from gas storage to catalysis, largely due to their high porosity and customizable architectures. However, most MOFs were electrically insulating, limiting their relevance for electronics or electrochemical energy storage. This perception began to shift in the early 2010 s when pioneering work demonstrated that certain MOFs could be rendered conductive through design strategies such as redox-active linkers, extended π-conjugation, or partial oxidation/reduction of the metal centers [64]. Notable breakthroughs included the discovery of layered, 2D frameworks like $Ni_3(HITP)_2$ that support long-range electron transport [5]. These so-called *conductive MOFs* (cMOFs) have since attracted significant interest for electrochemical applications, including supercapacitor electrodes.

10.3.2.1 Predicting Conductive MOFs

A key challenge in harnessing cMOFs for supercapacitors is the sheer diversity of potential metal-linker combinations and structural topologies, which can number in the tens of thousands. High-throughput computational screening thus becomes essential for identifying promising cMOFs without resorting to trial-and-error synthesis.

One prominent example is the QMOF project [65], a large-scale initiative that compiles quantum-chemical data for thousands of MOFs to facilitate the discovery of electronically conductive and electrochemically active materials. This database is built upon high-throughput DFT calculations, systematically applied to a diverse set of MOFs, providing key electronic descriptors such as band gaps, partial atomic charges, density of states, and spin densities (Fig. 10.12a). By focusing on experimentally synthesized MOFs, the QMOF database ensures practical relevance while maintaining computational accuracy through rigorous geometry optimizations and careful curation of the structures.

Beyond serving as a static database, the QMOF framework integrates machine-learning techniques to accelerate materials discovery. Machine-learning models, particularly crystal graph convolutional neural networks (CGCNNs), are trained on the dataset to predict quantum-chemical properties with high accuracy at a fraction of the computational cost of full DFT calculations (Fig. 10.12b). These models enable rapid screening of thousands of hypothetical or newly synthesized MOFs, identifying candidates with high electronic conductivity, optimal charge mobility, or tunable redox states. Furthermore, by leveraging unsupervised learning techniques such as uniform manifold approximation and projection (UMAP) (Fig. 10.12c), researchers can analyze structure-property relationships, revealing design rules for tailoring MOF electronic properties.

In the context of energy storage applications, such as supercapacitors, the QMOF project provides a foundational tool for identifying high-performance conductive

MOFs. The open-access nature of the database ensures that it remains a dynamic and continuously expanding resource, fostering further integration of quantum chemistry, machine learning, and experimental validation in the search for next-generation electrode materials.

10.3.2.2 Modeling MOF-Based Supercapacitors

Bi et al. used constant potential molecular dynamics simulations to investigate charge storage in cMOF electrodes [66]. Their study examined symmetric c-MOF electrodes in an ionic liquid ([Emim][BF_4]), as shown in Fig. 10.13a. Three cMOFs with different pore sizes–$Ni_3(HHB)_2$ (0.8 nm), $Ni_3(HITP)_2$ (1.5 nm), and $Ni_3(HITN)_2$ (2.4 nm)–were studied. Capacitance-potential curves revealed a strong dependence on pore size: the smallest pores exhibited a double-humped profile, while larger ones showed bell-shaped curves near the potential of zero charge (PZC) [66]. Simulated capacitances closely matched experimental values, with $Ni_3(HITP)_2$ achieving 9μ F.cm^{-2}, comparable to the experimental 11μ F.cm^{-2} [66].

A transmission line model was used to interpret the charging dynamics, showing how charge propagates through the porous electrodes (Fig. 10.13c). They found that the conductivity of in-pore electrolyte ions increased with temperature, enhancing power density. Also, equivalent circuit simulations predicted device-scale performance, identifying the largest-pore MOF among the studied ones as the most efficient in terms of energy and power density (Fig. 10.13d). These findings highlight the potential of cMOF-based supercapacitors for high-performance energy storage.

Integrating such simulations with in situ techniques like electrochemical quartz crystal microbalance (EQCM) and nuclear magnetic resonance (NMR) spectroscopy could deepen our understanding of cMOF-based supercapacitors. Given the vast diversity of MOF structures, this approach enables high-throughput screening for next-generation supercapacitors with optimized properties.

10.3.3 2D Layered Nanomaterials

Two-dimensional (2D) layered materials have emerged as promising electrode candidates for supercapacitors due to their high surface area, tunable interlayer spacing, and excellent electronic conductivity. Among these materials, metallic 1T-phase molybdenum disulfide (1T-MoS_2) and MXenes stand out for their superior electrochemical performance. Unlike the semiconducting 2H-MoS_2, the 1T polymorph exhibits metallic conductivity, making it highly effective for charge-storage applications. Its layered structure enables efficient ion intercalation, while its intrinsic flexibility allows for mechanical stability during cycling. Similarly, MXenes–a family of transition metal carbides and nitrides–offer exceptional conductivity, hydrophilicity, and surface functionalization tunability, making them ideal for both electrical double-layer and pseudocapacitive charge storage. Their high volumetric

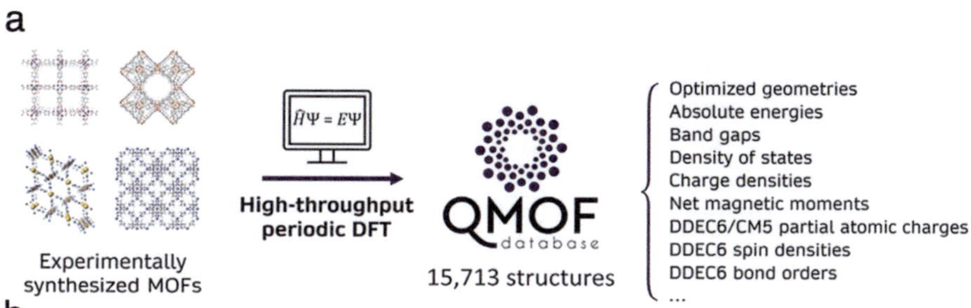

b

ML method	MAE (eV)	R^2	ρ
Constant mean model	0.973	–	–
Sine Coulomb matrix	0.529 ± 0.008	0.643 ± 0.012	0.787 ± 0.008
Stoichiometric-45	0.437 ± 0.004	0.743 ± 0.006	0.842 ± 0.004
Stoichiometric-120	0.433 ± 0.010	0.750 ± 0.009	0.847 ± 0.005
Orbital field matrix	0.417 ± 0.008	0.763 ± 0.010	0.863 ± 0.003
SOAP	0.357 ± 0.008	0.822 ± 0.010	0.910 ± 0.003
CGCNN	0.274 ± 0.008	0.876 ± 0.011	0.932 ± 0.005

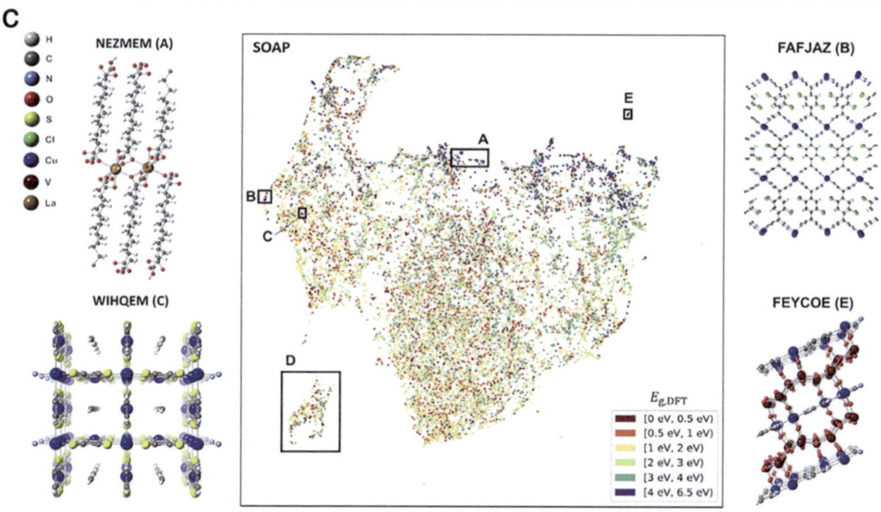

Fig. 10.12 QMOF database. **a** Overview of the QMOF database. **b** Summary of the testing-set mean absolute error (MAE), coefficient of determination (R^2), and Spearman rank-order correlation coefficient (ρ) for several machine-learning methods to predict the computed band gaps of MOFs from their deposited crystal structures with free solvent removed. **c** Unsupervised structural dimensionality reduction performed using uniform manifold approximation and projection (UMAP), with a distance matrix obtained from the SOAP average similarity kernel of the unrelaxed structures in the QMOF dataset. Reproduced with permission from [65]

capacitance, coupled with fast ion transport, enables MXene-based supercapacitors to achieve remarkable power and energy densities. From a modeling perspective, these 2D materials provide an ideal platform for theoretical investigations, as their

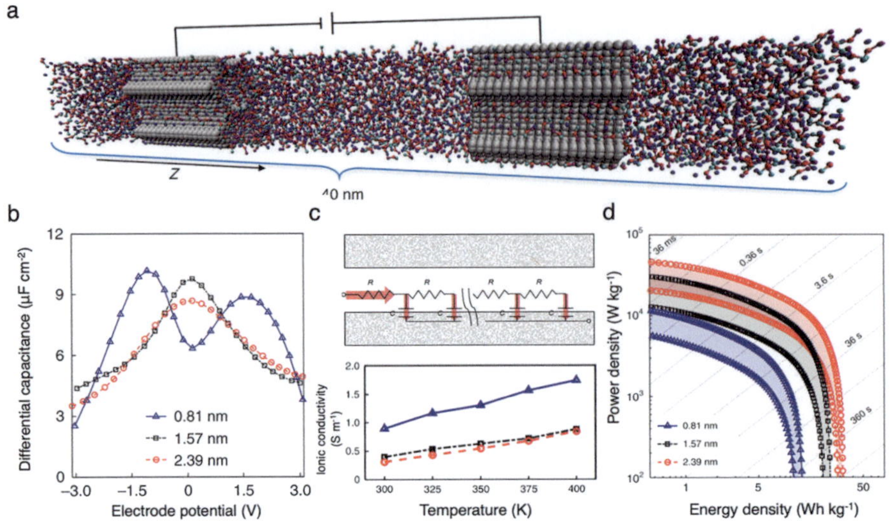

Fig. 10.13 Molecular dynamics simulations on three c-MOFs in ionic liquids. **a** Schematics of molecular simulations of c-MOF-based supercapacitor. **b** Differential areal capacitance of c-MOFs. **c** Intrinsic relaxation time and ionic conductivity of ILs in the MOF pores at different temperatures obtained by using transmission line model. **d** Gravimetric Ragone plot for c-MOFs-based supercapacitors from 300 K to 400 K. Reproduced with permission from [66]

well-defined crystalline structures allow for accurate simulations of nanoconfinement effects, ion intercalation mechanisms, and electrostatic screening behavior at electrified interfaces. This makes them valuable model systems for advancing our understanding of supercapacitor charge storage at the atomic scale.

10.3.3.1 Heterogeneous Electrode Model

Unlike the vast chemical diversity found in MOFs, where high-throughput screening is essential for identifying conductive candidates, 2D layered materials such as metallic 1T-MoS_2 and MXenes occupy a more constrained chemical space. However, their electrochemical behavior is inherently complex due to the presence of multiple atomic species (e.g., transition metals, chalcogenides, carbides, nitrides) and the diverse intercalation processes they undergo. This complexity necessitates a more detailed and accurate electrode model rather than a broad screening approach. In particular, conventional constant potential models often oversimplify the charge distribution across these heterogeneous systems. More advanced modeling frameworks, such as multi-component constant potential methods or electronic structure-informed approaches, are needed for describing these effects accurately.

A refined extension of the original constant potential method was recently proposed for multi-component electrodes such as MoS_2, known as constant chemical potential method (CPM_χ) [43]. This approach allows independent control over the electrochemical potential of different elements, providing a more accurate representation of electrode behavior. A detailed description of this method can be found in [43]. Unlike the standard CPM, CPM_χ assigns distinct Gaussian charge distributions to different elements within the MoS_2 electrode and modifies the electrical potential constraint to account for their varying electronegativities (Fig. 10.14a–b). This refinement enhances the accuracy of simulations involving multi-component electrodes, making CPM_χ particularly useful for systems where charge distribution and polarization effects play a crucial role.

The authors investigated the 1T-MoS_2/[EMIM][BF_4] supercapacitor using the CPM_χ model. Despite the significant difference in the electronegativities between Mo and S atoms, the computed capacitance (40.9F.g^{-1}) remains comparable to previous reports based on conventional CPM. However, CPM_χ simulations uncovered a distinct charging mechanism dominated by co-ion desorption in 1T-MoS_2, a behavior not observed in nanoporous carbon electrodes and undetectable without the enhanced model (Fig. 10.14c).

In parallel, a similar advancement over the conventional CPM method, named heteroatomic constant potential method (HCPM) was also proposed to model heteroatomic electrodes such as MXene [44]. Similar to the CPM_χ method, this method extends CPM by incorporating two additional parameters for each element: Gaussian charge widths and hardness offsets, which are optimized using DFT calculations (Fig. 10.15a–b). This allows HCPM to more accurately reproduce the charge distribution observed in quantum simulations. The optimization process involves fitting these parameters to induced charges derived from DFT simulations of a charged ion near an MXene electrode. The methodology employs derivative-free optimization to minimize the difference between DFT-calculated and HCPM-predicted charge distributions.

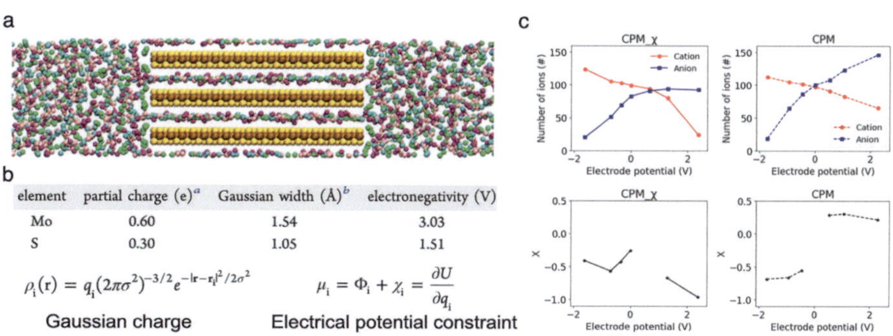

Fig. 10.14 Molecular dynamics simulations on 1T-MoS_2 in ionic liquids using constant chemical potential method (CPM_χ method). **a** Molecular dynamics simulations setup. **b** Partial Charge, Gaussian Width of Mo and S Atoms. **c** The number of ions inside the interlayers for the CPM_χ simulations (left) and the conventional CPM simulations (right). **d** Charging mechanism parameter (X) obtained by the two methods: negative value indicates co-ion desorption mechanism whereas positive value indicates counter-ion adsorption. Adapted with permission from ref [43]

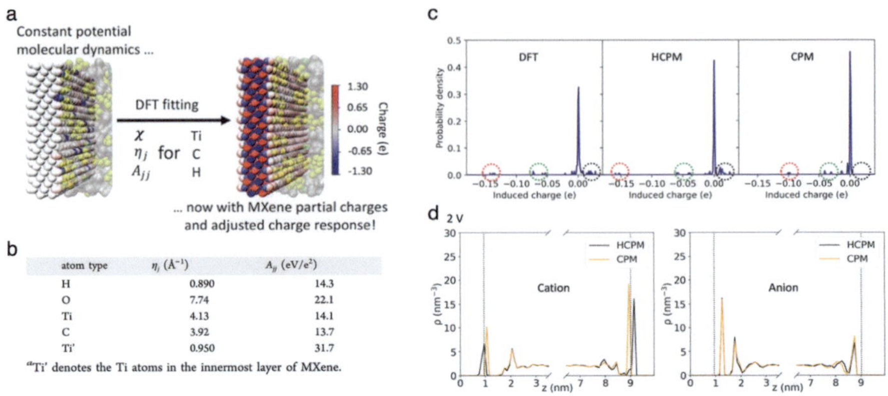

Fig. 10.15 Molecular dynamics simulations on MXene in LiTFSI/acetonitrile using HCPM method. **a** Schematics of HCPM method. **b** HCPM Parameters for Ti3$C_2(OH)_2$ obtained from the derivative-free optimization. **c** Probability density functions for induced charges on MXene using different methods. Red dashed circles highlight H atoms near Li+; red dashed circles also mark O and Ti atoms near Li+; black dashed circles indicate MXene atoms with small positive induced charges, present in DFT and HCPM but absent in CPM. **d** Ion number density distribution along the z direction by HCPM, CPM, or CCM under 2 V. Reproduced with permission from [44]

Compared to standard CPM, HCPM provides a more accurate charge distribution in molecular dynamics simulations of MXene electrodes with LiTFSI/acetonitrile solvent-in-salt electrolytes, preserving partial charges on MXene atoms in agreement with DFT results (Fig. 10.15c). This leads to enhanced cation adsorption (Fig. 10.15d), as Li+ ions exhibit stronger attraction to the MXene surface due to the improved treatment of electronic interactions. A key distinction between HCPM and CPM lies in the electrode charge response, particularly in the first few atomic layers: while CPM predicts a largely neutral electrode surface at zero voltage, HCPM maintains realistic partial charges on atoms, aligning better with experimental and DFT studies. Additionally, HCPM captures essential interfacial effects, such as variations in the electrostatic potential and ionic structure near the electrode, which standard CPM fails to resolve.

10.3.3.2 Charging Mechanisms in 2D Layered Materials

2D layered materials, including MXenes and TMDs, often exhibit pseudocapacitance, particularly in acidic aqueous electrolytes, where Faradaic surface redox reactions contribute significantly to charge storage alongside conventional electric double-layer capacitance. This dual mechanism presents both opportunities and challenges in understanding their electrochemical performance. Classical molecular dynamics simulations, while effective in capturing ion dynamics and double-layer formation, typically lack the ability to model surface redox processes due to their reliance on empirical force fields that do not account for electronic structure changes.

Consequently, joint DFT approaches, which explicitly describe electronic states and charge transfer, are frequently employed to model these systems. Such methods allow researchers to quantify the interplay between EDLC and pseudocapacitive mechanisms, offering deeper insights into the role of ion intercalation, solvent interactions, and electrode surface chemistry in determining overall capacitance.

Zhan et al. developed a theoretical model using joint DFT to capture the competition between surface redox chemistry and double-layer charging [67]. They applied this approach to predict the pseudocapacitive performance of $Ti_3C_2T_x$ (T = O, OH) in an acidic H_2SO_4 electrolyte (Fig. 10.16a). The Gibbs free energy, $G(x, \phi)$, was computed for different H coverages at fixed electrode potentials, determining the most probable H coverage (Fig. 10.16b). This allowed them to establish the relationship between Faradaic charge, surface net charge (from EDL capacitance), and the applied potential. It was found that the total charge storage is dominated by redox charging, while EDL charge counteracts this process, leading to a consistently positive surface net charge (Fig. 10.16c). This indicates that part of the proton transfer charge contributes to the net surface charge.

In addition, the specific differential capacitance was computed based on the total charge-potential relationship (Fig. 10.16d), obtaining a capacitance of approximately 230 F g^{-1} in the redox-dominant region (below 0.5 V vs. SHE), which aligns well with experimental measurements [68]. The total and local density of states (DOS) of the studied MXene indicate that the total DOS near the Fermi level is primarily contributed by Ti, suggesting that electron transfer predominantly occurs on Ti atoms (Fig. 10.16e). Moreover, they found that the surface charge responsible for EDL capacitance is mainly localized on the surface oxygen layer, which directly interacts with the electrolyte to form the double layer (Fig. 10.16f).

10.4 Challenges

Despite significant advancements, computational modeling of supercapacitors faces multiple technical and methodological challenges and limitations. Classical molecular dynamics, while computationally efficient, are usually incapable of capturing complex interfacial chemistry that involves explicit redox reactions. Quantum-mechanical methods, such as density functional theory (DFT) and hybrid quantum mechanics/molecular mechanics (QM/MM) approaches, in principle can account for these electronic effects but suffer from severe computational limitations, restricting their practical use to small system sizes and short timescales. Although recent hybrid methods like joint density functional theory provide a balance between computational feasibility and electronic accuracy, they often oversimplify electrolyte behavior through continuum approximations, missing critical molecular-level details of solvent structuring and ion correlations. Machine-learning-based approaches such as machine-learning force fields (MLFFs) have emerged as promising tools, offering near-quantum accuracy at classical computational cost, yet their development remains a challenge by the lack of diverse and systematically constructed

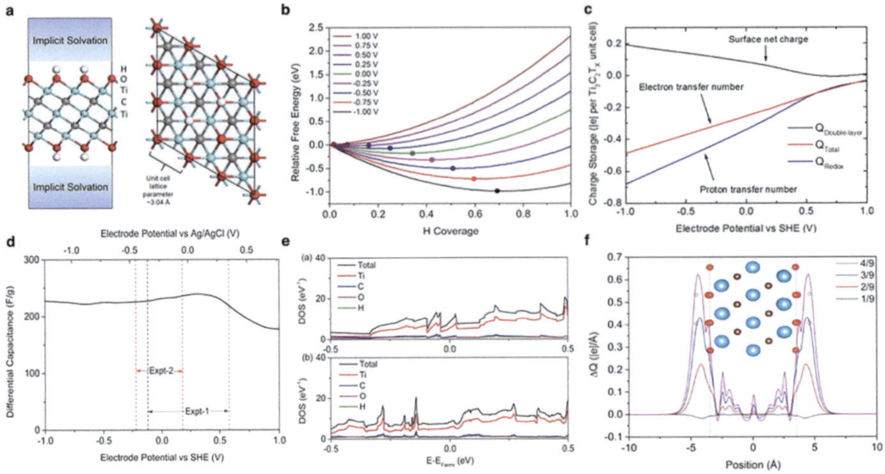

Fig. 10.16 Joint-DFT simulations on MXenes-based pseudocapacitors. **a** Schematics of the simulation on MXenes in the H_2SO_4 electrolyte. **b** Gibbs free energy of Ti_3C_2Tx (T = O, OH) as a function of H coverage x at different electrode potentials, ϕ, relative to SHE. **c** Faradaic charge (blue), EDL charge (black), and total charge (red) stored at different electrode potentials. **d** Differential capacitance of Ti_3C_2Tx in 1 M H_2SO_4. **e** Total electronic DOS and local DOS of the Ti_3C_2Tx electrode with a H coverage of 1/9 (upper) and a H coverage 2/9 (bottom). **f** Planar charge density change (ΔQ) of the Ti_3C_2Tx electrode after charging from a neutral surface with different H coverages. Reproduced with permission from ref [67]

training datasets that cover the broad chemical variability present in electrode-electrolyte systems. In future, developing multiscale simulation methods that integrate quantum-level accuracy, robust molecular mechanics approaches, and data-driven machine-learning strategies will be essential for addressing these challenges, ultimately enhancing the reliability and predictive capability of simulations for realistic electrochemical systems.

10.5 Conclusion

Over the past decade, computational modeling has become a cornerstone in advancing our fundamental understanding and practical development of advanced supercapacitors utilizing emerging nanomaterials. Molecular simulations have provided critical microscopic insights of charge storage, including ion confinement and quantum capacitance effects in nanoporous carbons, electronic conductivity and porosity-capacitance relationships in conductive metal-organic frameworks, and complex charging mechanisms in two-dimensional layered electrodes such as MXenes and molybdenum disulfide. These theoretical insights have established clear connections between atomistic-scale phenomena and macroscopic electrochemical performance, driving informed experimental exploration and optimization. Going forward, the continued synergy between advanced computational methods and targeted experimental validation will be essential to unlock innovative supercapacitor materials with superior performance, cost-effectiveness, and sustainability.

References

1. S. Fleischmann, Y. Zhang, X. Wang, P.T. Cummings, J. Wu, P. Simon, Y. Gogotsi, V. Presser, V. Augustyn, Continuous transition from double-layer to faradaic charge storage in confined electrolytes. Nat. Energy **7**(3), 222–228 (2022)
2. P. Simon, Y. Gogotsi, Perspectives for electrochemical capacitors and related devices. Nat. Mater. **19**(11), 1151–1163 (2020)
3. M.R. Lukatskaya, O. Mashtalir, C.E. Ren, Y. Dall'Agnese, P. Rozier, P.L. Taberna, M. Naguib, P. Simon, M.W. Barsoum, Y. Gogotsi, Cation intercalation and high volumetric capacitance of two-dimensional titanium carbide. Science **341**(6153), 1502–1505 (2013)
4. H. Wang, H. Feng, J. Li, Graphene and graphene-like layered transition metal dichalcogenides in energy conversion and storage. Small **10**(11), 2165–2181 (2014)
5. D. Sheberla, J.C. Bachman, J.S. Elias, C.-J. Sun, Y. Shao-Horn, M. Dinca, Conductive MOF electrodes for stable supercapacitors with high areal capacitance. Nat. Mater.**16**(2), 220–224 (2017)
6. A. Halder, M. Ghosh, A.M. Khayum, S. Bera, M. Addicoat, H.S. Sasmal, S. Karak, S. Kurungot, R.Banerjee, Interlayer hydrogen-bonded covalent organic frameworks as high-performance supercapacitors. J. Am. Chem. Soc. **140**(35), 10941–10945 (2018)
7. B.E. Conway, W.G. Pell, Double-layer and pseudocapacitance types of electrochemical capacitors and their applications to the development of hybrid devices. J. Solid State Electrochem. **7**(9), 637–644 (2003)
8. T. Brousse, D. Bélanger, J.W. Long, To be or not to be pseudocapacitive?. J. Electrochem. Soc. **162**(5), A5185–A5189 (2015)
9. J. Chmiola, Anomalous increase in carbon capacitance at pore sizes less than 1 nanometer. Science **313**(5794), 1760–1763 (2006)
10. G. Feng, P.T. Cummings, Supercapacitor capacitance exhibits oscillatory behavior as a function of nanopore size. J. Phys. Chem. Lett. **2**(22), 2859–2864 (2011)
11. C. Merlet, B. Rotenberg, P.A. Madden, P.-L. Taberna, P. Simon, Y. Gogotsi, M. Salanne, On the molecular origin of supercapacitance in nanoporous carbon electrodes. Nat. Mater. **11**(4), 306–310 (2012)
12. D.T.L. Galhena, B.C. Bayer, S. Hofmann, G.A.J. Amaratunga, Understanding capacitance variation in sub-nanometer pores by in-situ tuning of interlayer constrictions. ACS Nano **10**, 747–754 (2016)
13. X. Liu, D. Lyu, C. Merlet, M.J.A. Leesmith, X. Hua, Z. Xu, C.P. Grey, A.C. Forse, Structural disorder determines capacitance in nanoporous carbons. Science **384**(6693), 321–325 (2024)
14. X. Kui, H. Shao, Z. Lin, C. Merlet, G. Feng, J. Zhu, P. Simon, Computational insights into charge storage mechanisms of supercapacitors. Energy Environ. Mater. **3**(3), 235–246 (2020)
15. G. Jeanmairet, B. Rotenberg, M. Salanne, Microscopic simulations of electrochemical double-layer capacitors. Chem. Rev. **122**(12), 10860–10898 (2022)
16. S. Bi, L. Knijff, X. Lian, A. Van Hees, C. Zhang, M. Salanne, Modeling of nanomaterials for supercapacitors: beyond carbon electrodes. ACS Nano. **18**(31), 19931–19949 (2024)
17. M. Salanne. Ionic liquids for supercapacitor applications. in *Topics in Current Chemistry*, vol. 375 (Springer, 2017), pp. 29–53
18. C. Zhang, J. Cheng, Y. Chen, M. Chan, Q. Cai, R.P. Carvalho, C.F.N. Marchiori, D. Brandell, C.M. Araujo, M. Chen, X. Ji, G. Feng, K. Goloviznina, A. Serva, M. Salanne, T. Mandai, T.

Hosaka, M. Alhanash, P. Johansson, Y. Qiu, H. Xiao, M.H .Eikerling, R. Jinnouchi, M.M. Melander, G. Kastlunger, A. Bouzid, A. Pasquarello, S.-J. Shin, M.M. Kim, H. Kim, K. Schwarz, R. Sundararaman, 2023 roadmap on molecular modelling of electrochemical energy materials. J. Phys.: Energy (2023)
19. C. Zhan, C. Lian, Y. Zhang, M.W. Thompson, Y. Xie, J. Wu, P.R.C. Kent, P.T. Cummings, D.E. Jiang, D.J. Wesolowski, Computational insights into materials and interfaces for capacitive energy storage. Adv. Sci. (Weinh.) **4**(7) (2017)
20. W.L. Jorgensen, D.S. Maxwell, J. Tirado-Rives, Development and testing of the OPLS all-atom force field on conformational energetics and properties of organic liquids. J. Am. Chem. Soc. **118**(45), 11225–11236 (1996)
21. ...A.D. MacKerell, D. Bashford, M. Bellott, R.L. Dunbrack, J.D. Evanseck, M.J. Field, S. Fischer, J. Gao, H. Guo, S. Ha, D. Joseph-McCarthy, L. Kuchnir, K. Kuczera, F.T.K. Lau, C. Mattos, S. Michnick, T. Ngo, D.T. Nguyen, B. Prodhom, W.E. Reiher, B. Roux, M. Schlenkrich, J.C. Smith, R. Stote, J. Straub, M. Watanabe, J. Wiórkiewicz-Kuczera, D. Yin, M. Karplus, All-atom empirical potential for molecular modeling and dynamics studies of proteins. J. Phys. Chem. B **102**(18), 3586–3616 (1998)
22. C.I. Bayly, K.M. Merz, D.M. Ferguson, W.D. Cornell, T. Fox, J.W. Caldwell, P.A. Kollman, P. Cieplak, I.R. Gould, D.C. Spellmeyer, A second generation force field for the simulation of proteins, nucleic acids, and organic molecules. J. Am. Chem. Soc. **117**(19), 5179–5197 (1995)
23. C.C.J. Roothaan, New developments in molecular orbital theory. Rev. Mod. Phys. **23**(2), 69 (1951)
24. W. Kohn, L.J. Sham, Self-consistent equations including exchange and correlation effects. Phys. Rev. **140**(4A), A1133–A1138 (1965)
25. X. Zhao, Y. Liu, Origin of selective production of hydrogen peroxide by electrochemical oxygen reduction. J. Am. Chem. Soc. **143**(25), 9423–9428 (2021)
26. C. Zhang, T. Sayer, J. Hutter, M. Sprik, Modelling electrochemical systems with finite field molecular dynamics. J. Phys. Ener. **2**, 032005 (2020)
27. V.W.D. Cruzeiro, M. Manathunga, K.M. Merz, A.W. Götz, Open-source multi-GPU-accelerated QM/MM simulations with AMBER and QUICK. J. Chem. Inf. Model. **61**(5), 2109–2115 (2021)
28. R. Sundararaman, W.A. Goddard, T.A. Arias, Grand canonical electronic density-functional theory: algorithms and applications to electrochemistry. J. Chem. Phys. **146**(11), 114104 (2017)
29. J. Behler, Four generations of high-dimensional neural network potentials. Chem. Rev. **121**(16), 10037–10072 (2021)
30. V.L. Deringer, A.P. Bartók, N. Bernstein, D.M. Wilkins, M. Ceriotti, G. Csányi, Gaussian process regression for materials and molecules. Chem. Rev. **121**(16), 10073–10141 (2021)
31. A.A. Kornyshev, Double-layer in ionic liquids: paradigm change? J. Phys. Chem. B **111**(20), 5545–5557 (2007)
32. M.Z. Bazant, B.D. Storey, A.A. Kornyshev, Double layer in ionic liquids: overscreening versus crowding. Phys. Rev. Lett. **106**(4), 046102 (2011)
33. K. Goloviznina, J.C. Lopes, M.C. Gomes, A.A.H. Pádua, Transferable, polarizable force field for ionic liquids. J. Chem. Theory Comput. **15**(11), 5858–5871 (2019)
34. A.C.T.V. Duin, S. Dasgupta, F. Lorant, W.A. Goddard, ReaxFF: a reactive force field for hydrocarbons. J. Phys. Chem. A **105**(41), 9396–9409 (2001)
35. J.N.C. Lopes, A.A.H. Pádua, CL&P: a generic and systematic force field for ionic liquids modeling. Theor. Chem. Acc. **131**(3), 1–11 (2012)
36. J.I. Siepmann, M. Sprik, Influence of surface topology and electrostatic potential on water/electrode systems. J. Chem. Phys. **102**(1), 511–524 (1995)
37. S.K. Reed, P.A. Madden, A. Papadopoulos, Electrochemical charge transfer at a metallic electrode: a simulation study. J. Chem. Phys. **128**(12) (2008)
38. J. Yang, Z. Bo, H. Yang, H. Qi, J. Kong, J. Yan, K. Cen, Reliability of constant charge method for molecular dynamics simulations on EDLCs in nanometer and sub-nanometer spaces. ChemElectroChem **4**(10), 2486–2493 (2017)

39. A. Marin-Laflèche, M. Haefele, L. Scalfi, A. Coretti, T. Dufils, G. Jeanmairet, S. Reed, A. Serva, R. Berthin, C. Bacon, S. Bonella, B. Rotenberg, P. Madden, M. Salanne, MetalWalls: a classical molecular dynamics software dedicated to the simulation of electrochemical systems. J. Open Source Softw. **5**(53), 2373 (2020)
40. L.J.V. Ahrens-Iwers, M. Janssen, S.R. Tee, R.H. Meißner, ELECTRODE: an electrochemistry package for atomistic simulations. J. Chem. Phys. **157**(8), 084801 (2022)
41. S. Bi, R. Wang, S. Liu, J. Yan, B. Mao, A.A. Kornyshev, G. Feng, Minimizing the electrosorption of water from humid ionic liquids on electrodes. Nat. Commun. **9**(1), 5222 (2018)
42. L. Zeng, X. Tan, X. Ji, S. Li, J. Zhang, J. Peng, S. Bi, G. Feng, Constant charge method or constant potential method: which is better for molecular modeling of electrical double layers? J. Energy Chem. **94**, 54–60 (2024)
43. S. Bi, M. Salanne, Co-ion desorption as the main charging mechanism in metallic 1T-MoS2 supercapacitors. ACS Nano **16**(11), 18658–18666 (2022)
44. X. Lin, S.R. Tee, P.R.C. Kent, D.J. Searles, P.T. Cummings, Development of heteroatomic constant potential method with application to MXene-based supercapacitors. J. Chem. Theory Comput. **20**(2), 651–664 (2024)
45. J.P. Perdew, K. Burke, M. Ernzerhof, Generalized gradient approximation made simple. Phys. Rev. Lett. **77**(18), 3865–3868 (1996)
46. J. Sun, A. Ruzsinszky, J.P. Perdew, Strongly constrained and appropriately normed semilocal density functional. Phys. Rev. Lett. **115**(3), 036402 (2015)
47. M. Finnis, Interatomic forces in materials. Prog. Mater Sci. **49**(1), 1–18 (2004)
48. C. Zhang, M. Sprik, Computing the dielectric constant of liquid water at constant dielectric displacement. Phys. Rev. B **93**(14), 144201 (2016)
49. S. Naserifar, Y. Chen, S. Kwon, H. Xiao, W.A. Goddard, Artificial intelligence and QM/MM with a polarizable reactive force field for next-generation electrocatalysts. Matter **4**(1), 195–216 (2021)
50. K. Letchworth-Weaver, T.A. Arias, Joint density functional theory of the electrode-electrolyte interface: application to fixed electrode potentials, interfacial capacitances, and potentials of zero charge. Phys. Rev. B Condens. Matter Mater. Phys. **86**(7), 075140 (2012)
51. J. Behler, Atom-centered symmetry functions for constructing high-dimensional neural network potentials. J. Chem. Phys. **134**(7), 074106 (2011)
52. A.P. Bartók, R. Kondor, G. Csányi, On representing chemical environments. Phys. Rev. B **87**(18), 184115 (2013)
53. O.T. Unke, S. Chmiela, H.E. Sauceda, M. Gastegger, I. Poltavsky, K.T. Schütt, A. Tkatchenko, K.R. Müller, Machine learning force fields. Chem. Rev. **121**(16), 10142–10186 (2021)
54. Y. Shao, M. Hellström, P.D. Mitev, L. Knijff, C. Zhang, PiNN: A python library for building atomic neural networks of molecules and materials. J. Chem. Inf. Model. **60**(3), 1184–1193 (2020)
55. T. Dufils, L. Knijff, Y. Shao, C. Zhang, PiNNwall: heterogeneous electrode models from integrating machine learning and atomistic simulation. J. Chem. Theory Comput. **19**(15), 5199–5209 (2023)
56. A. Coretti, C. Bacon, R. Berthin, A. Serva, L. Scalfi, I. Chubak, K. Goloviznina, M. Haefele, A. Marin-Laflèche, B. Rotenberg, S. Bonella, M. Salanne, MetalWalls: simulating electrochemical interfaces between polarizable electrolytes and metallic electrodes. J. Chem. Phys. **157**, 184801 (2022)
57. A. Grisafi, A. Bussy, M. Salanne, R. Vuilleumier, Predicting the charge density response in metal electrodes. Phys. Rev. Mater. **7**(12), 125403 (2023)
58. A.M. Lewis, P. Lazzaroni, M. Rossi, Predicting the electronic density response of condensed-phase systems to electric field perturbations. J. Chem. Phys. **159**(1), 014103 (2023)
59. K.T. Thomson, K.E. Gubbins, Modeling structural morphology of microporous carbons by reverse Monte Carlo. Langmuir **16**(13), 5761–5773 (2000)
60. M. Qamar, M. Mrovec, Y. Lysogorskiy, A. Bochkarev, R. Drautz, Atomic cluster expansion for quantum-accurate large-scale simulations of carbon. J. Chem. Theory Comput. **19**(15), 5151–5167 (2023)

61. J. Xia, F. Chen, J. Li, N. Tao, Measurement of the quantum capacitance of graphene. Nat. Nanotechnol. **4**(8), 505–509 (2009)
62. L. Scalfi, T. Dufils, K.G. Reeves, B. Rotenberg, M. Salanne, A semiclassical thomas-fermi model to tune the metallicity of electrodes in molecular simulations. J. Chem. Phys. **153**(17) (2020)
63. Z. Wang, M. Chen, J. Wu, X. Ji, L. Zeng, J. Peng, J. Yan, A.A. Kornyshev, B. Mao, G. Feng, Constant-potential modeling of electrical double layers accounting for electron spillover. Phys. Rev. Lett. **134**(4) (2025)
64. S. Takaishi, M. Hosoda, T. Kajiwara, H. Miyasaka, M. Yamashita, Y. Nakanishi, Y. Kitagawa, K. Yamaguchi, A. Kobayashi, H. Kitagawa, Electroconductive porous coordination polymer Cu[Cu(pdt)2] composed of donor and acceptor building units. Inorg. Chem. **48**(19), 9048–9050 (2009)
65. A.S. Rosen, S.M. Iyer, D. Ray, Z. Yao, A. Aspuru-Guzik, L. Gagliardi, J.M. Notestein, R.Q. Snurr, Machine learning the quantum-chemical properties of metal-organic frameworks for accelerated materials discovery. Matter **4**(5), 1578–1597 (2021)
66. S. Bi, H. Banda, M. Chen, L. Niu, M. Chen, W. Taizheng, J. Wang, R. Wang, J. Feng, T. Chen, Molecular understanding of charge storage and charging dynamics in supercapacitors with MOF electrodes and ionic liquid electrolytes. Nat. Mater. **19**(5), 552–558 (2020)
67. C. Zhan, M. Naguib, M. Lukatskaya, P.R.C. Kent, Y. Gogotsi, D.E. Jiang, Understanding the MXene pseudocapacitance. J. Phys. Chem. Lett. **9**(6), 1223–1228 (2018)
68. M.R. Lukatskaya, S.-M. Bak, X. Yu, X.-Q. Yang, M.W. Barsoum, Y. Gogotsi, Probing the mechanism of high capacitance in 2D titanium carbide using in situ X-ray absorption spectroscopy. Adv. Energy Mater. **5**(15), 1500589 (2015)

Index

© The Editor(s) (if applicable) and The Author(s), under exclusive license to Springer Nature Switzerland AG 2026

K. Raju et al. (eds.), *Hybrid Nanostructures for Advanced Supercapacitors*, Modern Aspects of Electrochemistry 63, https://doi.org/10.1007/978-3-032-04593-5

MIX
Papier aus verantwortungsvollen Quellen
Paper from responsible sources
FSC® C105338

If you have any concerns about our products,
you can contact us on
ProductSafety@springernature.com

In case Publisher is established outside the EU,
the EU authorized representative is:
Springer Nature Customer Service Center GmbH
Europaplatz 3, 69115 Heidelberg, Germany

Printed by Libri Plureos GmbH
in Hamburg, Germany